Theoretical Mechanics

THEORETICAL MECHANICS

T. C. Bradbury

Department of Physics
California State College at Los Angeles

John Wiley & Sons, Inc.   *New York   London   Sydney*

Preface

This text is designed for the intermediate mechanics course primarily intended for physics majors.

In it I have emphasized the Lagrange formulation of mechanics, but without undue neglect of Newtonian concepts. I introduce Hamiltonian equations, but leave the more advanced aspects of Hamiltonian theory, such as contact transformations and Hamilton-Jacobs theory, to later courses. Covariance principles are stressed; consequently, there is a fuller development than usual, in texts at this level, of the theory of generalized coordinates.

In my use of vector and tensor notation, I have developed side by side the conventional boldface-type approach and component or tensor notation. I use the summation convention freely, and emphasize the transformation properties of the various quantities.

I hope the treatment of vectors, tensors, and matrices will prove useful to students in courses they will take later, such as quantum mechanics, general relativity, and the mechanics of continuous media. The simple nonlinear differential equations which occur in conservative mechanical systems are given considerably more attention in this text than in most others. We develop the approximate methods of solving these equations based on Fourier series as well as the exact solutions in terms of Jacobian elliptic functions.

The chapter on accelerated reference frames and rigid body dynamics is, I think, worthy of note on two counts. First, we treat the mechanics of a particle with respect to an accelerated reference frame by the Lagrange formulation. Second, we develop the theory of the motion of a rigid body with respect to a noninertial frame of reference.

The substantial introduction to the special theory of relativity should, I hope, be helpful to the students lacking familiarity with this important subject.

There is too much material in the text to cover in a one-year course. Chapters 1 through 9 represent a substantial basic course in mechanics. Some instructors may wish to spend less time on the first three chapters, which are substantially mathematical, if it is felt that these topics are being adequately covered in mathematics courses. Also, some of the material in Chapter 7 on the motion of a charged particle in an electro-magnetic field can be left out. Additional topics can be selected from Chapters 10 through 13, although most teachers will want to cover Chapter 13 which is on special relativity.

T. C. Bradbury

Los Angeles, CA
November 1967

Contents

1

Introduction to Vectors, Tensors, and Matrices

1.1. ALGEBRA OF VECTORS

It is assumed that the reader is familiar with the concept of a vector on an elementary basis and has had some experience with the use of vectors to represent kinematical quantities such as displacement, velocity, and acceleration.* Vectors and scalars are frequently introduced into elementary physics courses by a statement such as, "A vector has both magnitude and direction whereas a scalar has magnitude only." The purpose of this chapter is to put the idea of a vector on a more rigorous foundation and, at the same time, introduce the concepts of tensors and matrices.

The first encounter which most people have with vectors is in their use to represent displacements. Thus in Fig. 1.1.1, the directed line segment from point A to point B is thought of as a vector and is represented by a boldface letter $\mathbf{a}$. (Boldface letters are used to represent vectors and, in some instances, tensors, to be introduced later in this chapter. It will be clear from the context which is intended.)

In Fig. 1.1.1, a second displacement vector $\mathbf{b}$ is drawn from B to C. The vector $\mathbf{c}$ which goes directly from A to C is called the sum of $\mathbf{a}$ and $\mathbf{b}$:

$$\mathbf{c} = \mathbf{a} + \mathbf{b} \tag{1.1.1}$$

---

* Resnick, Robert, and David Halliday, *Physics*, Part I, John Wiley and Sons, 1966, is an excellent basis on which to build the material to be presented here.

The above idea is stated independently of any particular coordinate system. If a coordinate system such as the rectangular Cartesian coordinate system indicated in Fig. 1.1.1 is introduced, the vectors can be represented in terms of their components along the three coordinate axes. Equation (1.1.1) is then represented by the component equations

$$c_x = a_x + b_x$$
$$c_y = a_y + b_y \qquad\qquad (1.1.2)$$
$$c_z = a_z + b_z$$

There are an infinite number of rectangular coordinate systems that could have been chosen and in each case equations such as (1.1.2) could

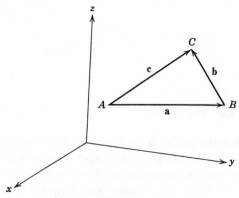

Fig. 1.1.1

be written down. In an abstract way, Eq. (1.1.1) represents all these sets of equations for all possible rectangular Cartesian coordinate systems.

Vectors obey the following basic laws:

I. Closure Law. If **a** and **b** are vectors, so is **a** + **b**.

II. Commutative Law of Addition. **a** + **b** = **b** + **a**

III. Associative Law Addition. (**a** + **b**) + **c** = **a** + (**b** + **c**)

IV. Existence of a zero vector. There is a unique vector **0** with the property that for every vector **a**, **a** + **0** = **a**.

V. For every vector **a** there is a unique vector (−**a**) such that **a** + (−**a**) = **0**.

These basic laws are easily verified by representing the vectors in a particular coordinate system and then using the component equations. For example, II is verified by noting that the components, being real numbers, satisfy

$$a_x + b_x = b_x + a_x$$
$$a_y + b_y = b_y + a_y$$
$$a_z + b_z = b_z + a_z$$

It is also possible to take the point of view that the above laws are basic *postulates* which vectors, considered as abstract mathematical entities, are to satisfy.

Two vectors are equal if and only if their components are equal; i.e., $\mathbf{a} = \mathbf{b}$ if and only if $a_x = b_x$, $a_y = b_y$, $a_z = b_z$. The following question might be asked: If it is known that the components of two vectors are equal in one particular coordinate system is it necessarily true that they are equal in all other coordinate systems? The answer is obviously yes if the vectors are displacement vectors. An algebraic proof must wait until

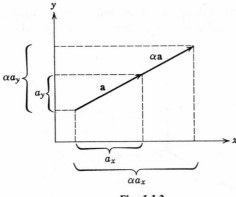

Fig. 1.1.2

we learn how to transform from one coordinate system to another. Note that it is not necessary for two vectors to be at the same location in space in order for them to be equal. For instance, two particles at quite different locations in space could have equal velocities.

The magnitude or length of a vector is denoted by $|\mathbf{a}|$ or sometimes a. In terms of components the Pythagorean theorem shows that

$$|\mathbf{a}| = (a_x^2 + a_y^2 + a_z^2)^{1/2} \tag{1.1.3}$$

If α is a positive scalar (a real number), then $\alpha\mathbf{a}$ is defined as a vector in the same direction as $\mathbf{a}$ but of magnitude $\alpha\,|\mathbf{a}|$. The components of $\alpha\mathbf{a}$ are

$$(\alpha a_x, \alpha a_y, \alpha a_z) \tag{1.1.4}$$

Figure 1.1.2 shows a two-dimensional geometrical interpretation of scalar multiplication. If α is negative, $\alpha\mathbf{a}$ is directed oppositely to $\mathbf{a}$. To the list of basic laws of vector algebra can be added the following which pertain specifically to scalar multiplication:

VI. $(\alpha_1 + \alpha_2)\mathbf{a} = \alpha_1\mathbf{a} + \alpha_2\mathbf{a}$

VII. $\alpha(\mathbf{a} + \mathbf{b}) = \alpha\mathbf{a} + \alpha\mathbf{b}$

VIII. $\alpha_1(\alpha_2\mathbf{a}) = (\alpha_1\alpha_2)\mathbf{a}$

In the developments of this section, vectors have been pictured as displacements in three-dimensional space. Note that *any* set of quantities which obey the same basic algebraic laws and which in addition have the correct transformation property (to be discussed) are by definition vectors.

The vector concept can be extended to spaces of an arbitrary number of dimensions. The discussion is confined for the present to ordinary three-dimensional space.

1.2. UNIT VECTORS AND SUMMATION CONVENTION

In a three-dimensional space, three independent numbers must be specified in order to define a vector. It is common practice to introduce

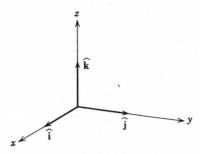

Fig. 1.2.1

unit vectors or basis vectors $\hat{\mathbf{i}}$, $\hat{\mathbf{j}}$, $\hat{\mathbf{k}}$ directed along the coordinate axes, Fig. 1.2.1. These vectors have components

$$\hat{\mathbf{i}}: \ (1, 0, 0)$$
$$\hat{\mathbf{j}}: \ (0, 1, 0)$$
$$\hat{\mathbf{k}}: \ (0, 0, 1)$$

and have magnitude unity. Unit vectors are denoted by lower-case boldface letters with the addition of a circumflex or hat ($\wedge$). An arbitrary vector can be expressed in terms of its components as

$$\mathbf{a} = a_x \hat{\mathbf{i}} + a_y \hat{\mathbf{j}} + a_z \hat{\mathbf{k}} \tag{1.2.1}$$

It is expedient in many calculations to label the coordinate axes x_1, x_2, and x_3 in place of x, y, and z. The unit vectors can be labeled $\hat{\mathbf{e}}_1$, $\hat{\mathbf{e}}_2$, and $\hat{\mathbf{e}}_3$ and the components of $\mathbf{a}$ as a_1, a_2, and a_3. The components of a vector are actually an *ordered triplet* of numbers. Equation (1.2.1) reads

$$\mathbf{a} = \sum_{i=1}^{3} \hat{\mathbf{e}}_i a_i = \hat{\mathbf{e}}_1 a_1 + \hat{\mathbf{e}}_2 a_2 + \hat{\mathbf{e}}_3 a_3 \tag{1.2.2}$$

To streamline the notation even more, the summation sign in (1.2.2) is dropped:

$$\mathbf{a} = \hat{\mathbf{e}}_i a_i \tag{1.2.3}$$

It should be understood, unless otherwise indicated, that repeated subscripts are summed from 1 to 3. This so-called *summation convention* was first introduced by Albert Einstein and is extensively employed in this book. Notice that it makes no difference what letter is used for the repeated subscript:

$$\hat{\mathbf{e}}_j a_j = \hat{\mathbf{e}}_k a_k = \hat{\mathbf{e}}_1 a_1 + \hat{\mathbf{e}}_2 a_2 + \hat{\mathbf{e}}_3 a_3 \tag{1.2.4}$$

By using the summation convention, the square of the length of a vector can be expressed

$$a^2 = a_i a_i \tag{1.2.5}$$

1.3. LINEAR DEPENDENCE

Three vectors are said to be *linearly dependent* if there exists scalars l, m, and n, not all zero, such that

$$l\mathbf{a} + m\mathbf{b} + n\mathbf{c} = 0 \tag{1.3.1}$$

If no such constants exist, the vectors are independent. Written out in component form, (1.3.1) is

$$la_x + mb_x + nc_x = 0$$
$$la_y + mb_y + nc_y = 0 \tag{1.3.2}$$
$$la_z + mb_z + nc_z = 0$$

If the vectors $\mathbf{a}$, $\mathbf{b}$, and $\mathbf{c}$ are given, then finding constants l, m, and n involves solving the three simultaneous homogeneous equations (1.3.2). As we know from algebra, such a set of homogeneous equations has a nonzero solution if and only if the determinant of the coefficients vanishes:

$$\begin{vmatrix} a_x & b_x & c_x \\ a_y & b_y & c_y \\ a_z & b_z & c_z \end{vmatrix} = 0 \tag{1.3.3}$$

Thus (1.3.3) provides a test for the dependence or independence of any three vectors.

In the case of the unit vectors $\hat{\mathbf{i}}$, $\hat{\mathbf{j}}$, $\hat{\mathbf{k}}$ (1.3.3) gives

$$\begin{vmatrix} 1 & 0 & 0 \\ 0 & 1 & 0 \\ 0 & 0 & 1 \end{vmatrix} = 1 \neq 0 \tag{1.3.4}$$

showing that they are independent.

Any three vectors which lie in the same plane are dependent. This is geometrically obvious. From Fig. 1.3.1 we see that scalars α and β can be found such that $\alpha\mathbf{a}$ and $\beta\mathbf{b}$ form the sides of a parallelogram, of which $\mathbf{c}$ is the diagonal:

$$\mathbf{c} = \alpha\mathbf{a} + \beta\mathbf{b} \qquad (1.3.5)$$

A plane can be thought of as a two dimensional world (subspace of the three-dimensional world) in which at most two vectors can be independent and all other vectors can be expressed in terms of these.

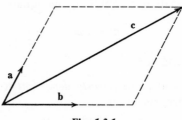

Fig. 1.3.1

Four vectors cannot be independent in a three-dimensional space. There are always numbers l, m, n, and p not all of which are zero such that

$$l\mathbf{a} + m\mathbf{b} + n\mathbf{c} + p\mathbf{d} = 0 \qquad (1.3.6)$$

for any four vectors $\mathbf{a}$, $\mathbf{b}$, $\mathbf{c}$, and $\mathbf{d}$. If (1.3.6) is written out in component form there results three homogeneous equations in four unknowns l, m, n, and p. Such a system always has nontrivial solutions. To put it in another way, given any three vectors which are independent (not coplanar), a fourth vector can always be expressed in terms of them. For this reason, we frequently speak of three linearly independent vectors as being a *complete set*. We also say that three linearly independent vectors *span* the three-dimensional space.

The unit vectors $\hat{\mathbf{i}}$, $\hat{\mathbf{j}}$, and $\hat{\mathbf{k}}$ are a *complete set* of mutually orthogonal unit vectors. For spaces of more dimensions than three, the number of vectors required to make a complete set is equal to the number of dimensions of the space.

As an example, suppose that two vectors $\mathbf{a}$ and $\mathbf{b}$ have a common origin O. Their tips determine a line AB. Under what conditions will a third vector $\mathbf{r}$, also with its origin at O, have its tip on this line? Clearly $\mathbf{a}$, $\mathbf{b}$, and $\mathbf{r}$ are dependent vectors. From Fig. 1.3.2 we see that the vectors $\mathbf{a} - \mathbf{r}$ and $\mathbf{r} - \mathbf{b}$ are colinear:

$$\mathbf{a} - \mathbf{r} = k(\mathbf{r} - \mathbf{b}) \qquad (1.3.7)$$

where k is a constant. Solution for **r** gives

$$\mathbf{r} = \frac{1}{k+1}\mathbf{a} + \frac{k}{k+1}\mathbf{b} \qquad (1.3.8)$$

As a second example, consider a body of weight w suspended by three strings as in Fig. 1.3.3. Let $\mathbf{F}_1$, $\mathbf{F}_2$, and $\mathbf{F}_3$ be the tensions in the three strings. Obviously, $F_3 = w$. The knot where the three strings come

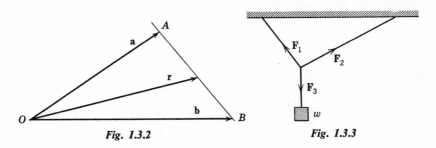

Fig. 1.3.2   *Fig. 1.3.3*

together is a small body in equilibrium. According to the principles of statics,

$$\mathbf{F}_1 + \mathbf{F}_2 + \mathbf{F}_3 = 0 \qquad (1.3.9)$$

showing that the three forces are linearly dependent, and hence coplanar.

1.4. ORTHOGONAL TRANSFORMATIONS

A basic principle of physics is that the fundamental laws are independent of any particular coordinate system or mode of description. In other words, if two observers use entirely different coordinate systems in the description of a physical system, they must come to the same conclusions as to the basic laws that govern the physical system under consideration. Coordinate systems are, of course, absolutely necessary in the description of physical phenomena, and so the best that can be done is to state the laws of physics in such a way that no particular coordinate system is implied. A law of physics that has been so stated is said to be in *covariant form*. A coordinate system is often called a *frame of reference*.

If we are to write the laws of physics in covariant form it is essential to know how the various quantities such as mass, force, velocity, etc., that enter our equation *transform* when a change of the reference frame is made. In this chapter, transformations between three-dimensional rectangular coordinate systems are considered. The transformations are *time independent*. In Chapter 2 is to be found an extension of the theory to arbitrary curvilinear coordinates in three-dimensional space. Coordinate systems which are moving relative to one another at constant velocity are taken up in Chapter 8. Chapter 10 considers the transformations between coordinate systems that are in arbitrary motion relative to one another. This involves *accelerated* frames of reference. Finally, Chapter 13 treats the *Lorentz Transformation* which is the four-dimensional space-time transformation of special relativity.

The possible transformations between rectangular Cartesian coordinate systems can be divided into three categories. Figure 1.4.1a shows a

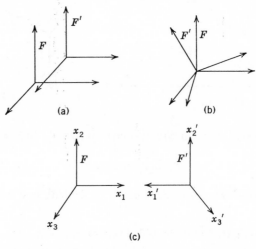

Fig. 1.4.1

simple translation of axes, Fig. 1.4.1b illustrates a *rotation* and Fig. 1.4.1c shows a *mirror reflection* in the $x_2 x_3$-plane resulting from the transformation $x_1' = -x_1$. The later transformation is termed *improper*. In the case of uniform translation and rotation, the new frame F' can be obtained from the original frame F by a continuous movement. Such is not the case with an improper transformation. In Fig. 1.4.1c it is not possible to move F in such a way that it coincides with F'. The most general transformation is a combination of the three basic types just discussed. Transformations between rectangular Cartesian coordinate systems that do not involve a change of scale (e.g., a transformation of the type $x_1' = kx_1$ where k is a constant) are called *orthogonal transformations*.

As an example of a rotation, consider a displacement vector **a** which lies in the $x_1 x_2$-plane. Let the new frame of reference be obtained by rotating the coordinate system by an angle θ about the x_3-axis, Fig. 1.4.2.

It is simple geometry to show that

$$a_1' = a_1 \cos \theta + a_2 \sin \theta$$
$$a_2' = -a_1 \sin \theta + a_2 \cos \theta \tag{1.4.1}$$

The *inverse transformation* is obtained by solving for a_1 and a_2:

$$a_1 = a_1' \cos \theta - a_2' \sin \theta$$
$$a_2 = a_1' \sin \theta + a_2' \cos \theta \tag{1.4.2}$$

Equation (1.4.1) can be thought of as a two-dimensional transformation. The transformation is *linear* because it involves only the first power of

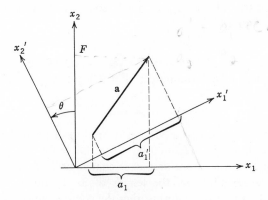

Fig. 1.4.2

a_1, a_2, a_1', etc. It is easy to see that if **a** does not lie in the x_1x_2 plane that a third equation $a_3' = a_3$ is to be added to (1.4.1) and (1.4.2).

A general three-dimensional orthogonal transformation can be represented by

$$a_1' = S_{11}a_1 + S_{12}a_2 + S_{13}a_3$$
$$a_2' = S_{21}a_1 + S_{22}a_2 + S_{23}a_3 \tag{1.4.3}$$
$$a_3' = S_{31}a_1 + S_{32}a_2 + S_{33}a_3$$

where $S_{11}, S_{12}, \ldots$ are the *transformation coefficients*. The simplest example of an orthogonal transformation is provided by a translation of axes as pictured in Fig. 1.4.1a. The components of a vector are unaltered, meaning that $a_1' = a_1$, $a_2' = a_2$, and $a_3' = a_3$. The values of the coefficients in (1.4.3), therefore, are $S_{11} = S_{22} = S_{33} = 1$ and all others zero. This is called the *identity* transformation. In the case of the mirror reflection illustrated in Fig. 1.4.1c, $S_{11} = -1$, $S_{22} = 1$, $S_{33} = 1$, and all others are zero. By using the summation convention (1.4.3) can be replaced by

$$a_i' = S_{ij}a_j \tag{1.4.4}$$

It is important that the reader understand exactly what is meant by (1.4.4). The subscript j is repeated in the same term and therefore a sum is implied:

$$a_i' = S_{i1}a_1 + S_{i2}a_2 + S_{i3}a_3$$

The subscript i is not repeated and can stand for any one of the three numbers 1, 2, or 3. These three values give the three equations (1.4.3). In other words, (1.4.4) is merely a convenient shorthand notation for the system (1.4.3).

It is possible to express the coefficients S_{ij} explicitly in terms of angles of rotation; see, e.g., Section 1.7. It is preferable, however, to derive the general properties of the S_{ij} without reference to a particular representation. The quantity

$$a_i a_i = a_1^2 + a_2^2 + a_3^2 \tag{1.4.5}$$

represents the square of the length of a vector, and a fundamental property of an orthogonal transformation is that it leaves the length of a vector unchanged:

$$a_i a_i = a_i' a_i' \tag{1.4.6}$$

This property is frequently used as a definition of an orthogonal transformation. It is obvious from Fig. 1.4.2 that only the *components* or *projections* of **a** on the coordinate axes change as F' is rotated, but this, of course, has no effect on the length of **a**. In other words, (1.4.6) expresses the fact that we can use either the components of **a** in F or in F' to compute its length.

Equation (1.4.4) can be substituted into (1.4.6) as follows: first write the two factors of a_i' as

$$a_i' = S_{ij}a_j \qquad a_i' = S_{ik}a_k$$

It is necessary to use different letters for the summed subscripts so that the two sums can be kept separate. Equation (1.4.6) is then

$$a_i a_i = S_{ij}S_{ik}a_j a_k \tag{1.4.7}$$

A triple sum is involved on the right side of (1.4.7)! This result must hold true for *all* vectors. In particular, it is true if the components of **a** are (1, 0, 0) in which case $a_i a_i = 1$ and only the terms for which $j = 1$ and $k = 1$ are nonzero:

$$1 = S_{i1}S_{i1} \tag{1.4.8}$$

By a similar procedure using the special vectors (0, 1, 0) and (0, 0, 1) one gets

$$1 = S_{i2}S_{i2}$$
$$1 = S_{i3}S_{i3} \tag{1.4.9}$$

As another example, consider the vector with components $(1, 1, 0)$. In $(1.4.7)$, $a_i a_i = 2$ and the sums over j and k work out to

$$2 = (S_{i1}a_1 + S_{i2}a_2 + S_{i3}a_3)(S_{i1}a_1 + S_{i2}a_2 + S_{i3}a_3)$$
$$= (S_{i1} + S_{i2})(S_{i1} + S_{i2})$$
$$= S_{i1}S_{i1} + S_{i2}S_{i2} + 2S_{i1}S_{i2}$$

By means of $(1.4.8)$ and $(1.4.9)$,

$$2 = 2 + 2S_{i1}S_{i2}$$
$$S_{i1}S_{i2} = 0 \qquad (1.4.10)$$

More generally,

$$S_{ij}S_{ik} = 0 \qquad \text{if } j \neq k \qquad (1.4.11)$$

The foregoing results can be stated compactly by using the *Kronecker delta symbol* defined as follows:

$$\delta_{ij} = 1 \qquad \text{if } i = j$$
$$\delta_{ij} = 0 \qquad \text{if } i \neq j \qquad (1.4.12)$$

Equations $(1.4.8)$, $(1.4.9)$, and $(1.4.11)$ are all contained in

$$S_{ij}S_{ik} = \delta_{jk} \qquad (1.4.13)$$

Equation $(1.4.13)$ is known as the *orthogonality condition*.

The *matrix* of the transformation is the square array

$$S = \begin{pmatrix} S_{11} & S_{12} & S_{13} \\ S_{21} & S_{22} & S_{23} \\ S_{31} & S_{32} & S_{33} \end{pmatrix} \qquad (1.4.14)$$

and is a particularly convenient way of tabulating the coefficients. Note carefully that the first subscript of an element of the matrix stands for the number of the row whereas the second subscript refers to the column. Equation $(1.4.11)$ can be seen to mean that a column of the matrix multiplied by itself is unity whereas one column multiplied by another gives zero. We can think of the columns of the transformation matrix as being unit vectors. In Section 1.6, it will be shown that when one column times another gives zero it means that these vectors are perpendicular to one another; hence, the term "orthogonal transformation." As an example, the matrix of the transformation $(1.4.1)$ is

$$S = \begin{pmatrix} \cos\theta & \sin\theta & 0 \\ -\sin\theta & \cos\theta & 0 \\ 0 & 0 & 1 \end{pmatrix} \qquad (1.4.15)$$

Equation (1.4.13) gives six independent equations connecting the nine transformation coefficients. An orthogonal transformation can be made to depend on three properly chosen independent parameters. See, for example, Section 10.5, especially Eq. (10.15.5), where a representation of an arbitrary rotation is obtained in terms of three angles which are known as the *Euler angles*.

To find the inverse transformation, it is necessary to solve (1.4.4) for the a_i in terms of $a_i{}'$. This can be done provided that the determinant of the transformation coefficients, i.e., the determinant formed from the array (1.4.14) and denoted det S_{ij}, does not vanish. By recalling the rules for multiplying determinants, we can construct the following equation from (1.4.13):

$$
\begin{vmatrix} S_{11} & S_{21} & S_{31} \\ S_{12} & S_{22} & S_{32} \\ S_{13} & S_{23} & S_{33} \end{vmatrix} \begin{vmatrix} S_{11} & S_{12} & S_{13} \\ S_{21} & S_{22} & S_{23} \\ S_{31} & S_{32} & S_{33} \end{vmatrix} = \begin{vmatrix} 1 & 0 & 0 \\ 0 & 1 & 0 \\ 0 & 0 & 1 \end{vmatrix} = 1
$$

This is conveniently abbreviated as

$$
(\det S_{ij})^2 = 1
$$
$$
\det S_{ij} = \pm 1 \tag{1.4.16}
$$

The negative sign occurs if the transformation is *improper*; (1.4.16) therefore provides a test for whether a given transformation is proper or improper. For instance, the transformation which gives the mirror reflection of Fig. 1.4.1c is

$$
S = \begin{pmatrix} -1 & 0 & 0 \\ 0 & 1 & 0 \\ 0 & 0 & 1 \end{pmatrix}
$$

and

$$
\det S_{ij} = -1
$$

The formal inversion of (1.4.4) is done by multiplying by S_{ik}:

$$
S_{ik}a_i{}' = S_{ik}S_{ij}a_j
$$
$$
= \delta_{kj}a_j = a_k
$$

where use is made of (1.4.13). Hence,

$$
a_k = S_{ik}a_i{}' \tag{1.4.17}
$$

Note carefully that in (1.4.17) the first subscript on S_{ik} is summed, whereas in (1.4.4) it is the second subscript that is summed. Equation

(1.4.17) written out in full is

$$a_1 = S_{11}a_1' + S_{21}a_2' + S_{31}a_3'$$
$$a_2 = S_{12}a_1' + S_{22}a_2' + S_{32}a_3' \qquad (1.4.18)$$
$$a_3 = S_{13}a_1' + S_{23}a_2' + S_{33}a_3'$$

The matrix of the inverse transformation (1.4.18) is

$$S^{-1} = \begin{pmatrix} S_{11} & S_{21} & S_{31} \\ S_{12} & S_{22} & S_{32} \\ S_{13} & S_{23} & S_{33} \end{pmatrix} = \tilde{S} \qquad (1.4.19)$$

Notice that (1.4.19) is obtained from (1.4.14) by an interchange of rows and columns. The matrix so obtained is called the *transpose* of S and is denoted $\tilde{S}$. The symbol S^{-1} is used to indicate that (1.4.19) is the *inverse* matrix of S. For the special case of orthogonal transformations the transpose and the inverse are identical, *but this is not true in general*!

For the moment, the arrays (1.4.14) and (1.4.19) are merely convenient ways of displaying the transformation coefficients. The matrices contain all the pertinent information about the transformation and, as will be seen, can be considered as mathematical entities in their own right. In Chapter 3, we take up the formal development of matrix algebra. By using (1.4.17) and (1.4.6) one can easily prove that

$$S_{ji}S_{ki} = \delta_{jk} \qquad (1.4.20)$$

In other words, the rows as well as the columns of the matrix (1.4.14) can be thought of as orthogonal unit vectors.

1.5. FORMAL DEFINITIONS OF VECTORS, SCALARS, AND INVARIANTS

A deeper meaning than was given in Section 1.1 can be given to the concept of a vector. A vector is a set of three functions of position in space that transform according to (1.4.4) under orthogonal transformations and, in addition, obeys the basic algebraic laws stated in Section 1.1. Of necessity the vector concept is based on the existence of co-ordinate systems or possibly other methods of representing vectors. The definition just given confines vectors to a particular class of coordinate systems. A vector is characterized by how its representations in the infinite number of possible rectangular Cartesian coordinate systems are related to one another. The vector concept can, of course, be extended to more general coordinate systems.

A *scalar* is a single function of position in space. Its value at a given point is independent of the particular coordinate system used to calculate it. For example, temperature has the same value at a point in space regardless of which coordinate system is used to calculate it, although its actual dependence on the coordinates might be quite different in different coordinate systems.

A scalar which has the same algebraic form in all coordinate systems is said to be an *invariant*. The magnitude of a vector has the *same* functional dependence on the components of the vector in all coordinate systems and, hence, is an *invariant*.

It is no longer necessary to think of a vector as a "directed line segment" drawn between two points in space. For instance, at each point of the trajectory of a particle there are associated three numbers which we call the three components of the velocity. Since the particle velocity changes from point to point, these three numbers are functions of the position. Whether or not they are to be regarded as the components of a vector is determined by how they transform. In other words, if two observers in entirely different Cartesian coordinate systems measure the three components of the velocity, then the components as measured in one frame should be related to those as measured in the other by means of (1.4.4) if velocity is rightfully to be regarded as a vector. That this is so is established in detail in Chapter two.

1.6. THE DOT PRODUCT

Let **a** and **b** be any two vectors. The scalar, inner, or dot product is indicated by the symbol $\mathbf{a} \cdot \mathbf{b}$ and is defined by

$$\mathbf{a} \cdot \mathbf{b} = a_i b_i = a_1 b_1 + a_2 b_2 + a_3 b_3 \tag{1.6.1}$$

The dot product is an invariant:

$$\begin{aligned} a_i' b_i' &= S_{ij} S_{ik} a_j b_k \\ &= \delta_{jk} a_j b_k \\ &= a_k b_k \end{aligned} \tag{1.6.2}$$

The square of the length of a vector is the dot product of the vector with itself. Given two vectors, there are essentially only three invariants which can be formed: namely, the two lengths and the dot product of the two vectors. All other apparently different invariants are functions of these three.

The dot product has a simple geometrical interpretation. Let **a** and **b** be two vectors emanating from a common point. Choose a coordinate system in such a way that the x_1x_2-plane coincides with the plane determined by **a** and **b** (Fig. 1.6.1). From the geometry one sees that

$$
\begin{aligned}
\mathbf{a} \cdot \mathbf{b} &= a_1 b_1 + a_2 b_2 \\
&= ab \cos \alpha \cos \beta + ab \sin \alpha \sin \beta \\
&= ab \cos (\alpha - \beta) \\
&= ab \cos \theta
\end{aligned}
\tag{1.6.3}
$$

Thus the dot product is an invariant the magnitude of which is the product of the lengths of the two vectors times the cosine of the included angle. If two vectors are orthogonal, the $\theta = 90°$ and their dot product is zero.

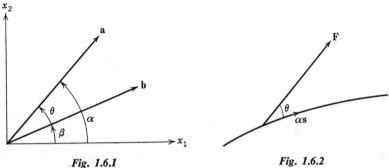

Fig. 1.6.1   *Fig. 1.6.2*

Hence, as we previously stated, the rows of the matrix (1.4.14) are mutually orthogonal. The choice of a special coordinate system to establish (1.6.3) is of no consequence since $\mathbf{a} \cdot \mathbf{b} = ab \cos \theta$ is an invariant expression.

As an example of the use of the dot product, recall from elementary physics that the work done on a particle by a force **F** in displacing it by $d\mathbf{s}$ (Fig. 1.6.2) is

$$
\begin{aligned}
dW &= F \, ds \cos \theta \\
&= \mathbf{F} \cdot d\mathbf{s} = F_1 \, dx_1 + F_2 \, dx_2 + F_3 \, dx_3
\end{aligned}
\tag{1.6.4}
$$

where dx_1, dx_2, and dx_3 are the components of $d\mathbf{s}$.

Unit vectors are of frequent use. In general, a unit vector $\hat{\mathbf{n}}$ has the property that

$$
\hat{\mathbf{n}} \cdot \hat{\mathbf{n}} = 1
\tag{1.6.5}
$$

Any vector **a** can be expressed

$$
\mathbf{a} = a\hat{\mathbf{n}}
\tag{1.6.6}
$$

where $\hat{\mathbf{n}}$ is a unit vector in the direction of **a** and a is the magnitude of **a**. Now **a** can be a physical quantity such as velocity or acceleration with definite units. Since

$$\hat{n} = \frac{a}{a} \tag{1.6.7}$$

$\hat{n}$ itself has no units or dimensions. It indicates direction only. The components of $\hat{n}$ are actually the cosines of the angles between $\hat{n}$ and the coordinate axes, i.e. the *direction cosines*.

As a further illustration of the use of the dot product, imagine a surface element of area $d\sigma$ in a fluid flowing with a velocity **u**. The problem is to calculate the volume of fluid per second which flows across

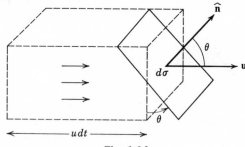

Fig. 1.6.3

this surface. The projection of the element of area perpendicular to the fluid flow is $d\sigma \cos \theta$, Fig. 1.6.3. In a time interval dt a volume of fluid $dV = u \, dt \, d\sigma \cos \theta$ will flow through the surface. The volume of fluid per second is

$$\frac{dV}{dt} = u \, d\sigma \cos \theta \tag{1.6.8}$$

Let $\hat{n}$ be a unit vector perpendicular to the surface. The element of surface area can be thought of as a vector defined by

$$d\boldsymbol{\sigma} = \hat{n} \, d\sigma \tag{1.6.9}$$

Equation (1.6.8) can then be expressed

$$\frac{dV}{dt} = \mathbf{u} \cdot d\boldsymbol{\sigma} = \mathbf{u} \cdot \hat{n} \, d\sigma = u_n \, d\sigma \tag{1.6.10}$$

where u_n represents the component of the velocity perpendicular to the surface.

An important concept in many calculations is that of solid angle. Consider a closed surface surrounding a point P, Fig. 1.6.4. A very narrow cone with its vertex at P intersects an element of area of magnitude

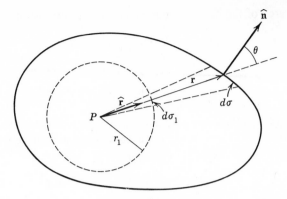

Fig. 1.6.4

$d\sigma$ on the surface. The element of solid angle subtended at P by $d\sigma$ is defined to be

$$d\Omega = \frac{\hat{\mathbf{r}} \cdot d\boldsymbol{\sigma}}{r^2} \qquad (1.6.11)$$

where $\mathbf{r}$ represents the position vector from P to $d\sigma$ and

$$\hat{\mathbf{r}} = \frac{\mathbf{r}}{r} \qquad (1.6.12)$$

is a unit vector in the direction of $\mathbf{r}$. We will show that

$$\int_{\text{surface}} d\Omega = 4\pi \qquad (1.6.13)$$

where the integral is taken over the entire closed surface. The result is independent of the shape of the surface.

Consider a spherical surface of radius r_1 surrounding P. The solid angle subtended by $d\sigma_1$ at P is

$$d\Omega_1 = \frac{d\sigma_1}{r_1^2} \qquad (1.6.14)$$

This is because $\hat{\mathbf{r}}$ is perpendicular to the surface of the sphere and

$$d\boldsymbol{\sigma}_1 \cdot \hat{\mathbf{r}} = d\sigma_1.$$

If (1.6.14) is integrated over the surface of the sphere, $1/r_1^2$ remains constant and can be taken outside the integral sign:

$$\int_{\text{sphere}} d\Omega_1 = \frac{1}{r_1^2} \int d\sigma_1 = \frac{1}{r_1^2} 4\pi r_1^2 = 4\pi \qquad (1.6.15)$$

Since the area of the base of a cone increases in proportion to r_1^2

$$d\Omega_1 = \frac{d\sigma_1}{r_1^2} = \frac{d\boldsymbol{\sigma}\cdot\hat{\mathbf{r}}}{r^2} = d\Omega \qquad (1.6.16)$$

Hence

$$\int_{\text{surface}} d\Omega = \int_{\text{sphere}} d\Omega_1 = 4\pi \qquad (1.6.17)$$

The basic mathematical properties of the dot product are established directly from the definition and are

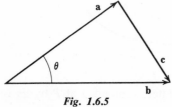

Fig. 1.6.5

VII $\quad$ $\mathbf{a}\cdot\mathbf{b} = \mathbf{b}\cdot\mathbf{a}$

VIII $\quad$ $\mathbf{a}\cdot(\mathbf{b}+\mathbf{c}) = \mathbf{a}\cdot\mathbf{b}+\mathbf{a}\cdot\mathbf{c}$

IX $\quad$ $\alpha(\mathbf{a}\cdot\mathbf{b}) = (\alpha\mathbf{a})\cdot\mathbf{b} = \mathbf{a}\cdot(\alpha\mathbf{b})$

where α is a real scalar.

Vectors can be used to establish the cosine law. Let $\mathbf{a}$, $\mathbf{b}$, and $\mathbf{c}$ be three vectors forming the sides of a triangle, Fig. 1.6.5. They are related by

$$\mathbf{b} - \mathbf{a} = \mathbf{c} \qquad (1.6.18)$$

If both sides of (1.6.18) are squared then

$$(\mathbf{b} - \mathbf{a})\cdot(\mathbf{b} - \mathbf{a}) = \mathbf{c}\cdot\mathbf{c} \qquad (1.6.19)$$

If properties VII and VIII are invoked,

$$\mathbf{a}\cdot\mathbf{a} - \mathbf{a}\cdot\mathbf{b} - \mathbf{b}\cdot\mathbf{a} + \mathbf{b}\cdot\mathbf{b} = \mathbf{c}\cdot\mathbf{c}$$

$$\mathbf{a}\cdot\mathbf{a} - 2\mathbf{a}\cdot\mathbf{b} + \mathbf{b}\cdot\mathbf{b} = \mathbf{c}\cdot\mathbf{c} \qquad (1.6.20)$$

$$a^2 - 2ab\cos\theta + b^2 = c^2$$

Note that squaring a vector equation such as (1.6.18) is very much like squaring an ordinary algebraic equation.

Important properties of the unit vectors $\hat{\mathbf{i}}$, $\hat{\mathbf{j}}$, $\hat{\mathbf{k}}$ are

$$\hat{\mathbf{i}}\cdot\hat{\mathbf{i}} = \hat{\mathbf{j}}\cdot\hat{\mathbf{j}} = \hat{\mathbf{k}}\cdot\hat{\mathbf{k}} = 1$$
$$\hat{\mathbf{i}}\cdot\hat{\mathbf{j}} = \hat{\mathbf{j}}\cdot\hat{\mathbf{k}} = \hat{\mathbf{k}}\cdot\hat{\mathbf{i}} = 0 \qquad (1.6.21)$$

The above properties are stated compactly by using the subscript notation:

$$\hat{\mathbf{e}}_i\cdot\hat{\mathbf{e}}_j = \delta_{ij} \qquad (1.6.22)$$

1.7. A REPRESENTATION OF AN ORTHOGONAL TRANSFORMATION

The vector $\mathbf{a}$ can be represented in either of the two coordinate systems illustrated in Fig. 1.7.1:

$$\mathbf{a} = a_i'\hat{\mathbf{e}}_i' = a_i\hat{\mathbf{e}}_i \qquad (1.7.1)$$

If one forms the dot product of both sides of (1.7.1) with the unit vector $\hat{\mathbf{e}}_j{}'$, the result is

$$a_i{}'(\hat{\mathbf{e}}_j{}' \cdot \hat{\mathbf{e}}_i{}') = a_i(\hat{\mathbf{e}}_j{}' \cdot \hat{\mathbf{e}}_i)$$

By means of (1.6.22) we are led to

$$a_j{}' = (\cos \alpha_{ji})a_i \qquad (1.7.2)$$

where α_{ji} are the angles between the new and the old axis. For instance, the angles illustrated in Fig. 1.7.1 are given by

$$\hat{\mathbf{e}}_1{}' \cdot \hat{\mathbf{e}}_1 = \cos \alpha_{11} \qquad \hat{\mathbf{e}}_1{}' \cdot \hat{\mathbf{e}}_2 = \cos \alpha_{12}$$
$$\hat{\mathbf{e}}_1{}' \cdot \hat{\mathbf{e}}_3 = \cos \alpha_{13} \qquad (1.7.3)$$

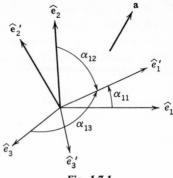

Fig. 1.7.1

The comparison of (1.7.2) and (1.4.4) shows that

$$S_{ji} = \cos \alpha_{ji} \qquad (1.7.4)$$

Thus the transformation coefficients in a proper orthogonal transformation or rotation are the *direction cosines* of the new axes with respect to the old. The direction cosines are also the components of the new basis vectors $\mathbf{e}_i{}'$ expressed in the old system:

$$\hat{\mathbf{e}}_i{}' = \cos \alpha_{ij}\hat{\mathbf{e}}_j = S_{ij}\hat{\mathbf{e}}_j \qquad (1.7.5)$$

which gives, incidentally, the transformation property of the basis vectors themselves. Moreover, the rows of the transformation matrix (1.4.14) are now seen to be the components of the new basis vectors expressed in the old system.

1.8. CARTESIAN TENSORS

A set of nine functions of position is said to be a *second rank Cartesian tensor* if they transform according to

$$A'_{ij} = S_{ik}S_{jl}A_{kl} \qquad (1.8.1)$$

under orthogonal transformations.

Second rank tensors are also called *diadics*. Just as we think of **a** as representing a vector, so we can think of a single boldface letter **A** as representing a tensor. In a particular coordinate system, **a** is represented by three components whereas the tensor **A** is represented by nine. The components of **A** in different coordinate systems can be conveniently displayed by means of matrices:

$$A = \begin{pmatrix} A_{11} & A_{12} & A_{13} \\ A_{21} & A_{22} & A_{23} \\ A_{31} & A_{32} & A_{33} \end{pmatrix}, \qquad A' = \begin{pmatrix} A'_{11} & A'_{12} & A'_{13} \\ A'_{21} & A'_{22} & A'_{23} \\ A'_{31} & A'_{32} & A'_{33} \end{pmatrix} \qquad (1.8.2)$$

Notice that ordinary italic letters are used for the matrix representations of a tensor in various coordinate systems. The tensor **A** itself has meaning independent of any particular coordinate system; or to put it another way, **A** stands for the sum total of all possible representations of a given tensor. According to the definition, these representations must be related to one another by transformations of the type (1.8.1). Other than the fact that there are nine components in a given coordinate system instead of three, the algebra of tensors is identical to that of vectors. A vector is actually a tensor of rank one and a scalar is a tensor of rank zero. Thus two tensors can be added together to give a third:

$$\mathbf{A} + \mathbf{B} = \mathbf{C} \tag{1.8.3}$$

In terms of components, (1.8.3) is understood to mean

$$\begin{pmatrix} A_{11} & A_{12} & A_{13} \\ A_{21} & A_{22} & A_{23} \\ A_{31} & A_{32} & A_{33} \end{pmatrix} + \begin{pmatrix} B_{11} & B_{12} & B_{13} \\ B_{21} & B_{22} & B_{23} \\ B_{31} & B_{32} & B_{33} \end{pmatrix}$$

$$= \begin{pmatrix} A_{11} + B_{11} & A_{12} + B_{12} & A_{13} + B_{13} \\ A_{21} + B_{21} & A_{22} + B_{22} & A_{23} + B_{23} \\ A_{31} + B_{31} & A_{32} + B_{32} & A_{33} + B_{33} \end{pmatrix} = \begin{pmatrix} C_{11} & C_{12} & C_{13} \\ C_{21} & C_{22} & C_{23} \\ C_{31} & C_{32} & C_{33} \end{pmatrix} \tag{1.8.4}$$

or, more simply,

$$A_{ij} + B_{ij} = C_{ij} \tag{1.8.5}$$

The list of the basic laws of vector algebra I through V that appears in Section 1 also applies to tensors. As with vectors, all components of the zero tensor are zero. Two tensors are equal if, and only if, their components are equal.

If α is a scalar, then the tensor $\alpha\mathbf{A}$ has components

$$\begin{pmatrix} \alpha A_{11} & \alpha A_{12} & \alpha A_{13} \\ \alpha A_{21} & \alpha A_{22} & \alpha A_{23} \\ \alpha A_{31} & \alpha A_{32} & \alpha A_{33} \end{pmatrix} \tag{1.8.6}$$

Every component of **A** is to be multiplied by α. Scalar multiplication of tensors obeys the rules VI through VIII of Section 1.1. The transformation (1.8.1) is inverted by forming the product of both sides with $S_{in}S_{jm}$:

$$\begin{aligned} S_{in}S_{jm}A'_{ij} &= S_{in}S_{jm}S_{ik}S_{jl}A_{kl} \\ &= (S_{in}S_{ik})(S_{jm}S_{jl})A_{kl} \\ &= \delta_{nk}\,\delta_{ml}A_{kl} = A_{nm} \end{aligned} \tag{1.8.7}$$

An example of a tensor is the Kronecker delta:

$$S_{ij}S_{kl}\,\delta_{jl} = S_{ij}S_{kj} = \delta_{ik} \tag{1.8.8}$$

The Kronecker delta is an unusual tensor because its components have the same numerical values in all coordinate systems. Tensors which have this property are called *isotropic tensors*. It can be shown that the Kronecker delta is the *only* second rank isotropic tensor. This tensor plays much the same role in tensor theory as the number *one* plays in the theory of ordinary numbers. Sometimes this tensor is represented by *I* and called the *identity tensor*:

$$I = \begin{pmatrix} 1 & 0 & 0 \\ 0 & 1 & 0 \\ 0 & 0 & 1 \end{pmatrix} \tag{1.8.9}$$

A second rank tensor can be constructed out of two vectors. Symbolically this is indicated by

$$\mathbf{A} = \mathbf{ab} \tag{1.8.10}$$

and called the *direct product* of **a** and **b**. The components of (1.8.10) are defined to be

$$\mathbf{A} = \begin{pmatrix} a_1b_1 & a_1b_2 & a_1b_3 \\ a_2b_1 & a_2b_2 & a_2b_3 \\ a_3b_1 & a_3b_2 & a_3b_3 \end{pmatrix} \tag{1.8.11}$$

or simply $A_{ij} = a_ib_j$. That A has the correct transformation property is readily verified:

$$A'_{ij} = a_i'b_j' = S_{ik}a_kS_{jl}b_l$$
$$= S_{ik}S_{jl}a_kb_l = S_{ik}S_{jl}A_{kl}$$

By writing out the components of the tensor **B** = **ba** and comparing with (1.8.11) one sees that they are not identical but that B is the transpose of A:

$$B = \tilde{A} \tag{1.8.12}$$

Notice that (1.8.12) can be replaced by the component equation

$$B_{ij} = A_{ji} \tag{1.8.13}$$

meaning that the element in row i and column j of B is found in row j and column i of A.

Whereas a second rank tensor can be constructed out of two vectors, the reverse is not necessarily true. A second rank tensor cannot, in general, be decomposed into two vectors.

The components of a *symmetric tensor* satisfy the condition

$$A_{ij} = A_{ji} \tag{1.8.14}$$

In matrix notation

$$\tilde{A} = A \tag{1.8.15}$$

For instance, such a tensor would result if the two vectors used in constructing (1.8.10) were the same.

Let us consider an example from physics. In the electrostatics of homogeneous, isotropic media, the relation between the displacement vector $\mathbf{D}$ and the electric field $\mathbf{E}$ is

$$\mathbf{D} = \epsilon \mathbf{E} \tag{1.8.16}$$

where ϵ is the dielectric constant. A medium such as crystal may have different properties in different directions and hence is not necessarily isotropic. The polarization, and hence $\mathbf{D}$, is not necessarily in the same direction as the applied field $\mathbf{E}$. The components of $\mathbf{D}$ are to be found from those of $\mathbf{E}$ by means of

$$
\begin{aligned}
D_1 &= \epsilon_{11}E_1 + \epsilon_{12}E_2 + \epsilon_{13}E_3 \\
D_2 &= \epsilon_{21}E_1 + \epsilon_{22}E_2 + \epsilon_{23}E_3 \\
D_3 &= \epsilon_{31}E_1 + \epsilon_{32}E_2 + \epsilon_{33}E_3
\end{aligned}
\tag{1.8.17}
$$

or, in shorthand notation $D_i = \epsilon_{ij}E_j$. Suppose it is definitely known that $\mathbf{D}$ is a vector for all possible values of the externally applied field $\mathbf{E}$:

$$D_i = S_{ji}D_j' \qquad E_j = S_{kj}E_k' \tag{1.8.18}$$

From this information it is possible to establish that ϵ_{ij} are the components of a second rank tensor. By substitution of (1.8.18) into (1.8.17),

$$S_{ji}D_j' = \epsilon_{ij}S_{kj}E_k' \tag{1.8.19}$$

Multiplication by S_{ni} and use of the orthogonality relation leads to

$$
\begin{aligned}
S_{ni}S_{ji}D_j' &= \epsilon_{ij}S_{ni}S_{kj}E_k' \\
D_n' &= S_{ni}S_{kj}\epsilon_{ij}E_k'
\end{aligned}
\tag{1.8.20}
$$

In the primed frame of reference, (1.8.17) reads

$$D_n' = \epsilon_{nk}'E_k' \tag{1.8.21}$$

By comparison of (1.8.21) and (1.8.20) and use of the arbitrariness of $\mathbf{E}$ we find

$$\epsilon_{nk}' = S_{ni}S_{kj}\epsilon_{ij} \tag{1.8.22}$$

thus establishing the tensor character of ϵ_{ij}.

The relation between **D** and **E** can be expressed

$$\mathbf{D} = \epsilon \mathbf{E} \qquad (1.8.23)$$

In this form, we think of the tensor ϵ as being an *operator* which transforms the vector **E** into the vector **D**. This is *not* a coordinate transformation. The components of all quantities, **D**, ϵ, and **E** are expressed in the same coordinate system. The action of ϵ on **E** is to produce a new vector **D**. Electromagnetic theory shows that ϵ is a symmetric tensor.

An *antisymmetric tensor* is one for which the components obey

$$A_{ij} = -A_{ji} \qquad (1.8.24)$$

or, in terms of a matrix representation,

$$\tilde{A} = -A \qquad (1.8.25)$$

The diagonal elements A_{11}, A_{22}, and A_{33} of an antisymmetric tensor are necessarily zero. If in (1.8.24) $i = j = 1$, then

$$A_{11} = -A_{11}, \qquad 2A_{11} = 0 \qquad (1.8.26)$$

Similarly,

$$A_{22} = A_{33} = 0 \qquad (1.8.27)$$

Symmetry and antisymmetry are invariant properties. If it is known that $A_{ij} = A_{ji}$ in one coordinate system, then in any other coordinate system $A'_{ij} = A'_{ji}$:

$$A'_{ij} = S_{ik}S_{jn}A_{kn} = S_{ik}S_{jn}A_{nk} \qquad (1.8.28)$$

k and n are dummy indices and may be interchanged, i.e., (1.8.28) means the same thing as

$$A'_{ij} = S_{in}S_{jk}A_{kn} = S_{jk}S_{in}A_{kn} = A'_{ji} \qquad (1.8.29)$$

A similar proof can be done for the case where A is antisymmetric. Any second rank tensor can be expressed as the sum of a symmetric and an antisymmetric part:

$$A_{ij} = \tfrac{1}{2}(A_{ij} + A_{ji}) + \tfrac{1}{2}(A_{ij} - A_{ji}) \qquad (1.8.30)$$

The components

$$B_{ij} = \tfrac{1}{2}(A_{ij} + A_{ji})$$

are symmetric whereas

$$C_{ij} = \tfrac{1}{2}(A_{ij} - A_{ji})$$

are antisymmetric. A symmetric tensor has six independent components; an antisymmetric tensor has only three.

Given a vector, it is possible to construct a single invariant, namely the length of the vector. All other apparently different invariants are functions of the length. With a second rank tensor there are associated three invariants,* one of which we now study. It is called the *trace* or *spur* of the tensor and is the sum of the diagonal elements

$$\text{tr } A = A_{ii} = A_{11} + A_{22} + A_{33} \tag{1.8.31}$$

The proof that (1.8.31) represents an invariant is left as a problem. Note that the trace of the tensor (1.8.10) is the scalar product of the vectors **a** and **b**.

The idea of an inner or dot product can be generalized. Given the components of a vector and a second rank tensor, there are two possible ways of forming an inner product:

$$b_i = A_{ij}a_j \qquad c_j = a_i A_{ij} \tag{1.8.32}$$

As is implied by the notation, b_i and c_j form the components of vectors. This fact must be established from the known transformation properties of A_{ij} and a_j:

$$\begin{aligned}
b_i' &= A_{ij}'a_j' = S_{ik}S_{jl}A_{kl}S_{jm}a_m \\
&= (S_{jl}S_{jm})S_{ik}A_{kl}a_m \\
&= \delta_{lm}S_{ik}A_{kl}a_m \\
&= S_{ik}A_{kl}a_l = S_{ik}b_k
\end{aligned} \tag{1.8.33}$$

A similar proof holds for the components c_j. Notice that **b** and **c** are not the same vector unless **A** is a symmetric tensor.

Let **A** and **B** be two second rank tensors. From them it is possible to construct other second rank tensors by forming inner products in various ways:

$$\begin{aligned}
C_{ik} = A_{ij}B_{jk} \qquad D_{ik} = A_{ij}B_{kj} \\
E_{ik} = A_{ji}B_{jk} \qquad F_{ik} = A_{ji}B_{kj}
\end{aligned} \tag{1.8.34}$$

The proof that $C_{ik}, D_{ik} \ldots$ form the components of tensors is left as a problem.

The following theorem is important. A tensor equation which is valid in one coordinate system is valid in any other. In other words, if it is known in one coordinate system that

$$\alpha A_{ij} + \beta B_{ij} + \cdots = 0 \tag{1.8.35}$$

where $\alpha, \beta, \ldots$ are scalars, then in any other coordinate system

$$\alpha A_{ij}' + \beta B_{ij}' + \cdots = 0 \tag{1.8.36}$$

---

* The number of invariants associated with a second rank tensor is equal to the dimension of the space in which it is defined.

starting with (1.8.35) one has

$$(\alpha A_{ij} + \beta B_{ij} + \cdots)S_{ki}S_{lj} = 0$$
$$(\alpha S_{ki}S_{lj}A_{ij} + \beta S_{ki}S_{lj}B_{ij} + \cdots) = 0 \qquad (1.8.37)$$
$$\alpha A'_{kl} + \beta B'_{kl} + \cdots = 0$$

The same statement can be made about a vector equation. The theorem just established allows us to write the fundamental equations of physics in such a way that they are automatically valid in all coordinate systems, and hence do not depend on any particular coordinate system. For example, Newton's second law of motion for a material particle of mass m can be expressed

$$F_i = ma_i \qquad (1.8.38)$$

We are then assured that in any other Cartesian coordinate system

$$F_i' = ma_i' \qquad (1.8.39)$$

We say that (1.8.38) expresses Newton's second law in *covariant form*. Of course, we have restricted ourselves to transformations between all possible Cartesian coordinate systems. In Chapter 2, we take up the problem of expressing Newton's second law for a particle in such a way that it is *generally covariant*, i.e., valid in all possible curvilinear coordinate systems.

The theorem just proved has the following consequence. If the basic laws of tensor algebra I to VIII of Section 1.1 are established in one coordinate system by using component equations, they are automatically valid in all other coordinate systems.

One of the main programs of special relativity is to write the basic equations of physics in *Lorentz covariant* fashion. As will be seen in Chapter 13, Lorentz transformations treat the three spatial coordinates and time on an equal footing, and hence are *four-dimensional* transformations which satisfy the basic postulates of special relativity. If we admit complex numbers, Lorentz transformations can be considered to be four-dimensional orthogonal transformations.

Tensors of higher rank can be defined. For example, a third rank tensor is a set of 27 functions of position which obey the transformation

$$A'_{ijk} = S_{il}S_{jm}S_{kn}A_{lmn} \qquad (1.8.40)$$

The rank of a particular tensor quantity is the number of free or unsummed subscripts, e.g.,

$$a_i B_{ij} \qquad B_{ij}A_{ijk} \qquad a_i A_{ijk}b_j$$

are all of rank one, meaning that their transformation property is that of a vector. The reader should verify this in detail. The number of free subscripts must be the same in all terms of a tensor equation, e.g.,

$$\alpha A_{ijk}b_k + \beta B_{ij} + \gamma C_{ik}D_{kj} = 0$$

is a possible tensor equation.

1.9. THE PERMUTATION SYMBOL

The permutation symbol, sometimes called the *Levi-Civita density*, is defined by

$\delta_{ijk} = +1$ if i, j, and k are an *even* permutation of 1, 2, and 3

$\delta_{ijk} = -1$ if i, j, and k are an *odd* permutation of 1, 2, and 3   (1.9.1)

$\delta_{ijk} = 0$ if any of the subscripts are equal

Other names for the permutation symbol are the *alternating tensor* or the *indicator*. A permutation is *even* or *odd* as determined by whether an *even* or *odd* number of exchanges are required to derive it from 1, 2, 3. For instance, 231 is an *even* permutation ($231 \rightarrow 321 \rightarrow 123$), whereas 132 is *odd* ($132 \rightarrow 123$). Thus,

$$\delta_{231} = +1 \qquad \delta_{132} = -1$$

Also from the definition

$$\delta_{122} = 0, \qquad \delta_{121} = 0, \qquad \delta_{333} = 0, \qquad \text{etc.}$$

The expansion of a determinant can be accomplished by using the permutation symbol:

$$\det a_{ij} = \begin{vmatrix} a_{11} & a_{12} & a_{13} \\ a_{21} & a_{22} & a_{23} \\ a_{31} & a_{32} & a_{33} \end{vmatrix} = a_{1i}a_{2j}a_{3k}\delta_{ijk} \qquad (1.9.2)$$

This can be proved by working out the indicated sums in detail:

$$a_{1i}a_{2j}a_{3k}\delta_{ijk}$$

$$= a_{1i}a_{2j}(a_{31}\delta_{ij1} + a_{32}\delta_{ij2} + a_{33}\delta_{ij3})$$

$$= a_{1i}(a_{22}a_{31}\delta_{i21} + a_{23}a_{31}\delta_{i31})$$

$$\quad + a_{1i}(a_{21}a_{32}\delta_{i12} + a_{23}a_{32}\delta_{i32})$$

$$\quad + a_{1i}(a_{21}a_{33}\delta_{i13} + a_{22}a_{33}\delta_{i23})$$ (1.9.3)

$$= a_{13}a_{22}a_{31}\delta_{321} + a_{12}a_{23}a_{31}\delta_{231} + a_{13}a_{21}a_{32}\delta_{312}$$

$$\quad + a_{11}a_{23}a_{32}\delta_{132} + a_{12}a_{21}a_{33}\delta_{213} + a_{11}a_{22}a_{33}\delta_{123}$$

$$= - a_{13}a_{22}a_{31} + a_{12}a_{23}a_{31} + a_{13}a_{21}a_{32}$$

$$\quad - a_{11}a_{23}a_{32} - a_{12}a_{21}a_{33} + a_{11}a_{22}a_{33}$$

A direct expansion of the determinant gives the same result.

From the properties of the permutation symbol it can be proved that exchanging two rows of a determinant changes its sign. If in (1.9.2) a_{1i} and a_{2j} are exchanged,

$$\det a_{ij} = a_{2j}a_{1i}a_{3k}\delta_{ijk} \qquad (1.9.4)$$

Interchange of i and j on the permutation symbol changes its sign. Therefore,

$$\det a_{ij} = -a_{2j}a_{1i}a_{3k}\delta_{jik} \qquad (1.9.5)$$

The name of a summed or "dummy" subscript can always be changed. Therefore, in the right side of (1.9.5) i can be called j and j can be called i:

$$\det a_{ij} = -a_{2i}a_{1j}a_{3k}\delta_{ijk} \qquad (1.9.6)$$

which is the equivalent of exchanging the first and second rows. The expansion of a determinant can be expressed

$$\delta_{lmn}(\det a_{ij}) = a_{li}a_{mj}a_{nk}\delta_{ijk} \qquad (1.9.7)$$

If the rows have been exchanged an odd or even number of times, then δ_{lmn} automatically provides the correct sign. If in (1.9.7) the a_{ij} are replaced by the coefficients of an orthogonal transformation,

$$\pm\delta_{lmn} = S_{li}S_{mj}S_{nk}\delta_{ijk} \qquad (1.9.8)$$

where $\det S_{ij} = \pm 1$ has been used. Recall that $\det S_{ij} = -1$ if the transformation is improper. From (1.9.8) we can say that δ_{ijk} is a third rank tensor under *proper* orthogonal transformations. If the transformation is *improper*, e.g., if it involves a mirror reflection, the minus sign must be used in (1.9.8). Such tensors are called *pseudo-tensors*. Like the Kronecker delta, the components of the permutation symbol have the same numerical values in all coordinate systems. The permutation symbol is thus an isotropic tensor and it can be shown that it is the *only* such third rank tensor.

An important identity is

$$\delta_{ijk}\delta_{ilm} = \delta_{jl}\delta_{km} - \delta_{jm}\delta_{kl} \qquad (1.9.9)$$

A general proof is difficult, but (1.9.9) can easily be checked for particular values of the subscripts. For instance, if $j = k = 1$, then the left side of (1.9.9) is zero by the definition of δ_{ijk}. It is easy to see that the right side is zero also. If $j = l = 1$ and $k = m = 2$

$$\delta_{i12}\delta_{i12} = \delta_{112}\delta_{112} + \delta_{212}\delta_{212} + \delta_{312}\delta_{312}$$
$$= 0 + 0 + 1 = 1;$$
$$\delta_{11}\delta_{22} - \delta_{12}\delta_{21} = 1 - 0 = 1, \qquad \text{etc.}$$

A further property of the permutation symbol is

$$\delta_{ijk} = \delta_{kij} = \delta_{jki} \qquad (1.9.10)$$

such rearrangements of the subscripts are called *cyclic permutations*. A cyclic permutation of three letters always involves an even number of exchanges

$$ijk \rightarrow ikj \rightarrow kij$$

Hence all three symbols in (1.9.10) have the same value.

1.10. THE CROSS PRODUCT

Given two vectors **a** and **b**, it is possible to form from them a *pseudo-vector* by using the permutation symbol:

$$\gamma_i = \delta_{ijk} a_j b_k \qquad (1.10.1)$$

The three-component entity so formed suffers from the same difficulty as does the permutation symbol under orthogonal transformations:

$$\gamma_i{}' = \pm S_{ij}\gamma_j \qquad (1.10.2)$$

where the minus sign must be employed if the transformation is improper. A pseudo-vector is also called an *axial vector* and an ordinary vector is called a *polar vector*. Provided that improper transformations are excluded, γ has all the properties of an ordinary vector.

The components of γ work out to be

$$\gamma_1 = \delta_{1jk} a_j b_k = \delta_{123} a_2 b_3 + \delta_{132} a_3 b_2 = a_2 b_3 - a_3 b_2 \qquad (1.10.3)$$

Similarly,

$$\gamma_2 = a_3 b_1 - a_1 b_3, \qquad \gamma_3 = a_1 b_2 - a_2 b_1 \qquad (1.10.4)$$

One frequently uses the notation

$$\gamma = \mathbf{a} \times \mathbf{b} \qquad (1.10.5)$$

and calls γ the *cross product* of **a** and **b**. The cross product is non-commutative. According to the definition (1.10.1), the components of **b** × **a** are

$$\delta_{ijk} b_j a_k = -\delta_{ikj} b_j a_k = -\delta_{ijk} b_k a_j$$
$$= -\delta_{ijk} a_j b_k$$

Hence,

$$\mathbf{b} \times \mathbf{a} = -\mathbf{a} \times \mathbf{b} \qquad (1.10.6)$$

A useful vector identity is

$$\mathbf{a} \times (\mathbf{b} \times \mathbf{c}) = \mathbf{b}(\mathbf{a} \cdot \mathbf{c}) - \mathbf{c}(\mathbf{a} \cdot \mathbf{b}) \qquad (1.10.7)$$

The identity is established by use of (1.9.9) and (1.9.10). The components of $\mathbf{a} \times (\mathbf{b} \times \mathbf{c})$ are understood to be obtained by first forming $\boldsymbol{\gamma} = (\mathbf{b} \times \mathbf{c})$ and then $\mathbf{a} \times \boldsymbol{\gamma}$. The ith component of $\mathbf{a} \times (\mathbf{b} \times \mathbf{c})$, therefore, is

$$
\begin{aligned}
[\mathbf{a} \times (\mathbf{b} \times \mathbf{c})]_i &= \delta_{ijk} a_j \delta_{klm} b_l c_m = \delta_{kij} \delta_{klm} a_j b_l c_m \\
&= (\delta_{il}\delta_{jm} - \delta_{im}\delta_{jl}) a_j b_l c_m \\
&= a_m b_i c_m - c_i a_l b_l \\
&= b_i (\mathbf{a} \cdot \mathbf{c}) - c_i (\mathbf{a} \cdot \mathbf{b})
\end{aligned}
$$

Equation (1.10.7) is sometimes called the "back minus cab rule." It becomes apparent from (1.10.6) and (1.10.7) that the cross product is nonassociative;

$$
(\mathbf{a} \times \mathbf{b}) \times \mathbf{c} = -\mathbf{c} \times (\mathbf{a} \times \mathbf{b}) = -\mathbf{a}(\mathbf{c} \cdot \mathbf{b}) + \mathbf{b}(\mathbf{c} \cdot \mathbf{a})
$$

Hence,

$$
\mathbf{a} \times (\mathbf{b} \times \mathbf{c}) - (\mathbf{a} \times \mathbf{b}) \times \mathbf{c} = \mathbf{a}(\mathbf{c} \cdot \mathbf{b}) - \mathbf{c}(\mathbf{a} \cdot \mathbf{b}) \qquad (1.10.8)
$$

which is not necessarily zero.

Note that the components of $\mathbf{a} \times (\mathbf{b} \times \mathbf{c})$ involve the permutation symbol *twice*, meaning that an improper transformation gives two factors of -1. If $\mathbf{a}, \mathbf{b}$, and $\mathbf{c}$ are polar vectors, then so is $\mathbf{a} \times (\mathbf{b} \times \mathbf{c})$. By a similar argument, if $\boldsymbol{\beta}$ is an axial vector and $\mathbf{a}$ is a *polar vector*, then $\mathbf{a} \times \boldsymbol{\beta}$ is a *polar vector*.

The cross product has a simple geometrical interpretation. Let two vectors $\mathbf{a}$ and $\mathbf{b}$ originate at the same point in space. Introduce a coordinate system such that the $x_1 x_2$-plane coincides with the plane of $\mathbf{a}$ and $\mathbf{b}$, Fig. 1.10.1. Since $a_3 = 0$ and $b_3 = 0$, it follows from (1.10.3) and (1.10.4)

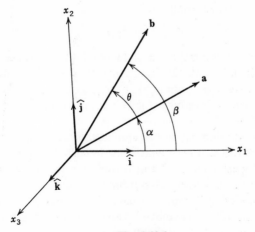

Fig. 1.10.1

that the cross product of **a** and **b** has only one component:

$$\gamma = \mathbf{a} \times \mathbf{b} = \hat{\mathbf{k}}(a_1 b_2 - a_2 b_1) \qquad (1.10.9)$$

From the geometry of Fig. 1.10.1,

$$a_1 b_2 - a_2 b_1 = ab(\cos \alpha \sin \beta - \sin \alpha \cos \beta)$$
$$= ab \sin (\beta - \alpha) = ab \sin \theta \qquad (1.10.10)$$

Thus **a** × **b** is in a direction perpendicular to the plane formed by **a** and **b** and has a magnitude equal to the product of the lengths of the vectors times the sine of the included angle. The magnitude of the cross product is the area of the parallelogram formed by **a** and **b**.

The cross product follows the *right-hand screw rule*. If **a** were turned into **b** with $\hat{\mathbf{k}}$ as axis, then a right-hand screw would advance in the direction of **a** × **b**. Note that

$$\mathbf{b} \times \mathbf{a} = -\hat{\mathbf{k}} ab \sin \theta \qquad (1.10.11)$$

If **b** is turned into **a**, then a right-hand screw advances in the direction of $-\hat{\mathbf{k}}$.

In the coordinate system of Fig. 1.10.1, the unit vectors $\hat{\mathbf{i}}$, $\hat{\mathbf{j}}$, $\hat{\mathbf{k}}$ are related by

$$\hat{\mathbf{i}} \times \hat{\mathbf{j}} = \hat{\mathbf{k}} \qquad \hat{\mathbf{j}} \times \hat{\mathbf{k}} = \hat{\mathbf{i}} \qquad \hat{\mathbf{k}} \times \hat{\mathbf{i}} = \hat{\mathbf{j}} \qquad (1.10.12)$$

$$\hat{\mathbf{i}} \times \hat{\mathbf{i}} = 0 \qquad \hat{\mathbf{j}} \times \hat{\mathbf{j}} = 0 \qquad \hat{\mathbf{k}} \times \hat{\mathbf{k}} = 0 \qquad (1.10.13)$$

Equations (1.10.13) follow from the fact that the angle between a vector and itself is zero.

The cross-product relations for the unit vectors can be stated compactly by means of the subscript notation:

$$\hat{\mathbf{e}}_i \times \hat{\mathbf{e}}_j = \delta_{ijk} \hat{\mathbf{e}}_k \qquad (1.10.14)$$

You should convince yourself that (1.10.14) is the equivalent of all the equations contained in (1.10.12) and (1.10.13). The coordinate system of Fig. 1.10.1 is *right-handed*, meaning that the first unit vector $\hat{\mathbf{i}}$ crossed into the second unit vector $\hat{\mathbf{j}}$ produces the third unit vector $\hat{\mathbf{k}}$ if a right-hand screw rule is followed.

The subscript notation x_1, x_2, x_3 emphasizes that there is a *definite ordering* of the coordinate axes which distinguishes a right- from a left-hand system. Proper orthogonal transformations are said to preserve the *chirality* (Greek: *cheir*, hand) of a coordinate system.

Is the existence of right- and left-hand coordinate systems merely a mathematical curiosity to be avoided by always using a right-hand system or does it have some deep lying significance?

Consider the improper transformation given by

$$S = \begin{pmatrix} -1 & 0 & 0 \\ 0 & 1 & 0 \\ 0 & 0 & 1 \end{pmatrix} \qquad (1.10.15)$$

and illustrated in Fig. 1.10.2. Suppose that there is a particle at point P that has velocity $\mathbf{u}$. According to (1.10.2), axial vectors reverse direction under improper transformations and of course we cannot set a particle

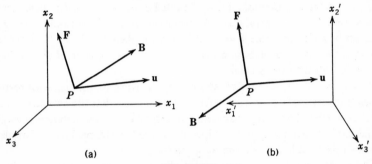

(a)   (b)

Fig. 1.10.2

moving in the opposite direction merely by changing the reference frame. Therefore, velocity must be a polar vector. It transformation is

$$u_1{}' = -u_1, \qquad u_2{}' = u_2, \qquad u_3{}' = u_3 \qquad (1.10.16)$$

Similarly, a force $\mathbf{F}$ which acts on the particle must be a polar vector.

Recall from electrodynamics that a particle of charge q and velocity $\mathbf{u}$ experiences a force in a magnetic field $\mathbf{B}$ given by the *Lorentz force*

$$\mathbf{F} = \frac{q}{c}(\mathbf{u} \times \mathbf{B}) \qquad (1.10.17)$$

where c is a constant. (Gaussian units are used in this book; c is the velocity of light and has the value 3.00×10^{10} cm per sec).* This is very curious because if $\mathbf{u}$ and $\mathbf{B}$ were both polar vectors, we should have to conclude that $\mathbf{F}$ is axial which cannot be. The inescapable conclusion is that the magnetic field is an axial vector! To find its transformation, we must use the *negative* sign in (1.10.2):

$$B_1{}' = B_1, \qquad B_2{}' = -B_2, \qquad B_3{}' = -B_3 \qquad (1.10.18)$$

---

* See Appendix B for a summary of units used in electromagnetic theory.

Equation (1.10.18) means that in the mirror-image coordinate system of Fig. 1.10.2b **B** is to be drawn in a direction opposite to that in Fig. 1.10.2a. How can it be that a mere change of reference frame reverses the physical direction of a magnetic field? The answer is that insofar as experiments with moving charges in a magnetic field are concerned, the direction of **B** is defined only up to ± 1. In other words, a "line of magnetic force" should be represented by just a line with no arrow on it!

The force and the velocity can be measured and have unambiguous directions. If a direction is to be assigned to **B**, an arbitrary choice must be made between the directions as illustrated by Figs. 1.10.2a and b. If Fig. 1.10.2b is adopted, formula (1.10.17) can be retained if a *left-hand* screw rule is used for cross products. Another possibility is to retain the right-hand screw rule and write $\mathbf{F} = (q/c)(\mathbf{B} \times \mathbf{u})$ in place of (1.10.17). In actual practice, the relation of **B** to **u** and **F** is chosen to be that of Fig. 1.10.2a and a right-hand rule is used.

It was long thought that the laws of physics were symmetric with respect to space inversion (i.e., mirror reflection), meaning that the choice between a right- and left-hand representation is arbitrary as, for instance, was just explained for the example of the Lorentz force. In the parlance of modern physics this is called *conservation of parity.*

There are recognized today essentially four types of forces or inter-actions between the fundamental particles of matter. The weakest of all forces are gravitational. Stronger than gravitational are the so-called weak interactions which are responsible for β-decay. Next in line are the electromagnetic forces. Finally, the strongest of all forces are nuclear forces. Of these four forces, three obey the principle of reflection symmetry. These are gravitational, electromagnetic, and nuclear. The conservation of parity breaks down in the case of weak-decay or beta-decay! In the original experimental confirmation of nonconservation of parity, the atoms of an isotope of cobalt that undergo beta decay were lined up by means of a magnetic field. With **B** in the upward direction, the electrons in the beta-decay were found to be emitted preferentially downward. Here is an experiment which allows us to assign a definite direction to a magnetic field!

It is really true ultimately that right or left symmetry is maintained if account is taken of the existence of *antimatter.* As it turns out, "right-handed" ordinary matter is symmetrical with "left-handed" antimatter. We cannot go further here into this interesting subject, but must leave it to advanced courses in quantum mechanics and particle theory. The forces of classical physics are produced by gravitation and electromagnetism and this book is primarily concerned with the treatment of the mechanics of particles in these two force fields by Newtonian or classical methods.

If **a**, **b**, and **c** represent polar vectors, then $(\mathbf{a} \times \mathbf{b}) \cdot \mathbf{c}$ must represent a *pseudo-scalar* the magnitude of which is the volume of the parallelopiped formed by the three vectors, Fig. 1.10.3. A pseudo-scalar changes sign under improper transformation. From the geometrical interpretations as a volume we can see that

$$(\mathbf{a} \times \mathbf{b}) \cdot \mathbf{c} = (\mathbf{c} \times \mathbf{a}) \cdot \mathbf{b} = (\mathbf{b} \times \mathbf{c}) \cdot \mathbf{a} \qquad (1.10.19)$$

This result can also be established algebraically:

$$\begin{aligned}
(\mathbf{a} \times \mathbf{b}) \cdot \mathbf{c} &= \delta_{ijk} c_i a_j b_k \\
&= \delta_{kij} c_i a_j b_k = (\mathbf{c} \times \mathbf{a}) \cdot \mathbf{b}
\end{aligned} \qquad (1.10.20)$$

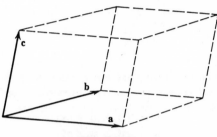

Fig. 1.10.3

The cyclic permutation of vectors in (1.10.19) is equivalent to the cyclic permutation of the subscripts of the permutation symbol in (1.10.20).

It is easily proved that the cross product obeys

$$\mathbf{a} \times (\mathbf{b} + \mathbf{c}) = \mathbf{a} \times \mathbf{b} + \mathbf{a} \times \mathbf{c} \qquad (1.10.21)$$

$$(\mathbf{b} + \mathbf{c}) \times \mathbf{a} = \mathbf{b} \times \mathbf{a} + \mathbf{c} \times \mathbf{a} \qquad (1.10.22)$$

PROBLEMS

The problems are numbered according to the following rule: The first numeral stands for the chapter, the second numeral indicates the section to which the problem is relevant, and the third is the number of the problem.

1.3.1   The tips of three independent vectors from a common origin O define a plane. Under what condition will a fourth vector, also with its origin at O, have its tip on this plane?

1.3.2   Verify that $\mathbf{a} = (1, 2, 3)$, $\mathbf{b} = (1, 1, 0)$, $\mathbf{c} = (2, 0, 3)$ are independent. Express $\mathbf{d} = (0, 1, 4)$ in terms of $\mathbf{a}$, $\mathbf{b}$, and $\mathbf{c}$.

1.3.3   The vectors

$$\mathbf{a} = \hat{\imath} + \hat{\jmath} + 2\hat{k}$$
$$\mathbf{b} = 2\hat{\imath} - 2\hat{\jmath} + 2\hat{k}$$

extend from the origin of coordinates. Show that the line joining the tips of these vectors is parallel to the xy-plane and find its length.

1.4.4   A vector **a** has components $(a_i) = (1, 1, 1)$. Find the components of **a** in a new coordinate system obtained by a 30° rotation about the x_2-axis.

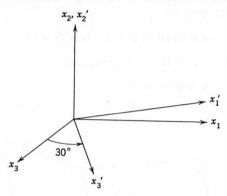

1.6.5   If **a** is a constant vector and **r** is a vector from the origin to a point (x_1, x_2, x_3), show that $(\mathbf{r} - \mathbf{a}) \cdot \mathbf{r} = 0$ represents the equation of a sphere.

1.6.6   Show that

$$\mathbf{a} = \tfrac{1}{7}(2\mathbf{\hat{i}} + 3\mathbf{\hat{j}} + 6\mathbf{\hat{k}})$$

$$\mathbf{b} = \tfrac{1}{7}(3\mathbf{\hat{i}} - 6\mathbf{\hat{j}} + 2\mathbf{\hat{k}})$$

$$\mathbf{c} = \tfrac{1}{7}(6\mathbf{\hat{i}} + 2\mathbf{\hat{j}} - 3\mathbf{\hat{k}})$$

are mutually orthogonal unit vectors.

1.6.7   If **a** is a constant vector and **r** is a vector from the origin to (x_1, x_2, x_3), what kind of a surface does $\mathbf{r} \cdot \mathbf{a} = $ constant represent?

1.6.8   If in Fig. 1.6.4, the point P lies outside the closed surface, prove that the total solid angle subtended at P is zero. The unit vector **n** is always drawn *outward* from the closed surface.

1.6.9   A point P is on the axis and at the center of a cylinder of length s and radius a. Calculate the solid angle subtended at P by the cylinder. The two ends of the cylinder are *open* so that it does not form a closed surface.

1.6.10   Prove that $\mathbf{a} \cdot (\mathbf{b} + \mathbf{c}) = \mathbf{a} \cdot \mathbf{b} + \mathbf{a} \cdot \mathbf{c}$ directly from the definition of the dot product.

1.6.11   If **a**, **b**, and **c** are three vectors, what conclusions can be drawn if $\mathbf{a} \cdot \mathbf{b} = \mathbf{a} \cdot \mathbf{c}$?

1.8.12   At a particular point in space the components of a second rank tensor are

$$A = \begin{pmatrix} 1 & 0 & 1 \\ 0 & 0 & 0 \\ 1 & 0 & 1 \end{pmatrix}$$

in a frame x_1, x_2, x_3. What are the components of this tensor in a frame x_1', x_2', x_3' obtained by the transformation of Problem 1.4.4?

1.8.13 If **a**, **b**, and **c** are three linearly independent vectors, show that

$$A_{ij}a_j = B_{ij}a_j \qquad A_{ij}b_j = B_{ij}b_j \qquad A_{ij}c_j = B_{ij}c_j$$

implies $A_{ij} = B_{ij}$.

1.8.14 Prove that the trace of a tensor is an invariant.

1.8.15 If **A** is a second rank tensor and **b** is a vector, prove that $A_{ij}b_k$ form the components of a third rank tensor.

1.8.16 Given that A_{ij} and B_{ij} are the components of second rank tensors, prove that $C_{ik} = A_{ij}B_{jk}$ are also components of a second rank tensor.

1.8.17 What is $A_{ij}B_{ij}$? Support your answer with an appropriate proof.

1.8.18 Given that

$$A = \begin{pmatrix} 0 & 1 & 2 \\ 4 & 2 & 0 \\ 1 & 0 & 3 \end{pmatrix} \qquad a_i = (1, 2, 0).$$

work out the numerical values of

$$c_i = A_{ij}a_j, \qquad b_i = A_{ji}a_j, \qquad A_{ii}, \qquad a_iA_{ij}a_j$$

1.4.19 Could

$$S = \begin{pmatrix} \dfrac{1}{2} & -\dfrac{1}{\sqrt{2}} & \dfrac{1}{2} \\ -\dfrac{1}{2} & -\dfrac{1}{\sqrt{2}} & -\dfrac{1}{2} \\ -\dfrac{1}{\sqrt{2}} & 0 & \dfrac{1}{\sqrt{2}} \end{pmatrix}$$

represent a rotation?

1.9.20 What kind of transformation property does $\delta_{ijk}\delta_{ilm}$ have?

1.9.21 Derive the formula

$$\frac{d}{dt}(\det A_{ij}) = \begin{vmatrix} \dot{A}_{11} & \dot{A}_{12} & \dot{A}_{13} \\ A_{21} & A_{22} & A_{23} \\ A_{31} & A_{32} & A_{33} \end{vmatrix} + \begin{vmatrix} A_{11} & A_{12} & A_{13} \\ \dot{A}_{21} & \dot{A}_{22} & \dot{A}_{23} \\ A_{31} & A_{32} & A_{33} \end{vmatrix} + \begin{vmatrix} A_{11} & A_{12} & A_{13} \\ A_{21} & A_{22} & A_{23} \\ \dot{A}_{31} & \dot{A}_{32} & \dot{A}_{33} \end{vmatrix}$$

1.9.22 Show that

$$\hat{e}_l \cdot \hat{e}_m \times \hat{e}_n = \delta_{lmn}$$

1.9.23 Show that $C_{jk} = a_jb_k - a_kb_j$ is a second rank antisymmetric tensor. Show that the cross product of **a** and **b** is represented by

$$C_i = \tfrac{1}{2}\delta_{ijk}C_{jk}.$$

1.9.24 Show that if any two rows (or any two columns) of a determinant are identical, the value of the determinant is zero.

1.9.25   Write the expansion of a 4 × 4 determinant in terms of a suitably defined permutation symbol.

1.9.26   If α is a scalar, show that

$$\alpha \det A_{ij} = \begin{vmatrix} \alpha A_{11} & A_{12} & A_{13} \\ \alpha A_{21} & A_{22} & A_{23} \\ \alpha A_{31} & A_{32} & A_{33} \end{vmatrix} = \begin{vmatrix} A_{11} & \alpha A_{12} & A_{13} \\ A_{21} & \alpha A_{22} & A_{23} \\ A_{31} & \alpha A_{32} & A_{33} \end{vmatrix} = \cdots$$

1.9.27   Show that the value of a determinant is not altered if a column (row) is multiplied by a scalar and the result added to another column (row), e.g.,

$$\begin{vmatrix} A_{11} & A_{12} & A_{13} \\ A_{21} & \cdot & \cdot \\ \cdot & \cdot & \cdot \end{vmatrix} = \begin{vmatrix} A_{11} & A_{12} + \alpha A_{11} & A_{13} \\ A_{21} & A_{22} + \alpha A_{21} & \cdot \\ \cdot & \cdot & \cdot & \cdot \end{vmatrix}$$

This result is very useful in simplifying the numerical evaluation of a determinant.

1.9.28   Show that

$$\begin{vmatrix} 1 & 2 & 3 \\ 2 & 1 & 2 \\ 3 & 3 & 2 \end{vmatrix} = \begin{vmatrix} 1 & 0 & 0 \\ 0 & -3 & -4 \\ 0 & -3 & -7 \end{vmatrix}$$

(See Problem 1.9.27.)

1.9.29   Prove the following vector identities

(a) $(\mathbf{a} \times \mathbf{b}) \times (\mathbf{c} \times \mathbf{d}) = (\mathbf{a} \times \mathbf{b} \cdot \mathbf{d})\mathbf{c} - (\mathbf{a} \times \mathbf{b} \cdot \mathbf{c})\mathbf{d}$
$= (\mathbf{a} \times \mathbf{c} \cdot \mathbf{d})\mathbf{b} - (\mathbf{b} \times \mathbf{c} \cdot \mathbf{d})\mathbf{a}$

(b) $(\mathbf{a} \times \mathbf{b}) \cdot (\mathbf{c} \times \mathbf{d}) = (\mathbf{a} \cdot \mathbf{c})(\mathbf{b} \cdot \mathbf{d}) - (\mathbf{a} \cdot \mathbf{d})(\mathbf{b} \cdot \mathbf{c})$

(c) $\mathbf{a} \times (\mathbf{b} \times \mathbf{c}) + \mathbf{b} \times (\mathbf{c} \times \mathbf{a}) + \mathbf{c} \times (\mathbf{a} \times \mathbf{b}) = 0$

1.9.30   The three vectors

$$\mathbf{a} = \hat{\mathbf{i}} + \hat{\mathbf{j}} - 2\hat{\mathbf{k}}$$
$$\mathbf{b} = 2\hat{\mathbf{i}} - \hat{\mathbf{j}} + \hat{\mathbf{k}}$$
$$\mathbf{c} = \hat{\mathbf{i}} + 3\hat{\mathbf{j}} - \hat{\mathbf{k}}$$

extend from the origin of coordinates. Their tips determine a plane. Find the perpendicular distance from the origin to this plane. (*Hint:* Find any vector perpendicular to the plane and then find the projection of $\mathbf{a}$ on it.)

1.10.31   Verify equation (1.10.2) using the known transformation properties of δ_{ijk} (Eq. 1.9.8) and the vector components a_i, b_i.

1.8.32   Show that if A is antisymmetric, then $a_i A_{ij} a_j = 0$ for any vector $\mathbf{a}$.

1.5.33   If x_i and x_i' refer to the coordinates of a point with respect to two rectangular Cartesian coordinate systems, why is it not correct in general to write $x_i' = S_{ij}x_j$? Is there a restricted class of transformations where this is true?

1.10.34   Show that

$$\mathbf{a} \cdot \mathbf{b} \times \mathbf{c} = \begin{vmatrix} a_1 & a_2 & a_3 \\ b_1 & b_2 & b_3 \\ c_1 & c_2 & c_3 \end{vmatrix}$$

REFERENCES

1. Temple, G., *Cartesian Tensors*, John Wiley and Sons, New York, 1960.
2. Phillips, J. B., *Vector Analysis*, John Wiley and Sons, New York, 1933.
3. Halliday, David, and Robert Resnick, *Physics for Students of Science and Engineering*, John Wiley and Sons, New York, 1962.
4. Feynman, Richard P., Robert B. Leighton, and M. L. Sands, *The Feynman Lectures On Physics*, Addison-Wesley, Palo Alto, 1963.
5. Ferrar, W. L., *Algebra, A Text Book of Determinants, Matrices, And Algebraic Forms*, Oxford University Press, New York, 1953.
6. Sokolnikoff, I. S., *Mathematical Theory of Elasticity*, McGraw-Hill Book Co., New York, 1956.

2

Vector and Tensor Calculus

2.1. DIFFERENTIATION WITH RESPECT TO A PARAMETER

The motion of a particle along a trajectory can be represented by a position vector $\mathbf{r}$ extending from an arbitrary fixed point O in space, Fig. 2.1.1. If a coordinate system is introduced, the three components of $\mathbf{r}$ are functions of the time. The derivative of $\mathbf{r}$ with respect to time is defined by means of

$$\frac{d\mathbf{r}}{dt} = \lim_{\Delta t \to 0} \frac{1}{\Delta t} [\mathbf{r}(t + \Delta t) - \mathbf{r}(t)] = \lim_{\Delta t \to 0} \frac{\Delta \mathbf{s}}{\Delta t} \qquad (2.1.1)$$

where $\Delta \mathbf{s}$ is the displacement vector between the points a and b on the particle trajectory. Since the time is a scalar, the transformation property

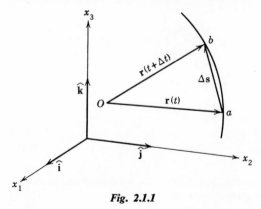

Fig. 2.1.1

38

of $d\mathbf{r}/dt$ is that of $\Delta\mathbf{s}$, i.e., $d\mathbf{r}/dt$ is a vector. In component form (2.1.1) is

$$\frac{d\mathbf{r}}{dt} = \lim_{\Delta t \to 0} \frac{\Delta x_1 \hat{\mathbf{i}} + \Delta x_2 \hat{\mathbf{j}} + \Delta x_3 \hat{\mathbf{k}}}{\Delta t} \tag{2.1.2}$$

where Δx_1, Δx_2, and Δx_3 represent the three components of $\Delta\mathbf{s}$.

It is common to use a dot to indicate the total derivative of a quantity with respect to the time. Equation (2.1.2) represents the *instantaneous velocity* of the particle:

$$\mathbf{u} = \dot{\mathbf{r}} = \dot{x}_1 \hat{\mathbf{i}} + \dot{x}_2 \hat{\mathbf{j}} + \dot{x}_3 \hat{\mathbf{k}} = u_1 \hat{\mathbf{i}} + u_2 \hat{\mathbf{j}} + u_3 \hat{\mathbf{k}} \tag{2.1.3}$$

A second differentiation yields the acceleration:

$$\mathbf{a} = \ddot{\mathbf{r}} = \ddot{x}_1 \hat{\mathbf{i}} + \ddot{x}_2 \hat{\mathbf{j}} + \ddot{x}_3 \hat{\mathbf{k}} = a_1 \hat{\mathbf{i}} + a_2 \hat{\mathbf{j}} + a_3 \hat{\mathbf{k}} \tag{2.1.4}$$

Thus in addition to the position coordinates there are associated with each point of the particle trajectory six functions of the single parameter t, namely, the three velocity and the three acceleration components, which describe the kinematics of the particle.

Vectors the components of which are functions of a parameter satisfy the following easily verified relations:

$$\frac{d}{dt}(\mathbf{a} \cdot \mathbf{b}) = \mathbf{a} \cdot \frac{d\mathbf{b}}{dt} + \mathbf{b} \cdot \frac{d\mathbf{a}}{dt} \tag{2.1.5}$$

$$\frac{d}{dt}(\mathbf{a} \cdot \mathbf{a}) = 2\mathbf{a} \cdot \frac{d\mathbf{a}}{dt} = \frac{d}{dt}a^2 = 2a\frac{da}{dt} \tag{2.1.6}$$

$$\frac{d}{dt}(\mathbf{a} \times \mathbf{b}) = \mathbf{a} \times \frac{d\mathbf{b}}{dt} + \frac{d\mathbf{a}}{dt} \times \mathbf{b} \tag{2.1.7}$$

$$\frac{d}{dt}(\alpha\mathbf{a}) = \alpha\frac{d\mathbf{a}}{dt} + \frac{d\alpha}{dt}\mathbf{a} \tag{2.1.8}$$

where α is a scalar function of t. In (2.1.7) it is important to retain the order of the factors due to the non-commutativity of $\mathbf{a} \times \mathbf{b}$. In (2.1.6) $a = |\mathbf{a}|$ stands for the magnitude of the vector. A *constant vector* can be defined by the relation $d\mathbf{c}/dt = 0$ meaning that the components of $\mathbf{c}$ always have the same fixed values in a given frame of reference.

It is important to recognize that the vectors are being referred to a fixed coordinate system. Later, moving coordinate systems are treated and a new complication arises because the unit vectors $\hat{\mathbf{i}}$, $\hat{\mathbf{j}}$, and $\hat{\mathbf{k}}$ also become functions of the time and cannot be treated as constant vectors as was done in equation (2.1.2).

If **A** is a second rank tensor which is a function of the time, then by the derivative of **A** with respect to t is understood

$$\frac{d\mathbf{A}}{dt} = \lim_{\Delta t \to 0} \frac{1}{\Delta t} [\mathbf{A}(t + \Delta t) - \mathbf{A}(t)] \qquad (2.1.9)$$

Just as in the case of a vector, the scalar $1/\Delta t$ multiplies *every* component of the tensor. Therefore, $d\mathbf{A}/dt$ is a second rank tensor with components dA_{ij}/dt. Similarly, the derivative of a third rank tensor has components dA_{ijk}/dt.

2.2. THE SERRET-FRENET FORMULAS

Referring again to Fig. 2.1.1, let us regard the vector **r** as a function of the arc length s measured along the particle trajectory. We then have

$$\mathbf{u} = \frac{d\mathbf{r}}{dt} = \frac{d\mathbf{r}}{ds}\frac{ds}{dt} = \hat{\boldsymbol{\lambda}}u \qquad (2.2.1)$$

where $\hat{\boldsymbol{\lambda}}$ is a unit vector tangent to the particle trajectory and $u = ds/dt$ represents the speed of the particle. It is of course possible to represent any vector as a unit vector multiplied by the magnitude of the vector. A second differentiation yields the acceleration:

$$\mathbf{a} = \frac{d\mathbf{u}}{dt} = u\frac{d\hat{\boldsymbol{\lambda}}}{ds}\frac{ds}{dt} + \hat{\boldsymbol{\lambda}}\frac{du}{dt}$$

$$= u^2\frac{d\hat{\boldsymbol{\lambda}}}{ds} + \hat{\boldsymbol{\lambda}}\frac{du}{dt} \qquad (2.2.2)$$

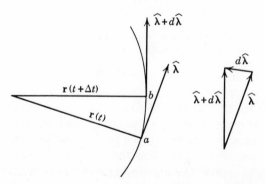

Fig. 2.2.1

The unit tangent vector is not constant but varies in *direction* along the particle trajectory, Fig. 2.2.1. The *curvature* k and the *radius of curvature* ρ at a given point on the curve are defined by

$$\frac{d\hat{\lambda}}{ds} = k\hat{n} \qquad k = \frac{1}{\rho} \tag{2.2.3}$$

where $\hat{n}$ is a unit vector. If one imagines taking the limit $b \to a$ in Fig. 2.2.1 it appears from the geometry that $d\hat{\lambda}$, and hence $\hat{n}$, are perpendicular to the trajectory at the point a where the derivative is evaluated. This fact is established by differentiation of $\hat{\lambda} \cdot \hat{\lambda} = 1$:

$$2\hat{\lambda} \cdot \frac{d\hat{\lambda}}{ds} = 0 \tag{2.2.4}$$

By combining (2.2.3) and (2.2.4),

$$\hat{\lambda} \cdot \hat{n} = 0 \tag{2.2.5}$$

showing that the angle between $\hat{\lambda}$ and $\hat{n}$ is 90°.

Equation (2.2.2) can be expressed

$$\mathbf{a} = \frac{u^2}{\rho} \hat{n} + \frac{du}{dt} \hat{\lambda} \tag{2.2.6}$$

which shows the acceleration resolved into two mutually perpendicular directions tangent and normal to the trajectory. The unit normal is called the *principal normal* to the curve. The normal component of **a** is the *centripetal acceleration*. If we compare (2.2.6) with elementary formulas for circular motion, we see that ρ becomes identical with the radius of the circle in this special case. Any small increment of an arbitrary trajectory can be approximated by a small portion of the arc of a circle. Since each successive portion has a slightly different curvature, the radius of curvature is a function of position along the curve. The tangential component of **a** is the time rate of change of the *speed* of the particle.

The plane determined at a given point on the trajectory of the particle by the tangent vector $\hat{\lambda}$ and the principle normal $\hat{n}$ is called the *osculating plane*.

A third unit vector can be defined which is perpendicular to $\hat{\lambda}$ and $\hat{n}$:

$$\hat{\nu} = \frac{1}{\tau} \frac{d\hat{n}}{ds} + \frac{1}{\rho\tau} \hat{\lambda} \tag{2.2.7}$$

The function τ is the *torsion* and its magnitude at a given point on the curve is fixed by the condition $\hat{\nu} \cdot \hat{\nu} = 1$. The perpendicularity of $\hat{\nu}$ to $\hat{n}$

and $\hat{\boldsymbol{\lambda}}$ is established by the following calculations:

$$\hat{\boldsymbol{\nu}} \cdot \hat{\mathbf{n}} = \frac{1}{\tau} \frac{d\hat{\mathbf{n}}}{ds} \cdot \hat{\mathbf{n}} + \frac{1}{\rho\tau} \hat{\boldsymbol{\lambda}} \cdot \hat{\mathbf{n}},$$

$$\hat{\mathbf{n}} \cdot \hat{\mathbf{n}} = 1, \qquad \frac{d\hat{\mathbf{n}}}{ds} \cdot \hat{\mathbf{n}} = 0, \qquad \therefore \ \hat{\boldsymbol{\nu}} \cdot \hat{\mathbf{n}} = 0 \qquad (2.2.8)$$

$$\hat{\boldsymbol{\nu}} \cdot \hat{\boldsymbol{\lambda}} = \frac{1}{\tau} \frac{d\hat{\mathbf{n}}}{ds} \cdot \hat{\boldsymbol{\lambda}} + \frac{1}{\rho\tau} \hat{\boldsymbol{\lambda}} \cdot \hat{\boldsymbol{\lambda}},$$

$$\hat{\mathbf{n}} \cdot \hat{\boldsymbol{\lambda}} = 0, \qquad \hat{\mathbf{n}} \cdot \frac{d\hat{\boldsymbol{\lambda}}}{ds} + \frac{d\hat{\mathbf{n}}}{ds} \cdot \hat{\boldsymbol{\lambda}} = 0, \qquad \hat{\boldsymbol{\lambda}} \cdot \frac{d\hat{\mathbf{n}}}{ds} = \frac{-1}{\rho},$$

$$\hat{\boldsymbol{\nu}} \cdot \hat{\boldsymbol{\lambda}} = - \frac{1}{\rho\tau} + \frac{1}{\rho\tau} = 0 \qquad (2.2.9)$$

The vector $\hat{\boldsymbol{\nu}}$ is known as the *unit bi-normal* to the curve. The vectors $\hat{\boldsymbol{\lambda}}$, $\hat{\mathbf{n}}$, and $\hat{\boldsymbol{\nu}}$ define at each point of the trajectory a Cartesian coordinate system, Fig. 2.2.2. The sign of τ can be chosen so as to make the coordinate system right-handed. Differentiation of $\hat{\boldsymbol{\nu}} = \hat{\boldsymbol{\lambda}} \times \hat{\mathbf{n}}$ yields

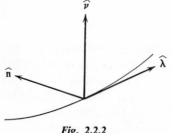

Fig. 2.2.2

$$\frac{d\hat{\boldsymbol{\nu}}}{ds} = \hat{\boldsymbol{\lambda}} \times \frac{d\hat{\mathbf{n}}}{ds} + \frac{d\hat{\boldsymbol{\lambda}}}{ds} \times \hat{\mathbf{n}}$$

If use is made of (2.2.7) and (2.2.3),

$$\frac{d\hat{\boldsymbol{\nu}}}{ds} = \hat{\boldsymbol{\lambda}} \times \left(\tau\hat{\boldsymbol{\nu}} - \frac{1}{\rho} \hat{\boldsymbol{\lambda}} \right) + \frac{1}{\rho} \hat{\mathbf{n}} \times \hat{\mathbf{n}}$$

$$= -\tau\hat{\mathbf{n}} \qquad (2.2.10)$$

Three of the formulas derived in this section, namely,

$$\frac{d\hat{\boldsymbol{\lambda}}}{ds} = \frac{1}{\rho} \hat{\mathbf{n}} \qquad (2.2.11)$$

$$\frac{d\hat{\mathbf{n}}}{ds} = \tau\hat{\boldsymbol{\nu}} - \frac{1}{\rho} \hat{\boldsymbol{\lambda}} \qquad (2.2.12)$$

$$\frac{d\hat{\boldsymbol{\nu}}}{ds} = -\tau\hat{\mathbf{n}} \qquad (2.2.13)$$

characterize, in the small, the geometrical properties of a space curve. They are known as the *Serret-Frenet* formulas.

2.3. VECTOR AND TENSOR FIELDS

If a small test body of mass m is placed in the gravitational field of a point mass M it experiences a force given by

$$\mathbf{F} = -G\frac{mM}{r^2}\,\hat{\mathbf{r}} \qquad (2.3.1)$$

where G is the gravitational constant and r is the distance between the masses, Fig. 2.3.1. The negative sign indicates that the force is one of

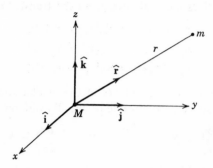

Fig. 2.3.1

attraction, i.e., opposite to the direction of the unit vector $\hat{\mathbf{r}}$. The *gravitational field* $\mathbf{g}$ at the location of the test body can be defined by

$$\mathbf{F} = m\mathbf{g} \qquad (2.3.2)$$

Equations (2.3.1) and (2.3.2) should be compared with Coulomb's law and the definition of the electric field in electrostatics.

If a coordinate system is introduced with its origin at the location of M, $\mathbf{g}$ can be expressed

$$\mathbf{g} = -GM\left(\frac{x\hat{\mathbf{i}}}{(x^2 + y^2 + z^2)^{3/2}} + \frac{y\hat{\mathbf{j}}}{(x^2 + y^2 + z^2)^{3/2}} + \frac{z\hat{\mathbf{k}}}{(x^2 + y^2 + z^2)^{3/2}}\right)$$

$$(2.3.3)$$

The components of the vector $\mathbf{g}$ are three functions of the coordinates. Such a vector is a *vector function* or a *vector field*.

2.4. THE GRADIENT

A tensor field of rank zero is a simple scalar function of position. From a scalar function $\psi(x_1, x_2, x_3)$, three partial derivatives can be formed:

$$\frac{\partial\psi}{\partial x_1},\quad \frac{\partial\psi}{\partial x_2},\quad \frac{\partial\psi}{\partial x_3} \qquad (2.4.1)$$

That the three partial derivatives represent the components of a vector is established by expressing the function ψ in terms of new coordinates x_i' obtained from the old by means of an orthogonal transformation:

$$\psi = \psi(x_1(x_1', x_2', x_3'), x_2(x_1', x_2', x_3'), x_3(x_1', x_2', x_3')) \qquad (2.4.2)$$

By the chain rule for partial differentiation

$$\frac{\partial \psi}{\partial x_1'} = \frac{\partial \psi}{\partial x_1}\frac{\partial x_1}{\partial x_1'} + \frac{\partial \psi}{\partial x_2}\frac{\partial x_2}{\partial x_1'} + \frac{\partial \psi}{\partial x_3}\frac{\partial x_3}{\partial x_1'} \qquad (2.4.3)$$

The partial derivatives are evaluated from the transformation equations:

$$\begin{aligned}
x_1 - x_1(P) &= S_{i1}[x_i' - x_i'(P)]\\
x_2 - x_2(P) &= S_{i2}[x_i' - x_i'(P)]\\
x_3 - x_3(P) &= S_{i3}[x_i' - x_i'(P)]
\end{aligned} \qquad (2.4.4)$$

where $x_i - x_i(P)$ are the components of the displacement vector $\mathbf{r}$ from a fixed point P in space to a variable point. By differentiation of (2.4.4),

$$\frac{\partial x_1}{\partial x_1'} = S_{11} \qquad \frac{\partial x_2}{\partial x_1'} = S_{12} \qquad \frac{\partial x_3}{\partial x_1'} = S_{13} \qquad (2.4.5)$$

If (2.4.5) is substituted in (2.4.3),

$$\frac{\partial \psi}{\partial x_1'} = S_{11}\frac{\partial \psi}{\partial x_1} + S_{12}\frac{\partial \psi}{\partial x_2} + S_{13}\frac{\partial \psi}{\partial x_3} \qquad (2.4.6)$$

A similar derivation can be done for the partial derivatives with respect to x_2' and x_3'. The results are summarized by

$$\partial_i'\psi = S_{ij}\,\partial_j\psi \qquad (2.4.7)$$

where the abbreviations

$$\partial_i' = \frac{\partial}{\partial x_i'}, \qquad \partial_j = \frac{\partial}{\partial x_j} \qquad (2.4.8)$$

are used.

Equation (2.4.7) establishes that $\partial_i\psi$ are the components of a vector. This vector is called the *gradient* of ψ and is written $\nabla\psi$. ∇ is a symbolic *vector operator*, sometimes called *del* or *nabla*:

$$\nabla = \hat{\mathbf{i}}\,\partial_1 + \hat{\mathbf{j}}\,\partial_2 + \hat{\mathbf{k}}\,\partial_3 \qquad (2.4.9)$$

The effect of del on a scalar function is

$$\nabla\psi = \operatorname{grad}\psi = \hat{\mathbf{i}}\,\partial_1\psi + \hat{\mathbf{j}}\,\partial_2\psi + \hat{\mathbf{k}}\,\partial_3\psi \qquad (2.4.10)$$

Let $\mathbf{r}$ be the displacement vector from a fixed point P to a variable point. A simple scalar function is the magnitude of $\mathbf{r}$:

$$r = ([x_1 - x_1(P)]^2 + [x_2 - x_2(P)]^2 + [x_3 - x_3(P)]^2)^{1/2} \qquad (2.4.11)$$

The effect of ∇ on various simple functions of r is

$$\nabla r = \frac{\mathbf{r}}{r} = \hat{\mathbf{r}} \tag{2.4.12}$$

$$\nabla r^n = nr^{n-2}\mathbf{r} \tag{2.4.13}$$

$$\nabla \frac{1}{r} = -\frac{\mathbf{r}}{r^3} \tag{2.4.14}$$

$$\nabla r^{-n} = -nr r^{-n-2} \tag{2.4.15}$$

The function

$$\psi = -\frac{GM}{r} \tag{2.4.16}$$

is the *gravitational scalar potential* of a point of mass M. The gravitational field (2.3.3) is given by

$$\mathbf{g} = -\nabla \psi \tag{2.4.17}$$

2.5. THE DIRECTIONAL DERIVATIVE

Let $\psi(x_1, x_2, x_3)$ be a given scalar function. The equation

$$\psi(x_1, x_2, x_3) = c = \text{const.}$$

determines a surface in that it can be solved for x_3 in terms of x_1 and x_2. Each value of the constant c gives a different surface as illustrated in Fig. 2.5.1. Suppose it is required to calculate the difference between the value of the function at point P and a nearby point Q. In the limit $Q \to P$ this is the total differential:

$$d\psi_{PQ} = \partial_1 \psi \, dx_1 + \partial_2 \psi \, dx_2 + \partial_3 \psi \, dx_3 \tag{2.5.1}$$

where dx_1, dx_2, and dx_3 are components of the displacement vector $d\mathbf{s}$ between P and Q. The ratio $d\psi_{PQ}/ds$, where ds is the magnitude of $d\mathbf{s}$ is known as the *directional derivative*. $d\mathbf{s}$ should be thought of as a differential vector which can be taken in any direction we please from point P. Equation (2.5.1) can be expressed,

$$d\psi_{PQ} = \nabla\psi \cdot d\mathbf{s} \tag{2.5.2}$$

If both P and Q were chosen to lie on the surface $\psi = c_1$ (Fig. 2.5.1), then $d\psi_{PQ} = 0$. Thus if the displacement vector $d\mathbf{s}$ *lies in the surface* $\psi = c_1$, it is perpendicular to the vector $\nabla\psi$. If (2.5.2) is written as

$$d\psi_{PQ} = |\nabla\psi| \, |d\mathbf{s}| \cos\theta \tag{2.5.3}$$

then it is apparent that $d\psi_{PQ}$ has its maximum value when $\theta = 0$. *The gradient vector* $\nabla\psi$ *evaluated at point* P *is perpendicular to the surface*

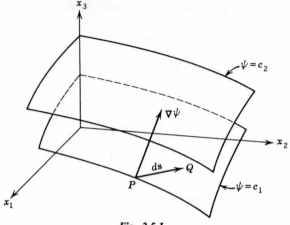

Fig. 2.5.1

$\psi = c_1$ *at this point and this is also the direction of the maximum rate of change of the function* ψ.

As an example

$$\psi = a_1 x_1 + a_2 x_2 + a_3 x_3 = c \qquad (2.5.4)$$

represents the equation of a plane and

$$\nabla \psi = a_1 \hat{\mathbf{i}} + a_2 \hat{\mathbf{j}} + a_3 \hat{\mathbf{k}} \qquad (2.5.5)$$

showing that a_1, a_2, and a_3 are the components of a vector perpendicular to the plane.

As a second example, let **a** and **b** be two constant vectors and let $\mathbf{r} = x_1 \hat{\mathbf{i}} + x_2 \hat{\mathbf{j}} + x_3 \hat{\mathbf{k}}$. Then if **a** and **b** are not parallel,

$$\phi = \mathbf{a} \cdot \mathbf{r} = c_1 \qquad \psi = \mathbf{b} \cdot \mathbf{r} = c_2 \qquad (2.5.6)$$

represent two planes which intersect along a line, Fig. 2.5.2. What is the equation of the line of intersection? In order to lie in the intersection, a

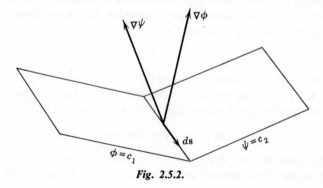

Fig. 2.5.2.

displacement vector $d\mathbf{s}$ must be perpendicular to both $\nabla\phi$ and $\nabla\psi$. This will be true if

$$\frac{d\mathbf{s}}{ds} = \frac{\nabla\phi \times \nabla\psi}{|\nabla\phi \times \nabla\psi|} = \frac{\mathbf{a} \times \mathbf{b}}{|\mathbf{a} \times \mathbf{b}|} \tag{2.5.7}$$

Let

$$m_1 = \frac{a_2 b_3 - a_3 b_2}{|\mathbf{a} \times \mathbf{b}|}, \qquad m_2 = \frac{a_3 b_1 - a_1 b_3}{|\mathbf{a} \times \mathbf{b}|}, \qquad m_3 = \frac{a_1 b_2 - a_2 b_1}{|\mathbf{a} \times \mathbf{b}|} \tag{2.5.8}$$

Then (2.5.7) expressed in component form is

$$dx_1 = m_1\, ds \qquad dx_2 = m_2\, ds \qquad dx_3 = m_3\, ds \tag{2.5.9}$$

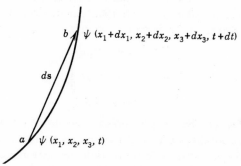

Fig. 2.5.3

Integration of (2.5.9) yields

$$x_1 = m_1 s + k_1 \qquad x_2 = m_2 s + k_2 \qquad x_3 = m_3 s + k_3 \tag{2.5.10}$$

which are recognized as the parametric equations of a straight line with distance measured along the line as the parameter. k_1, k_2, and k_3 are constants of integration and represent the coordinates of the point on the intersection from which s is measured.

Very often, a particle moves through a field of some kind that is changing as a function of time. For instance, the bodies that are setting up a gravitational field at the location of the particle may themselves be moving. Suppose that the field is described by a scalar potential $\psi(x_1, x_2, x_3, t)$. As far as determining the dynamical behavior of the particle, the important question is: how rapidly does the field change from the point of view of the particle? The potential difference between points a and b on the particle trajectory, Fig. 2.5.3, is

$$\begin{aligned}
d\psi &= \psi(x_1 + dx_1, x_2 + dx_2, x_3 + dx_3, t + dt) - \psi(x_1, x_2, x_3, t) \\
&= \partial_1\psi\, dx_1 + \partial_2\psi\, dx_2 + \partial_3\psi\, dx_3 + \partial_t\psi\, dt \tag{2.5.11} \\
&= \nabla\psi \cdot d\mathbf{s} + \partial_t\psi\, dt
\end{aligned}$$

The change $d\psi$ is the sum of the change brought about by displacing the particle by $d\mathbf{s}$ and a term which takes into account the variation of ψ with time while the displacement takes place. Since dt is the actual time required to displace the particle from a to b, the particle velocity is $\mathbf{u} = d\mathbf{s}/dt$ and

$$\frac{d\psi}{dt} = \nabla\psi \cdot \mathbf{u} + \partial_t \psi \qquad (2.5.12)$$

which gives the variation in ψ with time that the particle experiences.

2.6. THE DIVERGENCE OF A VECTOR

If $\mathbf{a}$ is a vector-field, the three components of $\mathbf{a}$ are functions of the coordinates. It is possible to form nine partial derivatives which are conveniently displayed in the form of a matrix:

$$\begin{pmatrix} \partial_1 a_1 & \partial_1 a_2 & \partial_1 a_3 \\ \partial_2 a_1 & \partial_2 a_2 & \partial_2 a_3 \\ \partial_3 a_1 & \partial_3 a_2 & \partial_3 a_3 \end{pmatrix} \qquad (2.6.1)$$

The proof that $\partial_i a_j$ are the components of a second rank tensor is formally the same as the proof that Eq. (1.8.11) represents a tensor. If more general than orthogonal coordinate transformations are considered, $\partial_i a_j$ are no longer the components of a second rank tensor. It is, however, possible to generalize the idea of partial differentiation so that the result is a second rank tensor. This generalization is known as the *covariant derivative*.

The trace of the tensor (2.6.1) is an invariant which is called the *divergence* of $\mathbf{a}$. It can be thought of as the dot product of the symbolic vector ∇ and $\mathbf{a}$:

$$\operatorname{div} \mathbf{a} = \nabla \cdot \mathbf{a} = \partial_1 a_1 + \partial_2 a_2 + \partial_3 a_3 \qquad (2.6.2)$$

If the vector $\mathbf{a}$ is the gradient of a scalar function ψ,

$$\operatorname{div} \operatorname{grad} \psi = \nabla \cdot \nabla \psi = \partial_1{}^2 \psi + \partial_2{}^2 \psi + \partial_3{}^2 \psi \qquad (2.6.3)$$

It is usual to define a symbolic differential operator by

$$\nabla \cdot \nabla = \nabla^2 = \partial_1{}^2 + \partial_2{}^2 + \partial_3{}^2 \qquad (2.6.4)$$

This operator is of frequent occurrence and is called the *Laplacian*.

2.7. EQUATIONS OF CONTINUITY

Consider a compressible fluid such as a gas flowing through a region of space. Flow is varying as a function of time and at a given point such as P in Fig. 2.7.1, the fluid density may be increasing or decreasing. The density ρ (mass per unit volume) is a scalar function of the four variables

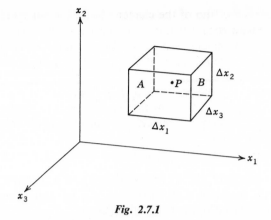

Fig. 2.7.1

x_1, x_2, x_3, and t and it is the purpose of this section to evaluate the rate of change of ρ at a given point i.e., $\partial \rho / \partial t$.

Constructed at P in Fig. 2.7.1 is a differential volume element in the form of a rectangular parallelepiped the edges of which are parallel to the coordinate axes and of length Δx_1, Δx_2, and Δx_3.* For the volume of the parallelepiped the notation

$$\Delta \Sigma = \Delta x_1 \, \Delta x_2 \, \Delta x_3$$

will be used. Previously, in Eq. (1.6.10) is was shown that $\mathbf{u} \cdot \hat{\mathbf{n}} \, d\sigma$ represents the volume per second of fluid which flows through a surface element of area $d\sigma$. Therefore, $\rho \mathbf{u} \cdot \hat{\mathbf{n}} \, d\sigma$ represents the *mass* per second which flows through the surface element. It is convenient to define a *mass current vector* $\mathbf{c} = \rho \mathbf{u}$ which has the dimensions of mass per second per unit area. The assumption is made that mass is not created or destroyed in the volume element. Any change in the mass inside the volume element must be accounted for by a flow of mass in or out through the six faces. At surface A in Fig. 2.7.1, the normal component of $\mathbf{c}$ is c_1. At B, the normal component has changed slightly due to the functional dependence of $\mathbf{c}$ on the coordinates but it can be represented by the first two terms of a Taylor's series expansion. The flow of mass out of the volume element due to fluid flowing across surfaces A and B, therefore, is

Outflow across B — Inflow across A

$$= \left(c_1 + \frac{\partial c_1}{\partial x_1} \Delta x_1 \right) \Delta x_2 \, \Delta x_3 - c_1 \, \Delta x_2 \, \Delta x_3$$

$$= \frac{\partial c_1}{\partial x_1} \Delta \Sigma \tag{2.7.1}$$

---

* In these heuristic derivations, Δx_i means a very small but still finite increment.

If account is taken of the remaining four faces, the net mass decrease in the volume element is

$$-\frac{d}{dt}(\rho\,\Delta\Sigma) = \left(\frac{\partial c_1}{\partial x_1} + \frac{\partial c_2}{\partial x_2} + \frac{\partial c_3}{\partial x_3}\right)\Delta\Sigma \qquad (2.7.2)$$

The volume element is constant in time and cancels out of (2.7.2). Since we are now talking about the time rate of change of the scalar function at a fixed point P in space [no displacement is involved such as was the case with the derivation of (2.5.12)],

$$\frac{d\rho}{dt} = \frac{\partial\rho}{\partial t} \qquad (2.7.3)$$

and

$$-\frac{\partial\rho}{\partial t} = \nabla \cdot \mathbf{c} \qquad (2.7.4)$$

This result expresses that the net outflow of mass is balanced by the decrease in matter in the volume element; it is an expression of the conservation of mass. For this reason, (2.7.4) is called an *equation of continuity*. "To diverge" means "to go out from." The divergence of the vector $\mathbf{c}$ gives the rate at which matter goes out from a volume element located at the point in space where the divergence is calculated.

If a fluid is incompressible, then necessarily $\partial\rho/\partial t = 0$ and

$$\nabla \cdot \mathbf{c} = 0 \qquad (2.7.5)$$

Any vector that has zero divergence in a region of space is said to be *solenoidal* in that region.

Equations of continuity exist in other branches of physics. If $\mathbf{j}$ is the electric current density (charge per second per unit area) then $\mathbf{j} \cdot \hat{\mathbf{n}}\,d\sigma$ represents the charge per second that flows across an element of area of magnitude $d\sigma$. If ρ is the charge density (charge per unit volume), then

$$\nabla \cdot \mathbf{j} + \frac{\partial\rho}{\partial t} = 0 \qquad (2.7.6)$$

expresses the conservation of charge.

An important concept in quantum mechanics is that of probability. It is possible to define a *probability density P* for a particle which has the significance that $P\,d\Sigma$ is the probability that a particle be found in the volume element $d\Sigma$. If $\mathbf{S}$ represents the probability current density (probability per second per unit area), then

$$\nabla \cdot \mathbf{S} + \frac{\partial P}{\partial t} = 0 \qquad (2.7.7)$$

expresses the fact that the particle is never created or destroyed.

2.8. GAUSS' DIVERGENCE THEOREM

If both sides of (2.7.4) are multiplied by a volume element $d\Sigma$ and the result integrated over a finite region of space such as that indicated in Fig. 2.8.1, the result is

$$-\int_{\text{volume}} \frac{\partial \rho}{\partial t} \, d\Sigma = \int_{\text{volume}} \nabla \cdot \mathbf{c} \, d\Sigma \tag{2.8.1}$$

Since time appears only through the explicit dependence of ρ on t,

$$\int_{\text{volume}} \frac{\partial \rho}{\partial t} \, d\Sigma = \frac{d}{dt} \int \rho \, d\Sigma = \frac{dm}{dt} \tag{2.8.2}$$

represents the total rate of change of the mass in the volume of integration. The total rate of decrease of mass must be equal to the rate at which mass

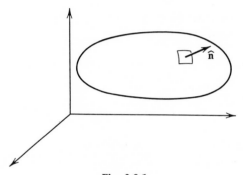

Fig. 2.8.1

flows outward across the closed surface of Fig. 2.8.1:

$$-\frac{dm}{dt} = \int_{\text{surface}} \mathbf{c} \cdot \hat{\mathbf{n}} \, d\sigma \tag{2.8.3}$$

where $\hat{\mathbf{n}}$ is the outward-drawn normal to the surface. Combining the preceding equations yields

$$\int_{\text{volume}} \nabla \cdot \mathbf{c} \, d\Sigma = \int_{\text{surface}} \mathbf{c} \cdot \hat{\mathbf{n}} \, d\sigma \tag{2.8.4}$$

which is Gauss' divergence theorem.

The proof of (2.8.4) has rested on the physical picture of fluid flow. The result, however, is a mathematical identity valid for any vector field $\mathbf{c}$. The quantity $\mathbf{c} \cdot \hat{\mathbf{n}} \, d\sigma$ is called the *flux* of the vector $\mathbf{c}$ across the element of area $d\sigma$, even though $\mathbf{c}$ may not represent an actual physical flow of anything.

If A is a second rank tensor field, the components of $\mathbf{c}$ in (2.8.4) can be replaced by any row of A to yield

$$\int_{\text{volume}} \partial_i A_{ji}\, d\Sigma = \int_{\text{surface}} A_{ji} n_i\, d\sigma \qquad (2.8.5)$$

A similar equation can be written for the columns of A. The integrands of (2.8.5) are not scalars, as in (2.8.4), but are the components of vectors. As long as the transformations are restricted to orthogonal, the integration of a vector yields another vector, but this is not true of more general transformations. Use will be found for (2.8.5) in the study of the mechanics of continuous media where A will be replaced by the symmetric stress tensor of the medium.

2.9. THE CURL OF A VECTOR

From the nine partial derivatives (2.6.1) it is possible to construct a pseudo-vector the components of which are

$$c_i = \delta_{ijk}\, \partial_j a_k \qquad (2.9.1)$$

Formally, $\mathbf{c}$ can be written as the cross product of the symbolic vector operator ∇ and $\mathbf{a}$:

$$\mathbf{c} = \nabla \times \mathbf{a} \qquad (2.9.2)$$

The vector $\mathbf{c}$ is called the *curl* of $\mathbf{a}$.

In fluid dynamics, a formal connection between curl and angular velocity can be established. Consider a particle of fluid which is in circular motion about an axis, (Fig. 2.9.1). Let $\mathbf{r}$ be the position vector of the particle measured from an arbitrary point P on the axis. The angular velocity of the particle is a pseudo-vector defined by

$$\mathbf{u} = \boldsymbol{\omega} \times \mathbf{r} \qquad (2.9.3)$$

where $\mathbf{u}$ is the velocity of the particle. As usual, a right-hand rule is used to define the direction of $\boldsymbol{\omega}$ which lies along the axis of rotation. The magnitude of $\mathbf{u}$ is

$$u = \omega r \sin \theta \qquad (2.9.4)$$

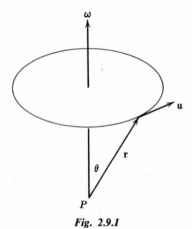

Fig. 2.9.1

Since $r \sin \theta$ is the radius of the circle on which the particle moves, (2.9.3) is consistent with the definition of angular velocity given in elementary texts.

The curl of the velocity vector (2.9.3) is

$$(\nabla \times \mathbf{u})_i = [\nabla \times (\boldsymbol{\omega} \times \mathbf{r})]_i$$
$$= \delta_{ijk} \, \partial_j (\delta_{klm} \omega_l r_m) \qquad (2.9.5)$$

The partial differentiations are to be performed on the components ω_l and r_m. The permutation symbol is a constant. Therefore,

$$(\nabla \times \mathbf{u})_i = [\delta_{il}\delta_{jm} - \delta_{im}\delta_{jl}][\omega_l(\partial_j r_m) + r_m(\partial_j \omega_l)] \qquad (2.9.6)$$

The components of $\mathbf{r}$ are $x_m - x_m(P)$ where $x_m(P)$ are the coordinates of the fixed point P and x_m are the coordinates of the variable point with respect to which the differentiations are being done. Appearing in (2.9.6) is

$$\partial_j r_m = \partial_j [x_m - x_m(P)] = \delta_{jm} \qquad (2.9.7)$$

Hence,

$$(\nabla \times \mathbf{u})_i = \omega_i \delta_{jj} + r_j \, \partial_j \omega_i - \omega_j \delta_{ji} - r_i \, \partial_l \omega_l$$
$$= 3\omega_i + (\mathbf{r} \cdot \nabla)\omega_i - \omega_i - r_i(\nabla \cdot \boldsymbol{\omega}) \qquad (2.9.8)$$
$$= 2\omega_i + (\mathbf{r} \cdot \nabla)\omega_i - r_i(\nabla \cdot \boldsymbol{\omega})$$

Use has been made of

$$\delta_{jm}\delta_{jm} = \delta_{jj} = 3 \qquad (2.9.9)$$

which is the trace of the Kronecker delta. The symbol $(\mathbf{r} \cdot \nabla)$ is a differential operator:

$$\mathbf{r} \cdot \nabla = r_j \, \partial_j = r_1 \, \partial_1 + r_2 \, \partial_2 + r_3 \, \partial_3 \qquad (2.9.10)$$

In (2.9.8) the differentiations indicated by $(\mathbf{r} \cdot \nabla)$ are done on the component ω_i:

$$(\mathbf{r} \cdot \nabla)\omega_i = r_1 \, \partial_1 \omega_i + r_2 \, \partial_2 \omega_i + r_3 \, \partial_3 \omega_i \qquad (2.9.11)$$

The component equation (2.9.8) is equivalent to the vector equation.

$$\nabla \times \mathbf{u} = 2\boldsymbol{\omega} + (\mathbf{r} \cdot \nabla)\boldsymbol{\omega} - \mathbf{r}(\nabla \cdot \boldsymbol{\omega}) \qquad (2.9.12)$$

In this equation, $(\mathbf{r} \cdot \nabla)$ operates on each component of $\boldsymbol{\omega}$. If the limit $\mathbf{r} \to 0$ is taken in (2.9.12), the result is

$$\boldsymbol{\omega} = \tfrac{1}{2}(\nabla \times \mathbf{u}) \qquad (2.9.13)$$

This represents a kind of limiting angular velocity as the point P, and hence the axis of rotation is taken closer and closer to the particle of fluid where the differentiations are performed. It was this result that led to the term "curl." In some texts, the curl is called the *rotation* and (2.9.13) is written

$$\boldsymbol{\omega} = \tfrac{1}{2} \operatorname{rot} \mathbf{u} \qquad (2.9.14)$$

In the example of a fluid flowing through space, the velocity is a *vector field* in the sense that at a given instant a velocity is associated with every point of the fluid. Even though the motion is not circular, it is still possible to calculate $\nabla \times \mathbf{u}$ and hence $\boldsymbol{\omega}$ given by (2.9.13) is defined. Since ∇ and $\mathbf{u}$ have the transformation properties of polar vectors, (2.9.13) shows that $\boldsymbol{\omega}$ is a pseudo-vector.

It is not surprising that a vector which satisfies the condition

$$\nabla \times \mathbf{a} = 0 \tag{2.9.15}$$

is called *irrotational*, a term which persists even if $\mathbf{a}$ has nothing to do with fluid flow.

2.10. DIFFERENTIATION FORMULAS

If ψ is a scalar field and $\mathbf{a}$ and $\mathbf{b}$ are vector fields, the following identities hold:

$$\nabla \cdot \psi \mathbf{a} = \psi \nabla \cdot \mathbf{a} + \mathbf{a} \cdot \nabla \psi \tag{2.10.1}$$

$$\nabla \times \psi \mathbf{a} = \psi \nabla \times \mathbf{a} + \nabla \psi \times \mathbf{a} \tag{2.10.2}$$

$$\nabla \cdot (\mathbf{a} \times \mathbf{b}) = \mathbf{b} \cdot (\nabla \times \mathbf{a}) - \mathbf{a} \cdot (\nabla \times \mathbf{b}) \tag{2.10.3}$$

$$\nabla \times (\mathbf{a} \times \mathbf{b}) = (\mathbf{b} \cdot \nabla)\mathbf{a} - (\mathbf{a} \cdot \nabla)\mathbf{b} + \mathbf{a}(\nabla \cdot \mathbf{b}) - \mathbf{b}(\nabla \cdot \mathbf{a}) \tag{2.10.4}$$

$$\nabla(\mathbf{a} \cdot \mathbf{b}) = (\mathbf{a} \cdot \nabla)\mathbf{b} + (\mathbf{b} \cdot \nabla)\mathbf{a} + \mathbf{a} \times (\nabla \times \mathbf{b}) + \mathbf{b} \times (\nabla \times \mathbf{a}) \tag{2.10.5}$$

$$\nabla \times \nabla \psi = 0 \tag{2.10.6}$$

$$\nabla \cdot \nabla \times \mathbf{a} = 0 \tag{2.10.7}$$

$$\nabla \times (\nabla \times \mathbf{a}) = \nabla(\nabla \cdot \mathbf{a}) - \nabla^2 \mathbf{a} \tag{2.10.8}$$

For example, the proof of (2.10.3) is

$$\begin{aligned}
\nabla \cdot (\mathbf{a} \times \mathbf{b}) &= \partial_i \delta_{ijk} a_j b_k \\
&= \delta_{ijk}(\partial_i a_j)b_k + \delta_{ijk} a_j (\partial_i b_k) \\
&= b_k \delta_{kij} \, \partial_i a_j - a_j \delta_{jik} \, \partial_i b_k \\
&= \mathbf{b} \cdot (\nabla \times \mathbf{a}) - \mathbf{a} \cdot (\nabla \times \mathbf{b})
\end{aligned}$$

The derivation of (2.9.12) illustrates the method of proof of (2.10.4). The proof of (2.10.8) is

$$\begin{aligned}
[\nabla \times (\nabla \times \mathbf{a})]_i &= \delta_{ijk} \, \partial_j (\delta_{klm} \, \partial_l a_m) \\
&= [\delta_{il}\delta_{jm} - \delta_{im}\delta_{jl}] \, \partial_j \, \partial_l a_m \\
&= \partial_m \, \partial_i a_m - \partial_j \, \partial_j a_i \\
&= \partial_i (\nabla \cdot \mathbf{a}) - \nabla^2 a_i
\end{aligned}$$

Note carefully the meaning of $\nabla^2 \mathbf{a}$:

$$\nabla^2 \mathbf{a} = \nabla^2 a_1 \hat{\mathbf{i}} + \nabla^2 a_2 \hat{\mathbf{j}} + \nabla^2 a_3 \hat{\mathbf{k}} \tag{2.10.9}$$

The last identity just proved is used in the derivation of wave equations in electromagnetic theory.

If **r** is the position vector from a fixed point P to a variable point,

$$\nabla \cdot \mathbf{r} = 3 \tag{2.10.10}$$

$$\nabla \times \mathbf{r} = 0 \tag{2.10.11}$$

$$(\mathbf{a} \cdot \nabla)\mathbf{r} = \mathbf{a} \tag{2.10.12}$$

The application of vector differentiation formulas is sometimes cumbersome. More headway can frequently be made by working directly with component equations and the permutation symbol.

2.11. LINE INTEGRALS. STOKES' THEOREM

As a particle moves along a trajectory, the amount of work done by a force **F** during an infinitesimal displacement $d\mathbf{s}$ is $\mathbf{F} \cdot d\mathbf{s}$. The total work done between two points a and b on the path of the particle is

$$W_{ab} = \int_a^b \mathbf{F} \cdot d\mathbf{s} \tag{2.11.1}$$

This is an example of a *line integral*. The line integral of any vector field is a scalar quantity. As a specific example, suppose that a vector field is given by

$$\mathbf{a} = (x + 5y + 4z)\hat{\mathbf{i}} + (5x + y + 3z)\hat{\mathbf{j}} + (4x + 3y + z)\hat{\mathbf{k}} \tag{2.11.2}$$

and suppose that the line integral $\int \mathbf{a} \cdot d\mathbf{s}$ along the straight line between $(0, 0, 0)$ and $(1, 1, 1)$ is to be calculated. The parametric equations of the path of integration are

$$x = ks \qquad y = ks \qquad z = ks \tag{2.11.3}$$

where k is a constant and s is the arc length measured from $(0, 0, 0)$. Since the value of s at $(1, 1, 1)$ is $\sqrt{3}$, the value of k must be $1/\sqrt{3}$. From (2.11.3),

$$dx = \frac{1}{\sqrt{3}} ds \qquad dy = \frac{1}{\sqrt{3}} ds \qquad dz = \frac{1}{\sqrt{3}} ds \tag{2.11.4}$$

Hence,

$$\mathbf{a} \cdot d\mathbf{s} = (x + 5y + 4z)\, dx + (5x + y + 3z)\, dy + (4x + 3y + z)\, dz$$

$$= \frac{10s\, ds}{3} + \frac{9s\, ds}{3} + \frac{8s\, ds}{3} = 9s\, ds$$

and

$$\int_0^{\sqrt{3}} \mathbf{a} \cdot d\mathbf{s} = \frac{27}{2} \tag{2.11.5}$$

In general, the value of the line integral of a vector between two points depends on the path of integration between the two points. There is a special class of vector fields for which the value of the line integral between two points is *independent* of the path of integration. Its value depends *only* on the coordinates of the initial and final points. These are called *conservative vector fields*. This concept is presented in more detail in Chapter 5.

The line integral around a *closed path* is denoted

$$\oint \mathbf{a} \cdot ds$$

and is called the *circulation* of the vector **a**. Figure 2.11.1 shows a closed rectangular path located in the x_1x_2-plane. The dimensions of the

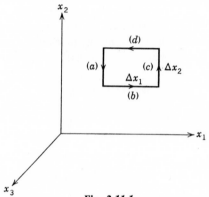

Fig. 2.11.1

rectangle are small so that the contribution to the circulation from side (*a*) is approximately

$$\mathbf{a}_a \cdot \Delta \mathbf{s}_a = -a_2 \, \Delta x_2 \tag{2.11.6}$$

Along side (*c*) the contribution is

$$\mathbf{a}_c \cdot \Delta \mathbf{s}_c = \left(a_2 + \frac{\partial a_2}{\partial x_1} \Delta x_1 \right) \Delta x_2 \tag{2.11.7}$$

Note that $\Delta \mathbf{s}_a = -\mathbf{j} \, \Delta x_2$ and $\Delta \mathbf{s}_c = \mathbf{j} \, \Delta x_2$. Taking account of all four sides of the rectangle gives

$$\oint \mathbf{a} \cdot ds = -a_2 \, \Delta x_2 + a_1 \, \Delta x_1 + \left(a_2 + \frac{\partial a_2}{\partial x_1} \Delta x_1 \right) \Delta x_2$$

$$- \left(a_1 + \frac{\partial a_1}{\partial x_2} \Delta x_2 \right) \Delta x_1$$

$$= \left(\frac{\partial a_2}{\partial x_1} - \frac{\partial a_1}{\partial x_2} \right) \Delta x_1 \, \Delta x_2 \tag{2.11.8}$$

If a coordinate system is chosen so that the rectangle does not lie in a coordinate plane, the result can be stated

$$\oint \mathbf{a} \cdot d\mathbf{s} = \nabla \times \mathbf{a} \cdot \hat{\mathbf{n}} \, \Delta\sigma \qquad (2.11.9)$$

where $\Delta\sigma$ is the magnitude of the area of the rectangle and $\hat{\mathbf{n}}$ is a unit vector normal to the surface and in the direction that a right-hand screw would advance if it were turned in the same direction that the circulation is calculated.

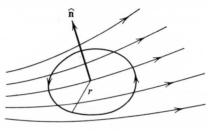

Fig. 2.11.2

The derivation of (2.11.9) does not pretend to mathematical rigor. It should be understood in the sense

$$\operatorname{curl} \mathbf{a} \cdot \hat{\mathbf{n}} = \lim_{\Delta\sigma \to 0} \frac{\oint \mathbf{a} \cdot d\mathbf{s}}{\Delta\sigma} \qquad (2.11.10)$$

which is frequently used as a definition of the curl.

Equation (2.11.9) can also be derived from fluid dynamics. Consider a closed path located in a flowing fluid which is an infinitesimal circle of radius r, Fig. 2.11.2. Since $\mathbf{u} \cdot d\mathbf{s}$ represents the tangential component of the velocity times an element of circular arc, the average velocity around the circle is

$$u_{\text{ave}} = \frac{1}{2\pi r} \oint \mathbf{u} \cdot d\mathbf{s} \qquad (2.11.11)$$

If we put $u_{\text{ave}} = r\omega$ and use (2.9.13), the result is

$$r\omega = \tfrac{1}{2} r(\nabla \times \mathbf{u}) \cdot \hat{\mathbf{n}} = \frac{1}{2\pi r} \oint \mathbf{u} \cdot d\mathbf{s}$$

$$\oint \mathbf{u} \cdot d\mathbf{s} = \pi r^2 (\nabla \times \mathbf{u} \cdot \hat{\mathbf{n}}) \qquad (2.11.12)$$

which is the same as (2.11.9).

The sum of the circulations around the two rectangles in Fig. 2.11.3a is the same as the circulation around the rectangle in Fig. 2.11.3b. This is because the contribution to the circulation of the common sides of the rectangles in Fig. 2.11.3a cancels.

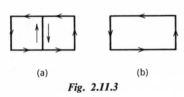

(a)   (b)

Fig. 2.11.3

If a surface σ of arbitrary shape bounded by a closed curve C (Fig. 2.11.4) is divided into a large number of small areas, then the sum of the circulations around the small areas is equal to the circulation around C, a result which is obtained by extending the argument of the last paragraph. Therefore,

$$\oint_C \mathbf{a} \cdot d\mathbf{s} = \sum_{k=1}^{N} [\boldsymbol{\nabla} \times \mathbf{a} \cdot \hat{\mathbf{n}} \, \Delta\sigma]_k \tag{2.11.13}$$

where N is the number of small areas into which the surface has been divided. If the number of small areas is taken larger and larger one obtains a surface integral:

$$\lim_{\substack{N \to \infty \\ \Delta\sigma \to 0}} \sum_{k=1}^{N} [\boldsymbol{\nabla} \times \mathbf{a} \cdot \hat{\mathbf{n}} \, \Delta\sigma]_k = \int_\sigma \boldsymbol{\nabla} \times \mathbf{a} \cdot \hat{\mathbf{n}} \, d\sigma \tag{2.11.14}$$

If (2.11.13) and (2.11.14) are combined, the result is *Stokes' theorem*:

$$\oint_C \mathbf{a} \cdot d\mathbf{s} = \int_\sigma \boldsymbol{\nabla} \times \mathbf{a} \cdot \hat{\mathbf{n}} \, d\sigma \tag{2.11.15}$$

Since $\mathbf{a} \cdot d\mathbf{s}$ is an ordinary scalar and $\boldsymbol{\nabla} \times \mathbf{a}$ is an axial vector, it is necessary to interpret $d\boldsymbol{\sigma} = \hat{\mathbf{n}} \, d\sigma$ as an axial vector in order that $\boldsymbol{\nabla} \times \mathbf{a} \cdot d\boldsymbol{\sigma}$ also be an ordinary scalar. If $d\mathbf{s}_1$ and $d\mathbf{s}_2$ are infinitesimal

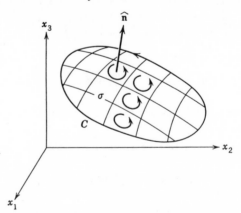

Fig. 2.11.4

displacement vectors lying in a surface, Fig. 2.11.5, then

$$d\boldsymbol{\sigma} = d\mathbf{s}_1 \times d\mathbf{s}_2 \qquad (2.11.16)$$

which represents $d\boldsymbol{\sigma}$ correctly as an axial vector. Note carefully that the direction of $d\boldsymbol{\sigma}$ is related to the direction of the circulation around the perimeter of the surface by the right-hand screw rule. The surface integral in the divergence theorem (2.8.4) is then also a pseudo-scalar. It is therefore necessary to represent the volume element as a pseudo-scalar. Three infinitesimal noncoplanar vectors $d\mathbf{a}$, $d\mathbf{b}$, and $d\mathbf{c}$ form the edges of a volume

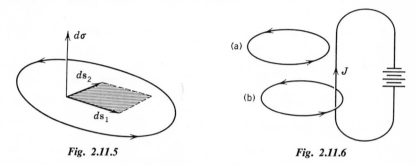

Fig. 2.11.5   Fig. 2.11.6

element in the shape of a parallelepiped. The volume is correctly represented as a pseudo-scalar by means of

$$d\Sigma = \delta_{ijk}\, da_i\, db_j\, dc_k \qquad (2.11.17)$$

If $da_i = (dx_1, 0, 0)$, $db_i = (0, dx_2, 0)$, and $dc_i = (0, 0, dx_3)$, then (2.11.17) is

$$d\Sigma = dx_1\, dx_2\, dx_3 \qquad (2.11.18)$$

The basic differential equation obeyed by a static magnetic field is

$$\boldsymbol{\nabla} \times \mathbf{B} = \frac{4\pi\mathbf{j}}{c} \qquad (2.11.19)$$

where $\mathbf{j}$ is the current density (charge per second per unit area) and c is a constant (Gaussian units are used). Let (2.11.19) be integrated over a surface bounded by a closed curve C:

$$\int \boldsymbol{\nabla} \times \mathbf{B} \cdot \hat{\mathbf{n}}\, d\sigma = \frac{4\pi}{c} \int \mathbf{j} \cdot \hat{\mathbf{n}}\, d\sigma = \frac{4\pi J}{c} \qquad (2.11.20)$$

where J represents the total current which flows through the surface. By Stokes' theorem

$$\oint_C \mathbf{B} \cdot d\mathbf{s} = \frac{4\pi}{c} J \qquad (2.11.21)$$

The circulation of the magnetic field about an arbitrary closed curve is equal to $4\pi/c$ times the total current which flows through the surface. In Fig. 2.11.6, a conducting wire carries a current J in a continuous loop.

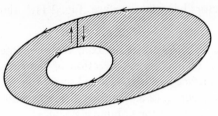

Fig. 2.11.7

The circulation of **B** about the closed curve (a) is zero whereas about (b) it is $4\pi J/c$.

Stokes' theorem is also valid when the surface is bounded by more than one curve as in Fig. 2.11.7. Such surfaces are said to be multiply connected. A cut can always be made which joins the two curves, making the surface effectively bounded by a single curve. The line integral along the cut is performed twice in opposite directions so that the contributions cancel.

2.12. INTEGRAL VECTOR IDENTITIES

There are four integral vector identities which are obtainable from Stokes' theorem and Gauss' divergence theorem. If ψ is a scalar and **a** is a vector field,

$$\oint_C \psi \, d\mathbf{s} = \int_\sigma \hat{\mathbf{n}} \times \nabla \psi \, d\sigma \tag{2.12.1}$$

$$\oint_C d\mathbf{s} \times \mathbf{a} = \int_\sigma (\hat{\mathbf{n}} \times \nabla) \times \mathbf{a} \, d\sigma \tag{2.12.2}$$

where, as in Stokes' theorem, C is a curve bounding the surface σ. In Stokes' theorem, the integrands are scalars whereas in (2.12.1) and (2.12.2) the integrands are vectors. For instance, (2.12.1) written out in full is

$$\hat{\mathbf{i}} \oint_C \psi \, dx_1 + \hat{\mathbf{j}} \oint_C \psi \, dx_2 + \hat{\mathbf{k}} \oint_C \psi \, dx_3$$

$$= \hat{\mathbf{i}} \int_\sigma (n_2 \, \partial_3 \psi - n_3 \, \partial_2 \psi) \, d\sigma + \hat{\mathbf{j}} \int_\sigma (n_3 \, \partial_1 \psi - n_1 \, \partial_3 \psi) \, d\sigma$$

$$+ \hat{\mathbf{k}} \int_\sigma (n_1 \, \partial_2 \psi - n_2 \, \partial_1 \psi) \, d\sigma$$

If σ is a closed surface bounding a volume Σ,

$$\int_\sigma \psi \hat{\mathbf{n}} \, d\sigma = \int_\Sigma \nabla \psi \, d\Sigma \tag{2.12.3}$$

$$\int_\sigma \hat{\mathbf{n}} \times \mathbf{a} \, d\sigma = \int_\Sigma \nabla \times \mathbf{a} \, d\Sigma \tag{2.12.4}$$

Equation (2.12.2) is proved by writing Stokes' theorem for $\mathbf{a} \times \mathbf{c}$ where $\mathbf{c}$ is an arbitrary *constant* vector:

$$\oint_C \mathbf{a} \times \mathbf{c} \cdot ds = \int_\sigma \hat{\mathbf{n}} \cdot \nabla \times (\mathbf{a} \times \mathbf{c}) \, d\sigma \qquad (2.12.5)$$

The line integral can be expressed

$$\oint_C \mathbf{a} \times \mathbf{c} \cdot ds = \oint_C ds \times \mathbf{a} \cdot \mathbf{c} = \mathbf{c} \cdot \oint_C ds \times \mathbf{a} \qquad (2.12.6)$$

and this suggests that an attempt be made to express the surface integral as the dot product of $\mathbf{c}$ with another vector:

$$\hat{\mathbf{n}} \cdot \nabla \times (\mathbf{a} \times \mathbf{c}) = n_i \delta_{ijk} \, \partial_j (\delta_{klm} a_l c_m) \qquad (2.12.7)$$

Since c_m is a constant, it is only a_l which is differentiated and the factors in (2.12.7) can be rearranged to give

$$\begin{aligned} \hat{\mathbf{n}} \cdot \nabla \times (\mathbf{a} \times \mathbf{c}) &= c_m \delta_{klm} n_i \delta_{ijk} \, \partial_j a_l \\ &= c_m \delta_{mkl} \delta_{kij} n_i \, \partial_j a_l \\ &= c_m \delta_{mkl} (\hat{\mathbf{n}} \times \nabla)_k a_l \\ &= \mathbf{c} \cdot (\hat{\mathbf{n}} \times \nabla) \times \mathbf{a} \end{aligned} \qquad (2.12.8)$$

Equation (2.12.5) can now be expressed

$$\mathbf{c} \cdot \oint ds \times \mathbf{a} = \mathbf{c} \cdot \int_\sigma (\hat{\mathbf{n}} \times \nabla) \times \mathbf{a} \, d\sigma \qquad (2.12.9)$$

Since $\mathbf{c}$ is an *arbitrary* constant vector, (2.12.9) implies

$$\oint_C ds \times \mathbf{a} = \int_\sigma (\hat{\mathbf{n}} \times \nabla) \times \mathbf{a} \, d\sigma \qquad (2.12.10)$$

To prove (2.12.3), write Gauss' theorem for the vector $\psi \mathbf{c}$ where $\mathbf{c}$ is an arbitrary constant vector:

$$\int_\sigma \psi \mathbf{c} \cdot \hat{\mathbf{n}} \, d\sigma = \int_\Sigma \nabla \cdot \psi \mathbf{c} \, d\Sigma$$

$$\mathbf{c} \cdot \int_\sigma \psi \hat{\mathbf{n}} \, d\sigma = \mathbf{c} \cdot \int_\Sigma \nabla \psi \, d\Sigma$$

$$\int_\sigma \psi \hat{\mathbf{n}} \, d\sigma = \int_\Sigma \nabla \psi \, d\Sigma$$

If Gauss' theorem is applied to the vector $\psi \nabla \phi$ where ψ and ϕ are scalar functions,

$$\begin{aligned} \int_\sigma \psi \nabla \phi \cdot \hat{\mathbf{n}} \, d\sigma &= \int_\Sigma \nabla \cdot \psi \nabla \phi \, d\Sigma \\ &= \int_\Sigma (\psi \nabla^2 \phi + \nabla \psi \cdot \nabla \phi) \, d\Sigma \end{aligned} \qquad (2.12.11)$$

a result which is known as *Green's first identity*. If the functions ψ and ϕ are interchanged,

$$\int_\sigma \phi \, \nabla \psi \cdot \hat{\mathbf{n}} \, d\sigma = \int_\Sigma (\phi \, \nabla^2 \psi + \nabla \phi \cdot \nabla \psi) \, d\Sigma \qquad (2.12.12)$$

Subtraction of (2.12.11) from (2.12.12) yields

$$\int_\Sigma (\phi \, \nabla^2 \psi - \psi \, \nabla^2 \phi) \, d\Sigma = \int_\sigma (\phi \, \nabla \psi - \psi \, \nabla \phi) \cdot \hat{\mathbf{n}} \, d\sigma \qquad (2.12.13)$$

which is *Green's second identity*, also referred to simply as "Green's theorem." Remember that $\hat{\mathbf{n}}$ is the *outward* directed normal to the surface bounding the volume Σ. Frequently, the notation

$$\nabla \psi \cdot \hat{\mathbf{n}} = \partial \psi / \partial n \qquad (2.12.14)$$

is used. Green's theorem then reads

$$\int_\Sigma (\phi \, \nabla^2 \psi - \psi \, \nabla^2 \phi) \, d\Sigma = \int_\sigma \left(\phi \frac{\partial \psi}{\partial n} - \psi \frac{\partial \phi}{\partial n} \right) d\sigma \qquad (2.12.15)$$

Green's identities are basic to the development of potential theory.

2.13. SPHERICAL POLAR COORDINATES

A few coordinate systems other than rectangular Cartesian are frequently used in physics. One of the most common systems is spherical polar coordinates illustrated in Fig. 2.13.1. The location of a point P can be specified by giving the length of the displacement vector $\mathbf{r}$ from the origin to the point along with two angles which give the orientation of $\mathbf{r}$. The angles used are a latitude angle θ measured between $\mathbf{r}$ and the x_3-axis and a longitude angle ϕ measured between the x_1-axis and the projection of $\mathbf{r}$ on the x_1x_2-plane. The three coordinates r, θ, and ϕ uniquely determine the location of the point P.

The *coordinate surfaces* are the surfaces determined by $r = $ const, $\theta = $ const, and $\phi = $ const. The surfaces $r = $ const are concentric spheres with their centers at the origin; the surfaces $\phi = $ const are planes which pass through the x_3-axis; the surfaces $\theta = $ const are cones with vertices at the origin. At a given point such as P in Fig. 2.13.1, these surfaces are mutually perpendicular. Coordinate systems which have this property are called *orthogonal*. The use of the term "orthogonal" in this connection is not to be confused with the term "orthogonal transformation" which specifically refers to a transformation between two rectangular Cartesian coordinate systems.

Any pair of the coordinate surfaces intersect along a continuous curve called a *coordinate line*. For instance, the cone $\theta = $ const intersects the

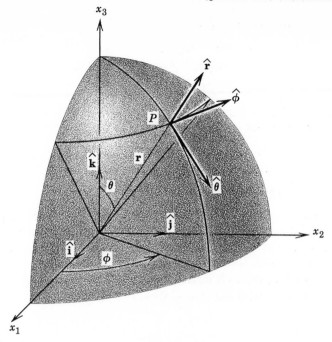

Fig. 2.13.1

sphere $r = $ const to form a circle. On this circle, only ϕ changes; r and θ are constant. Through P in Fig. 2.13.1 pass three mutually perpendicular coordinate lines which can be pictured as curved coordinate axes. Along each such axis only one variable changes. Generalized coordinates are sometimes refered to as *general curvilinear coordinates.*

It is possible to construct at any point three mutually perpendicular unit vectors $\hat{r}$, $\hat{\theta}$, and $\hat{\phi}$ which are tangent to the coordinate lines and perpendicular to the coordinate surfaces. These vectors, unlike $\hat{i}$, $\hat{j}$, and $\hat{k}$, change direction from point to point and hence are *functions of the coordinates.*

From the geometry of Fig. 2.13.1 the transformation from rectangular to spherical coordinates is found to be

$$x_1 = r \sin \theta \cos \phi$$
$$x_2 = r \sin \theta \sin \phi \qquad (2.13.1)$$
$$x_3 = r \cos \theta$$

The range of the new variables is

$$0 \leq r < \infty \qquad 0 \leq \theta \leq \pi \qquad 0 \leq \phi < 2\pi$$

The inverse transformation is

$$\phi = \tan^{-1}(x_2/x_1)$$
$$\theta = \cos^{-1}[x_3(x_1^2 + x_2^2 + x_3^2)^{-1/2}] \qquad (2.13.2)$$
$$r = (x_1^2 + x_2^2 + x_3^2)^{1/2}$$

The reader who is clever at geometry can show that the unit vectors $\hat{\mathbf{r}}$, $\hat{\boldsymbol{\theta}}$, and $\hat{\boldsymbol{\phi}}$ are related to $\hat{\mathbf{i}}$, $\hat{\mathbf{j}}$, $\hat{\mathbf{k}}$ by

$$\hat{\mathbf{r}} = \sin\theta \cos\phi \hat{\mathbf{i}} + \sin\theta \sin\phi \hat{\mathbf{j}} + \cos\theta \hat{\mathbf{k}}$$
$$\hat{\boldsymbol{\theta}} = \cos\phi \cos\theta \hat{\mathbf{i}} + \sin\phi \cos\theta \hat{\mathbf{j}} - \sin\theta \hat{\mathbf{k}} \qquad (2.13.3)$$
$$\hat{\boldsymbol{\phi}} = -\sin\phi \hat{\mathbf{i}} + \cos\phi \hat{\mathbf{j}}$$

These relations give the explicit functional dependence of the unit vectors on the coordinates. A derivation of (2.13.3) is postponed until Section 14, where the question of the transformation properties of basis vectors for arbitrary curvilinear coordinate systems is discussed.

The expression for small increments of displacement measured in the directions of $\hat{\mathbf{r}}$, $\hat{\boldsymbol{\theta}}$, and $\hat{\boldsymbol{\phi}}$ is obtained from Fig. 2.13.2:

$$ds(1) = dr$$
$$ds(2) = r\,d\theta \qquad (2.13.4)$$
$$ds(3) = r\sin\theta\,d\phi$$

A differential displacement vector measured in an arbitrary direction is expressible as

$$d\mathbf{s} = dr\hat{\mathbf{r}} + r\,d\theta\hat{\boldsymbol{\theta}} + r\sin\theta\,d\phi\hat{\boldsymbol{\phi}} \qquad (2.13.5)$$

The square of the magnitude of $d\mathbf{s}$ is

$$ds^2 = dr^2 + r^2\,d\theta^2 + r^2\sin^2\theta\,d\phi^2 \qquad (2.13.6)$$

This expression is called the *metric form* and is a basic expression which determines the geometry of the coordinate system. $d\mathbf{s}$ itself is referred to as the *line element*.

A differential volume element can be expressed

$$d\Sigma = ds_1\,ds_2\,ds_3 = r^2\sin\theta\,dr\,d\theta\,d\phi \qquad (2.13.7)$$

An element of area on the spherical surface $r = $ const is

$$d\sigma = ds_2\,ds_3 = r^2\sin\theta\,d\theta\,d\phi \qquad (2.13.8)$$

The velocity of a particle can be expressed in terms of its components in the $\hat{\mathbf{r}}$, $\hat{\boldsymbol{\theta}}$, and $\hat{\boldsymbol{\phi}}$ directions by dividing (2.13.5) by dt:

$$\mathbf{u} = d\mathbf{s}/dt = \dot{r}\hat{\mathbf{r}} + r\dot{\theta}\hat{\boldsymbol{\theta}} + r\sin\theta\dot{\phi}\hat{\boldsymbol{\phi}} \qquad (2.13.9)$$

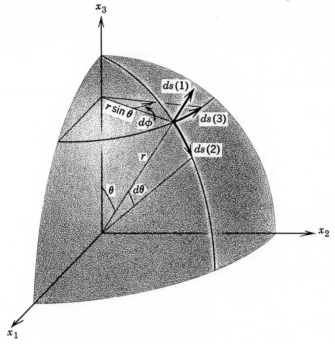

Fig. 2.13.2

The kinetic energy of a particle of mass m is

$$T = \tfrac{1}{2}m(\mathbf{u} \cdot \mathbf{u}) = \tfrac{1}{2}m(\dot{r}^2 + r^2\dot{\theta}^2 + r^2 \sin^2 \theta \dot{\phi}^2) \qquad (2.13.10)$$

In the treatment of a given problem, a great simplification can frequently be made by an appropriate choice of coordinate system. For instance, the gravitational field of a point mass M located at the origin is simply

$$\mathbf{g} = -GM\,\frac{\hat{\mathbf{r}}}{r^2} \qquad (2.13.11)$$

in spherical polar coordinates whereas in rectangular coordinates it is given by the more complicated equation (2.3.3).

2.14. GENERAL COORDINATE TRANSFORMATIONS

All of the foregoing developments of vector and tensor calculus have been based essentially on representations in rectangular Cartesian coordinate systems. We now take up the problem of generalizing these concepts to arbitrary curvilinear coordinate systems. Our central aim is the development of a covariant representation of particle mechanics, but we will obtain at the same time many results which are valuable in other areas of theoretical physics.

The functions relating the curvilinear coordinates to rectangular coordinates can be indicated by

$$q_1 = q_1(x_1, x_2, x_3)$$
$$q_2 = q_2(x_1, x_2, x_3) \qquad (2.14.1)$$
$$q_3 = q_3(x_1, x_2, x_3)$$

In the example of spherical coordinates, $q_1 = r$, $q_2 = \theta$, $q_3 = \phi$, and the transformations are given by (2.13.2).

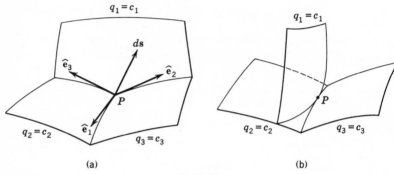

(a)   (b)

Fig. 2.14.1

A fundamental requirement of a coordinate system is that it must provide three independent parameters in terms of which distance can be measured as illustrated by Eq. (2.13.5) for the case of spherical coordinates. It must be possible to find three coordinate surfaces $q_1 = c_1, q_2 = c_2$, and $q_3 = c_3$ which pass through a given point. These coordinate surfaces must meet along coordinate lines which are such that if three unit tangent vectors are constructed at a given point, e.g., point P of Fig. 2.14.1a, they will form a linearly independent set. This guarantees that an infinitesimal displacement vector constructed at P can always be expressed in terms of components along the coordinate lines:

$$d\mathbf{s} = ds(1)\hat{\mathbf{e}}_1 + ds(2)\hat{\mathbf{e}}_2 + ds(3)\hat{\mathbf{e}}_3 \qquad (2.14.2)$$

The unit vectors are not necessarily mutually orthogonal in the most general coordinate system. Figure 2.14.1b illustrates a point where the basic requirement is not met. In rectangular coordinates, displacement vectors of finite length can be represented in the manner of (2.14.2), but due to the curvature of the coordinate lines, only *infinitesimal* displacement vectors can be so represented in a curvilinear coordinate system.

The gradient vectors, ∇q_1, ∇q_2, and ∇q_3 calculated at point P are normal to the coordinate surfaces at P. They are not in the same direction as the unit coordinate vectors except in the special case of orthogonal coordinate systems. Since the cross products of any two unit coordinate vectors is perpendicular to a coordinate surface, we may write

$$\nabla q_1 = \mathbf{b}^1 = k_1(\hat{\mathbf{e}}_2 \times \hat{\mathbf{e}}_3)$$
$$\nabla q_2 = \mathbf{b}^2 = k_2(\hat{\mathbf{e}}_3 \times \hat{\mathbf{e}}_1) \qquad (2.14.3)$$
$$\nabla q_3 = \mathbf{b}^3 = k_3(\hat{\mathbf{e}}_1 \times \hat{\mathbf{e}}_2)$$

The gradient vectors are given the special designation $\mathbf{b}^1$, $\mathbf{b}^2$, and $\mathbf{b}^3$ and called *reciprocal base vectors*. The superscripts are not powers, but are used to distinguish the reciprocal base vectors from coordinate base vectors to be introduced presently. The proportionality factors k_1, k_2, and k_3 will be determined in what follows.

The *Jacobian* of the transformation (2.14.1) is defined to be

$$J = \mathbf{b}^1 \cdot (\mathbf{b}^2 \times \mathbf{b}^3) = \nabla q_1 \cdot (\nabla q_2 \times \nabla q_3) = \det\left(\frac{\partial q_i}{\partial x_j}\right) \qquad (2.14.4)$$

By means of (2.14.3) and the "back minus cab rule" (1.10.7),

$$\begin{aligned} J &= k_1 k_2 k_3 (\hat{\mathbf{e}}_2 \times \hat{\mathbf{e}}_3) \cdot [(\hat{\mathbf{e}}_3 \times \hat{\mathbf{e}}_1) \times (\hat{\mathbf{e}}_1 \times \hat{\mathbf{e}}_2)] \\ &= k_1 k_2 k_3 (\hat{\mathbf{e}}_2 \times \hat{\mathbf{e}}_3) \cdot [\hat{\mathbf{e}}_1(\hat{\mathbf{e}}_3 \times \hat{\mathbf{e}}_1 \cdot \hat{\mathbf{e}}_2) - \hat{\mathbf{e}}_2(\hat{\mathbf{e}}_3 \times \hat{\mathbf{e}}_1 \cdot \hat{\mathbf{e}}_1)] \qquad (2.14.5) \\ &= k_1 k_2 k_3 (\hat{\mathbf{e}}_1 \cdot \hat{\mathbf{e}}_2 \times \hat{\mathbf{e}}_3)^2 \end{aligned}$$

Since $\hat{\mathbf{e}}_1 \cdot \hat{\mathbf{e}}_2 \times \hat{\mathbf{e}}_3 = 0$ if the three unit vectors are coplanar, the unit vectors at a given point form a linearly independent set if, and only if, the Jacobian calculated at that point is not zero. In courses on advanced calculus, it is proved that the transformation functions (2.14.1) can be solved for the x_i in terms of the q_i if, and only if, the Jacobian calculated at that point is not zero. A coordinate transformation will be called *admissible* in a region of space if $J \neq 0$ everywhere in that region.

Associated with an infinitesimal displacement vector $d\mathbf{s}$ are coordinate differentials which are computed from

$$dq^i = \frac{\partial q^i}{\partial x^j}\, dx^j \qquad (2.14.6)$$

This linear relation between the coordinate differentials is to be regarded as a generalization of the vector transformation law. The transformation coefficients are partial derivatives which are themselves functions of the coordinates. Equation (2.14.6) is called the *contravariant* transformation law and superscripts are used to distinguish quantities which transform in this way from others which transform according to the *covariant* law to be introduced presently. The coefficients themselves do not have any particular transformation property and so it dosen't really matter whether they are written as

$$\frac{\partial q^i}{\partial x^j} \quad \text{or} \quad \frac{\partial q_i}{\partial x_j}$$

Common practice dictates that we use the notation (2.14.6). Frequently, the functions (2.14.1) giving the actual relation between new and old coordinates are called "transformations", but they are *not* in general vector transformations.

The Jacobian (2.14.4) is just the determinant of the coefficients of (2.14.6); since $J \neq 0$, the inverse of (2.14.6) exists. By first expressing the x_i in terms of the q_i and taking the total differentials, we get

$$dx^j = \frac{\partial x^j}{\partial q^k} dq^k \tag{2.14.7}$$

By substituting (2.14.7) into (2.14.6),

$$dq^i = \frac{\partial q^i}{\partial x^j}\frac{\partial x^j}{\partial q^k} dq^k \tag{2.14.8}$$

The infinitesimal displacement vector ds pictured in Fig. 2.14.1a can be taken in any direction that we please, meaning that the coordinate differentials can take on arbitrary values independently of one another. By exactly the same argument that we used to get the orthogonality relations (1.4.13), we get

$$\frac{\partial q^i}{\partial x^j}\frac{\partial x^j}{\partial q^k} = \delta_k^{\ i} \tag{2.14.9}$$

where $\delta_k^{\ i}$ is the Kronecker delta. The reason for using one superscript and one subscript on the Kronecker delta when general coordinate transformations are being considered will soon become apparent. Similarly, we can show that

$$\frac{\partial q^i}{\partial x^j}\frac{\partial x^k}{\partial q^i} = \delta_j^{\ k} \tag{2.14.10}$$

The Jacobian of the inverse transformation (2.14.7) is

$$K = \det\left(\frac{\partial x^j}{\partial q^k}\right) \tag{2.14.11}$$

From (2.14.9) it can be determined that

$$JK = 1 \tag{2.14.12}$$

Coordinate systems used in applications frequently have isolated singular points, i.e., points where one or the other of J and K is zero. For instance,

$K = 0$ at the origin for the transformation from rectangular to spherical coordinates. This means simply that the transformation equations (2.13.1) cannot be solved for unique values of r, θ, and ϕ if $x_1 = x_2 = x_3 = 0$. We can only tell from this information that $r = 0$ and cannot assign any values to θ and ϕ. On the other hand, the transformation is all right in the other direction at the origin. If, for instance, it is known that $r = 0$, $\theta = \pi/2$, $\phi = \pi/4$, then there is only one solution for x_1, x_2, and x_3; namely, $x_1 = x_2 = x_3 = 0$.

It is not necessary that a new coordinate system be defined in terms of rectangular coordinates. Let q_i and q_i' be two admissible coordinate systems connected by

$$q_i' = q_i'(q_1, q_2, q_3) \tag{2.14.13}$$

Then

$$dq^{i'} = \frac{\partial q^{i'}}{\partial q^j} dq^j \qquad dq^j = \frac{\partial q^j}{\partial q^{k'}} dq^{k'} \tag{2.14.14}$$

We will frequently abbreviate the partial derivatives by the notation

$$P_j^{i'} = \frac{\partial q^{i'}}{\partial q^j} \qquad P_{k'}^j = \frac{\partial q^j}{\partial q^{k'}} \tag{2.14.15}$$

The reader should prove that

$$P_j^{i'} P_{k'}^j = \delta_k^{\ i} \qquad P_j^{i'} P_{i'}^k = \delta_j^{\ k} \tag{2.14.16}$$

We close this section with the following important definition: *Any* set of three functions which transform as

$$A^{i'} = P_j^{i'} A^j \tag{2.14.17}$$

under a general coordinate transformation are said to be the *contravariant* components of a vector. As the theory develops, examples of contravariant vector components other than the coordinate differentials will be encountered. There are also sets of three functions which transform as

$$A_i' = P_{i'}^j A_j \tag{2.14.18}$$

and are said to be the *covariant* components of a vector. We use *subscripts* to distinguish the *covariant* components of a vector from the contravariant components.

2.15. THE METRIC TENSOR

The discussion in Section 2.14 centers around the infinitesimal displacement vector pictured in Fig. 2.14.1. We now consider the problem of computing the magnitude of this vector. This is easy in rectangular coordinates:

$$ds^2 = dx^i \, dx^i \tag{2.15.1}$$

To see how to do the calculation in a general coordinate system, we use the transformations (2.14.7) for the coordinate differentials:

$$ds^2 = \frac{\partial x^i}{\partial q^j} \frac{\partial x^i}{\partial q^k} \, dq^j \, dq^k = g_{jk} \, dq^j \, dq^k \tag{2.15.2}$$

The set of nine quantities

$$g_{jk} = \frac{\partial x^i}{\partial q^j} \frac{\partial x^i}{\partial q^k} \tag{2.15.3}$$

are the covariant components of a second rank tensor called the *metric tensor*. In the example of spherical coordinates discussed in Section 2.13, it is seen from Eq. (2.13.6) that the components of the metric tensor are functions of the coordinates r, θ, and ϕ which can be conveniently displayed in matrix form as

$$(g_{ij}) = \begin{pmatrix} 1 & 0 & 0 \\ 0 & r^2 & 0 \\ 0 & 0 & r^2 \sin^2 \theta \end{pmatrix} \tag{2.15.4}$$

provided that $dq^1 = dr$, $dq^2 = d\theta$, and $dq^3 = d\phi$. In this example, we were able to infer the metric form (2.15.2) from simple geometry. Formally, the components of the metric tensor can be computed from (2.15.3). For example,

$$\begin{aligned} g_{11} &= \left(\frac{\partial x^1}{\partial r}\right)^2 + \left(\frac{\partial x^2}{\partial r}\right)^2 + \left(\frac{\partial x^3}{\partial r}\right)^2 \\ g_{12} &= \frac{\partial x^1}{\partial r} \frac{\partial x^1}{\partial \theta} + \frac{\partial x^2}{\partial r} \frac{\partial x^2}{\partial \theta} + \frac{\partial x^3}{\partial r} \frac{\partial x^3}{\partial \theta} \end{aligned} \tag{2.15.5}$$

and so on. The reader is invited to compute the partial derivatives from the transformations (2.13.1) and verify that (2.15.4) follows from (2.15.5).

From (2.15.3) it is seen that the metric tensor is *symmetric:*

$$g_{jk} = g_{kj} \tag{2.15.6}$$

meaning that it has only six, rather than nine, independent components.

The magnitude of ds can be computed in either of two admissible coordinate systems:

$$ds^2 = g_{ij} \, dq^i \, dq^j = g'_{ij} \, dq^{i'} \, dq^{j'} \tag{2.15.7}$$

By using the known transformation properties of the coordinate differentials (2.14.14) and the fact that the displacement vector ds is arbitrary, it is proved that

$$g'_{ij} = P^k_{i'} P^l_{j'} g_{kl} \tag{2.15.8}$$

Any set of nine functions which transform as

$$A'_{ij} = P^k_{i'}P^l_{j'}A_{kl} \tag{2.15.9}$$

are said to be the *covariant* components of a second rank tensor. A set of nine functions which transform as

$$A^{ij'} = P^{i'}_k P^{j'}_l A^{kl} \tag{2.15.10}$$

are said to be the *contravariant* components of a second rank tensor.

There exists also so called *mixed* second rank tensor components. They are given one covariant index (subscript) and one contravariant index (superscript) and transform as

$$A^{i'}_j = P^{i'}_k P^l_{j'} A^k_l \tag{2.15.11}$$

We already have an example of such a tensor; it is the Kronecker delta symbol. It is readily proved by using (2.14.16) that

$$\delta^{i'}_j = P^{i'}_k P^l_{j'}\, \delta^k_l \tag{2.15.12}$$

The Kronecker delta has the unique property of being *isotropic*, i.e., its components have the same numerical values in all coordinate systems:

$$\delta^{i'}_j = \delta^{\ i}_j \tag{2.15.13}$$

No other second rank tensor has this property.

The great importance of the metric or fundamental tensor, as it is sometimes called, is that it completely determines the geometry of a coordinate system. Once the metric tensor is known, further reference to any other coordinate system is unnecessary.

2.16. BASIS VECTORS

Referring once more to the infinitesimal displacement vector $d\mathbf{s}$ of Fig. 2.14.1, let us consider the problem of computing its components $ds(1)$, $ds(2)$, and $ds(3)$ as expressed by Eq. (2.14.2). To compute $ds(1)$, suppose that we start at point P and move a differential distance along the coordinate line to which $\hat{\mathbf{e}}_1$ is tangent. Only q_1 changes; $dq^2 = dq^3 = 0$. From the metric form (2.15.2) we get

$$ds(1)^2 = g_{11}(dq^1)^2, \qquad ds(1) = \sqrt{g_{11}}\, dq^1 \tag{2.16.1}$$

Similarly, by considering displacements along the other two coordinate lines,

$$ds(2) = \sqrt{g_{22}}\, dq^2 \qquad ds(3) = \sqrt{g_{33}}\, dq^3 \tag{2.16.2}$$

For a specific example, the reader can refer to Eq. (2.13.4). The components $ds(i)$ are the *physical* components of the displacement and have the actual dimensions of length, whereas the coordinate differentials in general do not have the dimensions of length.

An infinitesimal displacement vector can therefore be expressed

$$d\mathbf{s} = dq^1\sqrt{g_{11}}\,\hat{\mathbf{e}}_1 + dq^2\sqrt{g_{22}}\,\hat{\mathbf{e}}_2 + dq^3\sqrt{g_{33}}\,\hat{\mathbf{e}}_3 \qquad (2.16.3)$$

It is convenient to introduce a set of basis vectors defined by

$$\mathbf{b}_1 = \sqrt{g_{11}}\,\hat{\mathbf{e}}_1 \qquad \mathbf{b}_2 = \sqrt{g_{22}}\,\hat{\mathbf{e}}_2 \qquad \mathbf{b}_3 = \sqrt{g_{33}}\,\hat{\mathbf{e}}_3 \qquad (2.16.4)$$

These basis vectors are, of course, *not* unit vectors. Then

$$d\mathbf{s} = dq^i\mathbf{b}_i \qquad (2.16.5)$$

It is possible to compute $d\mathbf{s}$ in any other admissible coordinate system so that

$$d\mathbf{s} = dq^i\mathbf{b}_i = dq^{j'}\mathbf{b}_j' \qquad (2.16.6)$$

The above relation implies, by arguments that should by now be familiar to the reader, that

$$\mathbf{b}_i' = P_{i'}^{j}\mathbf{b}_j \qquad \mathbf{b}_i = P_i^{j'}\mathbf{b}_j' \qquad (2.16.7)$$

In other words, the basis vectors themselves have a simple *covariant* transformation property.

Recall the definition (2.14.3) of the *reciprocal* base vectors. The proportionality factors k_1, k_2, and k_3 can now be found. Let q_1 be expressed in terms of rectangular coordinates:

$$q_1 = q_1(x_1, x_2, x_3) \qquad (2.16.8)$$

It is possible to compute the total differential of (2.16.8) as

$$dq^1 = \boldsymbol{\nabla} q_1 \cdot d\mathbf{s} \qquad (2.16.9)$$

By (2.14.3) and (2.16.3)

$$dq^1 = k_1(\hat{\mathbf{e}}_2 \times \hat{\mathbf{e}}_3) \cdot \hat{\mathbf{e}}_1\sqrt{g_{11}}\,dq^1, \qquad k_1 = \frac{1}{\sqrt{g_{11}}\,\hat{\mathbf{e}}_1 \cdot \hat{\mathbf{e}}_2 \times \hat{\mathbf{e}}_3} \qquad (2.16.10)$$

Similarly,

$$k_2 = \frac{1}{\sqrt{g_{22}}\,\hat{\mathbf{e}}_1 \cdot \hat{\mathbf{e}}_2 \times \hat{\mathbf{e}}_3} \qquad k_3 = \frac{1}{\sqrt{g_{33}}\,\hat{\mathbf{e}}_1 \cdot \hat{\mathbf{e}}_2 \times \hat{\mathbf{e}}_3} \qquad (2.16.11)$$

The Jacobian (2.14.5) is expressible as

$$J = \mathbf{b}^1 \cdot \mathbf{b}^2 \times \mathbf{b}^3 = \frac{1}{\mathbf{b}_1 \cdot \mathbf{b}_2 \times \mathbf{b}_3} \qquad (2.16.12)$$

Note that $J < 0$ if the base vectors are a left-hand triad, i.e., if an improper transformation has been done. By means of (2.15.3),

$$g = \det g_{jk} = \left[\det\left(\frac{\partial x^i}{\partial q^j}\right)\right]^2 = \frac{1}{J^2} \qquad (2.16.13)$$

Hence

$$\mathbf{b}_1 \cdot \mathbf{b}_2 \times \mathbf{b}_3 = \sqrt{g} \qquad (2.16.14)$$

The reciprocal base vectors (2.14.3) can be expressed as

$$\mathbf{b}^1 = \frac{1}{\sqrt{g}} \mathbf{b}_2 \times \mathbf{b}_3 \qquad \mathbf{b}^2 = \frac{1}{\sqrt{g}} \mathbf{b}_3 \times \mathbf{b}_1 \qquad \mathbf{b}^3 = \frac{1}{\sqrt{g}} \mathbf{b}_1 \times \mathbf{b}_2 \qquad (2.16.15)$$

These relations can be solved for the base vectors in terms of the reciprocal base vectors:

$$\mathbf{b}_1 = \sqrt{g}\, \mathbf{b}^2 \times \mathbf{b}^3 \qquad \mathbf{b}_2 = \sqrt{g}\, \mathbf{b}^3 \times \mathbf{b}^1 \qquad \mathbf{b}_3 = \sqrt{g}\, \mathbf{b}^1 \times \mathbf{b}^2 \qquad (2.16.16)$$

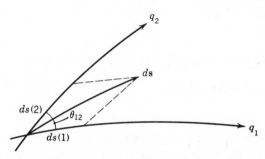

Fig. 2.16.1

A consequence of (2.16.15) is

$$\mathbf{b}^i \cdot \mathbf{b}_j = \delta_j{}^i \qquad (2.16.17)$$

By computing the magnitude of $d\mathbf{s}$ directly from (2.16.5),

$$ds^2 = (\mathbf{b}_i \cdot \mathbf{b}_j)\, dq^i\, dq^j \qquad (2.16.18)$$

The comparison of (2.16.18) with the metric form (2.15.2) shows that

$$g_{ij} = \mathbf{b}_i \cdot \mathbf{b}_j \qquad (2.16.19)$$

This result reveals that a coordinate system is orthogonal if, and only if, the metric tensor is diagonal, i.e., $g_{ij} = 0$ for $i \neq j$. See, for example, the metric tensor (2.15.4) of spherical coordinates. In general, the angle between the q_1 and the q_2 coordinate lines is found from

$$g_{12} = \sqrt{g_{11}}\,\sqrt{g_{22}}\,\hat{\mathbf{e}}_1 \cdot \hat{\mathbf{e}}_2 = \sqrt{g_{11}}\,\sqrt{g_{22}}\,\cos\theta_{12} \qquad (2.16.20)$$

In order to emphasize the meaning of the physical components of displacement, we have pictured in Fig. 2.16.1 an infinitesimal displacement vector which is tangent to the coordinate surface $q_3 = $ const. Since dq^3

is then zero, the metric form is

$$ds^2 = g_{11}(dq^1)^2 + g_{22}(dq^2)^2 + 2g_{12}\,dq^1\,dq^2$$

$$= (\sqrt{g_{11}}\,dq^1)^2 + (\sqrt{g_{22}}\,dq^2)^2 + 2\,\frac{g_{12}}{\sqrt{g_{11}}\,\sqrt{g_{22}}}\,(\sqrt{g_{11}}\,dq^1)(\sqrt{g_{22}}\,dq^2)$$

$$= ds(1)^2 + ds(2)^2 + 2\cos\theta_{12}\,ds(1)\,ds(2) \qquad (2.16.21)$$

This is the cosine law and shows that the physical components of displacement are to be combined by the familiar parallelogram rule to find the "resultant."

The base vectors and the reciprocal base vectors each form sets of three linearly independent vectors. It must, therefore, be possible to express the base vectors as linear combinations of the reciprocal base vectors:

$$\mathbf{b}_i = a_{ij}\mathbf{b}^j \qquad (2.16.22)$$

By forming the dot product of (2.16.22) with $\mathbf{b}_k$ we find

$$\mathbf{b}_k \cdot \mathbf{b}_i = a_{ij}\mathbf{b}_k \cdot \mathbf{b}^j, \qquad g_{ki} = a_{ij}\,\delta_k{}^j = a_{ik}$$

Hence,

$$\mathbf{b}_i = g_{ij}\mathbf{b}^j \qquad (2.16.23)$$

Since the transformation properties of $\mathbf{b}_i$ and g_{ij} are known to be covariant, the above result implies a contravariant transformation property for the reciprocal base vectors:

$$\mathbf{b}^{j'} = P_k^{j'}\mathbf{b}^k \qquad (2.16.24)$$

The proof of this is left as a problem. We will, however, obtain (2.16.24) in another way in the next section.

2.17. THE GRADIENT

The gradient of a scalar function expressed in rectangular Cartesian coordinates is

$$\mathbf{\nabla}\psi = \frac{\partial\psi}{\partial x^1}\,\hat{\mathbf{i}} + \frac{\partial\psi}{\partial x^2}\,\hat{\mathbf{j}} + \frac{\partial\psi}{\partial x^3}\,\hat{\mathbf{k}} = \frac{\partial\psi}{\partial x^j}\,\hat{\mathbf{i}}_j \qquad (2.17.1)$$

where we represented the unit Cartesian vectors as $\hat{\mathbf{i}}_1$, $\hat{\mathbf{i}}_2$, and $\hat{\mathbf{i}}_3$. How shall we represent the operator $\mathbf{\nabla}$ in an arbitrary curvilinear coordinate system? Let ψ be expressed in terms of the curvilinear coordinates. Then

$$\mathbf{\nabla}\psi = \frac{\partial\psi}{\partial q^k}\frac{\partial q^k}{\partial x^j}\,\hat{\mathbf{i}}_j = \frac{\partial\psi}{\partial q^k}\,\mathbf{\nabla}q_k \qquad (2.17.2)$$

But $\mathbf{\nabla}q_k$ are the reciprocal base vectors. Thus,

$$\mathbf{\nabla}\psi = \mathbf{b}^k\,\partial_k\psi \qquad (2.17.3)$$

meaning that the operator ∇ can be represented symbolically as

$$\nabla = \mathbf{b}^k \, \partial_k \qquad (2.17.4)$$

The partial derivatives themselves transform covariantly,

$$\partial_{k'} \psi = P_{k'}^j \, \partial_j \psi \qquad (2.17.5)$$

and are called the covariant components of the gradient. Since the gradient vector can be expressed in any curvilinear coordinate system

$$\nabla \psi = \mathbf{b}^k \, \partial_k \psi = \mathbf{b}^{i'} \, \partial_{i'} \psi \qquad (2.17.6)$$

the covariance of the partial derivatives implies the contravariance of the reciprocal base vectors as expressed by (2.16.24).

From the reciprocal base vectors we define

$$\mathbf{b}^i \cdot \mathbf{b}^j = g^{ij} \qquad (2.17.7)$$

It follows from (2.16.24) that

$$g^{kl'} = P_i^{k'} P_j^{l'} g^{ij} \qquad (2.17.8)$$

meaning that the g^{ij} are the *contravariant* components of a second rank tensor. Note that $g^{ij} = g^{ji}$. By dotting (2.16.23) with $\mathbf{b}^k$ we can prove

$$\delta_i{}^k = g_{ij} g^{kj} \qquad (2.17.9)$$

It is now easy to prove that

$$\mathbf{b}^i = g^{ij} \mathbf{b}_j \qquad (2.17.10)$$

If you will now look back over the various relations connecting the reciprocal base vectors with the base vectors, specifically equations (2.16.17), (2.16.19), (2.17.7), (2.17.9), and (2.17.10), you will realize that there is complete symmetry between a covariant representation and a contravariant representation. The g^{ij} are called the contravariant components of the metric tensor.

As an application, we will derive the relations (2.13.3) connecting the unit vectors of spherical coordinates with the Cartesian unit vectors. The base vectors of spherical coordinates are given by

$$\mathbf{b}_i = \frac{\partial x^k}{\partial q^i} \hat{\mathbf{i}}_k \qquad (2.17.11)$$

For instance

$$\mathbf{b}_2 = \frac{\partial x^1}{\partial \theta} \hat{\mathbf{i}} + \frac{\partial x^2}{\partial \theta} \hat{\mathbf{j}} + \frac{\partial x^3}{\partial \theta} \hat{\mathbf{k}}$$

$$= r \cos \theta \cos \phi \, \hat{\mathbf{i}} + r \cos \theta \sin \phi \, \hat{\mathbf{j}} - r \sin \theta \hat{\mathbf{k}} \qquad (2.17.12)$$

The base vector in terms of the unit vector is

$$\mathbf{b}_2 = \sqrt{g_{22}} \, \hat{\boldsymbol{\theta}} = r \hat{\boldsymbol{\theta}} \qquad (2.17.13)$$

Hence,

$$\hat{\theta} = \cos\theta\cos\phi\hat{\imath} + \cos\theta\sin\phi\hat{\jmath} - \sin\theta\hat{k} \qquad (2.17.14)$$

The formulas for $\hat{r}$ and $\hat{\phi}$ can be similarly obtained.

Finally, let us compute the physical components of the gradient in spherical coordinates. Since the coordinate system is orthogonal, (2.17.10) and (2.17.9) give

$$\mathbf{b}^1 = g^{11}\mathbf{b}_1 = \frac{1}{g_{11}}\mathbf{b}_1 = \frac{1}{\sqrt{g_{11}}}\hat{r}$$

$$\mathbf{b}^2 = \frac{1}{\sqrt{g_{22}}}\hat{\theta}, \qquad \mathbf{b}^3 = \frac{1}{\sqrt{g_{33}}}\hat{\phi} \qquad (2.17.15)$$

Hence,

$$\nabla\psi = \frac{1}{\sqrt{g_{11}}}\frac{\partial\psi}{\partial r}\hat{r} + \frac{1}{\sqrt{g_{22}}}\frac{\partial\psi}{\partial\theta}\hat{\theta} + \frac{1}{\sqrt{g_{33}}}\frac{\partial\psi}{\partial\phi}\hat{\phi}$$

$$= \frac{\partial\psi}{\partial r}\hat{r} + \frac{1}{r}\frac{\partial\psi}{\partial\theta}\hat{\theta} + \frac{1}{r\sin\theta}\frac{\partial\psi}{\partial\phi}\hat{\phi} \qquad (2.17.16)$$

2.18. FORMAL PROPERTIES OF VECTORS AND TENSORS

Given a vector expressed as

$$\mathbf{a} = A^i\mathbf{b}_i \qquad (2.18.1)$$

we can re-express it by means of (2.16.23) as

$$\mathbf{a} = A^i g_{ij}\mathbf{b}^j = A_j\mathbf{b}^j \qquad (2.18.2)$$

where

$$A_j = g_{ij}A^i \qquad (2.18.3)$$

a process which is known as lowering the index. That A_j transforms covariantly is readily proved. Write

$$A_{k'} = g'_{lk}A^{l'} = P_{l'}^i P_{k'}^j g_{ij}P_n^{l'}A^n$$
$$= \delta_n{}^i P_{k'}^j g_{ij}A^n = P_{k'}^j g_{ij}A^i = P_{k'}^j A_j \qquad (2.18.4)$$

By means of (2.17.9) it is possible to reverse the process and *raise* an index:

$$A_j g^{jk} = g^{jk}g_{ij}A^i = \delta_i{}^k A^i = A^k \qquad (2.18.5)$$

Thus any vector has both covariant and contravariant components, the connections being given by the process of raising or lowering an index. It should be stressed that any vector also has *physical* components given by

$$a(1) = \sqrt{g_{11}}\,A^1 \qquad a(2) = \sqrt{g_{22}}\,A^2 \qquad a(3) = \sqrt{g_{33}}\,A^3 \quad (2.18.6)$$

In rectangular Cartesian coordinates, the covariant, contravariant, and physical components of a vector all become one and the same. Thus it was unnecessary to make any distinction between covariance and contravariance in our development of the theory of Cartesian tensors.

Tensors of arbitrary rank and type can be defined. For instance, if

$$A_{jk}^{i'} = P_r^{i'} P_{j'}^s P_{k'}^t A_{st}^r \qquad (2.18.7)$$

then the set of 27 functions A_{st}^r are the components of a third rank tensor with one contravariant and two covariant indices.

It is possible to combine tensors by forming inner products. For instance,

$$A_{jk}^i B_l^k = C_{jl}^i \qquad A_j{}^i B^j = A^i \qquad A^i A_i = \psi \qquad (2.18.8)$$

In all cases, covariant indices are summed with contravariant indices. Quantities such as A_{ii}, $A^i B^i$, etc., have no simple transformation property.

Tensors of the same type and rank can be added together, e.g.,

$$\alpha A_j{}^i + \beta A_{jk}^i B^k + \gamma A^{ik} B_{jk} = 0 \qquad (2.18.9)$$

is a possible tensor equation. A tensor equation which is valid in one coordinate system is automatically valid in *all* coordinate systems. For example Eq. (2.18.9) implies that in a new coordinate system

$$\alpha A_j{}^{i'} + \beta A_{jk}^{i'} B^{k'} + \gamma A^{ik'} B_{jk}' = 0 \qquad (2.18.10)$$

The proof is left as a problem.

2.19. DOT AND CROSS PRODUCT

Consider any two vectors

$$\mathbf{a} = A^i \mathbf{b}_i \qquad \mathbf{b} = B^j \mathbf{b}_j \qquad (2.19.1)$$

The dot product of $\mathbf{a}$ and $\mathbf{b}$ is

$$\mathbf{a} \cdot \mathbf{b} = (\mathbf{b}_i \cdot \mathbf{b}_j) A^i B^j = g_{ij} A^i B^j \qquad (2.19.2)$$

Other forms are

$$\mathbf{a} \cdot \mathbf{b} = A_i B^i = A^i B_i = g^{ij} A_i B_j \qquad (2.19.3)$$

Notice that (2.16.15) and (2.16.16) can be expressed

$$\mathbf{b}_i \times \mathbf{b}_j = \sqrt{g}\, \delta_{ijk} \mathbf{b}^k \qquad \mathbf{b}^i \times \mathbf{b}^j = \frac{1}{\sqrt{g}}\, \delta_{ijk} \mathbf{b}_k \qquad (2.19.4)$$

The third rank pseudo-tensors

$$\epsilon_{ijk} = \sqrt{g}\, \delta_{ijk} \qquad \epsilon^{ijk} = \frac{1}{\sqrt{g}}\, \delta_{ijk} \qquad (2.19.5)$$

are the covariant generalizations of the permutation symbols and are called ϵ-tensors. The cross product of any two vectors is

$$\mathbf{a} \times \mathbf{b} = A^i B^j (\mathbf{b}_i \times \mathbf{b}_j) = \epsilon_{ijk} A^i B^j \mathbf{b}^k \qquad (2.19.6)$$

Or, we may write

$$\mathbf{a} = A_i \mathbf{b}^i \qquad \mathbf{b} = B_j \mathbf{b}^j \qquad (2.19.7)$$

$$\mathbf{a} \times \mathbf{b} = A_i B_j (\mathbf{b}^i \times \mathbf{b}^j) = A_i B_j \epsilon^{ijk} \mathbf{b}_k$$

Thus the covariant and contravariant components of $\mathbf{a} \times \mathbf{b}$ are

$$C_k = \epsilon_{ijk} A^i B^j \qquad C^k = \epsilon^{ijk} A_i B_j \qquad (2.19.8)$$

The ϵ-tensors obey

$$\epsilon^{ijk} \epsilon_{ilm} = \delta_l^{\ j} \delta_m^{\ k} - \delta_m^{\ j} \delta_l^{\ k} \qquad (2.19.9)$$

2.20. DIFFERENTIATION OF VECTORS

To calculate the velocity of a particle, merely divide the expression (2.16.5) for an infinitesimal displacement vector by dt:

$$\mathbf{u} = \frac{d\mathbf{s}}{dt} = \dot{q}^i \mathbf{b}_i \qquad (2.20.1)$$

The three quantities $\dot{q}^1$, $\dot{q}^2$, and $\dot{q}^3$ are the contravariant components of the velocity, but they are usually called *generalized velocity components*. The calculation of acceleration is not so easy. As is made clear by (2.17.12) for the example of spherical coordinates, the basis vectors are *vector functions*. Formally, acceleration is calculated by differentiating (2.20.1):

$$\frac{d\mathbf{u}}{dt} = \ddot{q}^i \mathbf{b}_i + \dot{q}^i \frac{d\mathbf{b}_i}{dt} \qquad (2.20.2)$$

Thus our problem is to learn how to differentiate the basis vectors. We write

$$\frac{d\mathbf{b}_i}{dt} = \frac{\partial \mathbf{b}_i}{\partial q^j} \dot{q}^j \qquad (2.20.3)$$

and compute the partial derivatives. It would be possible in any specific coordinate system to express the basis vectors in terms of the Cartesian vectors $\hat{\mathbf{i}}$, $\hat{\mathbf{j}}$, and $\hat{\mathbf{k}}$, e.g., equation (2.17.12), and then differentiate. The price to be paid for this approach is loss of covariance since results will be obtained that require the new coordinates to be expressed in terms of Cartesian coordinates. We will therefore do the calculation in such a way that all quantities are expressed in terms of generalized coordinates.

We begin with the premise that the partial derivatives of the basis vectors are themselves linear combinations of the basis vectors:

$$\frac{\partial \mathbf{b}_i}{\partial q^j} = \left\{ \begin{matrix} k \\ ij \end{matrix} \right\} \mathbf{b}_k \qquad (2.20.4)$$

The coefficients of the basis vectors in (2.20.4) are called *Christoffel symbols of the second kind*. In some texts they are written Γ_{ij}^k. The reciprocal base vectors also constitute a complete set of three linearly independent vectors so that a similar relation exists for them:

$$\frac{\partial \mathbf{b}^i}{\partial q^j} = c_{ijk}\mathbf{b}^k \tag{2.20.5}$$

We will show that the coefficients in (2.20.5) are also Christoffel symbols. The differentiation of $\mathbf{b}^i \cdot \mathbf{b}_k = \delta_k{}^i$ yields

$$\mathbf{b}^i \cdot \frac{\partial \mathbf{b}_k}{\partial q^j} + \frac{\partial \mathbf{b}^i}{\partial q^j} \cdot \mathbf{b}_k = 0 \tag{2.20.6}$$

By (2.20.3) and (2.20.4)

$$\mathbf{b}^i \cdot \mathbf{b}_n \begin{Bmatrix} n \\ kj \end{Bmatrix} + c_{ijn}\mathbf{b}^n \cdot \mathbf{b}_k = 0$$

$$\delta_n{}^i \begin{Bmatrix} n \\ kj \end{Bmatrix} + \delta_k{}^n c_{ijn} = 0, \qquad c_{ijk} = -\begin{Bmatrix} i \\ kj \end{Bmatrix} \tag{2.20.7}$$

Thus,

$$\frac{\partial \mathbf{b}^i}{\partial q^j} = -\begin{Bmatrix} i \\ kj \end{Bmatrix} \mathbf{b}^k \tag{2.20.8}$$

Another important property of the Christoffel symbols is that they are symmetric with respect to the lower two indices. This is most easily proved by expressing the basis vectors in terms of the Cartesian unit vectors:

$$\mathbf{b}_i = \frac{\partial x^k}{\partial q^i}\mathbf{i}_k \tag{2.20.9}$$

Differentiating

$$\frac{\partial \mathbf{b}_i}{\partial q^j} = \frac{\partial^2 x^k}{\partial q^j \partial q^i}\mathbf{i}_k \tag{2.20.10}$$

The order of taking the partial derivatives can be reversed, meaning that

$$\frac{\partial \mathbf{b}_i}{\partial q^j} = \frac{\partial \mathbf{b}_j}{\partial q^i}, \qquad \begin{Bmatrix} k \\ ij \end{Bmatrix} = \begin{Bmatrix} k \\ ji \end{Bmatrix} \tag{2.20.11}$$

We must now evaluate the Christoffel symbols explicitly. The differentiation of $\mathbf{b}_i \cdot \mathbf{b}_k = g_{ik}$ yields

$$\mathbf{b}_i \cdot \frac{\partial \mathbf{b}_k}{\partial q^j} + \frac{\partial \mathbf{b}_i}{\partial q^j} \cdot \mathbf{b}_k = \frac{\partial g_{ik}}{\partial q^j} \tag{2.20.12}$$

By a relabeling of indices, two more similar equations can be obtained:

$$\mathbf{b}_j \cdot \frac{\partial \mathbf{b}_k}{\partial q^i} + \frac{\partial \mathbf{b}_j}{\partial q^i} \cdot \mathbf{b}_k = \frac{\partial g_{jk}}{\partial q^i} \tag{2.20.13}$$

$$\mathbf{b}_i \cdot \frac{\partial \mathbf{b}_j}{\partial q^k} + \frac{\partial \mathbf{b}_i}{\partial q^k} \cdot \mathbf{b}_j = \frac{\partial g_{ij}}{\partial q^k} \tag{2.20.14}$$

By adding together (2.20.12) and (2.20.13), subtracting (2.20.14) and using the symmetry property (2.20.11), we get

$$2 \frac{\partial \mathbf{b}_i}{\partial q^j} \cdot \mathbf{b}_k = \frac{\partial g_{ik}}{\partial q^j} + \frac{\partial g_{jk}}{\partial q^i} - \frac{\partial g_{ij}}{\partial q^k} \qquad (2.20.15)$$

From (2.20.4),

$$\begin{Bmatrix} n \\ ij \end{Bmatrix} \mathbf{b}_n \cdot \mathbf{b}_k = \frac{1}{2} \left[\frac{\partial g_{ik}}{\partial q^j} + \frac{\partial g_{jk}}{\partial q^i} - \frac{\partial g_{ij}}{\partial q^k} \right] \qquad (2.20.16)$$

Then, since $\mathbf{b}_n \cdot \mathbf{b}_k = g_{nk}$ and $g_{nk}g^{lk} = \delta_n{}^l$

$$\begin{Bmatrix} l \\ ij \end{Bmatrix} = g^{lk} \frac{1}{2} \left[\frac{\partial g_{ik}}{\partial q^j} + \frac{\partial g_{jk}}{\partial q^i} - \frac{\partial g_{ij}}{\partial q^k} \right] = g^{lk}[ij, k] \qquad (2.20.17)$$

The symbol $[ij, k]$ defined above is called the Christoffel symbol of the first kind. Thus the Christoffel symbols are evaluated explicitly in terms of the metric tensor of the particular curvilinear coordinate system we happen to be using.

The acceleration of a particle can now be expressed

$$\mathbf{a} = \frac{d\mathbf{u}}{dt} = A^k \mathbf{b}_k \qquad (2.20.18)$$

where

$$A^k = \ddot{q}^k + \begin{Bmatrix} k \\ ij \end{Bmatrix} \dot{q}^i \dot{q}^j \qquad (2.20.19)$$

are the *contravariant* components of the acceleration. The second derivatives of the coordinates, $\ddot{q}^k$, are *not* vector components with respect to transformations between arbitrary curvilinear coordinates. The Christoffel symbols themselves are not tensors. The combination of quantities (2.20.19) *are* vector components and obey the contravariant vector transformation law. The covariant components of acceleration are obtained by lowering the index:

$$A_k = g_{ki}\ddot{q}^i + [ij, k]\dot{q}^i \dot{q}^j \qquad (2.20.20)$$

Note that

$$\frac{d}{dt}(g_{ki}\dot{q}^i) = \frac{\partial g_{ki}}{\partial q^j} \dot{q}^j \dot{q}^i + g_{ki}\ddot{q}^i \qquad (2.20.21)$$

Moreover, by a simple relabeling of dummy indices

$$\frac{\partial g_{ik}}{\partial q^j} \dot{q}^i \dot{q}^j = \frac{\partial g_{jk}}{\partial q^i} \dot{q}^i \dot{q}^j \qquad (2.20.22)$$

Thus (2.20.20) can be reduced to

$$A_k = \frac{d}{dt}(q_{ki}\dot{q}^i) - \frac{1}{2}\frac{\partial g_{ij}}{\partial q^k}\dot{q}^i\dot{q}^j \tag{2.20.23}$$

This form is especially convenient since the Christoffel symbols themselves are tedious to evaluate.

2.21. COVARIANT FORM OF NEWTON'S SECOND LAW

By using the covariant acceleration components (2.20.23), it is possible to state Newton's second law for a single particle as

$$Q_k = mA_k \tag{2.21.1}$$

The components Q_k are the covariant components of force, but they are usually called *generalized force components*. Since (2.21.1) is a vector equation, it is valid in *all* coordinate systems. One of our main goals, namely, that of expressing Newton's second law in general covariant form, is now achieved.

By means of the line element as given by (2.15.7), the kinetic energy of the particle can be expressed as a quadratic form in the generalized velocities:

$$T = \tfrac{1}{2}m\left(\frac{ds}{dt}\right)^2 = \tfrac{1}{2}mg_{ij}\dot{q}^i\dot{q}^j \tag{2.21.2}$$

The kinetic energy is a function of the generalized coordinates through the dependence of the metric tensor on the coordinates. Hence,

$$\frac{\partial T}{\partial q^k} = \tfrac{1}{2}m\frac{\partial g_{ij}}{\partial q^k}\dot{q}^i\dot{q}^j \tag{2.21.3}$$

Since the metric tensor itself does not depend on the velocities,

$$\frac{\partial T}{\partial \dot{q}^k} = mg_{ki}\dot{q}^i \tag{2.21.4}$$

If you have difficulty seeing (2.21.4), write out the double sum in (2.21.2) and differentiate with respect to one of the generalized velocities, say, $\dot{q}^1$. Remember that $g_{ij} = g_{ji}$! It is now possible to express Newton's second law in the especially convenient form

$$\frac{d}{dt}\left(\frac{\partial T}{\partial \dot{q}^k}\right) - \frac{\partial T}{\partial q^k} = Q_k \tag{2.21.5}$$

As an example of how (2.21.5) can be used, we will work out the physical components of acceleration in spherical coordinates. The kinetic energy of a particle is

$$T = \tfrac{1}{2}m(\dot{r}^2 + r^2\dot{\theta}^2 + r^2 \sin^2 \theta \dot{\phi}^2) \qquad (2.21.6)$$

With $q_1 = r$,

$$\frac{\partial T}{\partial r} = mr\dot{\theta}^2 + mr \sin^2 \theta \dot{\phi}^2, \qquad \frac{\partial T}{\partial \dot{r}} = m\dot{r},$$

$$\frac{d}{dt}\left(\frac{\partial T}{\partial \dot{r}}\right) - \frac{\partial T}{\partial r} = Q_1 = mA_1$$

Hence,

$$A_1 = \ddot{r} - r\dot{\theta}^2 - r \sin^2 \theta \dot{\phi}^2 \qquad (2.21.7)$$

Similarly,

$$A_2 = r^2\ddot{\theta} + 2r\dot{r}\dot{\theta} - r^2 \sin \theta \cos \theta \dot{\phi}^2$$

$$A_3 = \frac{d}{dt}(r^2 \sin^2 \theta \dot{\phi}) \qquad (2.21.8)$$

The contravariant components of the acceleration are

$$A^1 = g^{11}A_1 = \frac{1}{g_{11}} A_1 \qquad A^2 = \frac{1}{g_{22}} A_2 \qquad A^3 = \frac{1}{g_{33}} A_3 \qquad (2.21.9)$$

Finally, the physical components of acceleration are

$$a_r = \sqrt{g_{11}}A^1 = \frac{1}{\sqrt{g_{11}}} A_1 \qquad a_\theta = \frac{1}{\sqrt{g_{22}}} A_2 \qquad a_\phi = \frac{1}{\sqrt{g_{33}}} A_3 \qquad (2.21.10)$$

Since $g_{11} = 1$, $g_{22} = r^2$, and $g_{33} = r^2 \sin^2 \theta$,

$$a_r = \ddot{r} - r\dot{\theta}^2 - r \sin^2 \theta \dot{\phi}^2$$

$$a_\theta = r\ddot{\theta} + 2\dot{r}\dot{\theta} - r \sin \theta \cos \theta \dot{\phi}^2$$

$$a_\phi = \frac{1}{r \sin \theta} \frac{d}{dt}(r^2 \sin^2 \theta \dot{\phi}) \qquad (2.21.11)$$

The physical components of acceleration have the actual dimensions of cm/sec² in the cgs system of units. The physical components of force are given by

$$F_r = ma_r \qquad F_\theta = ma_\theta \qquad F_\phi = ma_\phi \qquad (2.21.12)$$

and are in dynes in the cgs system of units.

2.22. DIVERGENCE IN GENERALIZED COORDINATES

In addition to a covariant representation of particle mechanics, there are several other valuable results that can be obtained from our development of tensor analysis. Let **a** be any vector field:

$$\mathbf{a} = A^i \mathbf{b}_i \qquad (2.22.1)$$

where A^i are functions of the coordinates. For example, **a** might be a gravitational field. Of frequent occurrence in practice is the expression for the divergence of a vector which we now calculate in an arbitrary curvilinear coordinate system, using the symbolic operator ∇, Eq. (2.17.4):

$$\nabla \cdot \mathbf{a} = (\mathbf{b}^k \partial_k) \cdot (A^i \mathbf{b}_i) = \mathbf{b}^k \cdot (\mathbf{b}_i \, \partial_k A^i + A^i \partial_k \mathbf{b}_i)$$

$$= \mathbf{b}^k \cdot \left(\mathbf{b}_i \, \partial_k A^i + A^i \begin{Bmatrix} j \\ ik \end{Bmatrix} \mathbf{b}_j \right) = \delta_i^{\ k} \, \partial_k A^i + A^i \begin{Bmatrix} j \\ ik \end{Bmatrix} \delta_j^{\ k} \qquad (2.22.2)$$

$$= \partial_i A^i + A^i \begin{Bmatrix} k \\ ik \end{Bmatrix}$$

The first term looks just like the familiar expression for divergence in a Cartesian coordinate system. There is added a new term involving the Christoffel symbol which results when the basis vectors are differentiated. In Cartesian coordinates, the basis vectors are, of course, constant vectors so that their derivatives are zero.

There is a contraction on two indices in the Christoffel symbol; the result can be expressed in a simple way as we now show:

$$\begin{Bmatrix} k \\ ik \end{Bmatrix} = g^{kj} \frac{1}{2} \left[\frac{\partial g_{ij}}{\partial q^k} + \frac{\partial g_{kj}}{\partial q^i} - \frac{\partial g_{ik}}{\partial q^j} \right] \qquad (2.22.3)$$

By the same reasoning that led to (2.20.22), the first and third terms cancel:

$$\begin{Bmatrix} k \\ ik \end{Bmatrix} = \tfrac{1}{2} g^{kj} \frac{\partial g_{kj}}{\partial q^i} \qquad (2.22.4)$$

Further simplification is possible. The determinant of the metric tensor is

$$g = g_{1j} g_{2k} g_{3l} \, \delta_{jkl} \qquad (2.22.5)$$

Differentiation yields

$$\frac{\partial g}{\partial q^i} = \frac{\partial g_{1j}}{\partial q^i} g_{2k} g_{3l} \, \delta_{jkl} + g_{1j} \frac{\partial g_{2k}}{\partial q^i} g_{3l} \, \delta_{jkl} + g_{1j} g_{2k} \frac{\partial g_{3l}}{\partial q^i} \, \delta_{jkl} \qquad (2.22.6)$$

The metric tensor obeys $g_{1j} g^{1j} = 1$. Thus

$$g_{1j} g g^{1j} = g \qquad (2.22.7)$$

The comparison of (2.22.7) and (2.22.5) shows that

$$g_{2k} g_{3l} \delta_{jkl} = g g^{1j} \qquad (2.22.8)$$

The other terms in (2.22.6) can be similarly treated giving

$$\frac{\partial g}{\partial q^i} = \frac{\partial g_{1j}}{\partial q^i} g g^{1j} + \frac{\partial g_{2k}}{\partial q^i} g g^{2k} + \frac{\partial g_{3l}}{\partial q^i} g g^{3l}, \qquad \frac{1}{g} \frac{\partial g}{\partial q^i} = g^{kj} \frac{\partial g_{kj}}{\partial q^i} \qquad (2.22.9)$$

The expression for the divergence of a vector can now be written

$$\mathbf{\nabla} \cdot \mathbf{a} = \frac{\partial A^i}{\partial q^i} + \frac{A^i}{2g} \frac{\partial g}{\partial q^i} = \frac{1}{\sqrt{g}} \frac{\partial}{\partial q^i} (\sqrt{g} A^i) \qquad (2.22.10)$$

and is valid in any curvilinear coordinate system.

For spherical coordinates $g = g_{11} g_{22} g_{33} = r^4 \sin^2 \theta$ giving

$$\mathbf{\nabla} \cdot \mathbf{a} = \frac{1}{r^2} \frac{\partial}{\partial r} (r^2 A^1) + \frac{1}{\sin \theta} \frac{\partial}{\partial \theta} (\sin \theta A^2) + \frac{\partial A^3}{\partial \phi} \qquad (2.22.11)$$

In terms of the physical components of $\mathbf{a}$,

$$\mathbf{\nabla} \cdot \mathbf{a} = \frac{1}{r^2} \frac{\partial}{\partial r} (r^2 a_r) + \frac{1}{r \sin \theta} \frac{\partial}{\partial \theta} (\sin \theta a_\theta) + \frac{1}{r \sin \theta} \frac{\partial a_\phi}{\partial \phi} \quad (2.22.12)$$

2.23. THE LAPLACIAN

The Laplacian of a scalar function is

$$\mathbf{\nabla} \cdot \mathbf{\nabla} \psi = \nabla^2 \psi \qquad (2.23.1)$$

where

$$\mathbf{\nabla} \psi = \mathbf{b}^k (\partial_k \psi) = \mathbf{b}_i (g^{ik} \partial_k \psi) \qquad (2.23.2)$$

Thus

$$A^i = g^{ik} \partial_k \psi \qquad (2.23.3)$$

are the contravariant components of the gradient. By means of (2.22.10),

$$\nabla^2 \psi = \frac{1}{\sqrt{g}} \frac{\partial}{\partial q^i} (\sqrt{g} \, g^{ik} \partial_k \psi) \qquad (2.23.4)$$

The double sum makes (2.23.4) quite a cumbersome expression except in orthogonal coordinate systems where

$$g = g_{11} g_{22} g_{33} \qquad g^{11} = \frac{1}{g_{11}} \qquad g^{22} = \frac{1}{g_{22}} \qquad g^{33} = \frac{1}{g_{33}} \quad (2.23.5)$$

The Laplacian is then

$$\nabla^2 \psi = \frac{1}{\sqrt{g}} \Bigg[\frac{\partial}{\partial q^1} \bigg(\sqrt{\frac{g_{22} g_{33}}{g_{11}}} \frac{\partial \psi}{\partial q^1} \bigg) + \frac{\partial}{\partial q^2} \bigg(\sqrt{\frac{g_{11} g_{33}}{g_{22}}} \frac{\partial \psi}{\partial q^2} \bigg)$$
$$+ \frac{\partial}{\partial q^3} \bigg(\sqrt{\frac{g_{11} g_{22}}{g_{33}}} \frac{\partial \psi}{\partial q^3} \bigg) \Bigg] \qquad (2.23.6)$$

In spherical coordinates

$$\nabla^2 \psi = \frac{1}{r^2} \frac{\partial}{\partial r} \bigg(r^2 \frac{\partial \psi}{\partial r} \bigg) + \frac{1}{r^2 \sin \theta} \frac{\partial}{\partial \theta} \bigg(\sin \theta \frac{\partial \psi}{\partial \theta} \bigg) + \frac{1}{r^2 \sin^2 \theta} \frac{\partial^2 \psi}{\partial \phi^2} \quad (2.23.7)$$

2.24. CURL IN GENERALIZED COORDINATES

To obtain the curl of a vector we may write

$$\mathbf{c} = \nabla \times \mathbf{a} = (\mathbf{b}^j \partial_j) \times (A_k \mathbf{b}^k)$$

$$= \mathbf{b}^j \times (\mathbf{b}^k \partial_j A_k + A_k \partial_j \mathbf{b}^k) = \mathbf{b}^j \times \left(\mathbf{b}^k \partial_j A_k - A_k \begin{Bmatrix} k \\ ji \end{Bmatrix} \mathbf{b}^i \right) \quad (2.24.1)$$

$$= \epsilon^{jki} \mathbf{b}_i \partial_j A_k - A_k \begin{Bmatrix} k \\ ji \end{Bmatrix} \epsilon^{jil} \mathbf{b}_l$$

where (2.19.4) and (2.19.5) are used. The Christoffel symbol is symmetric with respect to its lower two indices and the ϵ-tensor is antisymmetric. Hence,

$$\begin{Bmatrix} k \\ ji \end{Bmatrix} \epsilon^{jil} = 0 \quad (2.24.2)$$

Thus the contravariant components of the curl are

$$C^i = \epsilon^{jki} \partial_j A_k \quad (2.24.3)$$

Written out, these are

$$C^1 = \frac{1}{\sqrt{g}} (\partial_2 A_3 - \partial_3 A_2) \quad C^2 = \frac{1}{\sqrt{g}} (\partial_3 A_1 - \partial_1 A_3)$$

$$C^3 = \frac{1}{\sqrt{g}} (\partial_1 A_2 - \partial_2 A_1) \quad (2.24.4)$$

If all vectors are expressed in terms of their physical components, the components of curl in spherical coordinates are

$$c_r = \frac{1}{r \sin \theta} \left(\frac{\partial}{\partial \theta} [\sin \theta a_\phi] - \frac{\partial a_\theta}{\partial \phi} \right) \quad (2.24.5)$$

$$c_\theta = \frac{1}{r \sin \theta} \left(\frac{\partial a_r}{\partial \phi} - \frac{\partial}{\partial r} [r \sin \theta a_\phi] \right) \quad (2.24.6)$$

$$c_\phi = \frac{1}{r} \left(\frac{\partial}{\partial r} [r a_\theta] - \frac{\partial a_r}{\partial \theta} \right) \quad (2.24.7)$$

2.25. VOLUME AND SURFACE INTEGRALS

The infinitesimal displacement vector (2.16.5) is made up of the three component vectors

$$dq^1 \mathbf{b}_1, \qquad dq^2 \mathbf{b}_2, \qquad dq^3 \mathbf{b}_3$$

which can be thought of as forming three edges of a differential parallelepiped with volume

$$d\Sigma = dq^1 \, dq^2 \, dq^3 \mathbf{b}_1 \cdot \mathbf{b}_2 \times \mathbf{b}_3 = \sqrt{g} \, dq^1 \, dq^2 \, dq^3 \qquad (2.25.1)$$

This is the appropriate form of the volume element if it is desired to calculate a volume integral in curvilinear coordinates.

If we wish to integrate over some surface, it is necessary to have an explicit formula for an element of surface area. A surface in three-dimensional space can be represented by the parametric equations

$$q_1 = q_1(u_1, u_2) \qquad q_2 = q_2(u_1, u_2) \qquad q_3 = q_3(u_1, u_2) \qquad (2.25.2)$$

For instance, if the surface is a sphere, (2.13.1) corresponds to (2.25.2) provided that $x_i = q_i$, $u_1 = \theta$, and $u_2 = \phi$. The radius of the sphere is to be regarded as a constant in this case. A line element can be represented by

$$ds^2 = g_{ij} \, dq^i \, dq^j \qquad (2.25.3)$$

but the requirement that it lie in the surface means that all three of the coordinate differentials are *not* independent. To see this, form the total differentials of (2.25.2)

$$dq^1 = \frac{\partial q^1}{\partial u^1} du^1 + \frac{\partial q^1}{\partial u^2} du^2, \qquad dq^2 = \frac{\partial q^2}{\partial u^1} du^1 + \frac{\partial q^2}{\partial u^2} du^2,$$
$$dq^3 = \frac{\partial q^3}{\partial u^1} du^1 + \frac{\partial q^3}{\partial u^2} du^2 \qquad (2.25.4)$$

If dq^1 and dq^2 are specified, then the first two equations determine du^1 and du^2 which then fix dq^3. Let us abbreviate (2.25.4) by

$$dq^i = \frac{\partial q^i}{\partial u^\alpha} du^\alpha \qquad (2.25.5)$$

where the Greek index is summed from 1 to 2. The line element is then

$$ds^2 = g_{ij} \frac{\partial q^i}{\partial u^\alpha} \frac{\partial q^j}{\partial u^\beta} du^\alpha \, du^\beta = h_{\alpha\beta} \, du^\alpha \, du^\beta \qquad (2.25.6)$$

where

$$h_{\alpha\beta} = g_{ij} \frac{\partial q^i}{\partial u^\alpha} \frac{\partial q^j}{\partial u^\beta} \qquad (2.25.7)$$

can be thought of as a metric tensor which determines the intrinsic (two-dimensional) geometry of the surface. The parameters u_1 and u_2 can be regarded as curvilinear coordinates on the surface.

Let $\mathbf{a}_1$ and $\mathbf{a}_2$ be two basis vectors tangent to the surface. An infinitesimal displacement vector lying in the surface can be represented

$$d\mathbf{s} = du^1 \mathbf{a}_1 + du^2 \mathbf{a}_2 \qquad (2.25.8)$$

An element of surface area is

$$d\boldsymbol{\sigma} = du^1\, du^2(\mathbf{a}_1 \times \mathbf{a}_2) \tag{2.25.9}$$

If θ is the angle between coordinate lines on the surface, the magnitude of $d\boldsymbol{\sigma}$ is

$$\begin{aligned}
d\sigma &= du^1\, du^2\sqrt{h_{11}h_{22}}\, \sin\theta = du^1\, du^2\sqrt{h_{11}h_{22}}\, \sqrt{1 - \cos^2\theta}\\
&= du^1\, du^2\sqrt{h_{11}h_{22}}\, \sqrt{1 - h_{12}^2/(h_{11}h_{22})} \tag{2.25.10}\\
&= du^1\, du^2\sqrt{h_{11}h_{22} - h_{12}^2} = du^1\, du^2\sqrt{h}
\end{aligned}$$

where h is the determinant of $h_{\alpha\beta}$. Thus the two-dimensional surface element (2.25.10) takes the same form as the three-dimensional volume element.

Since the integrands of Gauss' divergence theorem (2.8.4) are scalar invariants, these identities automatically retain their validity if the integrands are expressed in arbitrary curvilinear coordinates. The integral vector identities of Section 2.12 which have vectors for integrands *do not* retain their validity in curvilinear coordinates. For one thing, there is trouble when the integrals are performed because the basis vectors themselves are vector functions. For another thing, the idea of a *constant vector* as used in the proof of (2.12.10) has no covariant meaning because a vector that has constant components in Cartesian coordinates does not, in general, have constant components in curvilinear coordinates.

PROBLEMS

2.1.1   If the components of $\mathbf{a}$ and $\mathbf{b}$ are regarded as functions of the single parameter t, prove that

$$\mathbf{a} \cdot \frac{d\mathbf{a}}{dt} = a\,\frac{da}{dt}$$

where $a = |\mathbf{a}|$ and that

$$\frac{d}{dt}(\mathbf{a} \times \mathbf{b}) = \mathbf{a} \times \frac{d\mathbf{b}}{dt} + \frac{d\mathbf{a}}{dt} \times \mathbf{b}$$

2.2.2   A wheel of radius 2 ft is mounted on an axle and has an angular acceleration of 5 radians per $\sec^2$. Calculate the normal (or radial) and tangential components of the acceleration of a particle on the rim of the wheel after 10 sec if the wheel starts from rest. (Reference: Resnick and Halliday, *Physics*, Part I, Chapter 11)

2.4.3   Prove that

$$\nabla(r^{-n}) = \frac{-n\mathbf{r}}{r^{n+2}}$$

2.4.4   Let P and Q be two points and let r represent the distance between them. Show that $\nabla_P f(r) = -\nabla_Q f(r)$ where $f(r)$ is any scalar function of r and ∇_P and ∇_Q refer to the gradient with respect to the coordinates of P and Q.

2.5.5   Find the unit vector normal to the surface $x^2 - xy + yz = 3$ at the point $(1, 2, 2)$.

2.5.6   If $\phi = x^2 + y^2 - z^2$, describe qualitatively the surfaces $\phi = 2$ and $\phi = 3$. What is the approximate change in the value of the function ϕ between the points $(1, 1, 1)$ and $(1.001, 0.999, 1.002)$?

2.5.7   What conditions must be placed on the vectors $\mathbf{a}$, $\mathbf{b}$, and $\mathbf{c}$ in order that the planes

$$\mathbf{a} \cdot \mathbf{r} = k_1 \qquad \mathbf{b} \cdot \mathbf{r} = k_2 \qquad \mathbf{c} \cdot \mathbf{r} = k_3$$

will have one and only one point in common? Under what conditions will the planes intersect along a common line?

2.6.8   A vector field is given by

$$\mathbf{a} = (x + 5y + 4z + 2)\hat{\mathbf{i}} + (5x + y + 3z + 5)\hat{\mathbf{j}} + (4x + 3y + z + 6)\hat{\mathbf{k}}$$

Find $\nabla \cdot \mathbf{a}$ and $\nabla \times \mathbf{a}$.

2.8.9   Given $\mathbf{a} = x^2\hat{\mathbf{i}} + xy\hat{\mathbf{j}} + xz\hat{\mathbf{k}}$, calculate $\int \mathbf{a} \cdot \hat{\mathbf{n}}\, d\sigma$ over the surface of a cube formed by the coordinate planes and $x = c$, $y = c$, $z = c$. Check the divergence theorem by calculating $\int \nabla \cdot \mathbf{a}\, d\Sigma$ over the volume of the cube.

2.9.10   Prove:

$$\nabla \times \nabla\psi = 0 \qquad \nabla \cdot (\nabla \times \mathbf{a}) = 0$$

2.10.11   Show that

$$(\mathbf{a} \times \nabla) \cdot \mathbf{r} = 0 \qquad (\mathbf{a} \times \nabla) \times \mathbf{r} = -2\mathbf{a}$$

2.10.12   Show that $\nabla\phi_1 \times \nabla\phi_2 = \nabla \times (\phi_1\nabla\phi_2)$ where ϕ_1 and ϕ_2 are scalar functions.

2.10.13   If $\phi = z(x^2 + y^2 + z^2)^{-3/2}$, find $\nabla\phi$, $\nabla^2\phi$, and $\nabla \times \nabla\phi$.

2.11.14   Given $\mathbf{a} = (x^2 + y^2)\hat{\mathbf{i}} + xy\hat{\mathbf{j}}$ calculate the line integral of $\mathbf{a}$ around the square bounded by the coordinate axes and $x = c$, $y = c$ in the xy-plane. Check Stokes' theorem by calculating the surface integral of $\nabla \times \mathbf{a}$ over the square.

2.11.15   In the figure, a current J flows in a continuous loop. Calculate the circulation of the magnetic field $\mathbf{B}$ around the paths (a) and (b).

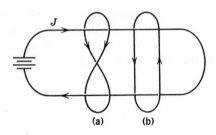

(a)   (b)

2.12.16 Evaluate $\oint_c \psi \, ds$ for $\psi = xy$ using the square path lying in the xy-plane formed by the coordinate axes and $x = c$, $y = c$. Check the identity

$$\oint_C \psi \, d\mathbf{s} = \int_\sigma \hat{\mathbf{n}} \times \nabla \psi \, d\sigma$$

2.12.17 If $\mathbf{c}$ is a constant vector, show that

$$\int_\sigma \hat{\mathbf{n}} \times (\mathbf{c} \times \mathbf{r}) \, d\sigma = 2\mathbf{c}\Sigma$$

where Σ is the volume enclosed by the surface σ.

2.12.18 Prove Green's theorem by first proving the identity

$$\phi\nabla^2\psi - \psi\nabla^2\phi = \nabla \cdot (\phi\nabla\psi - \psi\nabla\phi)$$

2.13.19 Two displacement vectors $\mathbf{r}_1$ and $\mathbf{r}_2$ project from the origin. The angle between $\mathbf{r}_1$ and $\mathbf{r}_2$ is α. Calculate $\cos \alpha$ in terms of the polar coordinate angles θ_1, ϕ_1, and θ_2, ϕ_2.

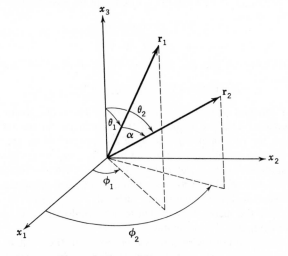

2.14.20 In what regions of space does

$$q_1 = x_1 \qquad q_2 = x_2 \qquad q_3 = \sqrt{x_1^2 + x_2^2 + x_3^2}$$

give an admissible coordinate transformation? Is the solution for x_i in terms of q_i unique? Where does the transformation break down?

2.14.21 If q_i and q_i' are admissible generalized coordinates meaning that

$$\det \frac{\partial q^i}{\partial x^j} \neq 0 \qquad \det \frac{\partial q^{i'}}{\partial x^j} \neq 0$$

prove that

$$\det \frac{\partial q^{i'}}{\partial q^j} \neq 0$$

2.15.22 Go through the steps of substituting (2.14.14) into (2.15.7). Make sure that you understand why it is necessary to assume that ds is an arbitrary vector in order to get (2.15.8). Invert the transformation (2.15.8), i.e., find the g_{kl} in terms of the g'_{ij}.

2.16.23 Prove directly from (2.16.23) that the reciprocal base vectors transform contravariantly. Do this by using the known transformation properties of the base vectors and the metric tensor.

2.19.24 Prove that in any *orthogonal* coordinate system the dot and cross products of any two vectors can be expressed

$$\mathbf{a} \cdot \mathbf{b} = a(1)b(1) + a(2)b(2) + a(3)b(3)$$

$$\mathbf{a} \times \mathbf{b} = [a(2)b(3) - a(3)b(2)]\hat{e}_1 + [a(3)b(1) - a(1)b(3)]\hat{e}_2$$
$$+ [a(1)b(2) - a(2)b(1)]\hat{e}_3$$

2.19.25 Prove that

$$\epsilon_{ijk} = g_{ir}g_{js}g_{kt}\epsilon^{rst}$$

2.20.26 Show that if curvilinear coordinates are expressed as functions of Cartesian coordinates, the Christoffel symbols can be expressed

$$\begin{Bmatrix} l \\ ji \end{Bmatrix} = \frac{\partial^2 x^k}{\partial q^i \, \partial q^j} \frac{\partial q^l}{\partial x^k}$$

2.21.27 Rectangular coordinates are given in terms of new coordinates by

$$x_1 = q_1 + q_2 \cos \theta \qquad x_2 = q_2 \sin \theta \qquad x_3 = q_3$$

where θ is a constant angle $0 < \theta < \pi/2$. Find the metric tensor of the new coordinate system. Draw a diagram of the coordinate system and show geometrically how the coordinates q_1 and q_2 of an arbitrary point are represented. Compute the covariant, contravariant, and physical components of acceleration. Find the magnitude of the acceleration. If the acceleration vector lies in the $q_1 q_2$-coordinate surface, sketch the vector and its physical components.

2.24.28 Cylindrical coordinates are defined by $x = r \cos \theta$, $y = r \sin \theta$, $z = z$. Calculate the metric tensor, divergence, Laplacian, curl, velocity, and acceleration in cylindrical coordinates.

2.24.29 A long straight wire carries a current J in the positive z-direction. The magnetic field lines are in the form of circles surrounding the wire. In cylindrical coordinates,

$$\mathbf{B} = \frac{2J}{rc}\theta$$

Compute the divergence and curl of this vector field at all points of space except $r = 0$. Find a vector field $\mathbf{A}$ which satisfies $\mathbf{B} = \nabla \times \mathbf{A}$.

2.24.30 Parabolic coordinates ξ, η, and ϕ are given in terms of spherical coordinates by $\xi = r(1 - \cos \theta)$, $\eta = (1 + \cos \theta)$, and $\phi = \phi$. Describe the coordinate surfaces qualitatively. Calculate the metric tensor and the Laplacian. (The solution of the hydrogen atom in parabolic coordinates finds application in quantum mechanics.)

2.25.31 Show that

$$\mathbf{a}_\alpha = \frac{\partial q^i}{\partial u^\alpha}\mathbf{b}_i \qquad \mathbf{a}_1 \times \mathbf{a}_2 = \frac{\partial q^i}{\partial u^1}\frac{\partial q^j}{\partial u^2}\sqrt{g}\,\delta_{ijk}\mathbf{b}_k$$

2.25.32 A surface is defined by

$$\frac{x}{a} + \frac{y}{b} + \frac{z}{c} = 1$$

where a, b, and c are positive constants. A vector field is given by

$$\mathbf{v} = \hat{\imath}x + \hat{\jmath}y$$

Find the surface integral $\int \mathbf{v}\,d\sigma$ over the region of the surface where $x > 0$, $y > 0$, and $z > 0$. Let the curvilinear coordinates of the surface be defined by $x = u_1$, $y = u_2$.

REFERENCES

1. Sokolnikoff, I. S., *Tensor Analysis*, 2nd ed., John Wiley and Sons, New York, 1964.
2. Wrede, Robert C., *Introduction to Vector and Tensor Analysis*, John Wiley and Sons, New York, 1963.

3

Matrices

On many occasions a transformation or a second rank tensor has been represented by a 3 by 3 square array called a matrix. In this chapter this idea will be formalized and a matrix algebra developed. The theory of orthogonal transformations will be extended to include the eigenvalue problem.

3.1. MATRIX MULTIPLICATION

All equations involving vectors, second rank tensors and transformations so far developed in this book can be represented by matrix equations provided that matrix multiplication is suitably defined. Let A and B be the two matrices

$$A = \begin{pmatrix} A_{11} & A_{12} & A_{13} \\ A_{21} & A_{22} & A_{23} \\ A_{31} & A_{32} & A_{33} \end{pmatrix} \qquad B = \begin{pmatrix} B_{11} & B_{12} & B_{12} \\ B_{21} & B_{22} & B_{23} \\ B_{31} & B_{32} & B_{33} \end{pmatrix} \qquad (3.1.1)$$

The product AB of the two matrices is defined to be a matrix C the components of which are

$$\begin{pmatrix} A_{11}B_{11} + A_{12}B_{21} + A_{13}B_{31} & A_{11}B_{12} + A_{12}B_{22} + A_{13}B_{32} & A_{11}B_{13} + A_{12}B_{23} + A_{13}B_{33} \\ A_{21}B_{11} + A_{22}B_{21} + A_{23}B_{31} & A_{21}B_{12} + A_{22}B_{22} + A_{23}B_{32} & A_{21}B_{13} + A_{22}B_{23} + A_{23}B_{33} \\ A_{31}B_{11} + A_{32}B_{21} + A_{33}B_{31} & A_{31}B_{12} + A_{32}B_{22} + A_{33}B_{32} & A_{31}B_{13} + A_{32}B_{23} + A_{33}B_{33} \end{pmatrix}$$

$$(3.1.2)$$

In component form this is

$$C_{ij} = A_{ik}B_{kj} \tag{3.1.3}$$

The component C_{11} is found by multiplying the first row of A by the first column B; C_{12} is the product of the first row of A and the *second* column of B, and so on.

Matrix multiplication is not commutative. The components of $BA = D$ are $D_{ik} = B_{ij}A_{jk}$ which in general are not equal to C_{ik}. This is perhaps best illustrated by a numerical example. Suppose the matrices A and B are

$$A = \begin{pmatrix} 1 & 2 & 0 \\ 3 & 1 & 0 \\ 0 & 0 & 1 \end{pmatrix} \qquad B = \begin{pmatrix} 0 & 1 & 1 \\ 0 & 2 & 3 \\ 1 & 2 & 0 \end{pmatrix}$$

The products AB and BA are

$$AB = \begin{pmatrix} 1 & 2 & 0 \\ 3 & 1 & 0 \\ 0 & 0 & 1 \end{pmatrix}\begin{pmatrix} 0 & 1 & 1 \\ 0 & 2 & 3 \\ 1 & 2 & 0 \end{pmatrix} = \begin{pmatrix} 0 & 5 & 7 \\ 0 & 5 & 6 \\ 1 & 2 & 0 \end{pmatrix}$$

$$BA = \begin{pmatrix} 0 & 1 & 1 \\ 0 & 2 & 3 \\ 1 & 2 & 0 \end{pmatrix}\begin{pmatrix} 1 & 2 & 0 \\ 3 & 1 & 0 \\ 0 & 0 & 1 \end{pmatrix} = \begin{pmatrix} 3 & 1 & 1 \\ 6 & 2 & 3 \\ 7 & 4 & 0 \end{pmatrix}$$

The rules for the addition of matrices and the multiplication of a matrix by a scalar are identical to the rules that apply to a second rank tensor developed in Chapter 1. Remember that in αA the scalar α multiplies every component of A! In addition we note the following rules that apply to the multiplication of matrices:

$$A(B + C) = AB + AC \tag{3.1.4}$$

$$(B + C)A = BA + CA \tag{3.1.5}$$

Equation (3.1.4) is proved by writing out the component equations:

$$[A(B + C)]_{ik} = A_{ij}(B + C)_{jk}$$
$$= A_{ij}(B_{jk} + C_{jk}) = A_{ij}B_{jk} + A_{ij}C_{jk}$$
$$= (AB)_{ik} + (AC)_{ik}$$

The matrix with all components equal to zero is called the *zero* or null matrix. Two matrices are equal if, and only if, each component of one matrix equals the corresponding component of the other.

Just as with second rank tensors, if the components of a matrix are functions of a parameter t, then the differentiation of a matrix proceeds by differentiating every component:

$$\frac{dA}{dt} = \begin{pmatrix} \dot{A}_{11} & \dot{A}_{12} & \dot{A}_{13} \\ \dot{A}_{21} & \dot{A}_{22} & \dot{A}_{23} \\ \dot{A}_{31} & \dot{A}_{32} & \dot{A}_{33} \end{pmatrix} \qquad (3.1.6)$$

Since

$$\frac{d}{dt}(A_{ik}B_{kj}) = A_{ik}\frac{dB_{kj}}{dt} + \frac{dA_{ik}}{dt}B_{kj} \qquad (3.1.7)$$

It follows that

$$\frac{d}{dt}(AB) = A\frac{dB}{dt} + \frac{dA}{dt}B \qquad (3.1.8)$$

If α is a scalar function of t,

$$\frac{d}{dt}(\alpha A) = \alpha\frac{dA}{dt} + \frac{d\alpha}{dt}A \qquad (3.1.9)$$

It is possible to define the integral of a matrix the components of which are function of a parameter t. Suppose the interval $t_a - t_b$ is broken up into sub-intervals $\Delta t_1, \Delta t_2, \ldots, \Delta t_n, \ldots, \Delta t_N$. Let t_n be a value of t chosen somewhere in the interval Δt_n. By $A(t_n)$ is understood the matrix obtained by evaluating the components of A for the special value t_n. The sum

$$\sum_{n=1}^{N} A(t_n)\,\Delta t_n = A(t_1)\,\Delta t_1 + \cdots + A(t_n)\,\Delta t_n + \cdots A(t_N)\,\Delta t_N \quad (3.1.10)$$

is defined since $A(t_n)\Delta t_n$ is the matrix $A(t_n)$ multiplied by the scalar Δt_n. Equation (3.1.10) merely involves adding together several matrices which we also know how to do. A typical component of (3.1.10) is

$$\sum_{n=1}^{N} A_{ij}(t_n)\,\Delta t_n \qquad (3.1.11)$$

The integral of the matrix A between the limits t_a and t_b is obtained by taking the limit $N \to \infty$ and $\Delta t_n \to 0$:

$$\lim_{\substack{N \to \infty \\ \Delta t_n \to 0}} \sum_{n=1}^{N} A(t_n)\,\Delta t_n = \int_{t_a}^{t_b} A(t)\,dt \qquad (3.1.12)$$

The components of the integral of $A(t)$ are simply

$$\int_{t_a}^{t_b} A_{ij}(t)\,dt \qquad (3.1.13)$$

A very special matrix is the identity matrix

$$I = \begin{pmatrix} 1 & 0 & 0 \\ 0 & 1 & 0 \\ 0 & 0 & 1 \end{pmatrix} \qquad (3.1.14)$$

It has the property that for every matrix A

$$IA = AI = A \qquad (3.1.15)$$

The identity matrix is like the number one in the algebra of ordinary numbers; it is also the matrix representation of the Kronecker delta.

It is possible to multiply together non-square matrices. For instance, the product

$$\begin{pmatrix} A_{11} & A_{12} \\ A_{21} & A_{22} \end{pmatrix} \begin{pmatrix} B_{11} & B_{12} & B_{13} \\ B_{21} & B_{22} & B_{23} \end{pmatrix} \qquad (3.1.16)$$

makes sense since the number of columns of the first matrix equals the number of rows of the second. The product of these matrices in reverse order is not defined. There are two types of non-square matrices which are extensively used. These are row and column matrices used to represent a vector:

$$X = \begin{pmatrix} a_1 \\ a_2 \\ a_3 \end{pmatrix} \qquad \tilde{X} = (a_1 \quad a_2 \quad a_3) \qquad (3.1.17)$$

The row and column matrices are transposes of each other. Unless otherwise specified, all matrices in this chapter are either square matrices denoted by capital letters $A, B, C, \ldots$ or column and row matrices denoted by $X, Y, \ldots, \tilde{X}, \tilde{Y}, \ldots$ and used to represent vectors.

If X and Y are matrix representations of two vectors $\mathbf{a}$ and $\mathbf{b}$, then the matrix product $\tilde{X} Y$ is equivalent to the dot product of $\mathbf{a}$ and $\mathbf{b}$:

$$\mathbf{a} \cdot \mathbf{b} = \tilde{X} Y = (a_1 \quad a_2 \quad a_3) \begin{pmatrix} b_1 \\ b_2 \\ b_3 \end{pmatrix} \qquad (3.1.18)$$

The dot product can also be represented by $\tilde{Y} X$. However, the products $X \tilde{Y}$ and $Y \tilde{X}$ are second rank tensors:

$$X \tilde{Y} = \begin{pmatrix} a_1 \\ a_2 \\ a_3 \end{pmatrix} (b_1 \quad b_2 \quad b_3) = \begin{pmatrix} a_1 b_1 & a_1 b_2 & a_1 b_3 \\ a_2 b_1 & a_2 b_2 & a_2 b_3 \\ a_3 b_1 & a_3 b_2 & a_3 b_3 \end{pmatrix} \qquad (3.1.19)$$

If $Y\tilde{X}$ is worked out, it is found to be the transpose of $X\tilde{Y}$:

$$Y\tilde{X} = \widetilde{X\tilde{Y}} \qquad (3.1.20)$$

To illustrate the use of matrix multiplication, several equations developed in Chapters 1 and 2 will be converted to matrix form. The orthogonal transformation of a vector can be represented by

$$\begin{pmatrix} a_1{}' \\ a_2{}' \\ a_3{}' \end{pmatrix} = \begin{pmatrix} S_{11} & S_{12} & S_{13} \\ S_{21} & S_{22} & S_{23} \\ S_{31} & S_{32} & S_{33} \end{pmatrix} \begin{pmatrix} a_1 \\ a_2 \\ a_3 \end{pmatrix} \qquad (3.1.21)$$

or simply

$$X' = SX \qquad (3.1.22)$$

The orthogonality condition which expresses the fact that the *rows* of S are orthogonal unit vectors can be replaced by the matrix equation

$$\begin{pmatrix} S_{11} & S_{12} & S_{13} \\ S_{21} & S_{22} & S_{23} \\ S_{31} & S_{32} & S_{33} \end{pmatrix} \begin{pmatrix} S_{11} & S_{21} & S_{31} \\ S_{12} & S_{22} & S_{32} \\ S_{13} & S_{23} & S_{33} \end{pmatrix} = \begin{pmatrix} 1 & 0 & 0 \\ 0 & 1 & 0 \\ 0 & 0 & 1 \end{pmatrix} \qquad (3.1.23)$$

or

$$S\tilde{S} = I \qquad (3.1.24)$$

The fact that the columns of S are also orthogonal unit vectors is expressed by

$$\tilde{S}S = I \qquad (3.1.25)$$

The quadratic form which represents the kinetic energy of a particle in generalized coordinates is*

$$T = \tfrac{1}{2}m(\dot{q}_1 \quad \dot{q}_2 \quad \dot{q}_3) \begin{pmatrix} g_{11} & g_{12} & g_{13} \\ g_{21} & g_{22} & g_{23} \\ g_{31} & g_{32} & g_{33} \end{pmatrix} \begin{pmatrix} \dot{q}_1 \\ \dot{q}_2 \\ \dot{q}_3 \end{pmatrix} \qquad (3.1.26)$$

If the components of a transformation matrix P are

$$P_{1'}^1 = \frac{\partial q_1}{\partial q_1{}'}, \qquad P_{2'}^1 = \frac{\partial q_1}{\partial q_2{}'}, \qquad P_{3'}^1 = \frac{\partial q_1}{\partial q_3{}'}, \dots \qquad (3.1.27)$$

then the transformation of the metric tensor from one coordinate system to another (2.15.8) in matrix form is

$$G' = \tilde{P}GP \qquad (3.1.28)$$

---

* Where no confusion will result, contravariant quantities are labelled with a subscript in place of a superscript.

So far, matrix equations have been used only as a convenient method of representing relations involving vectors, second rank tensors and transformations. Many calculations are more easily done by using matrices. For instance, Eq. (3.1.28) is much more convenient to use than the component equations

$$g'_{ij} = P^k_i P^l_j g_{kl} \qquad (3.1.29)$$

for the calculation of the components of the metric tensor in a new coordinate system. For a specific example, refer to Problem (2.21.27). From the given transformation equations found there the coefficients $P_j{}^i = \partial x^i / \partial q^j$ can be calculated and displayed in matrix form as

$$P = \begin{pmatrix} 1 & \cos\theta & 0 \\ 0 & \sin\theta & 0 \\ 0 & 0 & 1 \end{pmatrix} \qquad (3.1.30)$$

The component equations

$$g_{ij} = \frac{\partial x_k}{\partial q_i} \frac{\partial x_k}{\partial q_j} \qquad (3.1.31)$$

are represented by the matrix equation

$$G = \tilde{P}P = \begin{pmatrix} 1 & 0 & 0 \\ \cos\theta & \sin\theta & 0 \\ 0 & 0 & 1 \end{pmatrix} \begin{pmatrix} 1 & \cos\theta & 0 \\ 0 & \sin\theta & 0 \\ 0 & 0 & 1 \end{pmatrix} = \begin{pmatrix} 1 & \cos\theta & 0 \\ \cos\theta & 1 & 0 \\ 0 & 0 & 1 \end{pmatrix}$$

$$(3.1.32)$$

As this chapter is developed, we want to emphasize more and more the idea that the letters used to represent the matrices can be considered as algebraic quantities in their own right. In the purest sense, matrices are arrays of numbers which we endow with certain algebraic properties. A given matrix need not actually represent a tensor or a transformation or anything else so far discussed in this book.

3.2. FORMAL PROPERTIES OF MATRICES

Matrix multiplication is associative:

$$A(BC) = (AB)C = ABC \qquad (3.2.1)$$

The proof can be done by using component notation. The component of $A(BC)$ in row i and column l is

$$[A(BC)]_{il} = A_{ij}(BC)_{jl} = A_{ij}B_{jk}C_{kl}$$
$$= (AB)_{ik}C_{kl} = [(AB)C]_{il} \qquad (3.2.2)$$

If $AB = C$, then the transpose of C is the product of the transposes of A and B in the reverse order:

$$\tilde{C} = \widetilde{AB} = \tilde{B}\tilde{A} \tag{3.2.3}$$

The ik component of AB is

$$(AB)_{ik} = A_{ij}B_{jk} \tag{3.2.4}$$

Since $\widetilde{AB}$ is obtained by interchanging the rows and columns of AB, the ik component of $\widetilde{AB}$ is the ki component of AB:

$$(\widetilde{AB})_{ik} = A_{kj}B_{ji} = B_{ji}A_{kj}$$
$$(\widetilde{AB})_{ik} = (\tilde{B})_{ij}(\tilde{A})_{jk} \tag{3.2.5}$$

which is (3.2.3) in component form. Notice that Eq. (3.1.20) is a special case of the theorem just proved.

The inner product of a tensor and a vector yields a vector. In component form this is expressed by

$$b_i = A_{ij}a_j \tag{3.2.6}$$

The matrix representation of such an equation is

$$Y = AX \tag{3.2.7}$$

The matrix A symbolizes an *operation* which is done on a given vector to produce another. This is frequently called a transformation even though no change of coordinates is involved. A given vector has been transformed into another, but both the original and the transformed vector, as well as the matrix operator itself, are expressed in the *same coordinate system*. An example of this is the transformation of the electric field vector into the displacement vector in the electrostatics of non-isotropic media, Eq. (1.8.17). If

$$AX = BX \tag{3.2.8}$$

for every vector X, then $A = B$. In component form

$$A_{ij}a_j = B_{ij}a_j$$

which, because of the arbitrariness of the vector X, implies that

$$A_{ij} = B_{ij}$$

Given the transformation $Y = AX$ is it possible to reverse the process and find X if Y is known? Since we are dealing here with a set of linear non-homogeneous equations we know from algebra that X can be found uniquely in terms of Y provided that det $A \neq 0$. It can be assumed under this circumstance that a matrix exists, which we designate by A^{-1}, that has the property that

$$A^{-1}Y = A^{-1}AX = X \tag{3.2.9}$$

The process is reversible. If (3.2.9) is multiplied by A,

$$AA^{-1}Y = AX = Y \tag{3.2.10}$$

Thus if det $A \neq 0$, a matrix A^{-1} exists with the property

$$AA^{-1} = A^{-1}A = I \tag{3.2.11}$$

A^{-1} is called the *inverse* of A. A matrix that has no inverse is said to be *singular*.

The formal construction of A^{-1} begins with the expression for the determinant of A:

$$\det A = \delta_{ijk}A_{1i}A_{2j}A_{3k}$$
$$= A_{1i}(\delta_{ijk}A_{2j}A_{3k}) \tag{3.2.12}$$

The factors

$$C_{1i} = \delta_{ijk}A_{2j}A_{3k} \tag{3.2.13}$$

are called the *cofactors* of A_{1i}. Written out they are

$$C_{11} = \begin{vmatrix} A_{22} & A_{23} \\ A_{32} & A_{33} \end{vmatrix} \qquad C_{12} = -\begin{vmatrix} A_{21} & A_{23} \\ A_{31} & A_{33} \end{vmatrix} \qquad C_{13} = \begin{vmatrix} A_{21} & A_{22} \\ A_{31} & A_{32} \end{vmatrix}$$
$$\tag{3.2.14}$$

Similarly, the cofactors of the components in the remaining two rows are

$$C_{2j} = \delta_{ijk}A_{1i}A_{3k} \qquad C_{3k} = \delta_{ijk}A_{1i}A_{2j} \tag{3.2.15}$$

The cofactor of a given component is $\pm$ the determinant formed by striking out the row and column of det A that the component is in. The signs of the cofactors are determined by the scheme

$$\begin{pmatrix} + & - & + \\ - & + & - \\ + & - & + \end{pmatrix} \tag{3.2.16}$$

If the cofactors C_{2j} of the elements in the second row are multiplied by the elements of the first row the result is

$$A_{1i}C_{2i} = A_{1i}(\delta_{lim}A_{1l}A_{3m})$$
$$= (A_{1l}A_{1i}\,\delta_{lim})A_{3m} = 0 \tag{3.2.17}$$

In general,

$$\delta_{ik}(\det A) = A_{ij}C_{kj} \tag{3.2.18}$$

In other words, a given row multiplied by its own cofactors is equal to det A whereas if it is multiplied by the cofactors of another row, the result is zero. Equation (3.2.18) is valid for any 3 by 3 matrix even if det $A = 0$.

If, however, det $A \neq 0$,

$$A_{ij} \frac{C_{kj}}{\det A} = \delta_{ik} \tag{3.2.19}$$

If the matrix C is given by

$$C = \begin{pmatrix} C_{11} & C_{12} & C_{13} \\ C_{21} & C_{22} & C_{23} \\ C_{31} & C_{32} & C_{33} \end{pmatrix} \tag{3.2.20}$$

then Eq. (3.2.19) is equivalent to

$$A \frac{1}{\det A} \tilde{C} = I \tag{3.2.21}$$

The inverse of a matrix exists if and only if det $A \neq 0$. The inverse is given by

$$A^{-1} = \frac{1}{\det A} \tilde{C} \tag{3.2.22}$$

A matrix which represents a *coordinate* transformation must have an inverse. For an orthogonal transformation $S\tilde{S} = \tilde{S}S = I$ which means $\tilde{S} = S^{-1}$.

If A and B are two non-singular matrices,

$$(AB)^{-1} = B^{-1}A^{-1} \tag{3.2.23}$$

This is proved by applying the transformation AB to a vector X to produce another vector Y:

$$Y = ABX \tag{3.2.24}$$

The inversion of this transformation can be done in two ways. If (3.2.24) is multiplied by $(AB)^{-1}$ the result is

$$(AB)^{-1}Y = X \tag{3.2.25}$$

If (3.2.24) is multiplied first by A^{-1} and then by B^{-1}, there results

$$A^{-1}Y = BX$$
$$B^{-1}A^{-1}Y = X = (AB)^{-1}Y \tag{3.2.26}$$

Since Y can be chosen arbitrarily, the result follows.

If A is a non-singular matrix, then

$$(\tilde{A})^{-1} = (\widetilde{A^{-1}}) \tag{3.2.27}$$

Let $Y = AX$ be first transposed and then inverted:

$$\tilde{Y} = \tilde{X}\tilde{A} \qquad \tilde{Y}(\tilde{A})^{-1} = \tilde{X} \tag{3.2.28}$$

If the inversion is done first followed by the transposition

$$A^{-1}Y = X \qquad \widetilde{Y}(\widetilde{A^{-1}}) = \widetilde{X} \qquad (3.2.29)$$

Comparison of (3.2.28) and (3.2.29) gives (3.2.27).
The following are easily verified:

$$\widetilde{\widetilde{A}} = A, \qquad (A^{-1})^{-1} = A,$$
$$\widetilde{AB} = \widetilde{B}\widetilde{A}, \qquad (A^{-1}B)^{-1} = B^{-1}A, \text{ etc.} \qquad (3.2.30)$$

If A is any matrix, then any power of A can be defined:

$$A^2 = AA, \qquad A^3 = AAA, \qquad A^2A^3 = AAAAA = A^5, \ldots \quad (3.2.31)$$

In general, if n and m are positive integers,

$$A^n A^m = A^{n+m} \qquad (3.2.32)$$

If the matrix is non-singular, the above theorem can be extended to include negative integer exponents. Since $A^{+1}A^{-1} = A^0$, it is necessary to interpret A^0 as I.

In handling expressions involving matrices, care must be used to preserve the order of the factors due to the possibility of the matrices being non-commutative. For example,

$$(A + B)^2 = (A + B)(A + B) = A(A + B) + B(A + B)$$
$$= A^2 + AB + BA + B^2 \qquad (3.2.33)$$

It is not permissible to write $(A + B)^2 = A^2 + 2AB + B^2$ unless A and B commute!

The rule for multiplying two determinants together is the same as that for matrices. Hence,

$$\det (ABC) = \det A \det B \det C \qquad (3.2.34)$$

Note also that

$$\det A = \det \widetilde{A}$$
$$\det \widetilde{A}B = \det A \det B, \text{ etc.} \qquad (3.2.35)$$

This result has already been used, e.g., in the proof that $\det S = \pm 1$ from $S\widetilde{S} = I$.

3.3. ORTHOGONAL TRANSFORMATIONS

Most of the results of this section have already been obtained. It will therefore constitute a review of the theory of orthogonal transformations using matrix notation. The transformation of the components of a vector from one rectangular Cartesian coordinate system to another can be expressed by the matrix equation

$$X' = SX \tag{3.3.1}$$

The fundamental property of an orthogonal transformation is that it leaves the length of a vector unaltered. This is expressed by the matrix equation

$$\tilde{X}X = \tilde{X}'X' \tag{3.3.2}$$

If (3.3.1) is substituted into (3.3.2) and the rule (3.2.3) for the transpose of the product of two matrices is applied the result is

$$\tilde{X}X = \tilde{X}\tilde{S}SX \tag{3.3.3}$$

Since the vector X can be arbitrarily chosen, (3.3.3) implies that

$$\tilde{S}S = I \tag{3.3.4}$$

or

$$\tilde{S} = S^{-1} \tag{3.3.5}$$

Since a matrix commutes with its own inverse it is also true that

$$S\tilde{S} = I \tag{3.3.6}$$

Given a square matrix A, it is possible to apply it to a vector X to obtain another vector Y:

$$Y = AX \tag{3.3.7}$$

The components of both X and Y are expressed in the same coordinate system. The same relation set up in another coordinate system is

$$Y' = A'X' \tag{3.3.8}$$

The vectors can be transformed by means of (3.3.1):

$$SY = A'SX \tag{3.3.9}$$

Multiplication by $\tilde{S}$ and use of (3.3.4) yields

$$Y = (\tilde{S}A'S)X = AX \tag{3.3.10}$$

Since X is arbitrary, (3.3.10) implies

$$A = \tilde{S}A'S \tag{3.3.11}$$

If (3.3.11) is multiplied from the left by S and from the right by $\tilde{S}$, the transformation is inverted:

$$SA\tilde{S} = S\tilde{S}A'S\tilde{S} = IA'I = A' \tag{3.3.12}$$

The component form of (3.3.12) is of course

$$A'_{ij} = S_{ik}S_{jl}A_{kl} \tag{3.3.13}$$

which is recognized as the transformation of the components of a second rank tensor. The proof that we have done here is identical to the proof that ϵ_{ij} (Eq. 1.8.17) which relate the components of the displacement vector **D** to those of the electric field **E** are components of a second rank tensor.

Students of matrix algebra generally do not use the term "tensor." Matrix theorists use the notion of a *similarity transformation* which is defined by saying that two matrices A and A' are *similar* if there exists a non-singular matrix S such that

$$A' = SAS^{-1} \qquad (3.3.14)$$

If it is true that $S^{-1} = \tilde{S}$, then (3.3.14) specializes to an *orthogonal similarity transformation*. The present discussion is restricted entirely to orthogonal similarity.

It is safe to say that many branches of mathematics and physics developed independently of one another. Formalisms were developed which seemed appropriate for the treatment of particular problems. In our desire to learn how to transform physical quantities which are represented by vectors from one Cartesian coordinate system to another we were led to construct a theory of orthogonal transformations. A mathematician, on the other hand, may decide to study similarity transformations as defined by (3.3.14), not caring in the least whether or not his matrices and transformations actually correspond to anything physical or geometrical. Quite naturally, there is a lot of overlap in the theories and, unfortunately, a difference in terminology.

Vectors such as velocity and acceleration in three-dimensional space have three components and the tensors and transformations which operate on them have nine. There is no reason why the mathematical theory cannot be formally generalized to spaces of more dimensions. All the results so far developed in this section are valid in an n-dimensional space. Thus in Eq. (3.3.1) the column matrices X and X' have n-components instead of three; the square matrix S used to represent an orthogonal transformation has n rows and n columns. In reverting to the subscript notation of Eq. (3.3.13), allow the sums to run from 1 to n instead of 1 to 3. Matrices of larger size than 3 by 3 come up in the theory of oscillating systems possessing many degrees of freedom. Quantum mechanics can be formulated in an abstract vector space of an *infinite number* of dimensions called *Hilbert space*. An n by n matrix is generally called an n-dimensional matrix.

If S and T are orthogonal transformations, then so are the products ST and TS. For instance,

$$(\widetilde{ST})(ST) = \tilde{T}\tilde{S}ST = \tilde{T}T = I \qquad (3.3.15)$$

This means that if two orthogonal transformations are done in succession as

$$X' = SX \qquad X'' = TX' \qquad (3.3.16)$$

then the final result X'' can be obtained directly from X by the transformation

$$X'' = (TS)X \qquad (3.3.17)$$

The transformations applied in the reverse order do not yield the same result due to the non-commutativity of T and S

$$X''' = (ST)X \neq X'' \qquad (3.3.18)$$

The non-commutativity of two orthogonal transformations is illustrated by an example. Let S represent a rotation by $90°$ about the x_2-axis and T represent a rotation by $90°$ about the x_1-axis. Applying these in the order S, and then T yields:

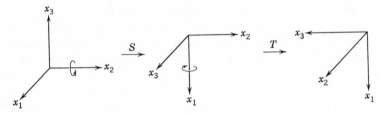

Application of S and T in the reverse order yields

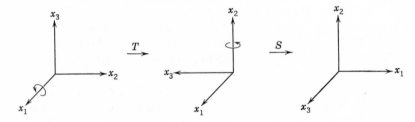

showing an obvious difference in the final result.

3.4. THE EIGENVALUE PROBLEM

Suppose a second rank tensor A represents some physical quantity. Equations in which this tensor appear could be greatly simplified if the coordinate system were chosen in such a way that some of the components were zero. We will prove that if A is a symmetric tensor it can be converted into diagonal form by an orthogonal transformation.

Again referring to the example of the relation between **D** and **E** for an anisotropic medium, a rectangular coordinate system can be oriented in the medium in such a way that the only non-zero elements of ϵ_{ij} are the three diagonal elements. The relation between **D** and **E** is then $D_1 = \epsilon_{11}E_1$, $D_2 = \epsilon_{22}E_2$, $D_3 = \epsilon_{33}E_3$. At a given point, the electrostatic properties of the most general medium are characterized by three numbers ϵ_{11}, ϵ_{22}, ϵ_{33}. The medium may, of course, be non-homogeneous in which case ϵ_{ij} is a *tensor field*. Still, at any given point it is possible to introduce a coordinate system in which ϵ_{ij} is diagonal.

The problem is to construct a transformation S such that $A' = SA\tilde{S}$ is diagonal A diagonal tensor is necessarily symmetric. Since symmetry is an invariant property of tensors, A expressed in the original system is necessarily symmetric. Thus the symmetry of A is a necessary condition for diagonalization by an orthogonal transformation to be possible. It will be shown that the condition is also sufficient. If A' is assumed to be in diagonal form, then the transformation in component form is

$$A'_{ij} = S_{ik}S_{jl}A_{kl} = \lambda(j)\delta_{ij} \tag{3.4.1}$$

The three numbers $\lambda(1)$, $\lambda(2)$, and $\lambda(3)$ are the diagonal components of A':

$$A' = \begin{pmatrix} \lambda(1) & 0 & 0 \\ 0 & \lambda(2) & 0 \\ 0 & 0 & \lambda(3) \end{pmatrix} \tag{3.4.2}$$

No sum over j is implied by $\lambda(j)\,\delta_{ij}$. It is necessary to cast the transformation (3.4.1) into a form from which the transformation coefficients S_{ij} can be inferred. The inner product of the transformation with S_{im} results in

$$S_{im}S_{ik}S_{jl}A_{kl} = \lambda(j)S_{im}\delta_{ij}$$
$$\delta_{mk}S_{jl}A_{kl} = \lambda(j)S_{jm} \tag{3.4.3}$$
$$S_{jl}A_{ml} = \lambda(j)\delta_{ml}S_{jl}$$

where in the last line S_{jm} is replaced by its equal $\delta_{ml}S_{jl}$. Continuing

$$(\lambda(j)\delta_{ml} - A_{ml})S_{jl} = 0 \tag{3.4.4}$$

Each value of j gives a different *row* of S. If X_1, X_2, and X_3 are the three vectors constructed from the rows of S,

$$X_1 = \begin{pmatrix} S_{11} \\ S_{12} \\ S_{13} \end{pmatrix} \qquad X_2 = \begin{pmatrix} S_{21} \\ S_{22} \\ S_{23} \end{pmatrix} \qquad X_3 = \begin{pmatrix} S_{31} \\ S_{32} \\ S_{33} \end{pmatrix} \tag{3.4.5}$$

Then (3.4.4) is equivalent to the three matrix equations

$$(\lambda(1)I - A)X_1 = 0 \qquad (3.4.6)$$

$$(\lambda(2)I - A)X_2 = 0 \qquad (3.4.7)$$

$$(\lambda(3)I - A)X_3 = 0 \qquad (3.4.8)$$

The three values $\lambda(1)$, $\lambda(2)$, and $\lambda(3)$ are called the *eigenvalues* of the tensor A, and X_1, X_2, and X_3 are called the *eigenvectors*.

The matrix $\lambda I - A$, with λ a variable, parameter is called the *characteristic matrix* of A. The function $f(\lambda) = \det(\lambda I - A)$ is the *characteristic function* of the matrix. If the characteristic matrix is allowed to operate on a vector X and the equation put equal to zero there results three homogeneous linear equations in three unknowns:

$$(\lambda I - A)X = 0 \qquad (3.4.9)$$

There can be a non-zero solution X if, and only if, the determinant of the coefficients vanishes:

$$f(\lambda) = \begin{vmatrix} \lambda - A_{11} & -A_{12} & -A_{13} \\ -A_{21} & \lambda - A_{22} & -A_{23} \\ -A_{31} & -A_{32} & \lambda - A_{33} \end{vmatrix} = 0 \qquad (3.4.10)$$

This is called the *characteristic equation* or, sometimes, the *secular equation*. It is a cubic equation in λ and has three roots, namely, the eigenvalues $\lambda(1)$, $\lambda(2)$, and $\lambda(3)$. After the eigenvalues have been found, Eqs. (3.4.6), (3.4.7), and (3.4.8) give, except for a scalar factor, the eigenvectors. The length of the vectors is not determined. If X is a solution of $(\lambda I - A)X = 0$, so is αX. If the eigenvectors are to be used as the rows of an orthogonal transformation S, then the orthogonality condition requires that they be unit vectors:

$$\tilde{X}_1 X_1 = \tilde{X}_2 X_2 = \tilde{X}_3 X_3 = 1 \qquad (3.4.11)$$

Eigenvectors which satisfy this condition are said to be normalized.

For *any* second rank tensor A three eigenvalues can be determined from $f(\lambda) = 0$. The corresponding eigenvectors are then found from $(\lambda I - A)X = 0$. It has been shown that a *necessary* condition for A to be diagonalized by an orthogonal transformation is that A be *symmetric*. We will now show that the condition is also *sufficient* by proving that the eigenvectors of a symmetric tensor are orthogonal, thus permitting the construction from them of a suitable transformation S. The eigenvalues and eigenvectors of a non-symmetric tensor exist, but the eigenvectors are not in general orthogonal so that an orthogonal transformation cannot be constructed from them.

If $AX_1 = \lambda(1)X_1$ is multiplied by $\tilde{X}_2$ and $AX_2 = \lambda(2)X_2$ is multiplied by $\tilde{X}_1$, the result is

$$\tilde{X}_2 A X_1 = \lambda(1)\tilde{X}_2 X_1 \qquad \tilde{X}_1 A X_2 = \lambda(2)\tilde{X}_1 X_2 \qquad (3.4.12)$$

The second equation is transposed and $\tilde{A} = A$ is used:

$$\tilde{X}_2 \tilde{A} X_1 = \tilde{X}_2 A X_1 = \lambda(2)\tilde{X}_2 X_1 \qquad (3.4.13)$$

Combining this with the first of equations (3.4.12) leads to

$$[\lambda(2) - \lambda(1)]\tilde{X}_2 X_1 = 0 \qquad (3.4.14)$$

If $\lambda(2) \neq \lambda(1)$, then $\tilde{X}_2 X_1 = 0$ which proves the orthogonality. It is quite possible that some of the roots $f(\lambda) = 0$ may be equal in which case the proof of the orthogonality of the eigenvectors just given fails.

The significance of equal eigenvalues is more easily appreciated after a discussion of the geometry of the eigenvalue problem. A transformation giving a translation of axes is trivial and the discussion will be confined to a pure rotation of axes. Let X represent a simple displacement vector projecting from the common origins of the new and old coordinate systems and having as its components the coordinates x_1, x_2, and x_3 in the old system and x_1', x_2', and x_3' in the new. The quadratic form

$$\phi = x_i A_{ij} x_j \qquad (3.4.15)$$

where $A_{ij} = A_{ji}$ represents a *central quadric*. If the axes are rotated so that A becomes diagonal,

$$\phi = x_1'^2\lambda(1) + x_2'^2\lambda(2) + x_3'^2\lambda(3) \qquad (3.4.16)$$

If all the eigenvalues are positive, $\phi = \text{const.}$ represents an ellipsoid, Fig. 3.4.1a. In this case, $\phi > 0$ for all x_1, x_2, and x_3 and is called *positive definite*. If one of the eigenvalues is negative, say, $\lambda(1)$, the surface is a hyperboloid of one sheet, Fig. 3.4.1b; and if two of the eigenvalues are negative, the surface as a hyperboloid of two sheets, Fig. 3.4.1c. It will be sufficient to discuss the case of the ellipsoid.

In the new coordinate system, the unit vectors $\hat{e}_1'$, $\hat{e}_2'$, and $\hat{e}_3'$ coincide with the *principal axes* of the ellipsoid. The transformation to a coordinate system in which A is diagonal is frequently called the *principal axis transformation* and $\hat{e}_i'$ are called the *principal directions* of the tensor A. The components of the unit vectors in the *old* system are just the rows of S (refer to equation 1.7.5 and Fig. 1.7.1). Thus $\hat{e}_i'$ are the eigenvectors. In the *new* coordinate system they are represented by the column matrices

$$X_1' = \begin{pmatrix} 1 \\ 0 \\ 0 \end{pmatrix} \qquad X_2' = \begin{pmatrix} 0 \\ 1 \\ 0 \end{pmatrix} \qquad X_3' = \begin{pmatrix} 0 \\ 0 \\ 1 \end{pmatrix} \qquad (3.4.17)$$

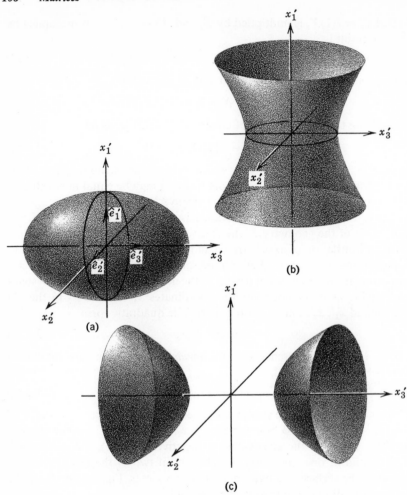

Fig. 3.4.1

Suppose that the two eigenvalues $\lambda(1)$ and $\lambda(2)$ are equal. This means simply that the cross section of the ellipsoid in the $x_1'x_2'$-plane is a *circle*. The directions $\hat{e}_1'$ and $\hat{e}_2'$ are no longer uniquely determined. Any further rotation about the x_3' axis would still leave A in a diagonal representation. *Any* vector X lying in the $x_1'x_2'$-plane satisfies $AX = \lambda(1)X = \lambda(2)X$, and hence represents an eigenvector. In this plane it is possible to choose two vectors X_1 and X_2 that are orthogonal. In this way, a set of mutually perpendicular vectors X_1, X_2, and X_3 can be chosen from which to construct a transformation S. The transformation is not unique due to the arbitrariness of X_1 and X_2. It is only the direction of X_3 which is unambiguously determined.

One further question remains. What guarantee is there that the roots of the characteristic equation $f(\lambda) = 0$ are real? The original tensor A is real. If A^* is used for the complex conjugate of A, then $A^* = A$. The complex conjugate of

$$AX = \lambda X \tag{3.4.18}$$

is

$$AX^* = \lambda^* X^*$$

which is transposed to yield

$$\tilde{X}^* A = \lambda^* \tilde{X}^* \tag{3.4.19}$$

Use has been made of $\tilde{A} = A$. Let (3.4.18) be multiplied by $\tilde{X}^*$ and (3.4.19) by X:

$$\tilde{X}^* A X = \lambda \tilde{X}^* X \qquad \tilde{X}^* A X = \lambda^* \tilde{X}^* X \tag{3.4.20}$$

Hence $\lambda = \lambda^*$ showing that the eigenvalues of a symmetric tensor are real. Non-symmetric tensors frequently do have complex eigenvalues.

Several theorems which relate to the eigenvalue problem will now be proved.

Any matrix which commutes with a diagonal matrix D is also diagonal provided that all the diagonal components of D are different. Assume $DB = BD$. In component form this is

$$\lambda(i)\delta_{ij}B_{jk} = B_{ij}\lambda(k)\delta_{jk}$$
$$\lambda(i)B_{ik} = B_{ik}\lambda(k) \tag{3.4.21}$$
$$[\lambda(i) - \lambda(k)]B_{ik} = 0$$

Since $\lambda(i) \neq \lambda(k)$, $B_{ik} = 0$ unless $i = k$ showing that B is diagonal.

If A is a symmetric tensor the eigenvalues of which are all different and B commutes with A, then B is symmetric and is diagonalized by the same principal axis transformation that diagonalizes A. Let S be the principal axis transformation that diagonalizes A. Since $AB = BA$ is a tensor equation, the same equation holds in the new coordinates: $A'B' = B'A'$. But A' is diagonal. If the eigenvalues of A are all different, B is necessarily also diagonal. If A has some eigenvalues equal, it is no longer possible to infer from $AB = BA$ that B is necessarily diagonalized by the same principal axis transformation that diagonalizes A. However, if B is known to be symmetric, a transformation can still be constructed that simultaneously diagonalizes both A and B. Suppose that A has been diagonalized and that $\lambda(1) = \lambda(2)$. Since $\lambda(1) \neq \lambda(3)$, Eq. (3.4.21) implies that B

is of the form

$$\begin{pmatrix} B_{11} & B_{12} & 0 \\ B_{21} & B_{22} & 0 \\ 0 & 0 & B_{33} \end{pmatrix} \tag{3.4.22}$$

The eigenvectors X_1 and X_2 of A can be chosen arbitrarily in the $x_1 x_2$-plane without disrupting the diagonal form of A, provided that $B_{21} = B_{12}$, X_1 and X_2 can be chosen in such a way that they are eigenvectors of B, thus completing its diagonalization.

This theorem is basic to the matrix representation of quantum mechanics. Measurable quantities can be represented by matrices, or matrix operators as they are called, the eigenvalues of which are the actual physically measurable quantities. Quantities which are simultaneously measurable are represented by commuting operators; that they are simultaneously measurable is expressed by the fact that there is a representation in which the commuting operators are all diagonal. The components of the matrices used in quantum mechanics are in general *complex numbers*. The generalization of the theory to cover this possibility is not difficult.

The eigenvalues of any tensor, symmetric or not, are invariants. The eigenvalue problem written in component form is

$$A_{ij}a_j = \lambda a_i \tag{3.4.23}$$

where a_i are components of the eigenvector corresponding to the eigenvalue λ. Formation of the inner product of (3.4.23) with a_i results in

$$a_i A_{ij} a_j = \lambda (a_i a_i) \tag{3.4.24}$$

Since $a_i A_{ij} a_j$ and $a_i a_i$ are invariants, so is λ.

The eigenvalues of any tensor are invariants. If the tensor is also symmetric, it is easily proved that all invariants are functions of the eigenvalues. The proof rests on the fact that a coordinate exists in which a symmetric tensor is diagonal, the diagonal components being the eigenvalues. For example, the invariant $\psi = A_{ij}A_{ij}$ can also be computed in any other coordinate system as $\psi = A'_{ij}A'_{ij}$. In particular, in the coordinate system in which A is diagonal,

$$\psi = \lambda_1{}^2 + \lambda_2{}^2 + \lambda_3{}^2$$

As has been previously noted, the trace of any tensor is an invariant. It is also easily proved that the determinant of any tensor is an invariant:

$$A' = SA\tilde{S}$$
$$\det A' = \det S \det A \det \tilde{S} = \det A \tag{3.4.25}$$

If the tensor is also symmetric, these invariants are easily related to the eigenvalues:

$$\text{trace } A = \lambda_1 + \lambda_2 + \lambda_3 \qquad (3.4.26)$$

$$\det A = \lambda_1 \lambda_2 \lambda_3 \qquad (3.4.27)$$

The notation λ_1, λ_2, λ_3 is used in place of $\lambda(1)$, $\lambda(2)$, $\lambda(3)$ where no confusion of the eigenvalues with the components of a vector will result.

In addition to the trace and the determinant, a third invariant can be constructed from the characteristic function. Since the characteristic function is a cubic equation in λ, the roots of which are λ_1, λ_2, and λ_3, it can be exhibited in the form

$$f(\lambda) = \det(\lambda I - A) = (\lambda - \lambda_1)(\lambda - \lambda_2)(\lambda - \lambda_3)$$
$$= \lambda^3 - [\lambda_1 + \lambda_2 + \lambda_3]\lambda^2 + [\lambda_1\lambda_2 + \lambda_1\lambda_3 + \lambda_2\lambda_3]\lambda - \lambda_1\lambda_2\lambda_3 \qquad (3.4.28)$$

Direct expansion of $\det(\lambda I - A)$ yields

$$f(\lambda) = \lambda^3 - [A_{11} + A_{22} + A_{33}]\lambda^2$$
$$+ \left(\begin{vmatrix} A_{11} & A_{12} \\ A_{21} & A_{22} \end{vmatrix} + \begin{vmatrix} A_{11} & A_{13} \\ A_{31} & A_{33} \end{vmatrix} + \begin{vmatrix} A_{22} & A_{23} \\ A_{32} & A_{33} \end{vmatrix} \right)\lambda - \det A \quad (3.4.29)$$

Comparison of the like powers of λ in (3.4.28) and (3.4.29) gives again (3.4.26) and (3.4.27) and the additional result

$$\lambda_1\lambda_2 + \lambda_1\lambda_3 + \lambda_2\lambda_3 = \begin{vmatrix} A_{11} & A_{12} \\ A_{21} & A_{22} \end{vmatrix} + \begin{vmatrix} A_{11} & A_{13} \\ A_{31} & A_{33} \end{vmatrix} + \begin{vmatrix} A_{22} & A_{23} \\ A_{32} & A_{33} \end{vmatrix} \quad (3.4.30)$$

Since no use was made of the symmetry of A in deriving (3.4.28) and (3.4.29), Eqs. (3.4.26), (3.4.27), and (3.4.30) hold for any tensor. The three invariants just derived find application in the mechanics of continuous media. The invariant (3.4.30) can also be expressed as

$$\lambda_1\lambda_2 + \lambda_1\lambda_3 + \lambda_2\lambda_3 = \tfrac{1}{2}(A_{ii})^2 - \tfrac{1}{2}A_{ij}A_{ji} \qquad (3.4.31)$$

The eigenvectors of a symmetric tensor are orthogonal and necessarily form a linearly independent set of vectors. It is possible to prove that the eigenvectors of a non-symmetric tensor form a linearly independent set provided that all the eigenvalues are distinct. The proof is done by showing that there exists no constants α, β, and γ such that

$$\alpha X_1 + \beta X_2 + \gamma X_3 = 0 \qquad (3.4.32)$$

where X_1, X_2, and X_3 are the eigenvectors. Multiplication of (3.4.32) by A yields

$$\alpha A X_1 + \beta A X_2 + \gamma A X_3 = 0$$

Since $A X_1 = \lambda_1 X_1$, $A X_2 = \lambda_2 X_2$, and $A X_3 = \lambda_3 X_3$,

$$\lambda_1(\alpha X_1) + \lambda_2(\beta X_2) + \lambda_3(\gamma X_3) = 0 \qquad (3.4.33)$$

A second multiplication by A yields

$$\lambda_1^2(\alpha X_1) + \lambda_2^2(\beta X_2) + \lambda_3^2(\gamma X_3) = 0 \qquad (3.4.34)$$

If (3.4.32), (3.4.33), and (3.4.34) are regarded as three homogeneous equations in three unknowns αX_1, βX_2, and γX_3, the determinant of the coefficients is

$$\begin{vmatrix} 1 & 1 & 1 \\ \lambda_1 & \lambda_2 & \lambda_3 \\ \lambda_1^2 & \lambda_2^2 & \lambda_3^2 \end{vmatrix} = (\lambda_1 - \lambda_2)(\lambda_2 - \lambda_3)(\lambda_3 - \lambda_1) \qquad (3.4.35)$$

Since the eigenvalues are all different, the determination of the coefficients does not vanish, meaning that the only solution of the homogeneous equations is the zero solution. Therefore, $\alpha = \beta = \gamma = 0$.

Determinants of the form (3.4.35) are known as *Vandermonde determinants*.

3.5. THE CAYLEY-HAMILTON THEOREM

Since all the powers of a square matrix A^p, A^{p-1}, $\ldots$, A, I are defined if p is a positive integer, it is possible to define a *matrix polynomial*

$$P(A) = \alpha_p A^p + \alpha_{p-1} A^{p-1} + \cdots + \alpha_1 A + \alpha_0 I \qquad (3.5.1)$$

If A is an n-dimensional matrix, the characteristic function

$$f(\lambda) = \det (\lambda I - A) \qquad (3.5.2)$$

is a polynomial of degree n in λ:

$$f(\lambda) = \lambda^n + \alpha_{n-1} \lambda^{n-1} + \cdots + \alpha_1 \lambda + \alpha_0 \qquad (3.5.3)$$

An important theorem in the algebra of matrices, known as the *Cayley-Hamilton theorem*, is that the matrix polynomial constructed from the characteristic function of the matrix by replacing λ by the matrix itself is equal to the zero matrix:

$$f(A) = A^n + \alpha_{n-1} A^{n-1} + \cdots + \alpha_1 A + \alpha_0 I = 0 \qquad (3.5.4)$$

The theorem is actually true for any matrix; we will prove its validity only for those matrices the eigenvectors of which form a linearly independent set. This includes all symmetric matrices and all non-symmetric matrices that have distinct eigenvalues. It is perhaps advisable to emphasize that given *any* matrix, symmetric or not, the eigenvalues are *defined* to be the roots of $\det (\lambda I - A) = 0$. The eigenvectors are then computed from $AX = \lambda X$. This is a purely formal procedure that can be carried out regardless of whether or not the eigenvalues and eigenvectors have any physical or geometrical significance. For non-symmetric matrices the eigenvalues and the components of the eigenvectors may be complex numbers. So far we have been able to guarantee that only symmetric matrices have real eigenvalues and eigenvectors with a simple geometrical interpretation.

If A is a three-dimensional matrix, the matrix polynomial formed from the characteristic function written in a factored form is

$$f(A) = (A - \lambda_1 I)(A - \lambda_2 I)(A - \lambda_3 I) \tag{3.5.5}$$

If X is an arbitrary vector, it can be written as a linear combination of the eigenvectors:

$$X = \alpha X_1 + \beta X_2 + \gamma X_3 \tag{3.5.6}$$

Since $(A - \lambda_1 I)X_1 = 0, (A - \lambda_2 I)X_2 = 0$, and $(A - \lambda_3 I)X_3 = 0$, and since $(A - \lambda_1 I), (A - \lambda_2 I)$, and $(A - \lambda_3 I)$ commute with one-another, it is true that

$$f(A)X = 0 \tag{3.5.7}$$

But X is an arbitrary vector. Hence, $f(A) = 0$. The theorem is easily extended to n-dimensional tensors.

The coefficients $\alpha_0, \alpha_1, \ldots$ that appear in (3.5.4) are invariants. This is proved as follows. Note that

$$SA\tilde{S} = A' \qquad SA^2\tilde{S} = SA\tilde{S}SA\tilde{S} = A'^2$$
$$SA^n\tilde{S} = A'^n \tag{3.5.8}$$

If $f(A) = 0$ is multiplied on the left by S and on the right by $\tilde{S}$, the result is

$$S(A^n + \alpha_{n-1}A^{n-1} + \cdots + \alpha_1 A + \alpha_0 I)\tilde{S} = 0$$
$$A'^n + \alpha_{n-1}A'^{n-1} + \cdots + \alpha_1 A' + \alpha_0 I = 0 \tag{3.5.9}$$

The invariants $\alpha_{n-1}, \ldots, \alpha_0$ can be related to the eigenvalues of A by the same procedure used to construct Eqs. (3.4.28) and (3.4.29) for the special case of a three-dimensional matrix. For example,

$$\alpha_{n-1} = -\text{trace } A = -[\lambda_1 + \lambda_2 + \cdots + \lambda_n] \tag{3.5.10}$$

$$\alpha_0 = (-1)^n \det A = (-1)^n [\lambda_1 \lambda_2 \cdots \lambda_n] \tag{3.5.11}$$

3.6. THE SPECTRAL RESOLUTION OF A MATRIX

Let $P(A)$ be any polynomial function of the matrix A. The discussion will be confined to three dimensional matrices. If X_1 is one of the eigenvectors of A, then

$$AX_1 = \lambda_1 X_1, \qquad A^2 X_1 = A\lambda_1 X_1 = \lambda_1^2 X_1,$$
$$A^3 X_1 = \lambda_1^3 X_1, \qquad A^n X_1 = \lambda_1^n X_1 \tag{3.6.1}$$

Therefore, if $P(A)$ is applied to X_1, the result is

$$P(A)X_1 = P(\lambda_1)X_1 \tag{3.6.2}$$

Similarly, for the other two eigenvectors we have

$$P(A)X_2 = P(\lambda_2)X_2 \qquad P(A)X_3 = P(\lambda_3)X_3 \tag{3.6.3}$$

Let A be any tensor that has distinct eigenvalues. The eigenvectors are then a linearly independent set of vectors and any vector X can be expressed as

$$X = \alpha X_1 + \beta X_2 + \gamma X_3 \tag{3.6.4}$$

If $P(A)$ is allowed to operate on X

$$P(A)X = \alpha P(\lambda_1)X_1 + \beta P(\lambda_2)X_2 + \gamma P(\lambda_3)X_3 \tag{3.6.5}$$

Note that

$$(\lambda_3 I - A)X_1 = \lambda_3 X_1 - AX_1 = \lambda_3 X_1 - \lambda_1 X_1 = (\lambda_3 - \lambda_1)X_1;$$
$$(\lambda_2 I - A)(\lambda_3 I - A)X_1 = (\lambda_2 - \lambda_1)(\lambda_3 - \lambda_1)X_1;$$
$$(\lambda_2 I - A)(\lambda_3 I - A)X_2 = (\lambda_2 - \lambda_2)(\lambda_3 - \lambda_2)X_2 = 0$$
$$(\lambda_2 I - A)(\lambda_3 I - A)X_3 = (\lambda_2 - \lambda_3)(\lambda_3 - \lambda_3)X_3 = 0$$

Thus if a matrix G_1 is defined as

$$G_1 = \frac{(\lambda_2 I - A)(\lambda_3 I - A)}{(\lambda_2 - \lambda_1)(\lambda_3 - \lambda_1)} \tag{3.6.6}$$

it has the property that

$$G_1 X_1 = X_1 \qquad G_1 X_2 = 0 \qquad G_1 X_3 = 0 \tag{3.6.7}$$

Similarly, the matrices

$$G_2 = \frac{(\lambda_3 I - A)(\lambda_1 I - A)}{(\lambda_3 - \lambda_2)(\lambda_1 - \lambda_2)} \tag{3.6.8}$$

$$G_3 = \frac{(\lambda_1 I - A)(\lambda_2 I - A)}{(\lambda_1 - \lambda_3)(\lambda_2 - \lambda_3)} \tag{3.6.9}$$

have the properties

$$G_2 X_1 = 0 \qquad G_2 X_2 = X_2 \qquad G_2 X_3 = 0 \tag{3.6.10}$$

$$G_3 X_1 = 0 \qquad G_3 X_2 = 0 \qquad G_3 X_3 = X_3 \tag{3.6.11}$$

With the matrices G_i it is possible to convert (3.6.5) into the form

$$P(A)X = [P(\lambda_1)G_1 + P(\lambda_2)G_2 + P(\lambda_3)G_3]X \tag{3.6.12}$$

But X is an arbitrary vector. Therefore,

$$P(A) = P(\lambda_1)G_1 + P(\lambda_2)G_2 + P(\lambda_3)G_3 \tag{3.6.13}$$

If the matrix A is symmetric, the eigenvectors remain a linearly independent set even if some of the eigenvalues are equal. Assume that $\lambda_1 = \lambda_2 = \lambda$. The matrices

$$G = \frac{\lambda_3 I - A}{\lambda_3 - \lambda} \qquad G_3 = \frac{\lambda I - A}{\lambda - \lambda_3} \qquad (3.6.14)$$

have the property that

$$GX_1 = X_1 \qquad GX_2 = X_2 \qquad GX_3 = 0$$
$$G_3 X_1 = G_3 X_2 = 0 \qquad G_3 X_3 = X_3 \qquad (3.6.15)$$

In this case (3.6.5) becomes

$$P(A)X = [P(\lambda)G + P(\lambda_3)G_3]X \qquad (3.6.16)$$

which because of the arbitrariness of X implies

$$P(A) = P(\lambda)G + P(\lambda_3)G_3 \qquad (3.6.17)$$

This result applies *only* to a symmetric tensor with two eigenvalues equal. If *all three* of the eigenvalues are equal, (3.6.5) reduces to

$$P(A)X = P(\lambda)X \qquad (3.6.18)$$

which implies

$$P(A) = P(\lambda)I \qquad (3.6.19)$$

This result is true only for a symmetric tensor with *all* eigenvalues equal. For the special case where $P(A) = A$, Eq. (3.6.13) is

$$A = \lambda_1 G_1 + \lambda_2 G_2 + \lambda_3 G_3 \qquad (3.6.20)$$

The eigenvalues are sometimes called the *spectrum* and (3.6.20) is referred to as the *spectral resolution* of the matrix. That the same terminology is used here as is used in spectroscopy is no accident. The eigenvalues of the appropriate energy operator in quantum mechanics represent the electronic energy levels of an atom. By considering $P(A) = A^0 = I$ in (3.6.13) we find that $G_1 + G_2 + G_3 = I$.

For a symmetric matrix with all eigenvalues equal, Eq. (3.6.19) gives

$$A = \lambda I \qquad (3.6.21)$$

showing that A is diagonal in *all* representations.

It is as though a tensor were a supervector and the three eigenvalues were the components. The matrices G_1, G_2, and G_3 are called the *projections* of the tensor A and obey a sort of orthogonality condition. By means of the Cayley-Hamilton theorem we establish that

$$G_1G_2 = G_1G_3 = G_2G_3 = 0 \tag{3.6.22}$$

If both G_1 and $G_1{}^2$ are allowed to operate on the arbitrary vector X given by Eq. (3.6.4) and (3.6.7) is used,

$$G_1X = \alpha X_1 \qquad G_1{}^2X = \alpha X_1 \tag{3.6.23}$$

Hence,

$$G_1 = G_1{}^2 \tag{3.6.24}$$

Similarly,

$$G_2 = G_2{}^2 \qquad G_3 = G_3{}^2 \tag{3.6.25}$$

Equation (3.6.23) reveals that the application of G_1 to any vector yields its component in the direction of the eigenvector X_1.

3.7. FUNCTIONS OF MATRICES

Suppose that the polynomial

$$P(A) = I + A + \frac{1}{2!}A^2 + \frac{1}{3!}A^3 + \cdots + \frac{1}{N!}A^N \tag{3.7.1}$$

is substituted into (3.6.13). In the limit as N becomes infinite

$$P(\lambda_1) = e^{\lambda_1} \qquad P(\lambda_2) = e^{\lambda_2} \qquad P(\lambda_3) = e^{\lambda_3} \tag{3.7.2}$$

In this way it becomes possible to define

$$e^A = e^{\lambda_1}G_1 + e^{\lambda_2}G_2 + e^{\lambda_3}G_3 \tag{3.7.3}$$

Since

$$e^\lambda = \sum_{n=0}^{\infty} \frac{1}{n!} \lambda^n \tag{3.7.4}$$

converges for all λ, the convergence of

$$e^A = \sum_{n=0}^{\infty} \frac{1}{n!} A^n \tag{3.7.5}$$

is guaranteed if the matrix A satisfies equation (3.6.13), i.e., if the eigenvectors of A form a linearly independent set. The series (3.7.5) may be used to define e^A for other matrices but some other means is then required to establish convergence. It is possible to prove that (3.7.5) converges for *all* matrices, but this will not be carried out here.

If A and B commute, then

$$e^A \cdot e^B = e^{A+B} \tag{3.7.6}$$

This can be proved from the series expansion:

$$e^A \cdot e^B = \left(I + A + \frac{1}{2!}A^2 + \frac{1}{3!}A^3 + \cdots\right)\left(I + B + \frac{1}{2!}B^2 + \frac{1}{3!}B^3 + \cdots\right)$$

$$= I + B + \frac{1}{2!}B^2 + \frac{1}{3!}B^3 + \cdots$$

$$+ A + AB + \frac{1}{2!}AB^2 + \cdots$$

$$+ \frac{1}{2!}A^2 + \frac{1}{2!}A^2B + \cdots \tag{3.7.7}$$

$$+ \frac{1}{3!}A^3 + \cdots$$

$$= I + (A + B) + \frac{1}{2!}(A + B)^2 + \frac{1}{3!}(A + B)^3 + \cdots$$

$$= e^{A+B}$$

If α is a scalar function of t and A is a constant matrix, then

$$\frac{d}{dt}e^{\alpha A} = \frac{d}{dt}\left(I + \alpha A + \frac{1}{2!}\alpha^2 A^2 + \frac{1}{3!}\alpha^3 A^3 + \cdots\right)$$

$$= \frac{d\alpha}{dt}A + \frac{1}{2!}2\alpha\frac{d\alpha}{dt}A^2 + \frac{1}{3!}3\alpha^2\frac{d\alpha}{dt}A^3 + \cdots$$

$$= \frac{d\alpha}{dt}A\left(I + \alpha A + \frac{1}{2!}\alpha^2 A^2 + \cdots\right) \tag{3.7.8}$$

$$= \frac{d\alpha}{dt}Ae^{\alpha A}$$

PROBLEMS

3.1.1 Write the matrix equivalent of all the following component equations:

$$b_i = A_{ij}a_j \qquad C_i = A_{ji}a_j \qquad C_{ik} = A_{ij}B_{jk}$$
$$D_{ik} = A_{ij}B_{kj} \qquad E_{ik} = A_{ji}B_{jk} \qquad F_{ik} = A_{ji}B_{jk}$$

3.1.2 Suppose $A_{ij}a_j = B_{ij}a_j$ is known to be true for vectors that have $a_1 = 0$ but otherwise are arbitrary. What can be said about the relation of A_{ij} to B_{ij}?

3.1.3 If A and B are the matrices

$$A = \begin{pmatrix} 1 & 0 & 2 \\ 3 & 1 & 0 \\ 0 & 1 & 4 \end{pmatrix} \qquad B = \begin{pmatrix} 1 & 0 & 0 \\ 2 & 0 & 1 \\ 0 & 1 & 3 \end{pmatrix}$$

and $\alpha = 3$, find AB, αA, αAB, $A(\alpha B)$.

3.1.4 Is it true that $\dfrac{d}{dt} A^2 = 2A \dfrac{dA}{dt}$?

3.1.5 Show that $I = \begin{pmatrix} 1 & 0 \\ 0 & 1 \end{pmatrix}$ and $A = \begin{pmatrix} 0 & 1 \\ -1 & 0 \end{pmatrix}$ have all the algebraic

properties of 1 and $i = \sqrt{-1}$. Hence, the matrix representation of a complex

number $z = x + iy$ is $Z = xI + yA = \begin{pmatrix} x & y \\ -y & x \end{pmatrix}$

3.1.6 Write down the three matrices that represent (a) a rotation of ϕ about the x_3-axis, (b) a rotation by θ about the x_1'-aixs of the coordinate system obtained in part (a), (c) a rotation by ψ about the x_3''-axis. Find the product $S = S_a S_b S_c$. S then represents a general proper orthogonal transformation depending on three parameters ϕ, θ, and ψ. These are called the *Euler angles*.

3.1.7 Show by matrix methods that symmetry and antisymmetry of second rank tensors are invariant properties.

3.1.8 Does a matrix commute with its own transpose?

3.2.9 Given a vector $(a_i) = (1, 1, 0)$, what vector $\mathbf{b}$ is produced from it by the singular matrix

$$A = \begin{pmatrix} 1 & 0 & 0 \\ 0 & 0 & 1 \\ 2 & 0 & 3 \end{pmatrix}$$

Suppose $\mathbf{b}$ is known beforehand. What information does $b_i = A_{ij} a_j$ give about $\mathbf{a}$? Does the transformation make sense for *any* prior choice of $\mathbf{b}$?

3.2.10 Prove that $\widetilde{ABC} = \tilde{C}\tilde{B}\tilde{A}$ and that $(ABC)^{-1} = C^{-1}B^{-1}A^{-1}$.

3.2.11 If α is a scalar and A is a 3×3 square matrix, show that $\det(\alpha A) = \alpha^3 \det A$. What is the corresponding result if A is $n \times n$?

3.2.12 If A is a second rank tensor and A^{-1} exists, prove that A^{-1} is also a second rank tensor under orthogonal transformation. What is the transformation property of the matrix C formed from the cofactors of A, Eq. (2.20)?

3.2.13 Show that any antisymmetric tensor of rank 2 and odd dimension is singular.

3.4.14 If A is a non-singular symmetric matrix with eigenvalues λ_1, λ_2, and λ_3, show that the eigenvalues of A^{-1} are $1/\lambda_1$, $1/\lambda_2$, and $1/\lambda_3$.

3.3.15 Show formally that two orthogonal transformations that represent rotations by θ and θ' about the *same axis* commute, i.e., are equivalent to a single rotation of $\theta + \theta'$.

3.4.16 Discuss the eigenvalue problem for the symmetric tensor A if the central quadric $\phi = x_i A_{ij} x_j$ represents a *sphere* when $\phi = $ const.

3.4.17 Consider the symmetric tensor

$$A = \begin{pmatrix} \frac{2}{3} & 1 & 0 \\ 1 & \frac{2}{3} & 0 \\ 0 & 0 & 1 \end{pmatrix}$$

(a) Calculate the eigenvalues. (b) Determine the normalized eigenvectors. (c) From the normalized eigenvectors, construct the orthogonal transformation which will diagonalize the tensor. (d) Find the angle of rotation induced by the transformation.

3.4.18 Find the eigenvalues and eigenvectors of

$$A = \begin{pmatrix} \frac{2}{3} & 1 & 0 \\ -1 & \frac{2}{3} & 0 \\ 0 & 0 & 1 \end{pmatrix}$$

Compute the inverse of A by using the Cayley-Hamilton theorem. *Hint:* Write the Cayley-Hamilton theorem as

$$A^2 + \alpha_2 A + \alpha_1 I + \alpha_0 A^{-1} = 0$$

Compute α_0, α_1, α_2 in terms of the eigenvalues.

3.4.19 Show that the matrices

$$A = \begin{pmatrix} 2 & 0 & 0 & 0 \\ 0 & 2 & 0 & 0 \\ 0 & 0 & 2 & 1 \\ 0 & 0 & 1 & 2 \end{pmatrix} \qquad B = \begin{pmatrix} 1 & 3 & 0 & 0 \\ 3 & 1 & 0 & 0 \\ 0 & 0 & 1 & 2 \\ 0 & 0 & 2 & 1 \end{pmatrix}$$

commute. Find the eigenvalues of each. Construct a four-dimensional orthogonal transformation which diagonalizes both A and B.

3.4.20 If S_{ij} are the coefficients of a proper orthogonal transformation, show that S_{ij} is equal to its own cofactor.

3.4.21 Excluding the trivial case of uniform translation, all proper orthogonal transformations are rotations. The product of two improper transformations is proper and, hence, is equivalent to a rotation. Show that the product of the reflections in the planes

$$x_2 \cos \alpha = x_1 \sin \alpha$$

$$x_2 \cos \beta = x_1 \sin \beta$$

is equivalent to a rotation by $\theta = 2(\alpha - \beta)$.

3.5.22 Show that the eigenvectors of

$$\begin{pmatrix} \frac{2}{3} & 1 & 0 \\ 0 & \frac{2}{3} & 0 \\ 0 & 0 & 1 \end{pmatrix}$$

do not form a linearly independent set of vectors. Verify that the Cayley-Hamilton theorem holds for this matrix.

3.5.23   Verify by direct calculation that the Cayley-Hamilton theorem holds for any matrix of the form

$$A = \begin{pmatrix} 0 & A_{12} & A_{13} \\ 0 & 0 & A_{23} \\ 0 & 0 & 0 \end{pmatrix}$$

3.6.24   If $A = \begin{pmatrix} \lambda & 0 & 0 \\ 0 & \lambda & 0 \\ 0 & 0 & \lambda \end{pmatrix}$, show that $e^A = \begin{pmatrix} e^\lambda & 0 & 0 \\ 0 & e^\lambda & 0 \\ 0 & 0 & e^\lambda \end{pmatrix}$

3.6.25   Compute the inverse of the matrix A of Problem 3.4.17 by using

$$A^{-1} = \frac{1}{\lambda_1} G_1 + \frac{1}{\lambda_2} G_2 + \frac{1}{\lambda_3} G_3$$

3.7.26   If $A = \begin{pmatrix} 0 & A_{12} & A_{13} \\ 0 & 0 & A_{23} \\ 0 & 0 & 0 \end{pmatrix}$, show that all three eigenvalues are zero.

Calculate e^A by working out the expansion $e^A = I + A + \frac{1}{2}A^2 + \cdots$. (All powers of A higher than A^2 are zero.) Show that Eq. (3.6.19) *does not* work in this case.

3.7.27   If $i = \sqrt{-1}$ prove that $e^{iA} = \cos A + i \sin A$.

3.7.28   Prove that $\cos^2 A + \sin^2 A = I$.

3.7.29   For the matrix A of Problem 3.4.17 find $A^{1/2}$ and $A^{1/3}$ and verify by matrix multiplication that $A^{1/2}A^{1/2} = A$.

3.7.30 $dx_i/dt = A_{ij}x_j$ is a system of three differential equations for $x_1(t)$, $x_2(t)$, and $x_3(t)$. Convert the system to an equivalent matrix differential equation and solve. Treat the A_{ij} as constants.

3.7.31   Let N be a unit vector, i.e., $\tilde{N}N = 1$. Any vector X can be expressed as $X = \alpha N + Y$ where Y is perpendicular to N. Show that the matrix $G = N\tilde{N}$ has the property that when applied to any vector it takes the projection of that vector in the direction of N. Note also that $G^2 = G$. The matrix G is called a *projection operator*.

3.7.32   If A is a symmetric tensor with eigenvectors X_1, X_2, X_3 and eigenvalues λ_1, λ_2, λ_3 show that A can be expressed

$$A = X_1\tilde{X}_1\lambda_1 + X_2\tilde{X}_2\lambda_2 + X_3\tilde{X}_3\lambda_3$$

and hence that $G_1 = X_1\tilde{X}_1$, $G_2 = X_2\tilde{X}_2$, $G_3 = X_3\tilde{X}_3$ provided that the eigenvectors are normalized:

$$\tilde{X}_1X_1 = \tilde{X}_2X_2 = \tilde{X}_3X_3 = 1$$

This won't work if A is non-symmetric even if the eigenvectors are linearly independent. Why not?

3.7.33   Given the matrix

$$A = \begin{pmatrix} 0 & 1 & 1 \\ 1 & 0 & 0 \\ 1 & 0 & 0 \end{pmatrix}$$

Solve the following problems:
(a) Find the eigenvalues and normalized eigenvectors.
(b) Find the matrices G_1, G_2, and G_3.
(c) Find the general solutions of the following simultaneous linear differential equations:

$$\dot{x}_1 = x_2 + x_3 \qquad \dot{x}_2 = x_1 \qquad \dot{x}_3 = x_1$$

3.7.34   If A is a 2×2 matrix with two linearly independent eigenvectors, define appropriate matrices G_1 and G_2 analogous to equations (3.6.6), (3.6.7), and (3.6.8). If A is symmetric, show that G_1 and G_2 can be written in terms of the eigenvectors as in Problem 3.7.32. If

$$A = \begin{pmatrix} 1 & 2 \\ 2 & 1 \end{pmatrix}$$

find G_1 and G_2. Find the general solution of the system of linear differential equations

$$\dot{x}_1 = x_1 + 2x_2$$
$$\dot{x}_2 = 2x_1 + x_2$$

3.7.35   Find the eigenvalues of the matrix

$$A = \begin{pmatrix} -a & 0 & 0 \\ a & -b & 0 \\ 0 & b & -c \end{pmatrix}$$

Compute the projection operators G_1, G_2, and G_3. Three radioactive nuclei decay successively so that the numbers $N_i(t)$ of the three types obey the equations

$$\frac{dN_1}{dt} = -aN_1 \qquad \frac{dN_2}{dt} = aN_1 - bN_2 \qquad \frac{dN_3}{dt} = bN_2 - cN_3$$

where the rate constants a, b, and c are all different. If initially $N_1 = N, N_2 = 0$, and $N_3 = n$, find $N_i(t)$ as functions of time.

REFERENCES

1. Perlis, Sam, *Theory of Matrices*, Addison-Wesley, Cambridge, Mass., 1952.
2. Micheal, A. D., *Matrix and Tensor Calculus*, John Wiley and Sons, New York, 1955.

4

The One-Dimensional Motion
of a Particle

This chapter examines the mechanics of a single particle of constant mass
that is restricted to move along a straight line and other physical systems
which obey identical differential equations.

4.1. NEWTON'S LAWS OF MOTION

It is assumed that from elementary courses in physics the student is
familiar with the basic concepts of force and mass. In Chapter 1 we
studied the problem of expressing Newton's second law of motion for a
single particle of mass m so that it is covariant with respect to an orthogonal
transformation, Eqs. (1.8.38) and (1.8.39). In Chapter 2 the covariance
idea was extended to included arbitrary coordinate transformation in
three-dimensional Euclidean space. Either Eq. (2.21.1) or Eq. (2.21.5)
can be considered as a general covariant statement of Newton's second
law. In other words, at this stage of development the kinematics of
particle motion and essential accompanying geometrical concepts have
been completely treated. We now concern ourselves with the problem of
integrating the equations of motion in the simplest cases.

In the strict sense, a particle is a mathematical point. A point particle
is assumed to have mass but no extension in space or structure. There is
no such thing in nature as a true point particle but the actual motion of
real particles closely approximates that of the mathematical idealization
in many circumstances. For instance, a proton has a dimension of
the order of 10^{-13} cm. If the proton is being accelerated in a machine the

dimensions of which are several meters, it is quite accurate to treat the proton as a point particle. We concern ourselves exclusively with particles of constant mass. The problems of particles with variable mass is taken up in Chapter 8.

The following table reviews the three common systems of mechanical units:

	Length	Mass	Time	Force
mks	meter	kilogram	second	newton
cgs	centimeter	gram	second	dyne
English	foot	slug	second	pound

The conversion factors are

$$1 \text{ slug} = 14.594 \text{ kg}$$
$$1 \text{ meter} = 3.2808 \text{ ft}$$
$$1 \text{ newton} = 0.22480 \text{ lb}$$

The *momentum* of a particle of mass m is defined to be

$$\mathbf{p} = m\mathbf{u} \qquad (4.1.1)$$

where $\mathbf{u}$ is the particle velocity. Newton's second law of motion can be stated as

$$\mathbf{F} = \frac{d\mathbf{p}}{dt} \qquad (4.1.2)$$

where $\mathbf{F}$ is the force acting on the particle. Newton's first law is a special case of the second law and says that the momentum of a particle is a constant if the net force on the particle is zero. In other words, a particle moving in a straight line with a constant speed will continue to do so in the absence of any force. A particle with constant momentum is said to be in equilibrium.

In elementary texts, Newton's third law is usually stated by saying that when a body exerts a force on a second body the second body exerts a force on the first which is equal in magnitude but opposite in direction. This statement of the third law suffices for most simple problems in mechanics but in complex situations involving the interactions of many particles it is better to reformulate it. This reformulation is taken up in Chapter 8.

Newton's laws are based on experiment and cannot be proved or derived. The basic quantities which appear in Newton's second law, namely, length, time, mass, and force are difficult if not impossible to define. Length and time are perhaps the easiest concepts to grasp, at least intuitively. Mass is a more subtile concept. We learn from our elementary mechanics course that mass is a measure of *inertia*, i.e., the ease or difficulty with which we can change the state of motion of a body. However, in the last analysis, no one can really say what mass *is*. Force can, of course, be defined in terms of mass, length, and time by means of Eq. (4.1.2). Any one of the four basic quantities could be defined in terms of the other three by means of Newton's second law.

From where do the forces come that act on a particle? There are actually only two fundamental kinds of forces which properly belong to the domain of classical, i.e., Newtonian, mechanics; namely, gravitational and electromagnetic. All other apparently different forces are manifestations of these basic two. So-called "contact forces" between two bodies ultimately result from the electromagnetic forces (mainly electrostatic in this case) between the charged particles that make up the atoms of the body. The viscous frictional force experienced by a body moving through a fluid must also result from the electromagnetic forces between the atomic particles. It is frequently difficult and impractical to derive a macroscopic force by averaging over the electromagnetic forces between submicroscopic particles; hence, certain simple functions for forces are either assumed or obtained from experiment. As examples, a particle sliding across a table top is found to experience a frictional force that is approximately proportional to the normal force between particle and table; a particle falling through the air experiences a viscous frictional force that is approximately proportional to the velocity of the particle; the force exerted by a stretched spring is approximately proportional to the amount the spring is stretched.

In all of physics there are only four basic forces that are at present known. In addition to the two already mentioned, there is the nuclear force responsible for holding together the nucleons in an atomic nucleus and the so-called weak interactions which are forces that are observed to exist between certain elementary particles. The weak interactions are intermediate in strength between gravitational and electromagnetic forces and nuclear forces are the strongest of all. These forces are entirely out of the realm of classical mechanics.

In the remainder of this chapter, the mechanics of a particle that is constrained to move in one dimension, i.e., along a straight line, will be treated. Since Newton's laws are covariant, it is always possible to orient a Cartesian coordinate system in such a way that the particle moves parallel to a coordinate axis. The motion, therefore, will be given in terms of a single variable x which depends on the time. This type of motion is referred to as *rectilinear*. If the motion of a system depends on only one coordinate, the system is said to have one degree of freedom.

The most general type of force which will be considered is one which depends on position, velocity, and time. The equation of motion

$$m\ddot{x} = F(x, \dot{x}, t) \tag{4.1.3}$$

is a second order differential equation for x as a function of time. The solution involves two arbitrary constants which can be determined from the initial position and velocity of the particle. For all functions $F(x, \dot{x}, t)$ which occur in practice, it can be shown that the solutions of (4.1.3) is unique once the initial position and velocity of the particle are given. For a mathematical proof, reference can be made to any text on differential equations.

4.2. THE ENERGY INTEGRAL

Let a and b refer to any two points on the trajectory of the particle. The work done between these two points by the net force acting on the particle is defined to be

$$W_{ab} = \int_a^b F \, dx = \int_a^b m\ddot{x} \, dx = \int_a^b \frac{d}{dt} (\tfrac{1}{2}mu^2) \, dt$$

$$= \tfrac{1}{2}mu_b^2 - \tfrac{1}{2}mu_a^2 = T_b - T_a \tag{4.2.1}$$

Thus the work done by the net force on the particle is equal to the increase in the kinetic energy of the particle. Since the particle is constrained to move in one dimension, the net force is necessarily either parallel or antiparallel to the direction of motion. Thus F could be either positive or negative and T_b could be either greater than or less than T_a in Eq. (4.2.1).

If the force is a function of x and t but not the velocity, it is possible to write

$$F(x, t) = -\frac{\partial V(x, t)}{\partial x} \tag{4.2.2}$$

The function $V(x, t)$ is known as the *potential energy* of the particle. Equation (4.2.2) does not uniquely determine $V(x, t)$. If

$$V'(x, t) = V(x, t) + C$$

where C is a constant, then $F = -\partial V'/\partial x$ gives exactly the same force. Only *differences* in potential ever appear in any problem, and at a given point the value of the potential can be arbitrarily chosen by appropriate adjustment of the constant C. If the force has no explicit time dependence, then

$$F(x) = -\frac{dV(x)}{dx} \tag{4.2.3}$$

and in terms of $V(x)$ the work done by the force $F(x)$ between the points a and b is

$$W_{ab} = \int_a^b F \, dx = -\int_a^b \frac{dV}{dx} \, dx = -V_b + V_a \qquad (4.2.4)$$

Combining this with the expression for W_{ab} in terms of the kinetic energy yields

$$T_a + V_a = T_b + V_b \qquad (4.2.5)$$

The total energy of the particle is defined as

$$\mathscr{E} = T + V = \tfrac{1}{2}mu^2 + V \qquad (4.2.6)$$

and (4.2.5) shows that $\mathscr{E}$ is a *constant* whenever the force does not depend on the time. Such forces are said to be *conservative*.

If the force is time dependent, then the total differential of the potential energy is

$$dV = \frac{\partial V}{\partial x} \, dx + \frac{\partial V}{\partial t} \, dt$$
$$= -F \, dx + \frac{\partial V}{\partial t} \, dt \qquad (4.2.7)$$

Thus (4.2.4) no longer holds because of the presence of the extra term $(\partial V/\partial t) \, dt$. The definition (4.2.6) for the total energy of the particle can still be used, but it is no longer a constant:

$$\frac{d\mathscr{E}}{dt} = \frac{d}{dt}(\tfrac{1}{2}mu^2) + \frac{\partial V}{\partial x}\frac{dx}{dt} + \frac{\partial V}{\partial t}$$
$$= m\dot{x}\ddot{x} + \frac{\partial V}{\partial x}\dot{x} + \frac{\partial V}{\partial t} \qquad (4.2.8)$$
$$= F\dot{x} - F\dot{x} + \frac{\partial V}{\partial t} = \frac{\partial V}{\partial t}$$

Thus the time rate of change of the total energy is the partial derivative of $V(x, t)$ with respect to t.

The units of energy are *joules* in the mks system, *ergs* in the cgs system, and *ft-lb* in the English system.

4.3. MOTION IN A UNIFORM FIELD

If the range of motion of a particle near the surface of the earth is not too great, the gravitational field can be considered as approximately uniform. If x measures the altitude of the particle above the surface of the earth and if air resistance is neglected the equations of motion for a particle moving in a vertical direction are

$$m\ddot{x} = -mg \qquad (4.3.1)$$

$$u = \dot{x} = -gt + u_0 \qquad (4.3.2)$$

$$x = -\tfrac{1}{2}gt^2 + u_0 t + x_0 \qquad (4.3.3)$$

where x_0 and u_0 are the initial ($t = 0$) values of x and u. A similar set of equations results if a charged particle is placed in the uniform electric field between two condenser plates. A discussion of these equations is to be found in any elementary mechanics text. In this section, the theory is extended to include the effects of the viscous frictional force experienced by the particle as it moves through the air.

If the velocity is not too great, the frictional force is approximately proportional to the velocity of the particle and is in a direction opposite to the direction of the velocity. Equation (4.3.1) is replaced by the more accurate equation

$$m\ddot{x} = -mg - b\dot{x} \qquad (4.3.4)$$

The constant b can be regarded as experimentally determined from observations on a body falling through the air. Its value will depend on the size and shape of the particular body. The value of b is very large for a thing like a parachute and small for a steel ball bearing. The integration can be carried out by means of an integrating factor:

$$e^{bt/m}\left(\ddot{x} + \frac{b}{m}\dot{x}\right) = -ge^{bt/m}$$

$$\frac{d}{dt}(\dot{x}e^{bt/m}) = -ge^{bt/m}$$

Suppose that observations on the motion start at $t = 0$. It is then convenient to integrate between $t = 0$ and some later time t:

$$\int_0^t d(\dot{x}e^{bt/m}) = -\int_0^t ge^{bt/m}\,dt$$

$$\dot{x}e^{bt/m}\Big|_0^t = -\frac{gm}{b}e^{bt/m}\Big|_0^t$$

$$\dot{x}e^{bt/m} - u_0 = -\frac{gm}{b}[e^{bt/m} - 1]$$

where u_0 is the velocity at $t = 0$. Solution for the velocity yields

$$\dot{x} = \frac{gm}{b}\left[e^{-bt/m}\left(1 + \frac{bu_0}{gm}\right) - 1\right] \qquad (4.3.5)$$

After a long period of time, $e^{-bt/m} \to 0$.

The velocity of a particle eventually reaches a *terminal value* given by

$$u_t = -\frac{gm}{b} \qquad (4.3.6)$$

If b is small, then over the initial part of the motion

$$\frac{bt}{m} \ll 1 \qquad (4.3.7)$$

The symbol $\ll$ means "much less than." Under these circumstances, the exponential in (4.3.5) can be approximated by a few terms of its Taylor's series expansion:

$$\dot{x} = \frac{gm}{b}\left[\left(1 - \frac{bt}{m} + \frac{1}{2}\frac{b^2t^2}{m^2}\right)\left(1 + \frac{bu_0}{gm}\right) - 1\right] \qquad (4.3.8)$$

A rearrangement of terms yields

$$\dot{x} = u_0 - gt + \frac{gbt^2}{2m} - \frac{bu_0t}{m} \qquad (4.3.9)$$

where all terms of order b^2 or higher have been discarded. Note that because of the term gm/b in front of the brackets in (4.3.8), terms to order b^2 must be carried in the expansion of the exponential in order finally to end up with an expansion correct to terms linear in b. Equation (4.3.9) shows that when $b = 0$, Eq. (4.3.2) is obtained

Equation (4.3.5) may be written

$$dx = \frac{gm}{b}\left[e^{-bt/m}\left(1 + \frac{bu_0}{mg}\right) - 1\right] dt \qquad (4.3.10)$$

and integrated between the limits 0 and t to get

$$x - x_0 = \frac{gm}{b}\left[\frac{m}{b}\left(1 - e^{-bt/m}\right)\left(1 + \frac{bu_0}{mg}\right) - t\right] \qquad (4.3.11)$$

In order to obtain an approximation from (4.3.11) which is linear in b, it is necessary to carry the expansion of the exponential to order b^3. The result is

$$x - x_0 = u_0t - \tfrac{1}{2}gt^2 + \frac{bt^3g}{6m} - \frac{bt^2u_0}{2m} \qquad (4.3.12)$$

This result is also obtainable by direct integration of (4.3.9). Note that for $b = 0$, (4.3.12) reduces to the friction-free case given by Eq. (4.3.3). It is important to recognize that (4.3.12) and (4.3.9) become inaccurate after a sufficiently long time, no matter how small b is. Clearly, the condition (4.3.7) will eventually be violated.

Given the initial velocity u_0 and the initial height x_0 equation (4.3.11) provides the height x of the particle at any time. Suppose instead the height is known and it is desired to know the time. Let the particle be dropped from a height x_0 with zero initial velocity. The quantity $\tau = m/b$ has the dimensions of time and is a characteristic constant for the problem sometimes called the *time constant*. The distance $h = x_0 - x$ through which the particle has dropped in time t is then

$$h = -g\tau[\tau(1 - e^{-t/\tau}) - t] \qquad (4.3.13)$$

Since it is not possible to solve for t as a simple function of h, it is necessary to resort to some approximate procedure. If the exponential is expanded and terms are retained to t^4, there results the series

$$h = \tfrac{1}{2}gt^2\left[1 - \frac{1}{3}\frac{t}{\tau} + \frac{1}{12}\frac{t^2}{\tau^2} - \cdots\right] \qquad (4.3.14)$$

which represents h accurately as long as terms of higher power than $(t/\tau)^2$ are small. The very crudest approximation is

$$h = \tfrac{1}{2}gt^2 \qquad (4.3.15)$$

which is solved for t to yield

$$t_1 = \sqrt{2h/g} \qquad (4.3.16)$$

where the subscript 1 has been used to indicate that this is the first approximation to the time in terms of the distance the particle has fallen.

A *second* approximation is obtained by retaining one more term in the expansion:

$$h = \tfrac{1}{2}gt^2\left[1 - \frac{1}{3}\frac{t_1}{\tau}\right] \qquad (4.3.17)$$

Since the second term is only a correction, no great error is committed by replacing t by the first approximation t_1. The second approximation is found by solving (4.3.17) for t:

$$t^2 = \frac{2h}{g}\frac{1}{1 - t_1/(3\tau)} \qquad (4.3.18)$$

$$t_2 = \sqrt{\frac{2h}{g}}\left(1 - \frac{t_1}{3\tau}\right)^{-1/2} = t_1\left(1 + \frac{t_1}{6\tau}\right) \qquad (4.3.19)$$

where the binomial expansion has been used.

To obtain a third approximation, three terms in the expansion for h are retained:

$$h = \tfrac{1}{2}gt^2\left[1 - \frac{1}{3\tau}\left(t_1 + \frac{t_1^2}{6\tau}\right) + \frac{t_1^2}{12\tau^2}\right] \qquad (4.3.20)$$

Here t in the second term has been replaced by its second approximation whereas only the first approximation is necessary for t in the third term. Solution of (4.3.20) for t yields

$$t_3 = \sqrt{\frac{2h}{g}} \left[1 - \frac{t_1}{3\tau} + \frac{1}{36} \frac{t_1^2}{\tau^2} \right]^{-1/2} \qquad (4.3.21)$$

By using the binomial expansion

$$
\begin{aligned}
t_3 &= t_1 \left[1 - \frac{1}{2} \left(-\frac{t_1}{3\tau} + \frac{1}{36} \frac{t_1^2}{\tau^2} \right) + \frac{3}{8} \left(-\frac{t_1}{3\tau} \right)^2 \right] \\
&= t_1 \left[1 + \frac{t_1}{6\tau} + \frac{t_1^2}{36\tau^2} \right]
\end{aligned}
\qquad (4.3.22)
$$

The process can be continued to higher approximations. Clearly, if the third term in (4.3.22) amounts to only a small percentage, there is no point in going any further. More terms are needed in the original expansion (4.3.14) if it is necessary to continue. The process just carried out is the inversion of a power series.

The system is not conservative due to the presence of the frictional force $-b\dot{x}$. It is nevertheless possible to obtain an energy integral. If the equation of motion (4.3.4) is multiplied by $\dot{x}$, the result is

$$m\ddot{x}\dot{x} + mg\dot{x} = -b\dot{x}^2$$

$$\frac{d}{dt}(\tfrac{1}{2}m\dot{x}^2 + mgx) = -b\dot{x}^2 \qquad (4.3.23)$$

If the same expression is used for the energy as is used in the case where the frictional force is absent, then

$$\frac{d\mathscr{E}}{dt} = -b\dot{x}^2 \qquad (4.3.24)$$

which gives the rate at which the energy of the particle is dissipated as heat due to the friction.

There are analogous results for a charged particle being accelerated by a uniform electric field. It is known from electrodynamics that such accelerated particles lose energy in the form of electromagnetic radiation. This radiated energy is at the expense of the mechanical energy of the charged particle. The effect on the motion is qualitatively similar to the one just studied for a particle encountering a viscous frictional force and is known as *radiation damping*. The effect is extremely small for many situations encountered in the laboratory.

4.4. THE SIMPLE HARMONIC OSCILLATOR

A simple example of a linear harmonic oscillator is a particle of mass m suspended from a spring of force constant k, Fig. 4.4.1. When the particle hangs motionless, Fig. 4.4.1b, the weight of the particle is exactly balanced by the force which the spring exerts. If the spring has been stretched an amount s,

$$ks = mg \qquad (4.4.1)$$

Suppose the particle has been set into vertical motion, Fig. 4.4.1c. At the instant when the displacement from the equilibrium position is x,

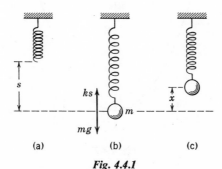

(a)   (b)   (c)

Fig. 4.4.1

the extension of the spring is $s - x$ and the equation of motion, neglecting any friction, is

$$F = k(s - x) - mg = m\ddot{x} \qquad (4.4.2)$$

By using Eq. (4.4.1), the equation of motion is reduced to

$$m\ddot{x} + kx = 0 \qquad (4.4.3)$$

When the particle is displaced from its equilibrium position, it experiences a *restoring force* that is proportional to the displacement. Motion which results from this type of force is called *simple harmonic*.

The potential function from which the force $F = -kx$ is derivable is

$$V(x) = \tfrac{1}{2}kx^2 \qquad F(x) = -\frac{dV}{dx} \qquad (4.4.4)$$

and the total energy is

$$\mathscr{E} = \tfrac{1}{2}mu^2 + \tfrac{1}{2}kx^2 \qquad (4.4.5)$$

and is a constant. The harmonic oscillator with no friction is a simple example of a *conservative system*. From (4.4.5) it is evident that $\mathscr{E} \geqslant 0$. Since

$$\tfrac{1}{2}mu^2 = \mathscr{E} - \tfrac{1}{2}kx^2 \qquad (4.4.6)$$

the potential energy can never be larger than the total energy; otherwise u^2 becomes negative which is not possible. Figure 4.4.2 shows a plot of $V(x)$. A possible total energy is marked on the $V(x)$ axis. The range of the motion is indicated by the horizontal dashed line. The particle cannot go beyond the points where this line intersects the curve $V(x)$, indicated by $\pm A$, since motion beyond these points would violate the condition that $V(x)$ never be larger than $\mathscr{E}$. These points are the *turning points* of the

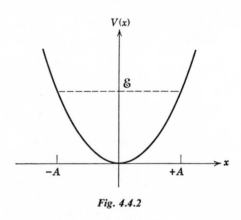

Fig. 4.4.2

motion and A is the *amplitude*. At $x = \pm A$ the potential and total energies are equal:

$$\mathscr{E} = \tfrac{1}{2}kA^2 \qquad (4.4.7)$$

meaning that the velocity of the particle is instantaneously zero at these points. The particle travels out to its maximum possible displacement A, stops, reverses direction, and travels to $-A$ where it again stops and reverses direction. The motion is *periodic*, repeating itself over and over again. A qualitative study of the motion of a particle based on the energy equation such as that just given is frequently very useful, especially in those cases where the exact solution of the differential equation is difficult.

The solution of the differential equation (4.4.3) containing two arbitrary constants of integration c_1 and c_2 is

$$x = c_1 \cos \omega_0 t + c_2 \sin \omega_0 t \qquad (4.4.8)$$

where

$$\omega_0 = \sqrt{k/m} \qquad (4.4.9)$$

is usually expressed in radians per second. The radian measure itself is a pure number having no units meaning that the actual units of ω_0 are

$\sec^{-1}$. The frequency ν_0 and period P_0 of the motion are given by

$$\omega_0 = 2\pi\nu_0 \qquad P_0 = \frac{1}{\nu_0}$$

$$\nu_0 = \frac{1}{2\pi}\sqrt{k/m} \qquad P_0 = 2\pi\sqrt{m/k} \tag{4.4.10}$$

Sometimes ω_0 is called the *circular frequency*. Quite often ω_0 is referred to simply as "the frequency" even though the actual frequency is really ν_0. The velocity of the particle is

$$\dot{x} = -\omega_0 c_1 \sin \omega_0 t + \omega_0 c_2 \cos \omega_0 t \tag{4.4.11}$$

and by setting $t = 0$ in the equations for x and $\dot{x}$, the constants c_1 and c_2 can be evaluated in terms of the initial position and velocity of the particle:

$$x_0 = c_1 \qquad u_0 = \omega_0 c_2 \tag{4.4.12}$$

The solution expressed in terms of x_0 and u_0 is

$$x = x_0 \cos \omega_0 t + \frac{u_0}{\omega_0} \sin \omega_0 t \tag{4.4.13}$$

The solution can be expressed in a different way. Let

$$A = \sqrt{x_0^2 + \frac{u_0^2}{\omega_0^2}} \tag{4.4.14}$$

and write (4.4.13) as

$$x = A\left[\frac{x_0}{A} \cos \omega_0 t + \frac{u_0}{A\omega_0} \sin \omega_0 t\right] \tag{4.4.15}$$

It is always possible to construct a right triangle with two sides of lengths x_0 and u_0/ω_0. From the geometry of Fig. 4.4.3,

$$\begin{aligned} x &= A[\cos \alpha \cos \omega_0 t + \sin \alpha \sin \omega_0 t] \\ &= A \cos(\omega_0 t - \alpha) \end{aligned} \tag{4.4.16}$$

α can be conveniently determined by

$$\tan \alpha = \frac{u_0}{x_0\omega_0} \tag{4.4.17}$$

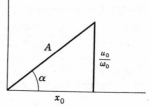

Fig. 4.4.3

This is an alternate expression for the solution in terms of the constants of integration A and α in place of c_1 and c_2. A represents the maximum value of x and is the amplitude of the motion. The quantity

$$\psi = \omega_0 t - \alpha \tag{4.4.18}$$

is called the *phase* and α is a phase-angle which is to be determined from the initial conditions. α can be any value, positive or negative, depending on the signs and magnitudes of x_0 and u_0.

4.5. THE HARMONIC OSCILLATOR WITH A SINUSOIDAL DRIVING FORCE

If a linear harmonic oscillator is driven by an external force which is a function of the time, the equation of motion is

$$m\ddot{x} + kx = F(t) \tag{4.5.1}$$

and is a linear nonhomogeneous, second order differential equation with constant coefficients. An important special case results if the driving force is given by

$$F(t) = F_0 \cos \omega t \tag{4.5.2}$$

where F_0 is a constant and ω is the frequency of the driving force. From differential equations one learns that if $x_2(t)$ is any particular integral of (4.5.1) and $x_1(t)$ is the general solution of the homogeneous equation, i.e., the equation that results if $F(t) = 0$, then a general solution of (4.5.1) is

$$x(t) = x_1(t) + x_2(t) \tag{4.5.3}$$

If the force on the particle is given by (4.5.2), a particular integral of (4.5.1) is

$$x_2 = c \cos (\omega t) \tag{4.5.4}$$

Substitution of (4.5.4) into (4.5.1) yields the constant c:

$$c = \frac{F_0}{m(\omega_0{}^2 - \omega^2)} \tag{4.5.5}$$

where $\omega_0 = \sqrt{k/m}$ is the natural frequency of the free oscillator. Hence the general solution is

$$x = \frac{F_0}{m(\omega_0{}^2 - \omega^2)} \cos \omega t + c_1 \cos \omega_0 t + c_2 \sin \omega_0 t \tag{4.5.6}$$

The relation of c_1 and c_2 to x_0 and u_0 is no longer given by (4.4.12) but must be determined anew from (4.5.6). The velocity of the particle is

$$\dot{x} = -\frac{F_0\omega}{m(\omega_0{}^2 - \omega^2)} \sin \omega t - c_1\omega_0 \sin \omega_0 t + c_2\omega_0 \cos \omega_0 t \tag{4.5.7}$$

Setting $x = x_0$ and $\dot{x} = u_0$ at $t = 0$ yields

$$c_1 = x_0 - \frac{F_0}{m(\omega_0{}^2 - \omega^2)} \qquad c_2 = \frac{u_0}{\omega_0} \tag{4.5.8}$$

As a specific example, suppose $k = 250$ dynes/cm, $m = 10$ gm, and $F_0 = 1600$ dynes. Let the driving frequency be $\omega = 3$ sec^{-1}. Then

$$\omega_0 = \sqrt{k/m} = 5 \text{ sec}^{-1} \qquad \frac{F_0}{m(\omega_0{}^2 - \omega^2)} = 10 \text{ cm} \qquad (4.5.9)$$

As initial conditions, let us take $x_0 = 0$ and $u_0 = 0$. Then

$$x = 10 \cos 3t - 10 \cos 5t \qquad (4.5.10)$$

which is a superposition of two cosine terms of different frequencies. A plot of this equation is shown in Fig. 4.5.1. Graphs of this type are most

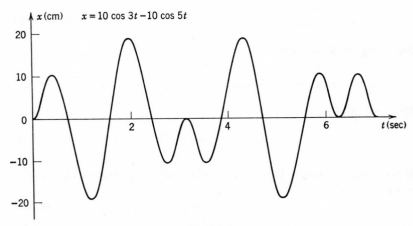

Fig. 4.5.1

easily obtained by first sketching the two cosine terms separately and then making a geometrical addition of the two curves.

As an example of a case where the driving frequency becomes somewhat higher than ω_0, suppose $\omega = 25$ sec^{-1} and all other numerical values are the same. Then

$$\frac{F_0}{m(\omega_0{}^2 - \omega^2)} = -0.267 \qquad (4.5.11)$$

Again using $x_0 = 0$ and $u_0 = 0$ we find

$$x = -0.267 \cos (25t) + 0.267 \cos (5t) \qquad (4.5.12)$$

The amplitude of the motion is considerably less. Equation (4.5.12) is illustrated in Fig. 4.5.2, and shows the effect of adding low and high frequency terms. The solution is always a combination of two simple harmonic vibrations one of which is the natural frequency of the system and the other the frequency of the driving force.

If the driving frequency is taken close to ω_0, the amplitude c of the particular integral (Eq. 4.5.5) becomes very large. This is a phenomenon known as *resonance* and the natural frequency ω_0 is therefore referred to as the *resonant frequency*. If $\omega = \omega_0$, c is actually *infinite* and the solution (4.5.6) is no longer valid. It is, however, possible to find a solution for this case by going back to the original differential equation which is now expressed

$$\ddot{x} + \omega_0{}^2 x = \frac{F_0}{m} \cos \omega_0 t \tag{4.5.13}$$

A particular integral is

$$x = \frac{F_0}{2m\omega_0} t \sin \omega_0 t \tag{4.5.14}$$

and the general solution is

$$x = \frac{F_0}{2m\omega_0} t \sin \omega_0 t + c_1 \cos \omega_0 t + c_2 \sin \omega_0 t \tag{4.5.15}$$

One finds that

$$c_1 = x_0 \qquad c_2 = \frac{u_0}{\omega_0} \tag{4.5.16}$$

If the same numerical values are used as in the previous examples, the result is

$$x = 16t \sin 5t \tag{4.5.17}$$

which is illustrated in Fig. 4.5.3. At resonance, a term $\sin \omega_0 t$ appears with an amplitude that is proportional to t and hence increases without limit as time goes on. Actually, the physical situation being described is

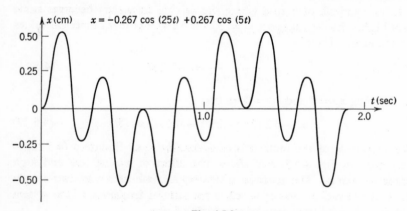

Fig. 4.5.2

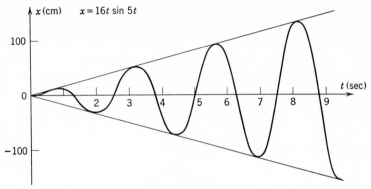

Fig. 4.5.3

not realistic. There are always present frictional forces which prevent such "run-away" solutions.

One important feature of the special driving force just considered lies in the fact that more complicated forces can be represented by a superposition of simple periodic forces of different frequencies.

4.6. THE DAMPED OSCILLATOR WITH NO DRIVING FORCE

If no driving force is present and a viscous frictional force proportional to the velocity is included, the simple harmonic oscillator equation (4.4.3) is replaced by

$$m\ddot{x} + b\dot{x} + kx = 0 \tag{4.6.1}$$

which is a general second order, linear, homogeneous differential equation with constant coefficients. The general solution is

$$x = c_1 e^{\alpha_1 t} + c_2 e^{\alpha_2 t} \tag{4.6.2}$$

Where α_1 and α_2 are the two roots of

$$m\alpha^2 + b\alpha + k = 0 \tag{4.6.3}$$

These roots are

$$\alpha_1 = -\frac{b}{2m} + \sqrt{\left(\frac{b}{2m}\right)^2 - \frac{k}{m}} \tag{4.6.4}$$

$$\alpha_2 = -\frac{b}{2m} - \sqrt{\left(\frac{b}{2m}\right)^2 - \frac{k}{m}} \tag{4.6.5}$$

If the frictional force is small it will be true that

$$\left(\frac{b}{2m}\right)^2 < \frac{k}{m} \tag{4.6.6}$$

in which case the roots α_1 and α_2 are complex numbers. It is then convenient to write them in the form

$$\alpha_1 = -\frac{b}{2m} + i\omega_1 \qquad \alpha_2 = -\frac{b}{2m} - i\omega_1 \qquad (4.6.7)$$

where

$$\omega_1 = \sqrt{\frac{k}{m} - \left(\frac{b}{2m}\right)^2} \qquad i = \sqrt{-1} \qquad (4.6.8)$$

The general solution is now expressed as

$$x = e^{-bt/2m}[c_1 e^{i\omega_1 t} + c_2 e^{-i\omega_1 t}] \qquad (4.6.9)$$

From the theory of complex numbers we obtain

$$e^{\pm i\omega_1 t} = \cos \omega_1 t \pm i \sin \omega_1 t \qquad (4.6.10)$$

which allows (4.6.9) to be written in the form

$$x = e^{-(b/2m)t}[(c_1 + c_2) \cos \omega_1 t + i(c_1 - c_2) \sin \omega_1 t] \qquad (4.6.11)$$

If new constants

$$a_1 = c_1 + c_2 \qquad a_2 = i(c_1 - c_2) \qquad (4.6.12)$$

are introduced, the resulting solution is

$$x = e^{-(b/2m)t}[a_1 \cos \omega_1 t + a_2 \sin \omega_1 t] \qquad (4.6.13)$$

The constants c_1 and c_2 can be expected to be complex, but since x is real a_1 and a_2 must also be real. The constants a_1 and a_2 can be evaluated in terms of x_0 and u_0 by the usual procedure:

$$a_1 = x_0 \qquad a_2 = \frac{u_0}{\omega_1} + \frac{bx_0}{2m\omega_1} \qquad (4.6.14)$$

by using the same procedure that lead to (4.4.16)

$$x = e^{-(b/2m)t}A \cos (\omega_1 t - \alpha) \qquad (4.6.15)$$

The solution is still oscillatory but the frictional force has introduced a decaying exponential which is called a *damping factor*. The amplitude of the oscillations is reduced continuously until the motion is finally damped out.

The frictional force also alters the natural frequency and to estimate this effect, equation (6.8) is written as

$$\omega_1 = \sqrt{\frac{k}{m}}\sqrt{1 - \frac{b^2}{4km}} \qquad (4.6.16)$$

If the dimensionless parameter $b^2/4km$ is small compared to one, two terms of a binomial expansion constitute a sufficient approximation for ω_1:

$$\omega_1 = \omega_0\left(1 - \frac{b^2}{8km}\right), \qquad \omega_0 = \sqrt{k/m} \qquad (4.6.17)$$

As a numerical example, suppose

$$\frac{m}{b} = 2 \text{ sec} \qquad k = 250 \text{ dynes/cm} \qquad m = 10 \text{ gm} \qquad (4.6.18)$$

This gives

$$\frac{b^2}{8km} = 1.25 \times 10^3 \qquad b = 5 \text{ gm/sec} \qquad (4.6.19)$$

In this example, the effect of the damping on the frequency of the oscillations is quite small and it is safe to use

$$\omega_1 = \omega_0 = 5 \text{ sec}^{-1}$$

Suppose that the initial displacement of the oscillator is 10 cm. Let the initial velocity be chosen such that a_2 in Eq. (4.6.14) is zero. The equation of motion is then

$$x = 10e^{-t/4} \cos 5t \qquad (4.6.20)$$

Figure 4.6.1 shows a plot of this equation. The dashed curves represent $\pm 10e^{-t/4}$ and show the effect of this factor on the motion. The solid curve represents x and touches the dashed curves whenever $\cos 5t = \pm 1$.

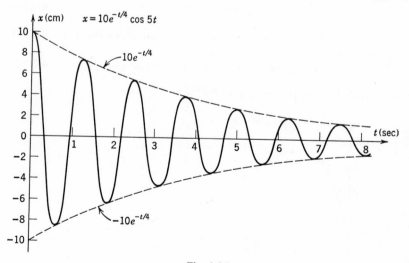

Fig. 4.6.1

If the frictional force is made to increase, a value of b is eventually reached where

$$\left(\frac{b}{2m}\right)^2 = \frac{k}{m} \tag{4.6.21}$$

in which case the two roots α_1 and α_2 are equal:

$$\alpha_1 = \alpha_2 = -\frac{b}{2m} \tag{4.6.22}$$

Equation (4.6.2) does not represent a general solution for this case but must be replaced by

$$x = (c_1 + c_2 t)e^{-(b/2m)t} \tag{4.6.23}$$

The constants c_1 and c_2 in terms of x_0 and u_0 are

$$c_1 = x_0 \qquad c_2 = u_0 + \frac{b x_0}{2m} \tag{4.6.24}$$

A critical value of b has been reached where the oscillatory motion disappears. The system is said to be *critically damped*.

As an example, let us again take the numerical values, $k = 250$ dynes/cm, $m = 10$ gm. We find from (4.6.21) that $b = 100$ gm/sec. Suppose this time that the initial position is zero and that the initial velocity is 50 cm/sec. Evaluation of the constants c_1 and c_2 yields

$$x = 50te^{-5t} \tag{4.6.25}$$

The velocity is given by

$$\dot{x} = 50(1 - 5t)e^{-5t} \tag{4.6.26}$$

Setting the velocity equal to zero yields the time at which the displacement is a maximum:

$$t = \tfrac{1}{5} \text{ sec} \qquad x_{\max} = 10e^{-1} = 3.68 \text{ cm} \tag{4.6.27}$$

Figure 4.6.2 shows a plot of Eq. (4.6.25).

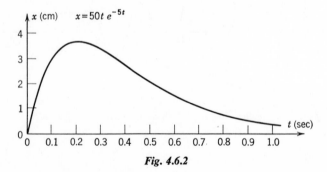

Fig. 4.6.2

If the friction force is sufficiently large,

$$\left(\frac{b}{2m}\right)^2 > \frac{k}{m} \tag{4.6.28}$$

The roots α_1 and α_2 are real and negative. The system is said to be *over damped* and the solution consists entirely of decaying exponentials.

An expression for the rate of change of the energy of the oscillator is obtained by multiplying the equation of motion (4.6.1) by $\dot{x}$ and writing the resulting equation in the form

$$\frac{d}{dt}\left(\tfrac{1}{2}m\dot{x}^2 + \tfrac{1}{2}kx^2\right) = -b\dot{x}^2 \tag{4.6.29}$$

The energy of the particle is the sum of the kinetic and potential energy and (4.6.29) gives the rate at which this energy is dissipated as heat due to friction.

If the motion is oscillatory and the damping is not large, there are two parameters which are used to measure the rate at which the oscillations of a system are damped out. One is the *logarithmic decrement* defined by

$$\delta = \ln \frac{x(t_1)}{x(t_2)} \tag{4.6.30}$$

where ln is the logarithm to the base e and the difference between t_1 and t_2 is one period of the motion:

$$\omega_1 t_2 - \omega_1 t_1 = 2\pi \tag{4.6.31}$$

The logarithmic decrement thus measures the decrease in the amplitude over 1 cycle of the motion. From Eq. (4.6.15)

$$\delta = \ln \left[\frac{e^{-(b/2m)t_1}\cos(\omega_1 t_1 - \alpha)}{e^{-(b/2m)t_2}\cos(\omega_1 t_2 - \alpha)}\right] \tag{4.6.32}$$

Since the phases of the two cosine factors differ by 2π, it is true that

$$\cos(\omega_1 t_1 - \alpha) = \cos(\omega_1 t_2 - \alpha) \tag{4.6.33}$$

Hence,

$$\delta = \ln e^{(b/2m)(t_2 - t_1)} = \frac{b}{2m}(t_2 - t_1) = \frac{\pi b}{m\omega_1} \tag{4.6.34}$$

The second parameter is the Q-factor defined by

$$Q = 2\pi \frac{\text{total energy}}{\text{decrease in energy per cycle}} \tag{4.6.35}$$

Let x_1 and x_2 be two positions 1 cycle apart where the velocity is zero. At these positions, all the energy of the particle is potential and

$$Q = 2\pi \frac{\tfrac{1}{2}kx_1{}^2}{\tfrac{1}{2}kx_1{}^2 - \tfrac{1}{2}kx_2{}^2} = 2\pi \frac{1}{1 - (x_2/x_1)^2} \qquad (4.6.36)$$

It is possible to express Q in terms of δ. From (4.6.30),

$$\frac{x_2}{x_1} = e^{-\delta} \qquad (4.6.37)$$

Hence,

$$Q = 2\pi \frac{1}{1 - e^{-2\delta}} \qquad (4.6.38)$$

If the damping is not too great, δ is small and

$$Q = 2\pi \frac{1}{1 - (1 - 2\delta)} = \frac{\pi}{\delta} \qquad (4.6.39)$$

The Q-factor is *large* if the damping is *small*.

4.7. THE DAMPED OSCILLATOR WITH A SINUSOIDAL DRIVING FORCE

Complex numbers will be found increasingly useful in the following developments. Any complex number $z = x + iy$ can be written in polar form. From the geometry of Fig. 4.7.1,

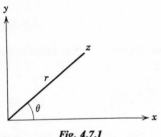

Fig. 4.7.1

$$r = \sqrt{x^2 + y^2}$$

$$z = r\cos\theta + ir\sin\theta = re^{i\theta} \qquad (4.7.1)$$

a result which will be used several times. The magnitude of the complex number is

$$r = |z| = \sqrt{zz^*} \qquad (4.7.2)$$

where z^* is the complex conjugate of z:

$$z^* = x - iy = re^{-i\theta} \qquad (4.7.3)$$

If the damped oscillator is driven by a force $F_0 \cos \omega t$, the equation of motion is

$$m\ddot{x} + b\dot{x} + kx = F_0 \cos \omega t \qquad (4.7.4)$$

The general solution of the homogeneous equation has been extensively investigated in Section 4.6 and it is only necessary to find a particular integral of (4.7.4). It is convenient to append to (4.7.4) the auxiliary equation

$$m\ddot{y} + b\dot{y} + ky = F_0 \sin \omega t \qquad (4.7.5)$$

If we multiply (4.7.5) by i, add (4.7.4) and (4.7.5), and set $z = x + iy$, the result is

$$m\ddot{z} + b\dot{z} + kz = F_0 e^{i\omega t} \qquad (4.7.6)$$

After the solution of (4.7.6) is found, it is necessary to take the *real part* in order to find a solution of the original equation (4.7.4). The solution of (4.7.6) is of the form

$$z = z_0 e^{i\omega t} \qquad (4.7.7)$$

Direct substitution of (4.7.7) into (4.7.6) yields

$$z_0 = \frac{F_0}{k - m\omega^2 + i\omega b} = \frac{F_0}{m(\omega_0{}^2 - \omega^2) + i\omega b} \qquad (4.7.8)$$

Where $\omega_0 = \sqrt{k/m}$ is the frequency of the oscillator with no damping and no driving force. By using (4.7.1) it is possible to write

$$z_0 = Be^{-i\beta} \qquad (4.7.9)$$

$$B = \frac{F_0}{\sqrt{m^2(\omega_0{}^2 - \omega^2)^2 + \omega^2 b^2}} \qquad \tan\beta = \frac{\omega b}{m(\omega_0{}^2 - \omega^2)} \qquad (4.7.10)$$

The general solution of (4.7.6) is

$$z = Be^{i(\omega t - \beta)} + z_1 \qquad (4.7.11)$$

where z_1 represents the general solution of the homogeneous equation. The real part of (4.7.11) is

$$x = B\cos(\omega t - \beta) + x_1 \qquad (4.7.12)$$

The solution of the homogeneous equation decays after a time, leaving only the particular integral. For this reason, x_1 is called a *transient* and $B\cos(\omega t - \beta)$ is called the *steady state solution*. If the damping is small enough so that the transient solution is oscillatory, then

$$x = B\cos(\omega t - \beta) + e^{-(b/2m)t}(a_1 \cos\omega_1 t + a_2 \sin\omega_1 t) \quad (4.7.13)$$

The constants a_1 and a_2 in terms of the initial position and velocity are found to be

$$a_1 = x_0 - B\cos\beta$$

$$a_2 = \frac{u_0}{\omega_1} - \frac{B\omega}{\omega_1}\sin\beta + \frac{b}{2m\omega_1}(x_0 - B\cos\beta) \qquad (4.7.14)$$

Let $k = 250$ dynes/cm, $m = 10$ gm, $b = 5$ gm/sec, and $F_0 = 1600$ dynes. If the driving frequency is $\omega = 3$ sec^{-1}, the following numerical values are found:

$$B = 10 \text{ cm} \qquad \beta = 0.094 \text{ radian} \qquad \cos\beta = 0.996$$

$$\omega_1 = \sqrt{\frac{k}{m}}\sqrt{1 - \frac{b^2}{4km}} = 5\sqrt{1 - \frac{1}{400}} \text{ sec}^{-1} = 5 \text{ sec}^{-1} \qquad (4.7.15)$$

By choosing $x_0 = 0$ and the initial velocity u_0 in (4.7.14) to make $a_2 = 0$,

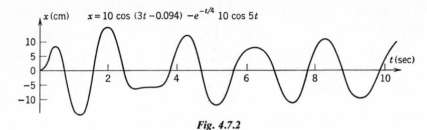

$x = 10 \cos (3t - 0.094) - e^{-t/4} 10 \cos 5t$

Fig. 4.7.2

the equation of motion is found to be approximately

$$x = 10 \cos (3t - 0.094) - e^{-t/4} 10 \cos (5t) \tag{4.7.16}$$

This equation is plotted in Fig. 4.7.2. At first, the motion is distorted by the transient. After a few cycles, the transient dies down and only the steady oscillation of the steady state solution is left.

If the driving frequency ω is 5 sec^{-1}, which is the natural frequency of the undamped oscillator, $F_0 = 1600$ dynes, and $b = 5$ gm/sec, the amplitude is not infinite as in the case of no damping but it is large:

$$B = \frac{F_0}{\omega b} = 64 \text{ cm}$$

$$\tan \beta = \infty \qquad \beta = 90° \tag{4.7.17}$$

The initial conditions $x_0 = 0$ and $u_0 = 0$ give

$$a_1 = 0 \qquad a_2 = -B \tag{4.7.18}$$

Then since $\cos (\omega t - 90°) = \sin \omega t$, the equation of motion is

$$x = B(1 - e^{-bt/2m}) \sin \omega t = 64(1 - e^{-t/4}) \sin 5t \tag{4.7.19}$$

If $bt/2m$ is small,

$$x = B\left[1 - \left(1 - \frac{bt}{2m}\right)\right] \sin \omega t = \frac{F_0}{2m\omega} t \sin \omega t \tag{4.7.20}$$

which is identical to Eq. (4.5.14) where no friction is present. However, as t becomes large, the exponential factor in (4.7.19) goes to zero, leaving a steady state oscillation. Figure 4.7.3 shows a plot of this equation.

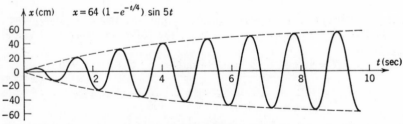

$x = 64 (1 - e^{-t/4}) \sin 5t$

Fig. 4.7.3

If the driving frequency is higher, say, $\omega = 25\,\text{sec}^{-1}$, the motion at first resembles Fig. 4.5.2; but as time progresses the low-frequency component is damped out, leaving only the high-frequency steady state component. As is typical of driving frequencies that are far from resonance, the amplitude of the steady state solution is small.

In the steady state there is a *phase difference* between the applied force and the displacement. This phase difference is the angle β given by Eq. (4.7.10). With $m = 10\,\text{gm}$ and $k = 250\,\text{dynes/cm}$, the angle is

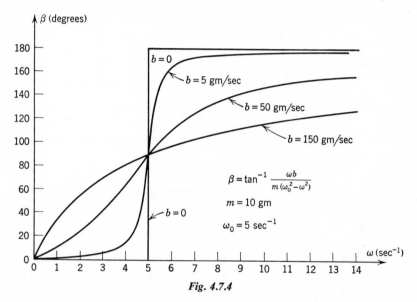

$$\beta = \tan^{-1}\frac{\omega b}{m(\omega_0^2 - \omega^2)}$$

$$m = 10\,\text{gm}$$

$$\omega_0 = 5\,\text{sec}^{-1}$$

Fig. 4.7.4

graphed as a function of the applied frequency for different values of the damping constant in Fig. 4.7.4.

If the damping is very small, the force and the displacement are nearly in phase up to the frequency ω_0 ($5\,\text{sec}^{-1}$ in this case) where there is a sharp change. Above this frequency, the force and the displacement are nearly 180° out of phase. For larger damping, the change is more gradual.

The amplitude of the steady state solution given by equation (4.7.10) is a function of the driving frequency. The *resonant frequency* for displacement is defined to be the frequency where $B(\omega)$ is a maximum. From $dB/d\omega = 0$ we find

$$\omega_{xr} = \sqrt{\omega_0^2 - (b^2/2m^2)} \qquad \omega_0 = \sqrt{k/m} \qquad (4.7.21)$$

where ω_{xr} stands for the resonant frequency for displacement. The resonant frequency is not identical to either ω_0 or ω_1, but all these frequencies are practically the same if the damping is not large. Figure 4.7.5 shows the behavior of $B(\omega)$ for different values of b. Again, $k = 250$ dynes/cm, $m = 10$ gm, and $F_0 = 1600$ dynes. The graphs in Fig. 4.7.5 are sometimes called *response characteristics* or *response curves* and are a measure of how the system responds to a driving force as a function of driving frequency.

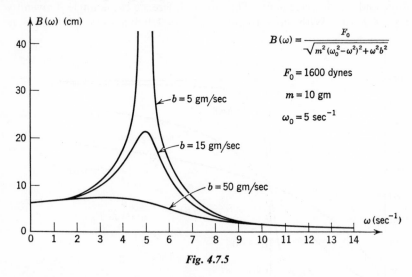

Fig. 4.7.5

If the damping is sufficiently large, ω_{xr} becomes imaginary meaning that there is no maximum. Thus a large frictional force will wipe out the resonance entirely.

The steady state velocity is given by

$$\dot{x} = -B\omega \sin(\omega t - \beta) \qquad (4.7.22)$$

The *velocity amplitude* is $A(\omega) = \omega B(\omega)$. The frequency of the *velocity resonance* is found from

$$\frac{dA(\omega)}{d\omega} = 0 \qquad (4.7.23)$$

and yields

$$\omega_{vr} = \omega_0 \qquad (4.7.24)$$

The velocity amplitude for various b is shown in Fig. 4.7.6. These graphs are *velocity response characteristics*.

Equation (4.7.22) can be written as

$$\dot{x} = A \cos\left(\omega t - \beta + \frac{\pi}{2}\right) \qquad (4.7.25)$$

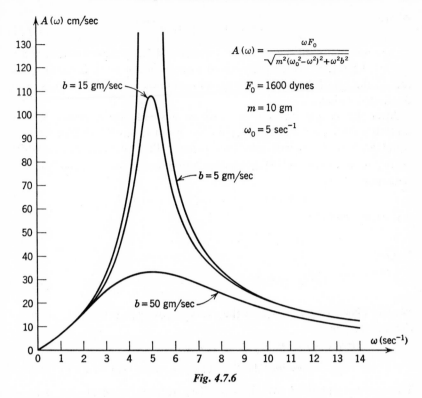

$$A(\omega) = \frac{\omega F_0}{\sqrt{m^2(\omega_0^2 - \omega^2)^2 + \omega^2 b^2}}$$

$F_0 = 1600$ dynes

$m = 10$ gm

$\omega_0 = 5$ sec^{-1}

Fig. 4.7.6

and shows that the velocity leads the displacement in phase by 90°. If the angle γ is defined by

$$\gamma = \beta - \frac{\pi}{2} \tag{4.7.26}$$

then

$$\tan \gamma = -\frac{1}{\tan \beta} = \frac{m(\omega^2 - \omega_0^2)}{\omega b} \tag{4.7.27}$$

$$\dot{x} = A \cos(\omega t - \gamma) \tag{4.7.28}$$

Thus the phase difference between velocity and force is γ.

As usual, the rate of energy dissipation is determined by multiplying the equation of motion by $\dot{x}$:

$$m\dot{x}\ddot{x} + b\dot{x}^2 + kx\dot{x} = \dot{x}F_0 \cos \omega t \tag{4.7.29}$$

This is then expressed as

$$\frac{d\mathscr{E}}{dt} + b\dot{x}^2 = \dot{x}F_0 \cos \omega t \tag{4.7.30}$$

where

$$\mathcal{E} = \tfrac{1}{2}m\dot{x}^2 + \tfrac{1}{2}kx^2 \tag{4.7.31}$$

is the total energy of the particle. Equation (4.7.30) expresses that at any instant the rate at which the applied force is doing work is equal to the rate of change of the particle energy plus the rate at which energy is being dissipated as heat.

After the system has settled down to a steady state, energy is continuously dissipated and the external force must, on the average, feed an equal amount of energy into the system. The average rate at which energy is dissipated is $\overline{b\dot{x}^2}$ where the bar denotes *time-average*. If it is remembered that

$$\cos^2(\omega t - \gamma) + \sin^2(\omega t - \gamma) = 1 \tag{4.7.32}$$

it is easily seen that

$$\overline{\cos^2(\omega t - \gamma)} = \overline{\sin^2(\omega t - \gamma)} = \tfrac{1}{2} \tag{4.7.33}$$

Hence,

$$\overline{b\dot{x}^2} = \tfrac{1}{2}B^2\omega^2 b = \frac{F_0^2\omega^2 b}{2[m^2(\omega^2 - \omega_0^2)^2 + b^2\omega^2]} \tag{4.7.34}$$

We can check that this is also the average power put into the system by the applied force:

$$\dot{x}F_0 \cos \omega t = B\omega F_0 \cos \omega t \cos(\omega t - \gamma)$$

$$= B\omega F_0[\cos^2 \omega t \cos \gamma + \cos \omega t \sin \omega t \sin \gamma] \tag{4.7.35}$$

The average value of $\cos^2 \omega t$ is $\tfrac{1}{2}$. The average value of $\cos \omega t \sin \omega t$ is found from

$$\overline{\cos \omega t \sin \omega t} = \frac{1}{2\pi}\int_{\omega t=0}^{2\pi} \cos \omega t \sin \omega t \, d(\omega t) = 0 \tag{4.7.36}$$

Hence,

$$\overline{\dot{x}F_0 \cos \omega t} = \tfrac{1}{2}B\omega F_0 \cos \gamma \tag{4.7.37}$$

But, as is readily determined from Eq. (4.7.27) and Fig. 4.7.7,

$$\cos \gamma = \frac{\omega b}{\sqrt{m^2(\omega^2 - \omega_0^2)^2 + \omega^2 b^2}} \tag{4.7.38}$$

Thus, in the steady state,

$$\overline{\dot{x}F_0 \cos \omega t} = \tfrac{1}{2}B^2\omega^2 b = \overline{b\dot{x}^2} \tag{4.7.39}$$

By computing

$$\frac{d}{d\omega}\overline{b\dot{x}^2} = 0 \tag{4.7.40}$$

from (4.7.34), we find that the *energy resonance* is

$$\omega_{er} = \omega_0 \tag{4.7.41}$$

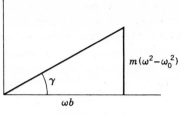

Fig. 4.7.7

that is, this is the frequency at which the system extracts energy from the driving force at the greatest rate. The value of this maximum power is

$$(\overline{b\dot{x}^2})_{\max} = F_0^2/(2b) \qquad (4.7.42)$$

and is large if the damping constant b is small. The power as a function of driving frequency ω is shown in Fig. 4.7.8. The power resonance becomes very narrow and high at the resonant frequency if b is small. As $b \to 0$ it becomes an infinitely high spike with no width.

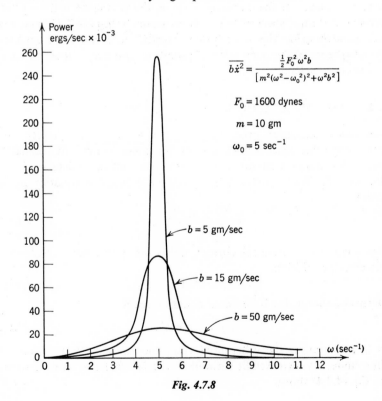

Fig. 4.7.8

A simple theory of the dispersion of electromagnetic radiation in a nonconducting material medium can be based on the theory developed in this section. We adopt the crude model that a material medium contains electrons which are bound by Hook's law restoring forces to the molecules. When a plane electromagnetic wave enters the medium, the electrons are drived by a simple periodic driving force. The equation of motion of the electrons is then

$$m\ddot{x} + b\dot{x} + kx = qE_0 \cos \omega t \tag{4.7.43}$$

where q is the charge on the electron and $E_0 \cos \omega t$ represents the electric field produced at the site of the electron by the plane electromagnetic wave. We neglect the magnetic forces which are typically much smaller. The electrons undergo harmonic oscillations of the same frequency as the incident wave and therefore are accelerated. The accelerated electrons reradiate electromagnetic energy of the same frequency as the incident radiation. This is why the damping term is present in equation (4.7.43). If the driving frequency is near the natural frequency of the electron, there will be a sharp increase in the absorption and scattering of the wave due to resonance. If the damping is small, the resonance will be a very narrow and sharp "spectral line," as is shown by Eq. (4.7.42) and Fig. 4.7.8. Something like this actually does happen, but of course a completely adequate theory of the interaction of radiation with matter must be based on quantum mechanics.

4.8. DYNAMICAL ANALOGIES

For every mechanical problem considered so far in this chapter there exists a simple electrical circuit that obeys an identical differential equation. Figure 4.8.1 shows a circuit with a battery of emf ϕ_0, resistance R and inductance L. When switch S is closed, the differential equation of the circuit is

$$L\frac{dJ}{dt} + RJ = \phi_0 \tag{4.8.1}$$

where J is the instantaneous current and is the charge per second which flows through the circuit:

$$J = \dot{q} \tag{4.8.2}$$

In terms of charge, the differential equation reads

$$L\ddot{q} + R\dot{q} = \phi_0 \tag{4.8.3}$$

and is identical to the differential equation for a particle of mass m falling under its own weight through a viscous medium, as comparison with Eq. (4.3.4) shows.

If the initial current is zero, Eq. (4.8.1) is integrated to yield

$$J = \frac{\phi_0}{R}[1 - e^{-Rt/L}] \tag{4.8.4}$$

and shows that the current rises from zero to the terminal value ϕ_0/R. In an analogous manner, a mass which is released from rest in a viscous medium reaches a terminal velocity given by mg/b.

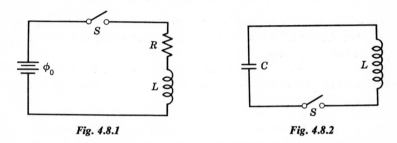

Fig. 4.8.1   *Fig. 4.8.2*

Figure 4.8.2 shows a circuit with capacitance C and inductance L. Suppose there is initially a charge q_0 on the capacitor. When switch S is closed, the circuit obeys

$$L\frac{dJ}{dt} + \frac{q}{C} = 0$$

or

$$L\ddot{q} + \frac{q}{C} = 0 \tag{4.8.5}$$

which is identical to the differential equation for a mass suspended from a spring. The general solution is

$$q = c_1 \cos \omega_0 t + c_2 \sin \omega_0 t$$
$$\omega_0 = \frac{1}{\sqrt{LC}} \tag{4.8.6}$$

The arbitrary constants of integration c_1 and c_2 are to be determined in terms of initial charge and initial current. In this case, the initial current is zero so that the solution is

$$q = q_0 \cos \omega_0 t \tag{4.8.7}$$

This gives the charge on the capacitor at any instant after the switch has been closed. The charge oscillates back and forth with a characteristic frequency, just as does a mass on a spring. The circuit just considered is known as a *tank circuit*.

Summarized below are the mks and Gaussian units and conversion factors for L, R, and C. Gaussian units are used in this book.

	mks	Gaussian
Inductance	1 henry	$\frac{1}{9} \times 10^{-11} \sec^2/cm$
Resistance	1 ohm	$\frac{1}{9} \times 10^{-11} \sec^2/cm$
Capacitance	1 farad	$9 \times 10^{-11} cm$

Fortunately, all the equations in this section are valid in all systems of units. See Appendix B for a discussion of electromagnetic equations in mks and Gaussian units.

From the two examples just considered, it is possible to identify the electrical analogues of various mechanical quantities. These are summarized in the following table:

Electrical		Mechanical	
Name	Symbol	Name	Symbol
Inductance	L	Mass	m
Resistance	R	Damping Force Constant	b
Reciprocal of Capacitance	$1/C$	Spring Constant	k
Charge	q	Coordinate	x
Current	J	Velocity	$\dot{x}$
Voltage	ϕ	Force	F

Some books use the analogue of capacitance in discussing mechanical systems. It is written $C_m = 1/k$ and called the *compliance*. The electrical analogues of kinetic and potential energies are

$$T_E = \tfrac{1}{2}LJ^2 \qquad V_E = \tfrac{1}{2}(1/C)q^2 \qquad (4.8.8)$$

and represent energy which is stored in the magnetic field of the inductor and the electric field of the capacitor.

Shown in Fig. 4.8.3 is the electric circuit analogue of the driven harmonic oscillator with damping. If the driving voltage is given by

$$\phi = \phi_0 \cos \omega t \qquad (4.8.9)$$

then the differential equation of the circuit is

$$L\ddot{q} + R\dot{q} + \frac{q}{C} = \phi_0 \cos \omega t \qquad (4.8.10)$$

One way of obtaining the steady state solution of this, or of any linear equation involving a sinusoidal driving force, is by the method of complex impedances. Just as in out treatment of Eq. (4.7.4) for a damped oscillator, we write in place of (4.8.10)

$$L\ddot{Q} + R\dot{Q} + \frac{1}{C}Q = V \qquad (4.8.11)$$

where Q is the complex charge and $V = \phi_0 e^{i\omega t}$ is the complex driving voltage. The actual steady state charge is the real part of the solution of

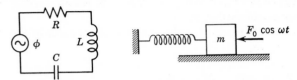

Fig. 4.8.3

(4.8.11). In the steady state, the complex charge and current are given by

$$Q = Q_0 e^{i\omega t} \qquad I = I_0 e^{i\omega t} \qquad (4.8.12)$$

where Q_0 and I_0 are complex amplitudes. The derivative of any such quantity is always $i\omega$ times itself:

$$\dot{Q} = i\omega Q \qquad \ddot{Q} = \dot{I} = i\omega I \qquad (4.8.13)$$

The differential equation (4.8.11) gives

$$Li\omega I + RI + \frac{1}{i\omega C}I = V \qquad (4.8.14)$$

If we introduce the *complex impedance* of the circuit defined by

$$Z = Li\omega + R + \frac{1}{i\omega C} \qquad (4.8.15)$$

the differential equation is reduced to

$$IZ = V \qquad (4.8.16)$$

which is just like Ohm's law for a direct current circuit! The solution of (4.8.16) for the complex current is trivial. The actual current is, of course, the real part I, and for this reason it is convenient to write the complex impedance Z as

$$Z = |Z| e^{i\gamma} \qquad (4.8.17)$$

Then

$$I = \frac{\phi_0}{|Z|} e^{i(\omega t - \gamma)} \qquad J = \frac{\phi_0}{|Z|} \cos(\omega t - \gamma) \qquad (4.8.18)$$

where J is the real part of I. The magnitude $|Z|$ and the phase angle γ are formed from (4.8.15):

$$|Z| = \frac{1}{\omega}\sqrt{R^2\omega^2 + L^2(\omega^2 - \omega_0{}^2)^2} \qquad \tan\gamma = \frac{L(\omega^2 - \omega_0{}^2)}{R\omega} \qquad (4.8.19)$$

where $\omega_0 = 1/\sqrt{LC}$ is the natural frequency of the free oscillations of the circuit when no resistance is present. The results we have obtained here are equivalent to the expression (4.7.28) for the steady rate velocity of the damped mechanical oscillator.

Equation (4.8.14) shows that the complex impedances of the individual circuit elements are

$$Z_L = i\omega L \qquad Z_R = R \qquad Z_C = \frac{1}{i\omega C} = -\frac{i}{\omega C} \qquad (4.8.20)$$

The complex voltages across the different circuit elements are

$$V_L = IZ_L \qquad V_R = IR \qquad V_C = IZ_C \qquad (4.8.21)$$

As usual, the actual instantaneous voltages are the real parts. Since

$$Z_L = \omega L e^{i\pi/2} \qquad Z_C = \frac{1}{\omega C} e^{-i\pi/2} \qquad (4.8.22)$$

it becomes apparent that the voltage across the inductor *leads* the current in phase by 90°, the voltage across the resistor is *in phase* with the current, and the voltage across the capacitor *lags* the current in phase by 90°. The phase difference between the current and the driving voltage is the angle γ of Eq. (4.8.19).

Complex impedances depend explicitly on the driving frequency and the simple idea of impedance developed here is valid only for a simple harmonic driving voltage. If a more general voltage is represented by a sum of such periodic voltages, each term in the series has its own characteristic impedance.

The rate at which the applied voltage does work on the circuit is

$$P = J\phi = \frac{\phi_0{}^2}{|Z|}\cos(\omega t - \gamma)\cos\omega t \qquad (4.8.23)$$

The average power delivered to the circuit is

$$\bar{P} = \frac{1}{2}\frac{\phi_0{}^2}{|Z|}\cos\gamma \qquad (4.8.24)$$

In the parlance of a-c circuit theory, $\cos \gamma$ is called the *power factor*. The average power can be written in terms of complex current, impedance, and driving voltage in the following different ways:

$$\bar{P} = \tfrac{1}{2}\,\mathrm{Re}\,(II^*Z^*) = \tfrac{1}{2}\,\mathrm{Re}\,(IV^*) = \tfrac{1}{2}\,\mathrm{Re}\left(\frac{VV^*}{Z}\right) \qquad (4.8.25)$$

where Re means real part.

By analogy, we can associate with the driven mechanical oscillator the impedances

$$Z_m = i\omega m \qquad Z_b = b \qquad Z_k = -\,i\,\frac{k}{\omega} \qquad (4.8.26)$$

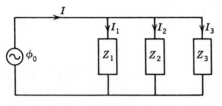

Fig. 4.8.4

Then the total impedance is

$$Z = Z_m + Z_b + Z_k \qquad (4.8.27)$$

and we may write the differential equation (4.7.6) as

$$UZ = F \qquad (4.8.28)$$

where $F = F_0 e^{i\omega t}$ is the complex driving force and U is the complex velocity. The solution now proceeds in precisely the same way as in the a-c circuit just completed. Remember that this method obtains only the steady state solution and not the transients.

One of the values of the complex impedance method lies in its use to analyze more complicated linear systems. The circuit of Fig. 4.8.4 has three branches with impedances Z_1, Z_2, and Z_3. What is the overall impedance of this circuit? The same voltage appears across each branch:

$$V = IZ = I_1 Z_1 = I_2 Z_2 = I_3 Z_3 \qquad (4.8.29)$$

where Z is the net impedance and

$$I = I_1 + I_2 + I_3 \qquad (4.8.30)$$

is the net current. On combining (4.8.29) and (4.8.30) we find

$$\frac{1}{Z} = \frac{1}{Z_1} + \frac{1}{Z_2} + \frac{1}{Z_3} \qquad (4.8.31)$$

Thus complex impedances behave in every respect like ordinary resistances in d-c circuits.

Let us consider an example. A circuit with parallel components and its rectilinear mechanical analogue are illustrated in Fig. 4.8.5. In Fig. 4.8.5b, the mass rests on a surface on which there is a viscous frictional force proportional to the velocity. The force is not applied directly to the mass, as in previous examples, but is applied to the opposite end of the spring to which the mass is attached. The velocity of the point where the

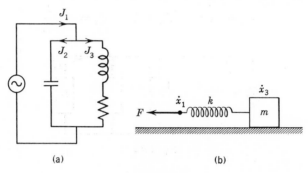

(a)   (b)

Fig.4.8.5

force is applied is $\dot{x}_1$, and $\dot{x}_3$ is the velocity of the mass itself. In the electrical circuit, the currents obey

$$J_1 = J_2 + J_3 \tag{4.8.32}$$

The analogous mechanical relation is

$$\dot{x}_1 = \dot{x}_2 + \dot{x}_3 \tag{4.8.33}$$

Hence, $\dot{x}_2$ is the relative velocity of opposite ends of the spring. If the mass of the spring itself is neglected and it is assumed that the applied force is transmitted instantaneously, then the same force appears on both ends of the spring and, therefore, is also the force which is applied to the mass. Analogously, the same voltage appears across both branches of the electrical circuit.

By using the electrical circuit as a guide, let us analyze the steady state motion of the mechanical system by the method of complex impedances. The net impedance is

$$Z_1 = \frac{Z_2 Z_3}{Z_2 + Z_3} \tag{4.8.34}$$

where

$$Z_2 = -i\frac{k}{\omega} \qquad Z_3 = b + i\omega m \tag{4.8.35}$$

Thus

$$Z_1 = \frac{m\omega_0^2(\omega m - ib)}{\omega b + im(\omega^2 - \omega_0^2)} \qquad (4.8.36)$$

where k has been eliminated using

$$k = m\omega_0^2 \qquad (4.8.37)$$

In Eq. (4.8.36), the complex numbers in the numerator and denominator can be written in polar form to yield

$$Z_1 = \frac{m\omega_0^2\sqrt{\omega^2 m^2 + b^2}e^{i\alpha}}{\sqrt{\omega^2 b^2 + m^2(\omega^2 - \omega_0^2)^2}e^{i\beta}} \qquad (4.8.38)$$

where

$$\tan \alpha = -\frac{b}{m\omega} \qquad \tan \beta = \frac{m(\omega^2 - \omega_0^2)}{\omega b} \qquad (4.8.39)$$

It is now possible to express (4.8.38) in the form

$$Z_1 = |Z_1| e^{i\gamma_1} \qquad (4.8.40)$$

where

$$|Z_1| = \frac{m\omega_0^2\sqrt{\omega^2 m^2 + b^2}}{\sqrt{\omega^2 b^2 + m^2(\omega^2 - \omega_0^2)^2}} \qquad \gamma_1 = \alpha - \beta \qquad (4.8.41)$$

By using the identity

$$\tan(\alpha - \beta) = \frac{\tan \alpha - \tan \beta}{1 + \tan \alpha \tan \beta} \qquad (4.8.42)$$

we find

$$\tan \gamma_1 = \frac{\omega b^2 + \omega m^2(\omega^2 - \omega_0^2)}{-mb\omega_0^2} \qquad (4.8.43)$$

The complex velocity of the point on the end of the spring where the force is applied is

$$U_1 = \frac{1}{Z_1} F_0 e^{i\omega t} = \frac{F_0}{|Z_1|} e^{i(\omega t - \gamma_1)} \qquad (4.8.44)$$

The actual velocity is the real part:

$$\dot{x}_1 = \frac{F_0}{|Z_1|} \cos(\omega t - \gamma_1) \qquad (4.8.45)$$

The *velocity amplitude* of the point where the force is applied is

$$A_1(\omega) = \frac{F_0}{|Z_1|} = \frac{F_0}{m\omega_0^2} \frac{\sqrt{\omega^2 b^2 + m^2(\omega^2 - \omega_0^2)^2}}{\sqrt{\omega^2 m^2 + b^2}} \qquad (4.8.46)$$

The motion of the mass itself can be computed from

$$U_3 Z_3 = F_0 e^{i\omega t} \tag{4.8.47}$$

where

$$Z_3 = b + i\omega m = |Z_3|\, e^{i\gamma_3}$$

$$|Z_3| = \sqrt{b^2 + \omega^2 m^2} \qquad \tan \gamma_3 = \frac{\omega m}{b} \tag{4.8.48}$$

The real part of U_3 is found to be

$$\dot{x}_3 = \frac{F_0}{|Z_3|} \cos\left(\omega t - \gamma_3\right) \tag{4.8.49}$$

The velocity amplitude of the mass, therefore, is

$$A_3(\omega) = \frac{F_0}{|Z_3|} = \frac{F_0}{\sqrt{b^2 + \omega^2 m^2}} \tag{4.8.50}$$

Referring to Fig. 4.8.5, we see that the spring is essentially a transmission device. The ratio of the *output amplitude* to the *input amplitude* is

$$\frac{A_3(\omega)}{A_1(\omega)} = \frac{m\omega_0^2}{\sqrt{\omega^2 b^2 + m^2(\omega^2 - \omega_0^2)^2}} \tag{4.8.51}$$

and is sometimes called the *magnification factor*. The function plotted in Fig. 4.7.6 is essentially the same as (4.8.51) and shows that the amplification factor is greatest at the resonant frequency ω_0. It is at this frequency that the greatest transmission of motion to the mass occurs.

The energy supplied to the system by the driving force $F_0 \cos \omega t$ must be entirely accounted for by the heat dissipated due to the friction. The average power, therefore, is,

$$\overline{b\dot{x}_3^2} = \overline{bA_3^2 \cos^2\left(\omega t - \gamma_3\right)}$$
$$= \tfrac{1}{2} b A_3^2 \tag{4.8.52}$$

By means of (4.8.50),

$$\overline{b\dot{x}_3^2} = \frac{bF_0^2}{2(b^2 + \omega^2 m^2)} \tag{4.8.53}$$

4.9. ACOUSTICAL SYSTEMS

There are simple acoustical systems which behave like the mechanical and electrical systems which have been treated in this chapter. Figure 4.9.1 shows a volume V of air in a container fitted with a small piston. If the piston is moved fairly rapidly, the compression will be adiabatic and the gas will obey

$$PV^\gamma = \text{const.} \tag{4.9.1}$$

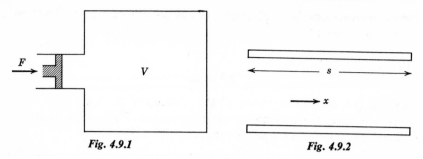

Fig. 4.9.1   **Fig. 4.9.2**

where P is the pressure and γ is the ratio of the heat capacity at constant pressure to the heat capacity at constant volume. For small changes in pressure and volume

$$\Delta P V^{\gamma} + P\gamma V^{\gamma-1}\Delta V = 0$$

$$\Delta P = -\gamma P \frac{\Delta V}{V}$$

(4.9.2)

The velocity of sound in air is

$$c = \sqrt{\gamma RT/W}$$

(4.9.3)

where R is the gas constant, T is the absolute temperature, and W is the molecular weight. According to the ideal gas law,

$$PV = nRT$$

(4.9.4)

where n is the number of moles of gas. Hence,

$$\Delta P = -\frac{c^2 W n \, \Delta V}{V^2}$$

(4.9.5)

The mass density of the gas is

$$\rho = \frac{nW}{V}$$

(4.9.6)

so

$$\Delta P = -\rho c^2 \frac{\Delta V}{V}$$

(4.9.7)

If ΔP is regarded as the analogue of force and ΔV as the analogue of displacement, then the *acoustical capacitance* is

$$C_A = \frac{V}{\rho c^2}$$

(4.9.8)

Consider the air which is in a small tube of cross sectional area σ and length s, Fig. 4.9.2. If there is a pressure difference ΔP across the tube, then according to Newton's second law

$$\Delta P \sigma = m\ddot{x}$$

(4.9.9)

where x represents the displacement of any particle of the air. The mass of the air in the tube is

$$m = \rho \sigma s \qquad (4.9.10)$$

Hence,

$$\Delta P = \rho s \ddot{x} \qquad (4.9.11)$$

Assume that at a given instant all the particles of air in the tube are in phase, i.e., are moving with the same velocity $\dot{x}$. The volume of air per second which flows through the tube is

$$U = \sigma \dot{x} \qquad (4.9.12)$$

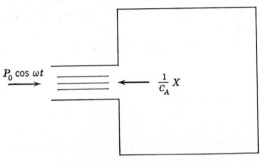

Fig. 4.9.3

Combining Eqs. (4.9.10), (4.9.11), and (4.9.12) yields

$$\Delta P = \frac{\rho s}{\sigma} \frac{dU}{dt} = \frac{m}{\sigma^2} \frac{dU}{dt} \qquad (4.9.13)$$

The *acoustical inertance* is defined by

$$M = m/\sigma^2 \qquad (4.9.14)$$

so that (4.9.13) reads

$$\Delta P = M\dot{U} \qquad (4.9.15)$$

Suppose now that the small tube is connected to a larger cavity and some fins are placed in it to provide a viscous frictional force as in Fig. 4.9.3. Suppose that a sound wave falls on the open end of the tube to provide a simple harmonic driving force. Since ΔV in Eq. (4.9.7) is the analogue of displacement (or charge in an electrical system), let us write

$$X = \Delta V \qquad (4.9.16)$$

The pressure increase in the large container is then X/C_A. The rate at which air flows through the tube is expressible as

$$U = \frac{d}{dt}(\Delta V) = \dot{X} \qquad (4.9.17)$$

At a given instant, the pressure difference across the tube is

$$\Delta P = P_0 \cos \omega t - \frac{X}{C_A} \qquad (4.9.18)$$

P_0 should be interpreted as the pressure amplitude of the incident sound wave and is the maximum pressure *above atmospheric* produced by the sound wave. Equation (4.9.18) includes both the effect of the cavity as given by Eq. (4.9.7) and the effect of the incident sound wave. Newton's second law for the air in the tube, including the effect of viscous friction is,

$$P_0 \cos \omega t - \frac{1}{C_A} X - b_A \dot{X} = M \ddot{X} \qquad (4.9.19)$$

where M is the acoustical inertance as defined by Eq. (4.9.14) and b_A is the acoustical resistance. The rearrangement of terms in (4.9.19) yields

$$M \ddot{X} + b_A \dot{X} + \frac{1}{C_A} X = P_0 \cos \omega t \qquad (4.9.20)$$

an equation which is identical in form to Eq. (4.7.4) and (4.8.10).

In order for the analysis that leads to Eq. (4.9.20) to be valid, the wavelength of the impinging sound must be long compared to the dimensions of the apparatus, otherwise the assumption that all the particles of the air in the tube move in phase is not valid. In other words, there is actually a finite time required for a pressure change to propagate from one end of the apparatus to the other which we neglect if the frequency of the incident sound wave is so small that one period of the motion is long by comparison. At higher frequencies, complications come into play due to the propagation of waves in the cavity and the consequent production of standing waves.

By analogy with electrical systems, it is evident that the steady state solutions of (4.9.20) can be analyzed by the use of *acoustical impedances* defined by

$$Z_C = -\frac{i}{\omega C_A} \qquad Z_b = b_A \qquad Z_M = i\omega M \qquad (4.9.21)$$

The device just discussed is known as a *Helmholz resonator*. An ordinary bottle forms a Helmholz resonator when oscillations are set up by blowing air across the mouth.

4.10. FORCES WHICH ARE PERIODIC FUNCTIONS OF TIME

In the previous sections, we have considered a linear system driven by a force which is periodic and varies harmonically with time. In this section, driving forces which are general periodic functions are treated.

A periodic force has the property that

$$F(t) = F(t + P) = F(t + 2P) = \cdots \qquad (4.10.1)$$

where P is the period. Such a force is pictured in Fig. 4.10.1. Just as in the case of a simple harmonic force, the circular frequency can be defined by means of

$$\omega P = 2\pi \qquad (4.10.2)$$

As is known from mathematical analysis, a periodic function can be expanded into a Fourier series as

$$F(t) = a_0 + \sum_{n=1}^{\infty}(a_n \sin n\omega t + b_n \cos n\omega t) \qquad (4.10.3)$$

A complete mathematical theory of Fourier series treating such problems as the existence and convergence of the series will not be presented here.

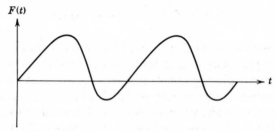

$F(t)$

t

Fig. 4.10.1

Reference can be made to books on analysis for this purpose. The coefficients a_n and b_n are formally obtained by first noting that

$$\int_0^{2\pi} \sin n'\omega t \sin n\omega t \, d(\omega t) = \delta_{n'n}\pi$$

$$\int_0^{2\pi} \cos n'\omega t \cos n\omega t \, d(\omega t) = \delta_{n'n}\pi \qquad (4.10.4)$$

$$\int_0^{2\pi} \sin n'\omega t \cos n\omega t \, d(\omega t) = 0$$

which, by analogy to the relations

$$\hat{\mathbf{e}}_i \cdot \hat{\mathbf{e}}_j = \delta_{ij} \qquad (4.10.5)$$

obeyed by the basis vectors of a rectangular Cartesian coordinate system, are called *orthogonality relations*. Note that the integrals (4.10.4) are over one complete cycle of the function $F(t)$; n and n' are integers. The functions

$$\sin n\omega t, \cos n\omega t, \quad n = 0, 1, 2, \ldots \qquad (4.10.6)$$

are said to be a *complete set* of orthogonal functions over the interval $0 \leq \omega t \leq 2\pi$ in the sense that any function defined over that interval can be represented in terms of them by a Fourier series. This is analogous to the fact that in ordinary three-dimensional space a vector can be represented as a linear combination of the basis vectors $\hat{e}_i$.

To evaluate the coefficients, let us write the Fourier series (4.10.3) as

$$F(t) = a_0 + \sum_{n'=1}^{\infty} (a_{n'} \sin n'\omega t + b_{n'} \cos n'\omega t) \qquad (4.10.7)$$

Upon multiplication by $\sin n\omega t$ and integration over one cycle of the function $F(t)$ we find

$$\int_0^{2\pi} F(t) \sin n\omega t \, d(\omega t) = \int_0^{2\pi} a_0 \sin n\omega t \, d(\omega t)$$

$$+ \sum_{n'=1}^{\infty} \left[a_{n'} \int_0^{2\pi} \sin n\omega t \sin n'\omega t \, d(\omega t) + \qquad (4.10.8) \right.$$

$$\left. b_{n'} \int_0^{2\pi} \sin n\omega t \cos n'\omega t \, d(\omega t) \right]$$

By application of the orthogonality relations (4.10.4) there results

$$a_n = \frac{1}{\pi} \int_0^{2\pi} F(t) \sin n\omega t \, d(\omega t) \qquad (4.10.9)$$

If the Fourier series (4.10.7) is multiplied by $d(\omega t)$ and integrated over 1 cycle, there results

$$a_0 = \frac{1}{2\pi} \int_0^{2\pi} F(t) \, d(\omega t) \qquad (4.10.10)$$

By a similar procedure

$$b_n = \frac{1}{\pi} \int_0^{2\pi} F(t) \cos n\omega t \, d(\omega t) \qquad (4.10.11)$$

As will be seen, the Fourier series representation of the applied force has the advantage that the series is a sum of terms of the type already considered in Section 4.7.

The Fourier series expansion of a function is made more reasonable by the consideration of an example. Suppose the force is of the form of a step function as pictured in Fig. 4.10.2 and defined by

$$F = F_0 \quad 0 \leq \omega t < \pi, 2\pi \leq \omega t < 3\pi, \ldots$$
$$F = 0 \quad \pi \leq \omega t < 2\pi, 3\pi \leq \omega t < 4\pi, \ldots \qquad (4.10.12)$$

For the coefficient a_0 we find

$$a_0 = \frac{1}{2\pi} \int_0^{2\pi} F(t) \, d(\omega t) = \tfrac{1}{2}F_0 \qquad (4.10.13)$$

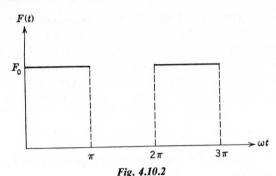

Fig. 4.10.2

Notice that a_0 has the significance that it is the *average value* of the function $F(t)$. The coefficients a_n are given by

$$a_n = \frac{1}{\pi} \int_0^{2\pi} F(t) \sin n\omega t \, d(\omega t)$$

$$= \frac{1}{\pi} \int_0^{\pi} F_0 \sin n\omega t \, d(\omega t)$$

$$= \frac{-F_0}{\pi n} \cos n\omega t \bigg|_0^{\pi} = \frac{F_0}{\pi n} [1 - (-1)^n]$$

$$a_n = \frac{2F_0}{n\pi} \qquad \text{if } n \text{ is odd} \qquad (4.10.14)$$

$$a_n = 0 \qquad \text{if } n \text{ is even}$$

The coefficient b_1 is

$$b_1 = \frac{1}{\pi} \int_0^{2\pi} F(t) \cos \omega t \, d(\omega t)$$

$$= \frac{1}{\pi} \int_0^{\pi} F_0 \cos \omega t \, d(\omega t) \qquad (4.10.15)$$

From Fig. 4.10.3 it is clear that $\cos \omega t$ is negative as much as it is positive over the interval $0 \leq \omega t < \pi$. Therefore, $b_1 = 0$. A similar argument

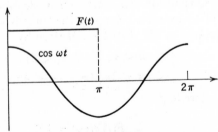

Fig. 4.10.3

can be made for $b_2, b_3, \ldots$. Thus

$$b_n = 0 \tag{4.10.16}$$

The Fourier series representation of $F(t)$ is therefore a *Fourier sine series:*

$$F(t) = \frac{F_0}{2} + \frac{2F_0}{\pi}(\sin \omega t + \tfrac{1}{3} \sin 3\omega t + \tfrac{1}{5} \sin 5\omega t + \cdots) \tag{4.10.17}$$

The coefficients a_n are sometimes called the *spectrum* of $F(t)$. They are pictured in Fig. 4.10.4 for the numerical value $F_0 = 1570$ dynes and form a

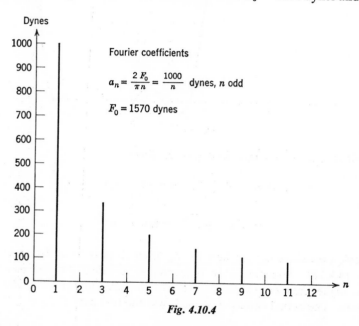

Fig. 4.10.4

discrete lime spectrum. The first four terms of the Fourier series can be expressed as

$$\frac{F}{F_0} = 0.5 + 0.637 \sin \omega t + 0.212 \sin 3\omega t + 0.127 \sin 5\omega t \tag{4.10.18}$$

and this function is plotted in Fig. 4.10.5 in order to show how the first few terms of the Fourier series approximate the original function. You may find it instructive to plot the terms in Eq. (4.10.18) separately and then add them geometrically.

Before proceding to solve the problem of a linear system driven by a general periodic force represented by a Fourier series, it is convenient

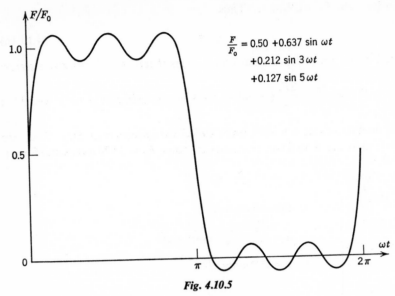

$$\frac{F}{F_0} = 0.50 + 0.637 \sin \omega t$$
$$+ 0.212 \sin 3\omega t$$
$$+ 0.127 \sin 5\omega t$$

Fig. 4.10.5

to rewrite the Fourier series (4.10.3) in the form

$$F(t) = a_0 + \sum_{n=1}^{\infty} F_n \cos(n\omega t - \alpha_n)$$

$$F_n = \sqrt{a_n^2 + b_n^2} \qquad \tan \alpha_n = \frac{a_n}{b_n}$$

(4.10.19)

The same procedure is used here as was used to pass from Eq. (4.4.13) to Eq. (4.4.16). Sometimes the frequency ω is called the *fundamental* and the higher frequency terms in (4.10.19) are called higher *harmonics* or *overtones*. The actual equation of motion is the real part of

$$m\ddot{z} + b\dot{z} + kz = a_0 + \sum_{n=1}^{\infty} F_n e^{i(n\omega t - \alpha_n)}$$

(4.10.20)

where $z = x + iy$. We will be concerned here with the steady state solution. The same procedure can be used as was used to solve equation (4.7.6). Assume a solution of the form

$$z = B_0 + \sum_{n=1}^{\infty} B_n e^{i(n\omega t - \gamma_n)}$$

(4.10.21)

Substitution into the differential equation yields

$$\sum_{n=1}^{\infty} (-mn^2\omega^2 + ibn\omega + k)B_n e^{i(n\omega t - \gamma_n)} + kB_0 = a_0 + \sum_{n=1}^{\infty} F_n e^{i(n\omega t - \alpha_n)}$$

(4.10.22)

In order for (4.10.22) to hold for all values of t, the coefficients of $e^{in\omega t}$ on the two sides of the equation must be equal:

$$kB_0 = a_0 \qquad (-mn^2\omega^2 + ibn\omega + k)B_n e^{-i\gamma_n} = F_n e^{-i\alpha_n} \quad (4.10.23)$$

This result can be expressed as

$$B_n\sqrt{m^2(\omega_0{}^2 - n^2\omega^2)^2 + b^2n^2\omega^2}\; e^{i(\beta_n - \gamma_n)} = F_n e^{-i\alpha_n}$$

$$\tan \beta_n = \frac{bn\omega}{m(\omega_0{}^2 - n^2\omega^2)} \qquad\qquad (4.10.24)$$

Since two complex numbers $re^{i\theta}$ and $r'r^{i\theta'}$ are equal if, and only if, $r' = r$ and $\theta' = \theta$, Eq. (4.10.24) implies

$$B_n = \frac{F_n}{\sqrt{m^2(\omega_0{}^2 - n^2\omega^2)^2 + b^2n^2\omega^2}}$$

$$\gamma_n = \alpha_n + \beta_n \qquad\qquad (4.10.25)$$

The actual steady state solution is the real part of (4.10.21):

$$x = \frac{a_0}{k} + \sum_{n=1}^{\infty} B_n \cos(n\omega t - \gamma_n) \qquad (4.10.26)$$

The solution is simply a sum of the solutions that would result if each term in the Fourier series expansion (4.10.19) of the driving force were present alone! This important *superposition principle* is valid for systems which obey *linear* differential equations. As will be demonstrated in a later section, the superposition principle breaks down for systems which obey *non linear* differential equations

The average rate of power dissipation is $b\dot{x}^2$. The velocity is given by

$$\dot{x} = -\sum_{n=1}^{\infty} B_n n\omega \sin(n\omega t - \gamma_n) \qquad (4.10.27)$$

Hence,

$$\overline{b\dot{x}^2} = \sum_{n=1}^{\infty} \sum_{n'=1}^{\infty} bB_n B_{n'} nn'\omega^2 \frac{1}{2\pi} \int_0^{2\pi} \sin(n\omega t - \gamma_n) \sin(n'\omega t - \gamma_{n'})\, d(\omega t)$$

$$(10.4.28)$$

By application of the identity $\sin(\alpha - \beta) = \sin\alpha\cos\beta - \cos\alpha\sin\beta$ and the orthogonality conditions (4.10.4) we obtain

$$\overline{b\dot{x}^2} = \sum_{n=1}^{\infty} \sum_{n'=1}^{\infty} bB_n B_{n'} nn'\omega^2 \frac{1}{2\pi} \delta_{nn'}\pi$$

$$= \sum_{n=1}^{\infty} \frac{\tfrac{1}{2}bn^2\omega^2 F_n{}^2}{m^2(\omega_0{}^2 - n^2\omega^2)^2 + b^2n^2\omega^2} \qquad (4.10.29)$$

An important point revealed by (4.10.29) is that each term in the solution (4.10.26) is responsible for the same power dissipation as if it were present alone.

As a specific example, suppose that a linear system is driven by the force (4.10.12). From (4.10.14),

$$a_n = F_n = \frac{2F_0}{\pi n} \qquad n \text{ odd}$$

$$B_n = \frac{2F_0}{\pi n \sqrt{m^2(\omega_0^2 - n^2\omega^2)^2 + b^2 n^2 \omega^2}} \qquad n \text{ odd} \qquad (4.10.30)$$

$$B_n = 0 \qquad n \text{ even}$$

For numerical values, choose $F_0 = 1570$ dynes, $m = 10$ gm, $b = 5$ gm/sec and $\omega_0 = 20$ sec^{-1}. The frequency spectrum of the force, F_n versus n, is already plotted in Fig. 4.10.4. For B_n is found

$$B_n = \frac{1000}{n} \frac{1}{\sqrt{100(400 - n^2\omega^2)^2 + 25n^2\omega^2}} \qquad (4.10.31)$$

The frequency spectrum of the displacement is found by plotting B_n versus n. This is done in Fig. 4.10.6 for two values of the driving frequency. For $\omega = 2.00$ sec^{-1} none of the frequencies of the harmonics of the Fourier series expansion of the driving force $F(t)$ exactly matches the resonant frequency ω. None the less, the displacement coefficients B_n show a distinct increase in the vicinity of $n\omega = \omega_0$. For $\omega = 2.22$ sec^{-1}, one of the harmonics in $F(t)$, namely, $n = 9$, exactly matches the resonant frequency $\omega_0 = 20$ sec^{-1} and this particular harmonic in the displacement has a much greater amplitude than any of the others.

If the driving frequency is made exactly equal to ω_0, the coefficient of the fundamental is $B_1 = 10$ cm whereas the coefficient of the next nonzero harmonic is $B_3 = 0.010$ cm. Thus, if the frequency of the driving force matches the resonant frequency, the exact shape of the periodic function $F(t)$ is of little importance. The system vibrates with essentially simple harmonic motion at its natural frequency.

Another measure of the response of the system is the *energy spectrum* which is the energy dissipation as a function of the term in Eq. (4.10.29). Using the same numerical values the total average power dissipated by the system is

$$\overline{b\dot{x}^2} = \sum_{n=1}^{\infty} \frac{2.5\omega^2 \times 10^6}{100(400 - n^2\omega^2)^2 + 25n^2\omega^2} \qquad (4.10.32)$$

Figure 4.10.7 shows a plot of the terms in Eq. (4.10.32) as a function of n for $\omega = 2$ sec^{-1}. The overall energy dissipation is quite small, but it is

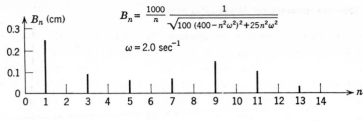

$$B_n = \frac{1000}{n} \frac{1}{\sqrt{100\,(400 - n^2\omega^2)^2 + 25n^2\omega^2}}$$

$\omega = 2.0 \text{ sec}^{-1}$

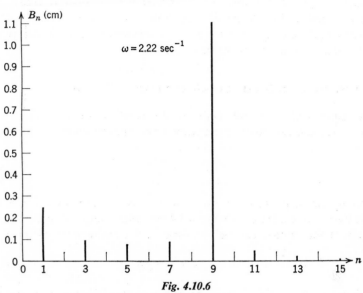

$\omega = 2.22 \text{ sec}^{-1}$

Fig. 4.10.6

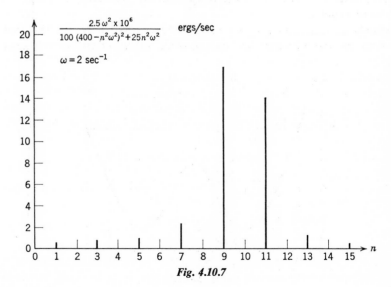

$$\frac{2.5\,\omega^2 \times 10^6}{100\,(400 - n^2\omega^2)^2 + 25n^2\omega^2} \quad \text{ergs/sec}$$

$\omega = 2 \text{ sec}^{-1}$

Fig. 4.10.7

evident that those frequency components of the driving force which are closest to ω_0 are responsible for most of the energy dissipation. For $\omega = 2.22$ sec^{-1}, $n = 9$ gives $\omega n = 20$ sec$^{-1} = \omega_0$. This time, the term $n = 9$ in Eq. (4.10.32) dissipates 1233 ergs per sec, and the other terms dissipate practically nothing by comparison. The value of the damping constant ($b = 5$ gm/sec) which has been used is fairly small resulting in a sharp resonance. The choice of a larger value of b would have the effect of spreading the energy dissipation out over more of the harmonics in the Fourier series expansion of the driving force.

4.11. ONE-DIMENSIONAL CONSERVATIVE SYSTEMS

Any force which depends only on the single position variable x is conservative and can be obtained from a potential function:

$$F(x) = -\frac{dV}{dx} \tag{4.11.1}$$

Following the same line of thought that was used in the discussion of the simple harmonic oscillator in Section 4.4, considerable qualitative information about the motion can be obtained by a consideration of the total energy:

$$\mathscr{E} = \tfrac{1}{2}m\dot{x}^2 + V(x) \tag{4.11.2}$$

This is especially useful in those cases where the differential equation is too hard to solve.

Shown in Fig. 4.11.1 is a conceptually possible potential function. Marked on the $V(x)$-axis are two possible values of the total energy,

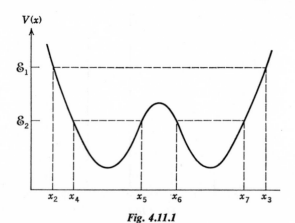

Fig. 4.11.1

$\mathscr{E}_1$ and $\mathscr{E}_2$. When the total energy is $\mathscr{E}_1$, the turning points are x_2 and x_3. When the total energy is $\mathscr{E}_2$, the particle is trapped in one of the two *potential sinks* or *potential wells*. In one of the sinks, the turning points are x_4 and x_5, and in the other they are x_6 and x_7. Once in one of the sinks, the particle cannot move to the other! In all cases illustrated by Fig. 4.11.1, the motion is periodic but *not* simple harmonic unless the sinks are parabolic in shape. Periodic motion which is not simple harmonic is called *anharmonic*.

Another type of potential function is illustrated by the example $V(x) = \mu x^3$ where μ is a constant, Fig. 4.11.2. If the particle has energy $\mathscr{E}$ and is

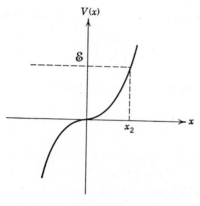

Fig. 4.11.2

initially moving in the positive x-direction, it moves up to the turning point x_2 where the velocity is instantaneously zero, reverses direction, and then moves indefinitely in the negative x-direction. Periodic motion does not occur with this type of potential function.

Consider a potential function which is sink-shaped as in Fig. 4.11.3 with a minimum at x_1. It is possible to expand $V(x)$ in a Taylor's series about x_1:

$$V(x) = V_1 + \left(\frac{dV}{dx}\right)_{x_1}(x - x_1) + \frac{1}{2!}\left(\frac{d^2V}{dx^2}\right)_{x_1}(x - x_1)^2 + \cdots \quad (4.11.3)$$

If the energy of the particle is only slightly larger than V_1, the maximum displacement of the particle from equilibrium will be small. There will be many potential function for which it is possible to neglect terms of order $(x - x_1)^3$ or higher, in which case

$$V(x) = V_1 + \tfrac{1}{2}k(x - x_1)^2 \quad (4.11.4)$$

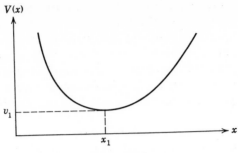

Fig. 4.11.3

where $(dV/dx)_{x_1} = 0$ since x_1 is the position of the minimum and

$$k = \left(\frac{d^2V}{dx^2}\right)_{x_1} > 0 \tag{4.11.5}$$

To this approximation, the force on the particle is

$$F = -\frac{dV}{dx} = -k(x - x_1) \tag{4.11.6}$$

The motion, therefore, is approximately simple harmonic with x_1 as the equilibrium position. If the particle has exactly the energy $\mathscr{E} = V_1$, it remains at rest in a condition of *stable equilibrium*. We call the equilibrium stable because if the particle is disturbed by the addition of a little energy, it begins to oscillate with a small amplitude about the point x_1.

In Fig. 4.11.4 is shown a potential function which is concave downward with its maximum of value V_1 located at x_1. If a particle is at x_1 with exactly the energy V_1 it is in equilibrium since the force is zero. The equilibrium is *unstable* because the particle continues to moves indefinitely away from x_1 once it is disturbed. It is again possible to expand $V(x)$

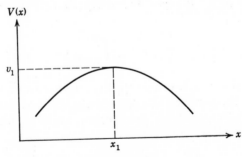

Fig. 4.11.4

about x_1 in a Taylor's series but this time the second derivative is *negative*.

$$k = - \left(\frac{d^2V}{dx^2}\right)_{x_1}$$

so that in place of (4.11.4),

$$V(x) = V_1 - \tfrac{1}{2}k(x - x_1)^2$$

approximately. If the particle is not too far from x_1 the force is given by

$$F = k(x - x_1) \qquad (4.11.7)$$

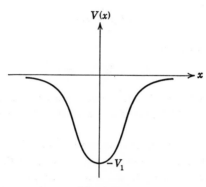

Fig. 4.11.5

and the equation of motion is

$$m\ddot{x} - k(x - x_1) = 0 \qquad (4.11.8)$$

The solution is not periodic but is

$$x - x_1 = c_1 e^{\sqrt{(k/m)}\,t} + c_2 e^{-\sqrt{(k/m)}\,t} \qquad (4.11.9)$$

The particle moves steadily away from x_1, once it is disturbed. When pushed, the particle slides down the potential hill!

Not all periodic motions are simple harmonic even if the amplitude is small. Consider the potential

$$V(x) = -V_1 e^{-ax^4} \qquad (4.11.10)$$

which is sketched in Fig. 4.11.5. The motion is clearly periodic for energies in the range $-V_1 \leq \mathscr{E} < 0$, but the first terms in the expansion of (4.11.10) are

$$V(x) = -V_1(1 - ax^4) \qquad (4.11.11)$$

To this approximation the force is

$$F(x) = -\frac{dV}{dx} = -4aV_1 x^3 \qquad (4.11.12)$$

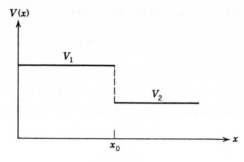

Fig. 4.11.6

The equation of motion is

$$m\ddot{x} + 4aV_1x^3 = 0 \tag{4.11.13}$$

showing that even for very small amplitudes the motion is not simple harmonic. Equation (4.11.13) is an example of a *nonlinear* differential equation.

Strictly speaking, a potential function should be continuous since it should possess a derivative at every point that gives the force. Nevertheless, it is convenient at times to use a discontinuous function such as the potential step function of Fig. 4.11.6. It should be understood that the step is an approximation to a physical situation where $V(x)$ changes rapidly over a short but finite distance. The question of what actually happens at the discontinuity can be avoided by using the principle of conservation of energy. If a particle is in the region where the potential energy is V_1 and is moving in the positive x-direction, then its velocity after it crosses the discontinuity at $x = x_0$ can be found from

$$\mathscr{E} = \tfrac{1}{2}mu_1{}^2 + V_1 = \tfrac{1}{2}mu_2{}^2 + V_2 \tag{4.11.14}$$

The force on the particle as given by $F = -dV/dx$ is not defined in the ordinary sense at the point of discontinuity. Nevertheless, the concept of derivative can be extended to include the point of discontinuity by the use of a *Dirac delta function* or *generalized function*. This concept can be arrived at by a limiting process. Consider the potential function in Fig. 4.11.7 and the force obtained from it:

$$F = 0 \quad x < x_1$$

$$F = -\frac{V_2 - V_1}{x_2 - x_1} = -\frac{\Delta V}{\Delta x} \quad x_1 < x < x_2 \tag{4.11.15}$$

$$F = 0 \quad x > x_2$$

The work done on the particle as it passes from $x < x_1$ to $x > x_2$ is

$$\int F \, dx = \tfrac{1}{2}mu_2{}^2 - \tfrac{1}{2}mu_1{}^2 = -(V_2 - V_1) \qquad (4.11.16)$$

This of course represents the area under the F versus x graph of Fig. 4.11.7. Imagine now that the limit $x_1 \to x_0$, $x_2 \to x_0$ is taken so that $V(x)$ pictured in Fig. 4.11.7 approaches the discontinuous function of Fig. 4.11.6. In

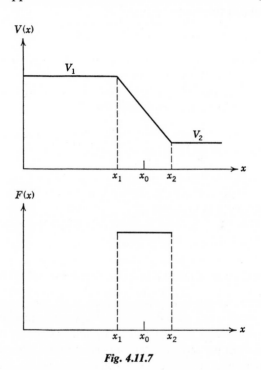

Fig. 4.11.7

order to preserve the conservation of energy, the limit must be taken in such a way that the area under the F versus x graph of Fig. 4.11.7 remains the same. This function therefore becomes very high and narrow in the limit. When the limit is actually reached, F has become a *Dirac delta function* which we write as

$$F(x) = C\delta(x - x_0) \qquad (4.11.17)$$

where C is a constant. By means of Eq. (4.11.16),

$$\int C\delta(x - x_0) \, dx = -(V_2 - V_1) \qquad (4.11.18)$$

It is the usual practice to normalize the δ-function so that

$$\int \delta(x - x_0)\, dx = 1 \qquad (4.11.19)$$

Therefore, the constant C is equal to the discontinuity in $V(x)$ at $x = x_0$

$$C = -(V_2 - V_1) \qquad (4.11.20)$$

The δ function has the property that

$$\delta(x - x_0) = 0 \qquad x \neq x_0 \qquad (4.11.21)$$

The integral of the δ-function is unity provided of course that the point $x = x_0$ is included in the range of integration.

Let $f(x)$ be a function which is continuous in the vicinity of x_0. The δ-function has the important property that

$$\int f(x)\, \delta(x - x_0)\, dx = f(x_0) \qquad (4.11.22)$$

The δ-function was first introduced by P. A. M. Dirac* in connection with certain analytical procedures used in quantum mechanics. Traditional mathematicians were at first outraged by the idea but physicists continued to use it. In recent years, the δ-function has attained respectability and a rigorous theory of it has been developed.†

4.12. EXAMPLES OF CONSERVATIVE NONLINEAR SYSTEMS

In this section, we study physical systems with one degree of freedom that are conservative but obey nonlinear differential equations. There is a problem in terminology which should be pointed out. In many texts the term linear motion is used to mean *one-dimensional motion*, i.e., motion along a straight line described by a single variable x. The differential equation, however, may be nonlinear in the mathematical sense. The most general linear second order differential equation is

$$\ddot{x} + f(t)\dot{x} + g(t)x = h(t) \qquad (4.12.1)$$

A differential equation which contains any other combination of x, $\dot{x}$, and $\ddot{x}$ such as x^2, $x\dot{x}$, $x^2\ddot{x}$, etc., is termed *nonlinear*. Actually, most physical systems obey nonlinear differential equations. The linear differential

---

* P. A. M. Dirac, *The Principles of Quantum Mechanics*, 3rd ed., Oxford University Press, London, 1947.

† M. J. Lighthill, *Fourier Analysis and Generalized Functions*, Cambridge University Press, London, 1958.

equations studied so far in this chapter represent in many instances only a first approximation. Before taking up the problem of solving nonlinear differential equations, several examples will be discussed.

In Fig. 4.12.1a is a particle held at rest by two identical springs. Assuming that there is no gravitational field, each spring will be stretched an amount s. Let the particle be pulled directly to the side as in Fig. 4.12.1b and released. When the displacement of the particle is x, the actual amount that the spring is stretched is

$$\sqrt{x^2 + s_0{}^2} - (s_0 - s)$$

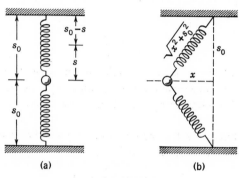

(a)   (b)

Fig. 4.12.1

The potential energy of the system is therefore

$$V(x) = 2 \cdot \tfrac{1}{2} k [\sqrt{x^2 + s_0{}^2} - (s_0 - s)]^2 \qquad (4.12.2)$$

The equation of motion is found from $m\ddot{x} = -dV/dx$ with the result

$$m\ddot{x} + \frac{2kx[s - s_0 + \sqrt{x^2 + s_0{}^2}]}{\sqrt{x^2 + s_0{}^2}} = 0 \qquad (4.12.3)$$

which is a difficult nonlinear differential equation. If x is less than s_0, it is possible to find an approximate equation of motion which is less formidable. The potential energy can be expressed

$$V(x) = k[x^2 + s_0{}^2 - 2(s_0 - s)\sqrt{x^2 + s_0{}^2} + (s_0 - s)^2] \qquad (4.12.4)$$

and, assuming $x < s_0$, the radical can be expanded by means of the binomial theorem to yield

$$\sqrt{x^2 + s_0{}^2} = s_0 \sqrt{1 + (x/s_0)^2} = s_0 \left(1 + \frac{1}{2}\frac{x^2}{s_0{}^2} - \frac{1}{8}\frac{x^4}{s_0{}^4} + \cdots\right) \qquad (4.12.5)$$

To this approximation, after rearrangement of terms, the potential energy is

$$V(x) = ks^2 + \frac{ks}{s_0}x^2 + \frac{k(s_0 - s)}{4s_0^{\,3}}x^4 \qquad (4.12.6)$$

The approximate equation of motion is

$$m\ddot{x} + \frac{2ks}{s_0}x + \frac{k(s_0 - s)}{s_0^{\,3}}x^3 = 0 \qquad (4.12.7)$$

Provided $s \neq 0$, the x^3 term can be neglected if the amplitude of the motion is very small. For example, suppose the maximum value of x is such that $x/s_0 = 0.1$. Then $(x/s_0)^3 = 0.001$. To a good approximation, the motion is simple harmonic of frequency

$$\omega_0 = \sqrt{\frac{2ks}{ms_0}} \qquad (4.12.8)$$

The x^3-term can be regarded as a small correction term or *perturbation* on the motion. After the consideration of some more examples we will take up the problem of computing the effect of the nonlinear x^3 term on the motion.

If the initial tension in the springs is made zero, then $s = 0$ and the *first* approximation to the motion is

$$m\ddot{x} + \frac{kx^3}{s_0^{\,2}} = 0 \qquad (4.12.9)$$

In this case, no linear approximation to the motion exists.

Recall from elementary mechanics that the equation of motion of a simple plane pendulum of length s (Fig. 4.12.2) is

$$\ddot{\theta} + \frac{g}{s}\sin\theta = 0 \qquad (4.12.10)$$

The motion is not one-dimensional, but the system has only one degree of freedom, i.e., is described by one coordinate θ. If the function $\sin\theta$ is expanded by means of a Taylor's series and two terms are retained

$$\ddot{\theta} + \frac{g}{s}\theta - \frac{g}{6s}\theta^3 = 0 \qquad (4.12.11)$$

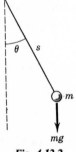

Fig. 4.12.2

which is a nonlinear differential equation of the same form as Eq. (4.12.7).

Figure 4.12.3 shows two straight wires carrying currents J_1 and J_2. One wire, carrying current J_1, is rigidly fixed and the other carrying current J_2 is suspended from two springs of force constant k. Let x be the position coordinate of the movable wire and let $x = 0$ be the equilibrium

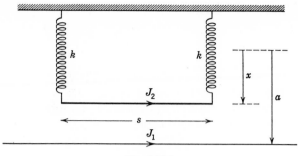

Fig. 4.12.3

position of this wire when no currents flow. Neglecting end effects, the magnetic force on the suspended wire is

$$F = \frac{J_2 sB}{c} \tag{4.12.12}$$

where B is the magnetic field due to the fixed wire and c is the velocity of light (Gaussian units). The magnetic field produced by the fixed wire is

$$B = \frac{2J_1}{c(a - x)} \tag{4.12.13}$$

The equation of motion of the suspended wire, therefore, is

$$m\ddot{x} + 2k\left[x - \frac{\alpha}{a - x}\right] = 0 \tag{4.12.14}$$

where the parameter α is

$$\alpha = \frac{sJ_1 J_2}{kc^2} \tag{4.12.15}$$

The motion can be studied qualitatively using the potential function

$$\frac{dV}{dx} = 2k\left[x - \frac{\alpha}{a - x}\right]$$

$$V(x) = k\left[x^2 + 2\alpha \log\left(1 - \frac{x}{a}\right)\right] \tag{4.12.16}$$

From $dV/dx = 0$ the maxima and minima of $V(x)$ are found to be

$$x_1 = \frac{a}{2} - \sqrt{(a/2)^2 - \alpha}$$

$$x_{11} = \frac{a}{2} + \sqrt{(a/2)^2 - \alpha} \tag{4.12.17}$$

Suppose, for example, that $a = 10\,\text{cm}$ and $\alpha = 16\,\text{cm}^2$. Then

$$\frac{V}{k} = x^2 + 32\log\left(1 - \frac{x}{10}\right)$$

$$x_1 = 2\,\text{cm} \qquad x_{11} = 8\,\text{cm}$$

(4.12.18)

This function is sketched in Fig. 4.12.4.

Oscillatory motion is possible for energy in the range $V_1 \leq \mathscr{E} < V_{11}$ where V_{11} is the value of $V(x)$ at its relative maximum at x_{11}. If the wire is

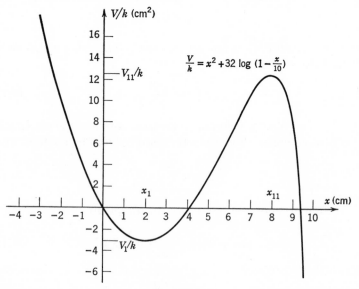

Fig. 4.12.4

placed at $x = 2.00\,\text{cm}$ with zero initial velocity, it remains there indefinitely in a condition of *stable equilibrium*. Physically, the force due to the springs exactly balances the force due to the magnetic field. If the wire is released at $x = 1.00\,\text{cm}$, the value of V/k is about $-2.36\,\text{cm}^2$. The motion would be very nearly simple harmonic between the turning points $x = 1\,\text{cm}$ and $x = 3\,\text{cm}$ as illustrated in Fig. 4.12.5a. If the wire is pulled back to $x = -2.35\,\text{cm}$ and released, its energy is only slightly less than V_{11} and its motion resembles Fig. 4.12.5b. It tends to linger somewhat longer at its maximum displacement. This is because the larger turning point is near to the point on $V(x)$ where $dV/dx = 0$. The particle approaches this point with diminishing velocity and the restoring force tending to accelerate it in the opposite direction is very small. If the initial displacement of the wire is made about $-2.40\,\text{cm}$, the energy is exactly V_{11}.

The motion is as sketched in Fig. 4.12.5c. The wire approaches asymptotically $x = 8.00$ cm and for this reason this special case is called the *sticking solution*. If the initial position is $x = -2.5$ cm, the energy is slightly above V_{11}. The wire will clear the peak at $x = 8.00$ cm with a small velocity (Fig. 4.12.5d) and then accelerate sharply toward the stationary wire until it finally collides with it. The curves drawn in Fig. 4.12.5 are only qualitative; as yet we do not know their exact shapes.

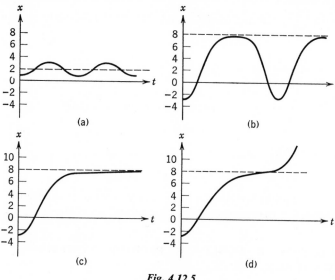

Fig. 4.12.5

The parameter α can be made sufficiently large, e.g., by making the currents large, so that

$$\left(\frac{a}{2}\right)^2 - \alpha < 0 \tag{4.12.19}$$

The roots (4.12.17) then become complex number, meaning that $V(x)$ no longer has any relative maxima or minima. Periodic motion can no longer occur. When the suspended wire is released, it moves toward the fixed wire until it collides with it.

If the motion is periodic, an approximate equation of motion can be obtained by expanding the potential function in a Taylor's series about the position of stable equilibrium x_1:

$$V(x) = V_1 + \frac{1}{2}\left(\frac{d^2V}{dx^2}\right)_{x_1}(x - x_1)^2 + \frac{1}{6}\left(\frac{d^3V}{dx^3}\right)_{x_1}(x - x_1)^3$$
$$+ \frac{1}{24}\left(\frac{d^4V}{dx^4}\right)(x - x_1)^4 + \cdots \tag{4.12.20}$$

It is found after evaluating the derivatives and using

$$a - x_1 = \frac{\alpha}{x_1} \tag{4.12.21}$$

that

$$V(x) = V_1 + \frac{k(\alpha - x_1{}^2)}{\alpha}(x - x_1)^2 - \frac{2kx_1{}^3}{3\alpha^2}(x - x_1)^3 - \frac{kx_1{}^4}{2\alpha^3}(x - x_1)^4 \tag{4.12.22}$$

Equation (4.12.21) follows from $(dV/dx)_{x_1} = 0$. The approximate equation of motion is

$$m\ddot{x} + \frac{2k(\alpha - x_1{}^2)}{\alpha}(x - x_1) - \frac{2kx_1{}^3}{\alpha^2}(x - x_1)^2 - \frac{2kx_1{}^4}{\alpha^3}(x - x_1)^3 = 0 \tag{4.12.23}$$

It is possible to change the dependent variable by letting

$$y = x - x_1 \tag{4.12.24}$$

The equation of motion is then

$$m\ddot{y} + k_1 y - k_2 y^2 - k_3 y^3 = 0 \tag{4.12.25}$$

Where k_1, k_2, and k_3 are constants. This amounts to choosing the origin of coordinates at the position of stable equilibrium of the system.

4.13. SOLUTION OF NONLINEAR PERIODIC SYSTEMS BY FOURIER SERIES

If the potential function is symmetric such as in the first two examples considered in Section 4.12,

$$V(x) = V(-x) \tag{4.13.1}$$

The Taylor's series expansion must contain only even powers of x:

$$V(x) = \tfrac{1}{2}k_1 x^2 + \tfrac{1}{4}k_3 x^4 + \cdots \tag{4.13.2}$$

There is no loss in generality incurred by setting any constant term in $V(x)$ equal to zero. The equation of motion is

$$m\ddot{x} + k_1 x + k_3 x^3 + \cdots = 0 \tag{4.13.3}$$

As examples, see Eq. (4.12.7) and (4.12.11). It is convenient to divide equation (4.13.3) by m and also to work with the potential energy per unit mass:

$$\ddot{x} + \alpha x + \gamma x^3 = 0 \tag{4.13.4}$$

$$\psi(x) = \frac{V}{m} = \tfrac{1}{2}\alpha x^2 + \tfrac{1}{4}\gamma x^4 \tag{4.13.5}$$

Where $\alpha = k_1/m$ and $\gamma = k_3/m$. Even though (4.13.4) generally represents only an approximate equation of motion, the problem of obtaining its solution will be treated in some detail. A variety of cases can occur depending on the signs of α and γ. If $\alpha > 0$ and $\gamma > 0$, $\psi(x)$ is as sketched in Fig. 4.13.1a. The only type of motion which can occur is periodic. This is sometimes referred to as a *hard spring*. If $\alpha > 0$ but $\gamma < 0$, the potential function $\psi(x)$ is like Fig. 4.13.1b. The spring is said to be *soft*. For a soft spring, both periodic and non-periodic motion can occur.

For the periodic motion which can occur for the potential functions of Figs. 4.13.1a and b, a Fourier series representation of the solution exists:

$$x(t) = a_0 + \sum_{n=1}^{\infty} (a_n \sin n\omega t + b_n \cos n\omega t) \qquad (4.13.6)$$

Many of the coefficients in (4.13.6) can be shown to be zero by using the *symmetry properties* of the motion. The symmetry of the potential function as given by equation (4.13.1) means that the average position of the particle is zero. Therefore,

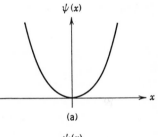

$$a_0 = \frac{1}{2\pi} \int_0^{2\pi} x(t) \, d(\omega t) = 0 \qquad (4.13.7)$$

The shape of $x(t)$ for $x < 0$ must be exactly the same as for $x > 0$.

Let us choose the initial velocity equal to zero. The qualitative form of $x(t)$ as a function of ωt is sketched in Fig. 4.13.2. The equation of motion (4.13.4) is not affected by a reversal of the direction of time. This symmetry with respect to a reversal of the direction of time is typical of systems which conserve energy. This means that if the system is started at $\omega t = 2\pi$ and run backwards, e.g., by running a motion picture of the particle

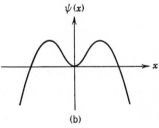

(a)

(b)

Fig. 4.13.1

backwards, exactly the same motion is obtained as by running the system forward from $\omega t = 0$. To put it another way, the solution for $\pi \leq \omega t < 2\pi$ is a mirror reflection of the solution for $0 \leq \omega t < \pi$ as illustrated in Fig. 4.13.2.

Figure 4.13.3 shows the first cycle of $x(t)$ and also $\sin \omega t$. Recall that

$$a_1 = \frac{1}{\pi} \int_0^{2\pi} x(t) \sin \omega t \, d(\omega t) \qquad (4.13.8)$$

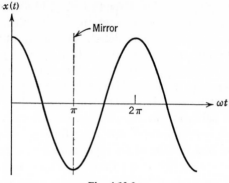

Fig. 4.13.2

It is evident that the integral over $\pi \leq \omega t < 2\pi$ cancels that over $0 \leq \omega t < \pi$. By a similar argument, *all* the coefficients of the sine terms in (4.13.6) can be shown to be zero.

Figure 4.13.4 shows the first cycle of $x(t)$ and also $\cos 2\omega t$. The coefficient b_2 is given by

$$b_2 = \frac{1}{\pi} \int_0^{2\pi} x(t) \cos 2\omega t \, d(\omega t) \tag{4.13.9}$$

and a moments consideration shows that the contribution to b_2 from $\pi/2 \leq \omega t < \pi$ cancels that from $0 \leq \omega t < \pi/2$. Thus $b_2 = 0$. Similarly, $b_4 = b_6 = \cdots = 0$. It can be concluded that the periodic solution of (4.13.4), for potential functions of the form given in Fig. 4.13.1, is of the form

$$x(t) = b_1 \cos \omega t + b_3 \cos 3\omega t + b_5 \cos 5\omega t + \cdots \tag{4.13.10}$$

Provided that $\dot{x} = 0$ at $t = 0$.

The remaining coefficients, as well as the frequency ω of the motion, are to be determined from the differential equation. Substituting the solution (4.13.10) into the differential equation and neglecting the fifth and higher

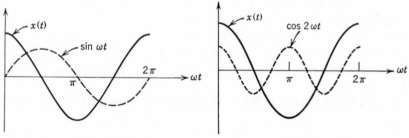

Fig. 4.13.3   Fig. 4.13.4

harmonics yields

$$b_1(\alpha - \omega^2) \cos \omega t + b_3(\alpha - 9\omega^2) \cos 3\omega t$$

$$+ \gamma (b_1 \cos \omega t + b_3 \cos 3\omega t)^3 = 0 \quad (4.13.11)$$

In working out the x^3 term we write

$$(b_1 \cos \omega t + b_3 \cos 3\omega t)^3 \cong b_1{}^3 \cos^3 \omega t + 3b_1{}^2 b_3 \cos^2 \omega t \cos 3\omega t$$

$$(4.13.12)$$

The coefficient b_1 is *first order* and b_3 is *third order*. A term such as $b_1{}^3$ is *third order* and $b_1{}^2 b_3$ is *fifth order*. The higher order terms $b_1 b_3{}^2$, and $b_3{}^3$ are neglected in (4.13.12). It is assumed that the magnitude of the terms decreases with increasing order. By means of the trigonometric identities

$$\cos^3 \omega t = \tfrac{1}{4} \cos 3\,\omega t + \tfrac{3}{4} \cos \omega t \quad (4.13.13)$$

$$\cos^2 \omega t \cos 3\omega t = \tfrac{1}{4} \cos 5\omega t + \tfrac{1}{2} \cos 3\omega t + \tfrac{1}{4} \cos \omega t \quad (4.13.14)$$

Eq. (4.13.11) can be rewritten as

$$[b_1(\alpha - \omega^2) + \tfrac{3}{4}\gamma b_1{}^3 + \tfrac{3}{4}\gamma b_1{}^2 b_3] \cos \omega t$$

$$+ [b_3(\alpha - 9\omega^2) + \tfrac{1}{4}\gamma b_1{}^3 + \tfrac{3}{2}\gamma b_1{}^2 b_3] \cos 3\omega t = 0 \quad (4.13.15)$$

where the fifth harmonic has been neglected. In Appendix A will be found several trigonometric identities which find use in this type of calculation. In order for (4.13.15) to be an identity for all values of t, it is necessary that the coefficients of harmonics $\cos \omega t$ and $\cos 3\omega t$ separately vanish. Therefore,

$$b_1(\alpha - \omega^2) + \tfrac{3}{4}\gamma b_1{}^3 + \tfrac{3}{4}\gamma b_1{}^2 b_3 = 0 \quad (4.13.16)$$

$$b_3(\alpha - 9\omega^2) + \tfrac{1}{4}\gamma b_1{}^3 + \tfrac{3}{2}\gamma b_1{}^2 b_3 = 0 \quad (4.13.17)$$

which gives two relations for the determination of b_3 and ω in terms of b_1. Neglecting the b_3 term in (4.13.16) gives for the frequency

$$\omega^2 = \alpha + \tfrac{3}{4}\gamma b_1{}^2 \quad (4.13.18)$$

and the use of this expression in (4.13.17) yields for the coefficient b_3

$$b_3 = \frac{\gamma b_1{}^3}{32\alpha \left[1 + \dfrac{21}{32\alpha}\gamma b_1{}^2\right]} \cong \frac{\gamma b_1{}^3}{32\alpha}\left[1 - \frac{21}{32\alpha}\gamma b_1{}^2\right] \cong \frac{\gamma b_1{}^3}{32\alpha} \quad (4.13.19)$$

This result can now be substituted back into (4.13.16) to yield a slightly more accurate expression for the frequency:

$$\omega^2 = \alpha + \tfrac{3}{4}\gamma b_1{}^2 + \frac{3\gamma^2 b_1{}^4}{128\alpha} \quad (4.13.20)$$

If the velocity at $t = 0$ is not zero, it is merely necessary to shift the origin in Fig. 4.13.2. To do this, we simply subtract a phase angle from ωt. Our approximate solution is then

$$x = b_1 \cos (\omega t - \phi) + \frac{\gamma b_1{}^3}{32\alpha} \cos (3\omega t - 3\phi) \qquad (4.13.21)$$

and contains two undetermined constants b_1 and ϕ which are to be found from initial conditions. Notice that no fifth order term, i.e., a term proportional to $b_1{}^5$, is included in (4.13.20) or (4.13.21). It does no good to include such terms since terms of the fifth order have already been excluded by neglecting the fifth harmonic. In deriving (4.13.20) and (4.13.21), it has been assumed that the linear term αx is relatively more important than the nonlinear term γx^3. In problems 4.13.30 and 4.13.31 we consider the case where the situation is reversed.

If $\gamma = 0$, the equation of motion (4.13.4) reduces to the equation of simple harmonic motion and (4.13.20) and (4.13.21) become simply

$$\omega^2 = \alpha = \frac{k_1}{m} \qquad x = b_1 \cos (\omega t - \phi) \qquad (4.13.22)$$

as is to be expected. The simple harmonic oscillator is *isochronous*, meaning that the frequency does not depend on the amplitude. Equation (4.13.20) reveals the important fact that the frequency of a nonlinear periodic system is *amplitude dependent*. By setting $\omega t - \phi = 0$, the actual amplitude of the motion is found to be

$$x_2 = b_1 + \frac{\gamma b_1{}^3}{32\alpha}$$

To second order, $x_2 = b_1$. The frequency is then approximately

$$\omega^2 = \alpha + \tfrac{3}{4}\gamma x_2{}^2 \qquad (4.13.23)$$

It is possible to evaluate the Fourier coefficients by another method. Writing

$$b_n = \frac{1}{\pi} \int_0^{2\pi} x \cos (n\omega t)\, d(\omega t)$$

$$= \frac{1}{\pi} \int_0^{2\pi} x \frac{1}{n} \frac{d}{d(\omega t)} \sin (n\omega t)\, d(\omega t)$$

and integrating by parts yields

$$b_n = \frac{1}{\pi n}\left[x \sin (n\omega t) \Big|_0^{2\pi} - \int_0^{2\pi} \frac{dx}{d(\omega t)} \sin (n\omega t)\, d(\omega t)\right]$$

$$= -\frac{1}{\pi n \omega} \int_0^{2\pi} \frac{dx}{dt} \sin (n\omega t)\, d(\omega t)$$

Integrating once more by parts gives

$$b_n = \frac{1}{\pi n^2 \omega} \left[\frac{dx}{dt} \cos (n\omega t) \Big|_0^{2\pi} - \frac{1}{\omega} \int_0^{2\pi} \frac{d^2 x}{dt^2} \cos (n\omega t) \, d(\omega t) \right]$$

Since the velocity is zero at 0 and 2π,

$$b_n = -\frac{1}{\pi n^2 \omega^2} \int_0^{2\pi} \ddot{x} \cos (n\omega t) \, d(\omega t) \qquad (4.13.24)$$

Substituting for the acceleration from the equation of motion yields

$$b_n = \frac{1}{\pi n^2 \omega^2} \int_0^{2\pi} (\alpha x + \gamma x^3) \cos (n\omega t) \, d(\omega t)$$

$$= \frac{\alpha b_n}{n^2 \omega^2} + \frac{\gamma}{\pi n^2 \omega^2} \int_0^{2\pi} x^3 \cos (n\omega t) \, d(\omega t)$$

Hence,

$$b_n = \frac{\gamma}{\pi (n^2 \omega^2 - \alpha)} \int_0^{2\pi} x^3 \cos (n\omega t) \, d(\omega t) \qquad (4.13.25)$$

By substituting $x = b_1 \cos \omega t + b_3 \cos 3\omega t + b_5 \cos 5\omega t + \cdots$ into the integral and performing the indicated integrations for $n = 1, 3, 5, \ldots$ a series of algebraic equations for the coefficients are found. For instance, in order to compute b_5 accurate to fifth order, Eq. (4.13.12) is substituted into the integrand and the identities (4.13.14) are used. Because of the orthogonality relations only the fifth harmonic contributes so that all other terms can be discarded. Thus

$$b_5 = \frac{\gamma}{\pi (25\omega^2 - \alpha)} \int_0^{2\pi} 3 b_1^2 b_3 \cdot \tfrac{1}{4} \cos (5\omega t) \cdot \cos (5\omega t) \, d(\omega t)$$

$$= \frac{3}{4} \frac{b_1^2 b_3 \gamma}{(25\omega^2 - \alpha)} \qquad (4.13.26)$$

Similarly, by letting $n = 1$ and 3 in (4.13.25), Eqs. (4.13.16) and (4.13.17) can be duplicated. In keeping all quantities accurate to fifth order, the approximation $\omega^2 = \alpha$ and b_3 as given by (4.13.19) are sufficient in (4.13.26):

$$b_5 = \frac{\gamma^2 b_1^5}{1024 \alpha^2} \qquad (4.13.27)$$

The equation of motion, accurate to fifth order terms, is

$$x = b_1 \cos \omega t + \frac{\gamma b_1^3}{32\alpha \left[1 + \dfrac{21\gamma b_1^2}{32\alpha} \right]} \cos 3\omega t + \frac{\gamma^2 b_1^5}{1024 \alpha^2} \cos 5\omega t \qquad (4.13.28)$$

The frequency is given by (4.13.20) and it is evident that it contains only even powers of b_1 so that the first missing term is of order 6. As mentioned previously, the point of view is taken that the linear term is more important than the nonlinear term since, obviously, if α is not too small, the coefficient of the fifth harmonic is small in Eq. (4.13.28). It has also been assumed that $\alpha > 0$. The Fourier series method, however, is quite versatile, applying in any situation where the motion is periodic. In studying new situations it is best to go back to the basic equations—(4.13.16), (4.13.17), and (4.13.26). See, for example, Problem 4.13.32. It may even be necessary to extend the calculation of x^3 to include more terms than was done in Eq. (4.13.12).

Equation (4.13.25) gives a method of estimating the order of magnitude of all coefficients. If the integrand is replaced by its maximum value,

$$|b_n| < \frac{|\gamma|}{\pi(n^2\omega^2 - \alpha)} \int_0^{2\pi} x_2^3 \, d(\omega t) = \frac{2\,|\gamma|\,x_2^3}{n^2\omega^2 - \alpha} \qquad (4.13.29)$$

where x_2 is the amplitude of the motion. The coefficients decrease roughly in proportion to $1/n^2$.

If the potential function is not symmetric, the Taylor's series expansion (4.13.2) will include a cube-term with a consequent appearance of a square-term in the differential equation:

$$\ddot{x} + \alpha x - \beta x^2 + \gamma x^3 = 0 \qquad (4.13.30)$$

(The minus sign is included for convenience; the constant β nearly always turns out to be positive in applications.) Both periodic and nonperiodic solutions of (4.13.30) exist; we consider here only the periodic solutions. The asymmetry of the potential function means that the Fourier series solution will contain both even and odd harmonics:

$$x = a_0 + b_1 \cos \omega t + b_2 \cos 2\omega t + b_3 \cos 3\omega t + \cdots \qquad (4.13.31)$$

The Fourier coefficients are found to obey

$$a_0 = \frac{1}{2\pi\alpha} \int_0^{2\pi} (\beta x^2 - \gamma x^3) \, d(\omega t) \qquad (4.13.32)$$

$$b_n = -\frac{1}{\pi(n^2\omega^2 - \alpha)} \int_0^{2\pi} x^2(\beta - \gamma x) \cos n\omega t \, d(\omega t) \qquad (4.13.33)$$

If third order accuracy is maintained, the above equations lead to

$$a_0 = \frac{1}{2}\frac{\beta}{\alpha}b_1^2$$

$$b_1(\omega^2 - \alpha) = \tfrac{3}{4}\gamma b_1^3 - 2\beta b_1 a_0 - \beta b_1 b_2$$

$$b_2(4\omega^2 - \alpha) = -\tfrac{1}{2}\beta b_1^2 \qquad (4.13.34)$$

$$b_3(9\omega^2 - \alpha) = \tfrac{1}{4}\gamma b_1^3 - \beta b_1 b_2$$

The approximate Fourier series solution is

$$x = \frac{\beta b_1^2}{2\alpha} + b_1 \cos \omega t - \frac{\beta b_1^2}{6\alpha} \cos 2\omega t$$

$$+ \left(\frac{\gamma b_1^3}{32\alpha} + \frac{\beta^2 b_1^3}{48\alpha^2} \right) \cos 3\omega t \qquad (4.13.35)$$

$$\omega^2 = \alpha + \tfrac{3}{4}\gamma b_1^2 - \frac{5}{6}\frac{\beta^2 b_1^2}{\alpha} \qquad (4.13.36)$$

There is an interesting and important feature of our solution which is often overlooked. The coefficient of the third harmonic has third order terms which result from the presence of both the square-term and the cube-term in (4.13.30). Frequently in applications these terms are both of the same order of magnitude! The same thing is true of the expression for the frequency which has two second order terms, one proportional to γ and the other to β^2. In many applications, it is not enough to include just one more term beyond the linear term in the differential equation. This leaves open the question of what happens if even higher terms are included in the differential equation (4.13.30). The interested reader may wish to pursue this question on his own.

PROBLEMS

4.2.1 If the force acting on a particle is conservative, show that a formal solution of the equation of motion is

$$t - t_0 = \int_{x_0}^{x} \frac{dx}{\sqrt{(2/m)[\mathscr{E} - V(x)]}}$$

where x_0 is the value of x at t_0.

4.3.2 A particle is released from rest and is observed to fall 320 ft in 5 sec. What is the value of the characteristic time τ? From equation (3.13),

$$\frac{h}{g} = 10 \sec^2 = -\tau^2(1 - e^{-5/\tau}) + 5\tau$$

This is not directly solvable for τ and the power series method is not convenient to use. Therefore, it is suggested that a plot of

$$f(\tau) = -\tau^2(1 - e^{-5/\tau}) + 5\tau$$

be made as a function of τ and the value of τ read from the graph which gives $f(\tau) = 10 \sec^2$.

4.3.3 If in Problem (4.3.2), the particle has a mass of 0.1 slug, how much energy is converted to heat over the 320-ft drop?

4.3.4 The frictional force on a body can frequently be represented by a series of the form

$$\text{frictional force} = \pm a - b\dot{x} \pm c\dot{x}^2 - \cdots$$

where the signs must be chosen so that the force is always opposite in direction to the velocity. The constants a, b, c, ... are experimentally determined. In elementary physics courses, the dry friction between surfaces is usually represented by just the constant term and written $\pm \mu N$ where μ is the coefficient of friction and N is the normal force. Suppose that a body moves vertically in a medium under the influence of a gravitational field and there is present a frictional force proportional to the square of the velocity. Integrate the equation of motion. Consider the case of upward motion and downward motion separately.

4.3.5 A particle moves upward in a medium in which the frictional force is proportional to the square of the velocity. Find the velocity as a function of height. *Hint*: In the equation of motion write

$$\ddot{x} = \frac{d\dot{x}}{dt} = \frac{d\dot{x}}{dx}\dot{x} = \frac{d}{dx}(\tfrac{1}{2}\dot{x}^2)$$

Solve the differential equation for $\dot{x}^2$ as a function of x. Find the maximum height to which the particle will rise for a given initial velocity.

4.6.6 Suppose that there is a *negative* frictional force proportional to the velocity on a harmonic oscillator so that the equation of motion is

$$m\ddot{x} - b\dot{x} + kx = 0$$

Without actually finding the solution of the equation of motion, state what the qualitative nature of the motion will be. The problem can be attacked as follows: What happens in the differential equation for positive damping,

$$m\ddot{x} + b\dot{x} + kx = 0$$

if the direction of time is reversed?

4.6.7 A mass of 10 gm is suspended from a spring of force constant 250 dynes/cm. The value of the viscous damping constant is $b = 50$ gm/sec. What is the approximate percentage of error made in calculating the frequency if the damping is neglected? If the initial conditions are $x_0 = 10$ cm and $u_0 = 0$, make a fairly accurate plot of the motion as a function of time. Are the points where $\cos(\omega_1 t - \alpha) = \pm 1$ the locations of relative maxima and minima?

4.6.8 A mass of 10 gm is suspended from a spring of force constant 250 dynes/cm. The viscous damping constant is $b = 120$ gm/sec. If $x_0 = 10$ cm and $u_0 = 0$, evaluate the constants of integration and make a fairly accurate plot of the motion as a function of time.

4.6.9 A damped oscillator is observed to have an initial amplitude of 6.00 cm. After 20 cycles the amplitude has reduced to 3.10 cm. If the frequency is $\nu_1 = 4$ cycles/sec, find the time constant $\tau = m/b$.

4.6.10 A particle of mass m is tied to a spring of force constant k and experiences a damping force which is that of dry friction, i.e., the friction force is constant but reverses direction when the velocity reverses direction. Discuss the motion in detail. Show that the decrease in amplitude per cycle is a constant and, therefore, that the envelope of the motion is two straight lines. Neglect any difference in the value of kinetic and static friction. Assume that the particle is released from rest at an initial displacement x_0. Plot the friction force as a function of velocity.

4.7.11 Figure 4.7.8 shows the average power dissipated by a linear system driven with a simple harmonic driving force of frequency ω. If $\omega_0 + \epsilon$ is defined to be the value of the driving frequency where the power is one-half its maximum value, show that $\epsilon \cong \pm b/(2m)$ if the damping is small. This is a measure of the width of the power resonance and shows it to be *narrow* if b is *small*.

4.7.12 A linear oscillator consists of a mass m attached to a spring of force constant k. The oscillator experiences a viscous damping proportional to the velocity. The oscillator is initially quiescent. At $t = 0$, a force of constant value F_0 is applied. Find the motion of the oscillator.

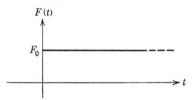

4.8.13 What combination of capacitors is equivalent to the following combinations of springs?

In each case, find the effective spring constant of the combination.

4.8.14 Construct the electric circuits which are the analogues of the following mechanical systems:

Write out the differential equations for both the mechanical and the electrical systems.

4.8.15   For the circuit shown in the figure, show that the complex current in impedance Z_3 is

$$I_3 = \frac{Z_2 V}{Z_1 Z_2 + Z_1 Z_3 + Z_2 Z_3}$$

where $V = \phi_0 e^{i\omega t}$ is the complex driving voltage.

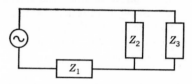

4.8.16   If a particle of mass m is suspended from a spring of force constant k, what is the electric circuit analogue of the constant gravitational field?

4.8.17   A mass is suspended from a spring that is attached to a ceiling. The ceiling vibrates with a *fixed displacement amplitude*. Find the average power

transmitted to the mass as a function of the driving frequency. Assume that there is a frictional force on the mass proportional to the velocity.

4.9.18   A simple Helmholtz resonator is made by blowing air into the mouth of a bottle. The volume of the bottle is 2,000 cm³ and the neck has a cross-sectional area of 10 cm² and a length of 5 cm. Calculate the frequency of the sound which is generated. Neglect the modification of the frequency due to

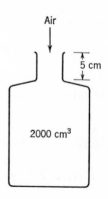

the presence of damping. The velocity of sound in air is $344 \times 10^2\,\text{cm/sec}$.

4.10.19 Prove from the Fourier series expansion equation (4.10.17) that

$$\frac{\pi}{4} = 1 - \frac{1}{3} + \frac{1}{5} - \frac{1}{7} + \frac{1}{9} - \cdots$$

4.10.20 Find the Fourier series expansion of the function

$$F(t) = \frac{F_0}{\pi}\,\omega t - F_0 \qquad 0 \le \omega t < 2\pi$$

Plot the first three terms of the expansion.

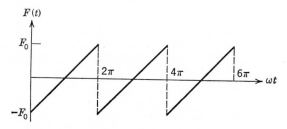

4.11.21 A particle of mass 0.2 gm moves in the positive x-direction in a potential

$$V(x) = -V_1 + k\,|x| \qquad -V_1/k < x < +V_1/k$$
$$V_1 = 10^4 \text{ ergs} \qquad k = 10^3 \text{ ergs/cm}$$
$$V(x) = 0 \text{ elsewhere}$$

The particle goes past $x = -20$ cm with an energy of 10^3 ergs. How long does it take it to get to $x = +20$ cm? Compare this with the case $V(x) = 0$ everywhere.

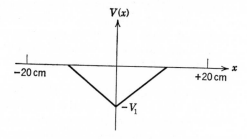

4.11.22 Find the one-dimensional motion of a particle if its potential energy is given by

$$V(x) = -k_0 x \qquad x < 0$$
$$V(x) = \tfrac{1}{2}k_1 x^2 \qquad x > 0$$

Let the initial displacement be x_2 and the initial velocity be zero. What is the period of the motion? Where are the turning points?

4.11.23   The potential energy of a particle moving in one dimension is

$$V(x) = \tfrac{1}{2}k_1 x^2 + \frac{k_2}{x}$$

$$k_1 > 0, \qquad k_2 > 0, \qquad x > 0$$

Show that the motion is periodic. What is the equilibrium position? What is the frequency of the motion if the amplitude of the vibrations is very small?

4.11.24   Three particles, each of mass 0.1 gm traveling in the positive x-direction with energies of 100, 1,000, and 10,000 ergs strike a potential barrier

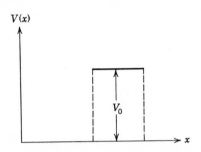

in the form of a step function of height $V_0 = 1,000$ ergs. Describe what happens to each particle.

4.11.25   A function is given by

$$V(x) = \tfrac{1}{2}kx^2 \qquad x \leq x_0$$
$$V(x) = 0 \qquad x > x_0$$

Find the derivative of this function everywhere.

4.12.26   Refer to the example of the attraction between current carrying wires in Section 4.12. If the initial displacement of the movable wire is 3 cm, what must its initial speed be in order that one of the turning points be 6 cm? Use the same numerical values that were used to construct Fig. 4.12.4 and the additional values $k = 1,000$ dynes/cm, $m = 10$ gm. What is the other turning point?

4.12.27   Continuing with the example of the current carrying wires, what is the critical value of α in Eq. (4.12.16) above which periodic motion is no longer possible? Let $a = 10$ cm. Sketch $V(x)$ for this case. Also sketch $V(x)$ for $\alpha = 50$ cm^2. Is there a qualitative difference between the two curves you have drawn?

4.13.28   Suppose that in equation (4.13.4) the linear term is missing so that the equation of motion is $\ddot{x} + \gamma x^3 = 0$. Find the Fourier series solution up to the fifth harmonic.

4.13.29   Suppose that in Eq. (4.13.4) the linear term is present but is very small. Find the Fourier series solution.

4.13.30   Suppose that the equation of motion of a particle is $\ddot{x} - \alpha x + \gamma x^3 = 0$ where $\alpha > 0$ and $\gamma > 0$. Sketch the potential function. Find a Fourier series solution for the case where the energy has a small positive value. *Suggestion:* Since this is near the sticking solution, the frequency will be small. Since

$$\omega^2 = -\alpha + \tfrac{3}{4}\gamma b_1{}^2 + \tfrac{3}{4}\gamma b_1 b_3$$

let

$$\alpha = \tfrac{3}{4}\gamma b_1{}^2$$

The remaining term will then give ω^2, a small positive value. Sketch the first and the third harmonic, add them geometrically, and interpret the result.

Without actually carrying out the details, suggest how the Fourier series solution for energy less than zero is to be obtained. Qualitatively, what do you expect the solution to look like if the energy is only slightly less than zero?

4.13.31   Show that the *exact* solution of $\ddot{x} - \alpha x + \gamma x^3 = 0$ is

$$x = \frac{x_2}{\cosh \sqrt{\alpha}\, t} \qquad x_2 = \sqrt{2}\sqrt{\alpha/\gamma}$$

for the special case where the energy is *exactly* zero. Sketch x as a function of t and interpret.

4.13.32   The equation of motion of a particle is

$$m\ddot{x} + k_1 x - k_2 x^2 = 0$$

with $m = 3$ gm, $k_1 = 3,000$ dynes/cm, and $k_2 = 300$ dynes/cm^2. If the initial displacement of the particle is 5 cm and the initial velocity is zero, find the numerical values of the Fourier coefficients up to and including the third harmonic. Find the approximate frequency of the motion. Locate the turning points. What is the energy of the particle? How does this compare with the maximum energy for which periodic motion is possible?

REFERENCES

1. Olson, H. F., *Dynamical Analogies*, Van Nostrand, New York, 1958.
2. McLachlan, N. W., *Theory of Vibrations*, Dover Publications, New York, 1951.
3. Stoker, J. J., *Non-Linear Vibrations*, Interscience Publishers, New York, 1963.
4. Minorsky, N., *Introduction to Non-Linear Vibrations*, J. W. Edwards, Ann Arbor, 1947.
5. Minorsky, N., *Non-Linear Oscillations*, Van Nostrand, Princeton, N.J. 1962.
6. Churchill, R. V., *Operational Mathematics*, McGraw-Hill Book Co., New York, 1958.
7. Lighthill, M. J., *An Introduction to Fourier Analysis and Generalized Functions*, Cambridge University Press, London, 1958.
8. Churchill, R. V., *Fourier Series and Boundary Value Problems*, McGraw-Hill Book Co., New York, 1963.
9. Irving, J., and N. Mullineux, *Mathematics in Physics and Engineering*, Academic Press, New York, 1959.

5

Introduction to Field Theory

As a preliminary to the study of the mechanics of a mass point moving in two and three dimensions, the study of the nature and origin of some of the forces which act on a particle is taken up in this chapter.

5.1. THE ENERGY INTEGRAL

Let a and b refer to any two points on a particle trajectory which is now to be regarded as a three-dimensional space curve. The work done on the particle between these two points by the *net* force acting on the particle is

$$
\begin{aligned}
W_{ab} &= \int_a^b \mathbf{F} \cdot d\mathbf{s} = \int_a^b m \frac{d\mathbf{u}}{dt} \cdot \frac{d\mathbf{s}}{dt}\, dt \\
&= \int_a^b m \frac{d\mathbf{u}}{dt} \cdot \mathbf{u}\, dt = \int_a^b \frac{d}{dt} \left(\tfrac{1}{2} m u^2 \right) dt \qquad (5.1.1) \\
&= \tfrac{1}{2} m u_b{}^2 - \tfrac{1}{2} m u_a{}^2 = T_b - T_a
\end{aligned}
$$

This is exactly the same result that was obtained in Section 4.2 for the case where the particle is constrained to move in one-dimension. The work done on a particle between two points on its trajectory by the net force is always equal to the change in the kinetic energy of the particle regardless of the nature of the force. It is to be emphasized that the net force is the vector sum of all forces which act on the particle.

196

5.2. CONSERVATIVE FORCES

Let **F** be a force which acts on a particle. If the particle is taken around a closed path, the net work done on the particle by the force is

$$W = \oint \mathbf{F} \cdot d\mathbf{s} \qquad (5.2.1)$$

where $\oint$ indicates an integral around a closed path. A region Σ of space is *simply connected* if every closed path in Σ can be continuously deformed into a point without any portion of it passing out of Σ. If in a simply connected region

$$\oint \mathbf{F} \cdot d\mathbf{s} = 0 \qquad (5.2.2)$$

for every closed path lying in Σ, and if **F** has no explicit time dependence, then **F** is said to be conservative in Σ.

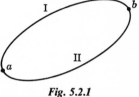

Fig. 5.2.1

As an example of a force which is *not* conservative, consider the velocity dependent force

$$\mathbf{F} = b\mathbf{u} \qquad (5.2.3)$$

Note that

$$\oint \mathbf{F} \cdot d\mathbf{s} = \oint b\mathbf{u} \cdot \frac{d\mathbf{s}}{dt} dt = \oint bu^2\, dt > 0$$

Let a and b be any two points on a closed path lying in Σ. The integral (5.2.2) can be divided into two parts:

$$\underset{\text{I}}{\int_a^b \mathbf{F} \cdot d\mathbf{s}} + \underset{\text{II}}{\int_b^a \mathbf{F} \cdot d\mathbf{s}} = 0 \qquad (5.2.4)$$

The limits of integration over portion II (Fig. 5.2.1) can be interchanged with the result

$$\underset{\text{I}}{\int_a^b \mathbf{F} \cdot d\mathbf{s}} = \underset{\text{II}}{\int_a^b \mathbf{F} \cdot d\mathbf{s}} \qquad (5.2.5)$$

Since any closed path lying in Σ and passing through the points a and b could have been chosen (5.2.5) implies that the value of the line integral of **F** between a and b is independent of the path of integration. Since it is possible to work backward from (5.2.5) to the original definition of a conservative force (5.2.2), (5.2.5) is a necessary and sufficient condition for a force to be conservative.

If Stokes' theorem, Eq. (2.11.15), is applied to (5.2.2), the result is

$$\oint \mathbf{F} \cdot d\mathbf{s} = \int_\sigma \mathbf{\nabla} \times \mathbf{F} \cdot \hat{\mathbf{n}} \, d\sigma \tag{5.2.6}$$

where σ denotes any surface lying entirely in the volume Σ and bounded by the closed path. Let the closed path of integration shrink so that it bounds a differential element of area. Equation (5.2.6) then gives

$$\mathbf{\nabla} \times \mathbf{F} \cdot \hat{\mathbf{n}} \, \Delta\sigma = 0 \tag{5.2.7}$$

Since the element of area is arbitrary and can be oriented in any direction, (5.2.7) implies

$$\mathbf{\nabla} \times \mathbf{F} = 0 \tag{5.2.8}$$

Thus if $\mathbf{F}$ is conservative in the simply connected region Σ, it also is irrotational in that region. It is easy to see that the steps leading from (5.2.2) to (5.2.8) are reversible so that (5.2.8) is a necessary and sufficient condition for $\mathbf{F}$ to be conservative.

Given that $\mathbf{F}$ is conservative in the simply connected region Σ it is possible to associate with $\mathbf{F}$ a scalar function $V(x, y, z)$ which is defined up to an arbitrary constant throughout Σ. Choose two points a and b in Σ. Let us arbitrarily assign the value V_a to V at the point a. Then since the value of the line integral of $\mathbf{F}$ between a and b is independent of the path chosen between these two points it is possible to assign a unique value to V at the point b by means of

$$\int_a^b \mathbf{F} \cdot d\mathbf{s} = -(V_b - V_a) \tag{5.2.9}$$

By choosing points in Σ other than the point b, it is possible to assign a unique value to V at every point of Σ once the value of V has been specified at a. The function $V(x, y, z)$ constructed in this way is the potential energy function associated with the conservative force $\mathbf{F}$.

Combining (5.2.9) and (5.1.1) results in

$$T_a + V_a = T_b + V_b \tag{5.2.10}$$

which implies that $\mathscr{E} = T + V$ has the same value at all points of the particle trajectory throughout the region Σ where F is conservative. $\mathscr{E}$ is recognized as the *total energy* of the particle and is defined up to an arbitrary constant. The conservation of energy theorem just derived is the origin of the term *conservative force*.

Consider two points a and b which are a differential distance $d\mathbf{s}$ apart. Equation (5.2.9) implies

$$\mathbf{F} \cdot d\mathbf{s} = -dV \tag{5.2.11}$$

showing that $\mathbf{F} \cdot d\mathbf{s}$ is an *exact differential*. Since the total differential of a scalar function can be expressed

$$dV = \nabla V \cdot d\mathbf{s} \tag{5.2.12}$$

and since the differential displacement $d\mathbf{s}$ can be chosen arbitrarily (5.2.11) and (5.2.12) imply

$$\mathbf{F} = -\nabla V \tag{5.2.13}$$

It is possible to start with equation (5.2.13) and derive the original definition (5.2.2) of a conservative force. Therefore, the existence of a scalar function from which the force can be obtained by means of (5.2.13) is a necessary and sufficient condition for the force to be conservative. The surfaces $V = $ const are called *equipotential surfaces*. The discussion of Section 2.5 shows that at any point the force $\mathbf{F}$ is always perpendicular to an equipotential surface.

As a specific example, suppose a force is given by

$$\mathbf{F} = -(2x + yz)\hat{\mathbf{i}} - (2y + xz)\hat{\mathbf{j}} - (2z + xy)\hat{\mathbf{k}} \tag{5.2.14}$$

By direct calculation it is found that $\nabla \times \mathbf{F} = 0$ showing that $\mathbf{F}$ is conservative. The potential function itself can be constructed by starting from

$$F_x = -\frac{\partial V}{\partial x} = -(2x + yz) \tag{5.2.15}$$

Integration with respect to x yields

$$V = x^2 + xyz + U(y, z) \tag{5.2.16}$$

where U can depend at most on y and z. Differentiation of (5.2.16) with respect to y yields

$$\frac{\partial V}{\partial y} = xz + \frac{\partial U}{\partial y} = -F_y = 2y + xz \tag{5.2.17}$$

$$\frac{\partial U}{\partial y} = 2y \tag{5.2.18}$$

Integration of (5.2.18) with respect to y gives

$$U = y^2 + W(z) \tag{5.2.19}$$

where W must depend only on the remaining variable z. Therefore,

$$V = x^2 + y^2 + xyz + W(z) \tag{5.2.20}$$

Differentiation of (5.2.20) with respect to z gives

$$\frac{\partial V}{\partial z} = xy + \frac{dW}{dz} = 2z + xy$$

$$\frac{dW}{dz} = 2z \qquad W = z^2 + c \tag{5.2.21}$$

where c is a constant. The potential function therefore, is

$$V = x^2 + y^2 + z^2 + xyz + c \qquad (5.2.22)$$

If $\mathbf{F}$ is time dependent, it may still be true that $\nabla \times \mathbf{F} = 0$ and $\mathbf{F} = -\nabla V$. However, $\oint \mathbf{F} \cdot d\mathbf{s} = 0$ would be true only if the integral were somehow done at a fixed instant of time. Such vector fields are referred to as *irrotational* or *lamellar*, but not conservative. The total mechanical energy of a particle is not a constant if $\mathbf{F}$ is time dependent, as can be verified by calculating the total time derivative of $\mathscr{E} = T + V$. This has already been done in Chapter 4 for the special case of one-dimensional motion (Eq. 4.2.8).

Let $\mathbf{a}$ be an arbitrary time-independent vector field which need not represent a force on a particle but may represent some other physical quantity. We can still speak of $\mathbf{a}$ as being conservative in the mathematical sense throughout a simply connected region Σ if the following conditions are obeyed:

I. $\oint \mathbf{a} \cdot d\mathbf{s} = 0$ for any closed path lying in Σ

II. $\displaystyle\int_a^b \mathbf{a} \cdot d\mathbf{s}$ is independent of the path of integration between a and b.

III. $\nabla \times \mathbf{a} = 0$ at all points of Σ.

IV. There exists a scalar function ψ such that at every point of Σ
$\mathbf{a} = -\nabla \psi$

Any one of the conditions I through IV is a necessary and sufficient condition for $\mathbf{a}$ to be conservative and any one implies the other three. The function ψ is called a *scalar potential* even though it may have nothing to do with an actual potential energy. It would obviously be possible to relax the condition that Σ be a simply connected region under certain circumstances. For example, Σ might be multiply connected but a sub-region of a larger simply connected region where the vector field is known to be conservative.

5.3. THE MAGNETOSTATIC FIELD

Suppose that a steady current J exists in the current loops of Fig. 5.3.1. At all points of space there will be a magnetostatic field that satisfies the differential equations

$$\nabla \times \mathbf{B} = \frac{4\pi \mathbf{j}}{c} \qquad \nabla \cdot \mathbf{B} = 0 \qquad (5.3.1)$$

where $\mathbf{j}$ is the current density. (See Appendix B for a discussion of Gaussian units.) The magnetostatic field is an example of a solenoidal vector field. Except in the wire which supports the current it will be true that

$$\nabla \times \mathbf{B} = 0 \tag{5.3.2}$$

The shaded region $\sum_a$ of Fig. 5.3.1a should be visualized as a toroid out of which a small slice has been cut. The current J lies entirely outside of $\sum_a$. Note that $\sum_a$ is *simply connected* and since (5.3.2) holds throughout $\sum_a$, $\mathbf{B}$ is conservative in $\sum_a$. The shaded region $\sum_b$ of Fig. 5.3.1b in the shape of a toroid is *not* simply connected. Even though (5.3.2) holds at every point

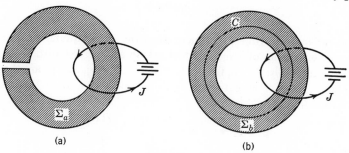

(a)   (b)

Fig. 5.3.1

of $\sum_b$, $\mathbf{B}$ is not conservative here. If the line integral of $\mathbf{B}$ around the closed path C is calculated, the result is

$$\oint_C \mathbf{B} \cdot d\mathbf{s} = \frac{4\pi J}{c} \tag{5.3.3}$$

where J is the total current enclosed by C. The curve C cannot be shrunk to a point without passing out of $\sum_b$.

5.4. THE STATIC GRAVITATIONAL FIELD

It is known from experiments that the gravitational field of a point particle of mass M is given by

$$\mathbf{g} = -\frac{GM\hat{\mathbf{r}}}{r^2} \tag{5.4.1}$$

where $\hat{\mathbf{r}}$ is a unit vector drawn outward from the particle. This concept was introduced in Section 2.3. The value of the gravitational constant is

$$G = 6.670 \times 10^{-8} \text{ cm}^3 \text{ gm}^{-1} \text{ sec}^{-2}$$
$$= 3.42 \times 10^{-8} \text{ ft}^3 \text{ slug}^{-1} \text{ sec}^{-2} \tag{5.4.2}$$

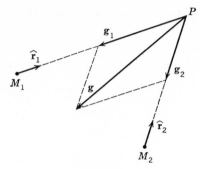

Fig. 5.4.1

It is also known from experiment that the gravitational field has the algebraic properties of a vector. For example, let P be a distance r_1, from mass M_1 and a distance r_2 from M_2, Fig. 5.4.1. If $\mathbf{g}_1$ and $\mathbf{g}_2$ are the gravitational fields at P due to M_1 and M_2 separately, then the resultant gravitational field at P is

$$\mathbf{g} = \mathbf{g}_1 + \mathbf{g}_2 = -\frac{GM_1\hat{\mathbf{r}}_1}{r_1^{\,2}} - \frac{GM_2\hat{\mathbf{r}}_2}{r_2^{\,2}} \qquad (5.4.3)$$

If there are n-point masses present, the net gravitational field is

$$\mathbf{g} = -\sum_{\alpha=1}^{n} \frac{GM_\alpha\hat{\mathbf{r}}_\alpha}{r_\alpha^{\,2}} \qquad (5.4.4)$$

To find the gravitational field of a continuous distribution of matter, the sum must be replaced by an integral. The point where the field is to be evaluated is known as the *field point* and its coordinates will be denoted x_i, Fig. 5.4.2. The variable point of integration is called the *source point*

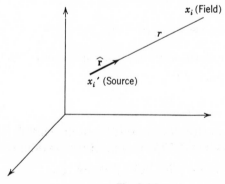

Fig. 5.4.2

and its coordinates will be labelled x_i'. If $\rho(x_i')$ is the mass density, the mass of a volume element located at the source point is

$$dm = \rho(x_i')\,d\Sigma' \tag{5.4.5}$$

The net gravitational field of the distribution of matter evaluated at the field point is

$$\mathbf{g} = -\int \frac{G\hat{\mathbf{r}}\,dm}{r^2} = -\int \frac{G\rho(x_i')\hat{\mathbf{r}}\,d\Sigma'}{r^2} \tag{5.4.6}$$

where

$$r = \sqrt{(x_1 - x_1')^2 + (x_2 - x_2')^2 + (x_3 - x_3')^2} \tag{5.4.7}$$

As an example, we will calculate the gravitational field at the point on the axis of the thin wire ring of radius a and mass M, illustrated in Fig. 5.4.3.

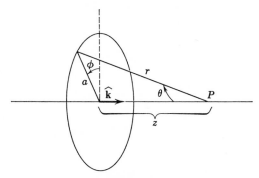

Fig. 5.4.3

The mass per unit length of the wire is

$$\mu = \frac{M}{2\pi a} \tag{5.4.8}$$

and the mass of an element of length ds is

$$dm = \mu\,ds = \mu a\,d\phi \tag{5.4.9}$$

Due to the symmetry, there will be no component of $\mathbf{g}$ perpendicular to the axis at point P. Since the component of $\hat{\mathbf{r}}$ in the z-direction is $\hat{\mathbf{k}}\cos\theta$, the expression for $\mathbf{g}$ at point P is

$$\mathbf{g} = -G\int_{\phi=0}^{2\pi} \frac{\hat{\mathbf{k}}\cos\theta\,\mu a\,d\phi}{r^2}$$

$$= -2\pi G\mu a\,\frac{\cos\theta}{r^2}\hat{\mathbf{k}} = -GM\frac{z}{r^3}\hat{\mathbf{k}} = \frac{-GMz\hat{\mathbf{k}}}{(a^2 + z^2)^{3/2}} \tag{5.4.10}$$

where z is the distance from the center of the ring to the point P.

Recall that

$$\nabla\left(\frac{1}{r}\right) = -\frac{\hat{\mathbf{r}}}{r^2} \tag{5.4.11}$$

where ∇ operates on the coordinates of the field point in Fig. 5.4.2. Hence, the gravitational field (5.4.6) of a continuous distribution of matter can be written

$$\mathbf{g} = \int G\,\nabla\left(\frac{1}{r}\right)\rho(x_i')\,d\Sigma' \tag{5.4.12}$$

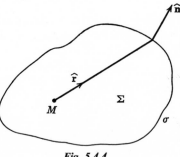

Fig. 5.4.4

Since ∇ does not operate on the variable of integration x_i' it can be factored out of the integral:

$$\mathbf{g} = \nabla\int \frac{G\rho(x_i')}{r}\,d\Sigma' \tag{5.4.13}$$

Thus $\mathbf{g}$ is derivable from a scalar potential:

$$\mathbf{g} = -\nabla\psi \qquad \psi = -\int \frac{G\rho(x_i')}{r}\,d\Sigma' \tag{5.4.14}$$

and this proves that the static gravitational field is a conservative vector field. It is frequently easier to calculate the gravitational potential by means of (5.4.14) rather than to calculate $\mathbf{g}$ directly from (5.4.6). If a particle of mass m is placed in a gravitational field, it experiences a force $\mathbf{F} = m\mathbf{g}$. Its potential energy can be taken as $V = m\psi$.

Let σ be a closed surface enclosing a volume Σ of space as in Fig. 5.4.4. Suppose that there is a single point mass M located in Σ. The net flux of the gravitational field across σ due to this point mass is

$$\int_\sigma \mathbf{g}\cdot\hat{\mathbf{n}}\,d\sigma = -GM\int_\sigma \frac{\hat{\mathbf{r}}\cdot\hat{\mathbf{n}}}{r^2}\,d\sigma = -GM\int d\Omega = -4\pi GM \tag{5.4.15}$$

where $d\Omega$ is the element of solid angle subtended at M by the element of area $d\sigma$ (Eq. 1.6.11 and 1.6.17). If M lies *outside* of the surface, $\int d\Omega = 0$ and

$$\int_\sigma \mathbf{g} \cdot \hat{\mathbf{n}} \, d\sigma = 0 \qquad (5.4.16)$$

These two cases are illustrated in Fig. 5.4.5. The two closed surfaces σ_1 and σ_2 are placed so that M is inside σ_1 but outside of σ_2. M acts as a "sink" for the "lines" of the gravitational field so that a total number

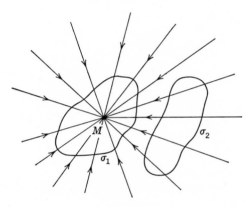

Fig. 5.4.5

$4\pi GM$ cross into σ_1. On the other hand, just as many lines leave σ_2 as enter it so that the net flux of $\mathbf{g}$ across σ_2 is zero.

If there are a number of point masses enclosed by a surface σ then

$$\int_\sigma \mathbf{g} \cdot \hat{\mathbf{n}} \, d\sigma = \int_\sigma (\mathbf{g}_1 + \mathbf{g}_2 + \cdots) \cdot \hat{\mathbf{n}} \, d\sigma$$
$$= -4\pi G(m_1 + m_2 + \cdots) \qquad (5.4.17)$$
$$= -4\pi G(\text{total mass enclosed by } \sigma)$$

If the mass enclosed by σ is in the form of a continuous distribution then (5.4.17) is to be replaced by

$$\int_\sigma \mathbf{g} \cdot \hat{\mathbf{n}} \, d\sigma = -4\pi G \int_\Sigma \rho \, d\Sigma \qquad (5.4.18)$$

where the volume integral is *only* over the region Σ enclosed by σ. Matter *outside* this region does not contribute to the net flux across the surface! Equation (4.18) is known as *Gauss' law*.

Gauss' law can be used directly to calculate the gravitational field of spherically symmetric mass distributions. For example, suppose that matter is uniformly distributed throughout a sphere of radius a. Figure 5.4.6 shows a Gaussian surface σ of radius $r > a$ surrounding the body. Due to the spherically symmetry, the gravitational field is everywhere perpendicular to σ and of the same magnitude. Therefore, $g_r = \mathbf{g} \cdot \hat{\mathbf{n}} = $ const and, by means of (5.4.18),

$$g_r \int_\sigma d\sigma = -4\pi G \int_\Sigma \rho \, d\Sigma = -4\pi GM$$

$$g_r \cdot 4\pi r^2 = -4\pi GM \qquad\qquad (5.4.19)$$

$$g_r = -\frac{GM}{r^2}$$

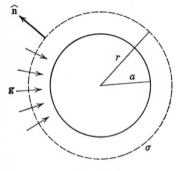

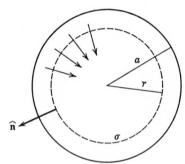

Fig. 5.4.6   **Fig. 5.4.7**

where M is the total mass of the sphere. Actually, this result is true if ρ is not a constant throughout the sphere but is a function of r. However, ρ must not depend on the polar coordinate angles θ and ϕ. Thus, at points *outside* a spherically symmetric distribution of matter, the gravitational field is the same as that of a point mass. To a first approximation the distribution of matter in the earth is spherically symmetric. In treating the mechanics of an artificial satellite it is permissible, as a first approximation, to consider the force on the satellite as identical to that produced by a point mass located at the center of the earth.

To find the gravitational field at points *inside* the sphere we compute the flux of $\mathbf{g}$ across a spherical Gaussian surface of radius $r < a$ as in Fig. 5.4.7. Assuming a constant density, the total mass enclosed by the surface of integration is

$$\rho \cdot \tfrac{4}{3}\pi r^3 = \frac{M}{\tfrac{4}{3}\pi a^3} \cdot \tfrac{4}{3}\pi r^3 = M \cdot \frac{r^3}{a^3} \qquad\qquad (5.4.20)$$

Gauss' law gives

$$g_r \cdot 4\pi r^2 = -4\pi GM \frac{r^3}{a^3}$$

$$g_r = -\frac{GM}{a^3} r \tag{5.4.21}$$

The field inside the sphere varies in direct proportion to r and, therefore, falls to zero at the center. The gravitational field for the sphere as a function of r is sketched in Fig. 5.4.8a.

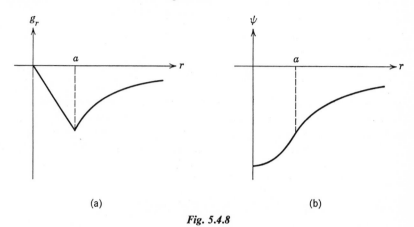

(a)   (b)

Fig. 5.4.8

The gravitational scalar potential of the sphere can be computed from

$$g_r = -\frac{\partial \psi}{\partial r} \tag{5.4.22}$$

Choosing the boundary condition $\psi = 0$ at $r = \infty$, we find

$$\psi(r) = -\frac{GM}{r} \qquad r > a \tag{5.4.23}$$

$$\psi(r) = \frac{GMr^2}{2a^3} + c \qquad r < a \tag{5.4.24}$$

Where c is a constant. The gravitational field is *finite* at every point of space and involves no δ-functions. The derivative of ψ, therefore, must be defined everywhere and, consequently, the constant c can be determined by the requirement that ψ be continuous at $r = a$:

$$-\frac{GM}{a} = \frac{GM}{2a} + c$$

$$c = -\frac{3}{2}\frac{GM}{a} \tag{5.4.25}$$

Therefore,

$$\psi(r) = -\frac{GM}{a}\left(\frac{3}{2} - \frac{r^2}{2a^2}\right) \qquad r < a \qquad (5.4.26)$$

The potential function is sketched in Fig. 5.4.8b.

By means of Eq. (2.8.4), Gauss' divergence theorem, Eq. (5.4.18) can be converted to

$$\int_{\Sigma} \nabla \cdot \mathbf{g} \, d\Sigma = -4\pi G \int_{\Sigma} \rho \, d\Sigma \qquad (5.4.27)$$

This result is an identity applying to any arbitrarily chosen region of integration implying that

$$\nabla \cdot \mathbf{g} = -4\pi G\rho \qquad (5.4.28)$$

The mass density ρ is therefore a *source function* for the gravitational field. More properly it should be called a "sink function" since the lines of $\mathbf{g}$ always *converge* toward the matter, as in Fig. 5.4.5. Since $\mathbf{g}$ is conservative, a second fundamental differential equation obeyed by $\mathbf{g}$ is

$$\nabla \times \mathbf{g} = 0 \qquad (5.4.29)$$

The substitution of $\mathbf{g} = -\nabla\psi$ into (5.4.28) leads to

$$\nabla^2\psi = 4\pi G\rho \qquad (5.4.30)$$

showing that the basic partial differential equation satisfied by the gravitational scalar potential is Poisson's equation.

The development of the theory of the electrostatic field parallels our development of the static gravitational field. The basic equations governing the electrostatic field are

$$\nabla \cdot \mathbf{E} = 4\pi\rho \qquad \nabla \times \mathbf{E} = 0 \qquad (5.4.31)$$

where ρ is the charge density. In contrast to mass density, charge density can be either positive or negative so that positive charge is a *source* and negative charge is a *sink* for the lines of $\mathbf{E}$.

There is a great body of literature on the solution of Poisson's equation. Since we are primarily concerned in this book with the mechanics of particles in given gravitational and electromagnetic fields, a further treatment of potential problems will not be given here. Because most students will take a separate course in electromagnetic theory, the development of this section is intended to show only the parallel between gravitation and the electrostatic field.

5.5. HEAT CONDUCTION

The development of the basic equations of heat conduction provide a good and important example of a vector field. Let us suppose that heat is flowing through some material medium. Let **h** be the *heat current vector* with units of ergs per second per square centimeter in the cgs system. The meaning of the divergence is explained in Section 2.7, and reveals that $\nabla \cdot \mathbf{h} \, d\Sigma$ represents the rate at which heat flows out of the volume element $d\Sigma$. Assume that energy in the form of heat is actually being created in the medium. This could happen if a chemical reaction were going on, or if an electric current were being passed through the medium. Let q be the density of heat in the medium (ergs cm^{-3}) and let f be the source density, i.e., the rate at which heat is being created per unit volume in the medium (ergs sec^{-1} cm^{-3}). Therefore,

$$\frac{d}{dt}(q \, d\Sigma) = \text{total rate of change of heat in volume element } d\Sigma$$

$$f \, d\Sigma = \text{rate at which heat is created in } d\Sigma$$

$$\nabla \cdot \mathbf{h} \, d\Sigma = \text{rate at which heat flows out of } d\Sigma$$

The principle of conservation of energy requires

$$\frac{d}{dt}(q \, d\Sigma) = f \, d\Sigma - \nabla \cdot \mathbf{h} \, d\Sigma \tag{5.5.1}$$

The heat density q is a scalar function of position and time in the medium: $q = q(x, y, z, t)$. In (5.5.1) we are talking about a volume element at a *fixed point* in the medium. Therefore,

$$\frac{dq}{dt} = \frac{\partial q}{\partial t} \tag{5.5.2}$$

Since $d\Sigma$ is not a function of time, Eq. (5.5.1) gives

$$\nabla \cdot \mathbf{h} + \frac{\partial q}{\partial t} = f \tag{5.5.3}$$

This is an equation of continuity and should be compared with the equations of continuity in Section 2.7. For instance, Eq. (2.7.6) expresses the fact that electric charge is conserved. In equation (5.5.3), we are dealing with a situation where energy in the form of heat is *not* conserved so a source function f appears.

The temperature of the medium can be introduced by remembering that if a body is raised in temperature by dT, it absorbs an amount of heat dq

per unit volume given by

$$dq = \rho c \, dT \tag{5.5.4}$$

where ρ is the mass density (gm cm^{-3}) and c is the specific heat (ergs gm^{-1} deg^{-1}) of the medium.* At a fixed point in the medium (5.5.4) can be replaced by

$$\frac{\partial q}{\partial t} = \rho c \, \frac{\partial T}{\partial t} \tag{5.5.5}$$

If (5.5.5) is substituted into the equation of continuity (5.5.3),

$$\nabla \cdot \mathbf{h} + \rho c \, \frac{\partial T}{\partial t} = f \tag{5.5.6}$$

In general, the density ρ will be a function of position in the medium.

Experimentally, it is known that the heat current in a medium is proportional to the gradient of the temperature:

$$\mathbf{h} = -K\nabla T \tag{5.5.7}$$

where K is the *thermal conductivity* of the medium (ergs sec^{-1} deg^{-1} cm^{-1}). If K is a constant throughout the medium, Eq. (5.5.7) implies

$$\nabla \times \mathbf{h} = 0 \tag{5.5.8}$$

Since $\mathbf{h}$ is time dependent, we have here an example of an *irrotational* but *not conservative* vector field. If K is not a constant but is a function of position in the medium, it is still true that

$$\nabla \times \frac{1}{K}\mathbf{h} = 0 \tag{5.5.9}$$

The vector $\mathbf{h}/K$ is irrotational. The function $1/K$ is an *integrating factor* for $\mathbf{h} \cdot d\mathbf{s}$, meaning that whereas $\mathbf{h} \cdot d\mathbf{s}$ is not an exact differential, $(1/K)\mathbf{h} \cdot d\mathbf{s}$ is exact (see Problem 5.4.10).

The combination of (5.5.7) and (5.5.6) results in a partial differential equation for the temperature:

$$\nabla \cdot (K \, \nabla T) = \rho c \, \frac{\partial T}{\partial t} - f \tag{5.5.10}$$

If K is a constant

$$\nabla^2 T = \frac{\rho c}{K} \frac{\partial T}{\partial t} - \frac{f}{K} \tag{5.5.11}$$

---

* Our theory is applicable to solids where there is little difference between the specific heat at constant pressure and that at constant volume. Therefore, it is possible to associate a definite quantity of heat with a given temperature change.

In steady state problems where there is no time dependence the temperature obeys Poisson's equation

$$\nabla^2 T = -\frac{f}{K} \tag{5.5.12}$$

The temperature plays the role of a potential function in the theory of heat conduction. As in all potential problems, only temperature differences appear so that T can be absolute temperature, centigrade temperature, or can be measured from any other convenient zero.

If the equation of continuity (5.5.6) is integrated over an arbitrary region Σ of the medium, the result is

$$\int_\Sigma \nabla \cdot \mathbf{h} \, d\Sigma + \frac{d}{dt} \int_\Sigma \rho c T \, d\Sigma = \int_\Sigma f \, d\Sigma \tag{5.5.13}$$

By the use of Gauss' divergence theorem and Eq. (5.5.4),

$$\frac{d}{dt} \int_\Sigma q \, d\Sigma = \int_\Sigma f \, d\Sigma - \int_\sigma \mathbf{h} \cdot \hat{\mathbf{n}} \, d\sigma \tag{5.5.14}$$

where σ is the closed surface bounding Σ. Equation (5.5.14) is an expression of conservation of energy in integral form and says that the total time rate of change of heat-energy in Σ equals the rate at which heat is created in Σ minus the rate at which heat flows out of Σ across the bounding surface. For a steady state (5.5.14) specializes to

$$\int_\sigma \mathbf{h} \cdot \hat{\mathbf{n}} \, d\sigma = \int_\Sigma f \, d\Sigma \tag{5.5.15}$$

which is Gauss' law for heat conduction. The source-function f plays the same role in heat conduction as does charge density in electrostatics and mass density in gravitational theory. Of course, f can be a negative source, or sink, if heat is being converted into some other form of energy in the medium.

We mention here very briefly that the diffusion of particles obeys the same kind of differential equation as does heat flow. If

$$n = \text{number of particles per unit volume}$$

$$\mathbf{c} = \text{particles per second per unit area}$$

then

$$\mathbf{c} = -D\nabla n \tag{5.5.16}$$

where D is the diffusion constant. The fact that the total number of particles is conserved is expressed by the equation continuity

$$\nabla \cdot \mathbf{c} + \frac{\partial n}{\partial t} = 0 \tag{5.5.17}$$

Assuming that D is a constant, (5.5.16) and (5.5.17) can be combined to yield the diffusion equation

$$\nabla^2 n = \frac{1}{D}\frac{\partial n}{\partial t} \tag{5.5.18}$$

Curiously, another area of physics where a similar partial differential occurs is in quantum mechanics. The Schrödinger equation for a free particle is

$$i\hbar\frac{\partial \psi}{\partial t} = -\frac{\hbar^2}{2m}\nabla^2\psi \tag{5.5.19}$$

where $\hbar$ is Planck's constant divided by 2π, $i = \sqrt{-1}$, m is the mass of the particle, and ψ is the wave-function. The occurrence of complex numbers in the Schrödinger equation makes an important difference; nevertheless, there are both qualitative and analytical similarities in the treatment of the wave equation and the equation of diffusion. The absolute value $|\psi|^2$ of the wave function is interpreted as the probability density for the position of the particle. Suppose that the position of the particle is measured with as great an accuracy as the uncertainty principle allows at $t = 0$. As time progresses, the probability density, as determined by the initial measurement, diffuses out in much the same way as an initial concentration of heat diffuses. In quantum-mechanical terminology, the "wave packet" spreads as time goes on and the position of the particle becomes less and less certain.

5.6. MAXWELL'S EQUATIONS

Maxwell's equations, which are the basic equations governing time-dependent electromagnetic fields, will be presented briefly in this section for reference purposes.

If the charges and currents which are the sources of an electromagnetic field are time dependent, the electric field is no longer conservative. According to Faraday's law, the line-integral of **E** around a closed path is equal to the time rate of change of the flux of the magnetic field calculated over the open surface bounded by the path of integration:

$$\oint_C \mathbf{E} \cdot d\mathbf{s} = -\frac{1}{c}\frac{d}{dt}\int_\sigma \mathbf{B} \cdot \hat{\mathbf{n}}\, d\sigma \tag{5.6.1}$$

If the surface of integration is fixed in space, the only time dependence comes from the time variation of **B** itself and, therefore, it is permissible to write (5.6.1) as

$$\oint \mathbf{E} \cdot d\mathbf{s} = -\frac{1}{c}\int_\sigma \frac{\partial \mathbf{B}}{\partial t} \cdot \hat{\mathbf{n}}\, d\sigma \tag{5.6.2}$$

The line integral of **E** can be transformed to a surface integral by means of Stokes' theorem:

$$\int_\sigma \left(\nabla \times \mathbf{E} + \frac{1}{c}\frac{\partial \mathbf{B}}{\partial t}\right) \cdot \hat{\mathbf{n}} \, d\sigma = 0 \tag{5.6.3}$$

Since the surface σ is arbitrary, (5.6.3) implies

$$\nabla \times \mathbf{E} + \frac{1}{c}\frac{\partial \mathbf{B}}{\partial t} = 0 \tag{5.6.4}$$

which is the appropriate generalization of (5.4.31) for time-dependent fields. The divergence equation

$$\nabla \cdot \mathbf{E} = 4\pi\rho \tag{5.6.5}$$

remains valid.

The static magnetic field obeys

$$\nabla \times \mathbf{B} = \frac{4\pi}{c}\mathbf{j} \tag{5.6.6}$$

but this equation cannot be valid for time dependent fields. If we take the divergence of (5.6.6) there results

$$0 = \nabla \cdot \nabla \times \mathbf{B} = \frac{4\pi}{c}(\nabla \cdot \mathbf{j}) \tag{5.6.7}$$

which is a valid equation of continuity for charge only if charge and current densities are independent of time. The correct generalization of (5.6.6) for the time-dependent case is

$$\nabla \times \mathbf{B} - \frac{1}{c}\frac{\partial \mathbf{E}}{\partial t} = \frac{4\pi}{c}\mathbf{j} \tag{5.6.8}$$

The added term is called *Maxwell's displacement current*. The divergence of (5.6.8) is

$$-\frac{1}{c}\frac{\partial}{\partial t}(\nabla \cdot \mathbf{E}) = \frac{4\pi}{c}(\nabla \cdot \mathbf{j}) \tag{5.6.9}$$

which, by (5.6.5), becomes

$$\nabla \cdot \mathbf{j} + \frac{\partial \rho}{\partial t} = 0 \tag{5.6.10}$$

and is the correct equation of continuity of charge. Finally, the divergence equation

$$\nabla \cdot \mathbf{B} = 0 \tag{5.6.11}$$

still holds for time-dependent fields. The four equations (5.6.4), (5.6.5), (5.6.8), and (5.6.11) determine the field quantities **E** and **B** in terms of the sources ρ and **j** and are known as *Maxwell's equations*.

In a later chapter, our development of the mechanics of a charged particle moving in a given electromagnetic field will require the use of a vector potential which can be defined in terms of the magnetic field by means of

$$\mathbf{B} = \nabla \times \mathbf{A} \tag{5.6.12}$$

Note that (5.6.12) satisfies one of the Maxwell equations (5.6.11) identically The substitution of (5.6.12) into (5.6.4) yields

$$\nabla \times \left(\mathbf{E} + \frac{1}{c} \frac{\partial \mathbf{A}}{\partial t} \right) = 0 \tag{5.6.13}$$

Thus (5.6.4) is satisfied if

$$\mathbf{E} + \frac{1}{c} \frac{\partial \mathbf{A}}{\partial t} = -\nabla \psi \tag{5.6.14}$$

where ψ is a scalar potential. Equations (5.6.12) and (5.6.14) give the field vectors $\mathbf{E}$ and $\mathbf{B}$ in terms of the vector potential $\mathbf{A}$ and the scalar potential ψ.

The potentials are not unique. New potentials $\mathbf{A}'$ and ψ' can be defined in terms of the old potentials by means of

$$\mathbf{A} = \mathbf{A}' + \nabla \phi \qquad \psi = \psi' - \frac{1}{c} \frac{\partial \phi}{\partial t} \tag{5.6.15}$$

where ϕ is an arbitrary scalar function.

One finds that the same electromagnetic field results from the new potentials, i.e.,

$$\mathbf{B} = \nabla \times \mathbf{A}' \qquad \mathbf{E} + \frac{1}{c} \frac{\partial \mathbf{A}'}{\partial t} = -\nabla \psi' \tag{5.6.16}$$

Frequently, the potentials are restricted further by requiring that they satisfy the *Lorentz gauge condition*

$$\nabla \cdot \mathbf{A} + \frac{1}{c} \frac{\partial \psi}{\partial t} = 0 \tag{5.6.17}$$

In order for the potentials to be of use in electromagnetic theory, it must be possible to find them in terms of the sources of the fields. The sought-after relations of the potentials to the source functions must follow from the remaining two Maxwell equations (5.6.5) and (5.6.8). Thus (5.6.5), (5.6.12), (5.6.14), and (5.6.17) yield

$$\nabla^2 \psi - \frac{1}{c^2} \frac{\partial^2 \psi}{\partial t^2} = -4\pi\rho \tag{5.6.18}$$

Similarly, the combination of (5.6.8), (5.6.12), (5.6.14), (5.6.17), and the vector identity (2.10.8) yields

$$\nabla^2 \mathbf{A} - \frac{1}{c^2}\frac{\partial^2}{\partial t^2}\mathbf{A} = -\frac{4\pi}{c}\mathbf{j} \qquad (5.6.19)$$

where

$$\nabla^2 \mathbf{A} = (\nabla^2 A_z)\hat{\mathbf{i}} + (\nabla^2 A_y)\hat{\mathbf{j}} + (\nabla^2 A_z)\hat{\mathbf{k}} \qquad (5.6.20)$$

Equations (5.6.18) and (5.6.19) are a set of four partial differential equations called *wave equations*, the solutions of which give the four functions A_x, A_y, A_z, and ψ. The fields are then to be determined by (5.6.12) and (5.6.14). In Chapter 12, we will solve the wave equation in detail for the special case where only one spatial coordinate and the time is involved, i.e., $\psi = \psi(x, t)$.

PROBLEMS

5.1.1   Suppose that a time dependent force can be expressed $\mathbf{F} = -\nabla V$ where V is a time-dependent potential function. By calculating the total derivative of $\mathscr{E} = T + V$ with respect to time, show that $\mathscr{E}$ is not a constant.

5.2.2   A force field is given by

$$F_x = y + z \qquad F_y = x + z \qquad F_z = x + y$$

Prove that $\mathbf{F}$ is conservative. Verify by direct calculation that the line integral of $\mathbf{F}$ around the triangular path in the xy-plane bounded by $y = 2$, $x = 6$, $y = 2x$ is zero. Find the potential function.

5.3.3   A current J flows from $-\infty$ to $+\infty$ along the z-axis. In cylindrical coordinates (r, θ, z) the magnetic field is given by

$$B_r = 0, \qquad B_\theta = \frac{2J}{rc}, \qquad B_z = 0$$

At all points of space except $r = 0$, $\mathbf{B}$ is irrotational, i.e., $\nabla \times \mathbf{B} = 0$. Find a scalar function ψ such that $\mathbf{B} = -\nabla\psi$ at all points except $r = 0$. Notice that ψ *is not* a single valued function.

5.4.4   Show that the gravitational field of a point mass is conservative by evaluating directly the line integral of $\mathbf{g}$ between two arbitrarily chosen points.

5.4.5   Compute the gravitational field at a point on the axis of a uniform thin disc of radius a and mass M.

5.4.6   A thin wire of mass M is bent into the form of a quarter of a circle of radius a. Calculate the gravitational field at the center of the circle.

5.4.7   Suppose that the density of a distribution of matter is given by

$$\rho(r) = \rho_0\left(1 - \frac{r}{a}\right) \qquad r < a, \qquad \rho(r) = 0, \qquad r > a$$

where ρ_0 is a constant. Find the gravitational field and the gravitational potential everywhere.

5.4.8   Find the gravitational field and the gravitational potential both inside and outside of a hollow spherical shell of radius a and mass M. Assume that the mass is uniformly distributed over the shell. If electric charge is placed on a hollow conductor, there is no electric field inside the conductor regardless of its shape. It is necessarily true that the gravitational field is always zero inside a hollow material body?

5.4.9   Verify by direct substitution that the gravitational potential of a point mass,

$$\psi = -\frac{GM}{r}$$

is a solution of Laplace's equation, $\nabla^2\psi = 0$ for $r \neq 0$. Do the calculation with ∇^2 expressed in spherical polar coordinates.

5.4.10   If $\mathbf{a} \cdot d\mathbf{s}$ is not an exact differential, it may be that an integrating factor exists, i.e., a function μ such that $\mu\mathbf{a} \cdot d\mathbf{s} = -d\psi$. Show that a necessary condition for the existence of an integrating factor is $\mathbf{a} \cdot \nabla \times \mathbf{a} = 0$. The condition is also sufficient, but this is harder to prove.

5.5.11   A cylindrical conductor of radius a, specific resistance ρ ohm-cm, thermal conductivity K_1 ergs sec^{-1} cm^{-1} deg^{-1} has a uniform electric current J amperes passing through it. Surrounding the conductor is an insulator of radius b and thermal conductivity K_2. The outside of the insulator is maintained at a temperature T_0. Find the steady-state temperature distribution throughout the insulator and the conductor. *Hint:* Use Gauss' law, Eq. (5.5.15). Verify that Poisson's equation $\nabla^2 T = -f/K_1$ is satisfied in the interior of the conductor.

5.5.12   Let $\mathbf{h}_1$ and $\mathbf{h}_2$ be the heat current vectors in two media of thermal conductivity K_1 and K_2 evaluated at a point on the interface between the two

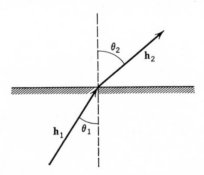

media. Find the relation between the normal and tangential components of $\mathbf{h}$ at the interface and show that $K_2 \tan \theta_1 = K_1 \tan \theta_2$ where θ_1 and θ_2 are the angles measured between the directions of $\mathbf{h}_1$ and $\mathbf{h}_2$ and the normal to the interface.

5.5.13   An electrical conductor in the form of a cylinder of length s and cross sectional area σ is thermally insulated except at its ends which are maintained at a temperature T_0. The specific resistance of the conductor ρ is ohm-cm and the current density is j amperes/cm^2. Find the steady state temperature as a function of position in the rod.

REFERENCES

1. Churchill, R. V., *Fourier Series and Boundary Value Problems*, McGraw-Hill Book Co., New York, 1963.
2. Jackson, J. D. *Classical Electrodynamics*, John Wiley and Sons, New York, 1962.

6

The Motion of a Particle
in Two and Three Dimensions

In this chapter, attention is concentrated on the development of the mechanics of a single mass point in a given force field. Both the Newtonian and the Lagrangian methods are used. Hamilton's equations are introduced. The solution of certain problems by means of Jacobian elliptic functions is also presented.

6.1. CONSTANTS OF THE MOTION

In solving mechanical problems, constants of the motion are frequently of great value. In problems that are not easily soluble, much qualitative information can be obtained by their use. Two constants of the motion have already been encountered: The linear momentum, $\mathbf{p} = m\mathbf{u}$, of a particle is constant if the force is zero and the total energy of a particle is constant if the force acting on the particle is conservative.

The motion of a particle is completely specified once its position $\mathbf{r}$, measured from some arbitrary origin O, and its linear momentum are known. The axial vector formed from the cross product of these two vectors,

$$\mathbf{l} = \mathbf{r} \times \mathbf{p} \tag{6.1.1}$$

is called the *moment of momentum* or *angular momentum* of the particle with respect to the point O. The angular momentum is a constant of the motion under certain circumstances. Its derivative with respect to time is

$$\frac{d\mathbf{l}}{dt} = \frac{d\mathbf{r}}{dt} \times \mathbf{p} + \mathbf{r} \times \frac{d\mathbf{p}}{dt} \tag{6.1.2}$$

Now

$$\frac{d\mathbf{r}}{dt} \times \mathbf{p} = \mathbf{u} \times m\mathbf{u} = 0 \tag{6.1.3}$$

and, therefore,

$$\frac{d\mathbf{l}}{dt} = \mathbf{r} \times \mathbf{F} = \mathbf{N} \tag{6.1.4}$$

The vector $\mathbf{N}$ is the moment of the force $\mathbf{F}$ about O or the *torque*. If the torque on the particle is zero, then the angular momentum is a constant of the motion.

All the possible vectors and scalars which can be constructed from $\mathbf{p}$ and $\mathbf{r}$ are

$$p^2 = \mathbf{p} \cdot \mathbf{p}, \qquad r^2 = \mathbf{r} \cdot \mathbf{r}, \qquad \mathbf{p} \cdot \mathbf{r}, \text{ and } \mathbf{r} \times \mathbf{p} \tag{6.1.5}$$

If $\mathbf{r}$ and $\mathbf{p}$ are not parallel, $\mathbf{l}$, $\mathbf{r}$, and $\mathbf{p}$ are a linearly independent set of vectors. All other apparently new vectors which can be formed from $\mathbf{r}$ and $\mathbf{p}$ are linear combinations of $\mathbf{l}$, $\mathbf{r}$, and $\mathbf{p}$.

There is a relation between the existence of constants of the motion and symmetry. There is a connection between rotational symmetry and conservation of angular momentum. Suppose, for example, that the force on a particle has spherical symmetry, i.e., suppose it is always directed toward a fixed origin and depends only on the separation r between the particle and the origin. This would be true of the gravitational force produced by the earth on an artificial satellite. In spherical coordinates the force can be expressed

$$\mathbf{F} = F(r)\hat{\mathbf{r}} \tag{6.1.6}$$

Such forces are called *central forces*. The torque about the origin produced by such a force is

$$\mathbf{N} = \mathbf{r} \times F(r)\hat{\mathbf{r}} = 0 \tag{6.1.7}$$

Thus particles acted on by central forces conserve angular momentum. Since $\mathbf{l}$ is a vector, it is constant both in direction and magnitude. A particle with a constant angular momentum must move in a plane. We may therefore treat the problem of a particle acted on by a central force in two dimensions without loss of generality. This results in a considerable simplification.

6.2. MOTION OF A PROJECTILE

If a particle moves in a uniform gravitational field, the motion can be treated in a plane without loss of generality. Let y be measured vertically, i.e., opposite in direction to the gravitational field, and let x be measured horizontally. From elementary mechanics, the integrated equations of motion are known to be

$$y = -\tfrac{1}{2}gt^2 + u_{0y}t + y_0 \qquad (6.2.1)$$

$$x = u_{0x}t + x_0 \qquad (6.2.2)$$

where (x_0, y_0) represents the coordinates of the initial position and (u_{0x}, u_{0y}) are the components of the initial velocity. The angle of projection can be introduced by means of

$$u_{0x} = u_0 \cos \theta \qquad u_{0y} = u_0 \sin \theta \qquad (6.2.3)$$

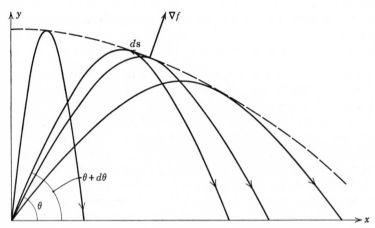

Fig. 6.2.1

where u_0 is the magnitude of the velocity of projection. If x_0 and y_0 are chosen to be zero and t is eliminated between (6.2.1) and (6.2.2),

$$y = - \frac{gx^2}{2u_0^2 \cos^2 \theta} + x \tan \theta \qquad (6.2.4)$$

If the velocity of projection u_0 is constant, then (6.2.4) represents a one-parameter family of curves, the variable parameter being the angle of projection θ.

Suppose it is desired to hit an airborne target. It is then necessary to compute the air-space which is accessible to the projectiles. Mathematically, we seek the *envelope* of the one-parameter family of curves which is the dotted curve of Fig. 6.2.1. A one-parameter family of curves is, in general, represented by an equation of the form

$$f(x, y, \theta) = 0 \qquad (6.2.5)$$

a different member of the family being obtained for each θ. Consider two points on the envelope which are a differential displacement ds apart as in

Fig. 6.2.1. Touching these points will be two parabolas that are found by varying θ by a differential amount. The total differential of (6.2.5) is

$$\frac{\partial f}{\partial x}\,dx + \frac{\partial f}{\partial y}\,dy + \frac{\partial f}{\partial \theta}\,d\theta = 0 \qquad (6.2.6)$$

which can also be expressed

$$\nabla f \cdot d\mathbf{s} + \frac{\partial f}{\partial \theta}\,d\theta = 0 \qquad (6.2.7)$$

where $d\mathbf{s}$ is tangent to the envelope. The vector ∇f is perpendicular to the curve $f(x, y, \theta) = 0$ and since the derivatives are evaluated at a point where this curve touches the envelope, ∇f is also perpendicular to the envelope. Hence, $\nabla f \cdot d\mathbf{s} = 0$ and

$$\frac{\partial f}{\partial \theta} = 0 \qquad (6.2.8)$$

which is the condition from which the equation of the envelope is to be computed.

For the example of the parabolic trajectories

$$f = y + \frac{gx^2}{2u_0^{\,2} \cos^2 \theta} - x \tan \theta = 0 \qquad (6.2.9)$$

$$\frac{\partial f}{\partial \theta} = \frac{gx^2 \sin \theta}{u_0^{\,2} \cos^3 \theta} - \frac{x}{\cos^2 \theta} = 0 \qquad (6.2.10)$$

Equation (6.2.10) simplifies to

$$gx \tan \theta = u_0^{\,2} \qquad (6.2.11)$$

Elimination of θ between (6.2.11) and (6.2.9) yields

$$y = \frac{u_0^{\,2}}{2g} - \frac{gx^2}{2u_0^{\,2}} \qquad (6.2.12)$$

which is the equation of the envelope. If the projectile can be fired in all possible directions, then the region of space in which a target can be hit is the paraboloid of revolution formed by (6.2.12) when it is rotated about the y-axis.

We have, incidentally, solved another problem. Given that the projectile lands at an elevation y above the starting point, then the maximum horizontal range is found by solving (6.2.12) for x. For instance, if $y = 0$, the maximum horizontal range is found to be

$$x = \frac{u_0^{\,2}}{g} \qquad (6.2.13)$$

If an air resistance proportional to the velocity is included, the equation of motion is

$$m\mathbf{g} - b\mathbf{u} = m\mathbf{a} \qquad (6.2.14)$$

Writing

$$\mathbf{g} = -\hat{j}g, \qquad \mathbf{u} = \dot{x}\hat{i} + \dot{y}\hat{j}, \qquad \mathbf{a} = \ddot{x}\hat{i} + \ddot{y}\hat{j} \qquad (6.2.15)$$

results in the component equations

$$m\ddot{x} + b\dot{x} = 0 \qquad (6.2.16)$$

$$m\ddot{y} + b\dot{y} + mg = 0 \qquad (6.2.17)$$

The treatment of these equations offers no difficulties beyond those discussed in Section 4.3. In actual practice, air resistance varies with altitude and becomes a more complex function of velocity at higher speeds.

6.3. THE PLANE PENDULUM

The plane pendulum consists of a particle of mass m on the end of a massless rigid rod pivoted at one end as in Fig. 6.3.1. This is a simple example of *constrained motion*. The particle moves in a vertical plane and is constrained to move on the arc of a circle. In addition to the uniform gravitational force, which is conservative, there is a *force of constraint* necessary to keep the particle on the arc of a circle which is the tension (or compression) in the rod. The force of constraint is *not conservative* as no scalar function exists from which it can be derived. However, neglecting friction, the conservation of energy principle is still valid because the force of constraint is always perpendicular to the direction of the motion of the particle and hence does no

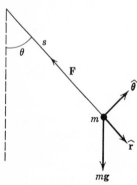

Fig. 6.3.1

work. This is an example of a *workless force of constraint*. We speak of such mechanical systems as being conservative even though, strictly speaking, nonconservative forces are involved.

The geometry of the problem makes polar coordinates especially suitable. In Fig. 6.3.1, $m\mathbf{g}$ is the weight of the particle and $\mathbf{F}$ is the force of constraint exerted on the particle by the rod. Resolving these forces into their components in the directions of $\hat{r}$ and $\hat{\theta}$ results in

$$F_r = mg\cos\theta - F \qquad (6.3.1)$$

$$F_\theta = -mg\sin\theta \qquad (6.3.2)$$

The kinematic expressions for the physical components of acceleration in polar coordinates are $a_r = \ddot{r} - r\dot{\theta}^2$ and $a_\theta = r\ddot{\theta} + 2\dot{r}\dot{\theta}$ as the reader can verify by means of Eq. (2.21.5). According to Newton's second law, $F_r = ma_r$ and $F_\theta = ma_\theta$. Since the equation of constraint is $r = s = $ const. we get from (6.3.1) and (6.3.2)

$$-ms\dot{\theta}^2 = mg\cos\theta - F \qquad (6.3.3)$$

$$ms\ddot{\theta} = -mg\sin\theta \qquad (6.3.4)$$

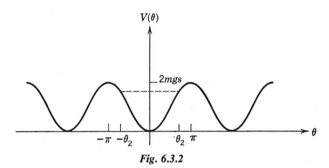

Fig. 6.3.2

The first equation allows the force of constraint to be calculated and the second equation can be written

$$\ddot{\theta} + \frac{g}{s}\sin\theta = 0 \qquad (6.3.5)$$

and is a nonlinear differential equation the solution of which yields θ as a function of t. Methods of obtaining opproximate solutions of nonlinear differential equations are discussed in Chapter 4. We will obtain here the exact solution of (6.3.5).

A first integral of the motion is obtained from the conservation of energy principle:

$$\tfrac{1}{2}ms^2\dot{\theta}^2 + mgs[1 - \cos\theta] = \mathscr{E} \qquad (6.3.6)$$

The qualitative features of the motion are obtained by a study of the potential function

$$V(\theta) = mgs[1 - \cos\theta] \qquad (6.3.7)$$

which is illustrated in Fig. 6.3.2. The maxima of $V(\theta)$ have the value

$$V_1 = 2mgs \qquad (6.3.8)$$

and as long as the total energy is less than this value the motion is periodic and symmetric about $\theta = 0$ with turning points at $\pm\theta_2$ given by

$$\mathscr{E} = mgs(1 - \cos\theta_2) = 2mgs\sin^2\frac{\theta_2}{2} \qquad (6.3.9)$$

If the energy is greater than $2mgs$, the motion is not periodic with respect to the coordinate θ and the magnitude of θ increases indefinitely. This is the case where the particle goes over the top and moves in a complete circle.

To complete the solution, the conservation of energy equation (6.3.6) is expressed

$$\dot{\theta}^2 = \frac{2\mathscr{E}}{ms^2}\left[1 - \frac{mgs}{\mathscr{E}}(1 - \cos\theta)\right] = \frac{2\mathscr{E}}{ms^2}\left[1 - \frac{2mgs}{\mathscr{E}}\sin^2\frac{\theta}{2}\right] \quad (6.3.10)$$

We treat first the case where the motion is periodic and introduce the dimensionless parameter

$$k^2 = \frac{\mathscr{E}}{2mgs} = \sin^2\frac{\theta_2}{2} < 1 \quad (6.3.11)$$

and the change of dependent variable

$$y = \frac{1}{k}\sin\frac{\theta}{2} \qquad \dot{\theta} = \frac{2k\dot{y}}{\sqrt{1 - k^2 y^2}} \quad (6.3.12)$$

In terms of y, the equation of motion as obtained from (6.3.10) is

$$\dot{y}^2 = \frac{g}{s}(1 - y^2)(1 - k^2 y^2) \quad (6.3.13)$$

The range of values of θ is $-\theta_2 \leq \theta \leq +\theta_2$ and from (6.3.11) and (6.3.12) the corresponding range of values of y is $-1 \leq y \leq +1$.

The first order differential Eq. (6.3.13) does not possess solutions in terms of the elementary functions, i.e., sines, cosines, exponentials, logarithms, or polynomials. It *does*, however, define a periodic function which, as we know from Chapter 4, must be symmetric with respect to time reversal $(t \rightarrow -t)$ and space inversion $(y \rightarrow -y)$. Qualitatively, the solution must resemble

$$y = \sin\left[\sqrt{g/s}\,(t - t_0)\right] \quad (6.3.14)$$

which is the solution for very small amplitudes. This can be checked by noting that as the energy becomes small, $k^2 \rightarrow 0$ and (6.3.13) becomes $\dot{y}^2 = (g/s)(1 - y^2)$ the solution of which is (6.3.14).

If in (6.3.13), the change of independent variable $u = \sqrt{g/s}\,t$ is made, the resulting differential equation is

$$\left(\frac{dy}{du}\right)^2 = (1 - y^2)(1 - k^2 y^2) \quad (6.3.15)$$

It is the conventional practice to write the solution of this differential equation as

$$y = sn(u - u_0) \quad (6.3.16)$$

where $u_0 = \sqrt{g/s}\, t_0$ is a constant of integration; t_0 represents the time when $y = 0$. Without loss of generality t_0 can be put equal to zero, and this will be the general practice from now on. The second constant of integration is the parameter k and is called the *modulus*. The modulus is related to the total energy and to the turning points of the motion through equation (6.3.11). The function $sn\,u$ is one of the *Jacobian elliptic functions*. It has the range of values $-1 \leq sn\,u \leq +1$ and is defined to have the value zero at $u = 0$.

Imagine that at $t = 0$ the pendulum passes through its equilibrium position, moving in the direction of positive θ. The angular velocity $\dot{\theta}$ and, therefore, dy/du is positive. Over this part of the motion, the positive square root of (6.3.15) applies:

$$du = \frac{dy}{\sqrt{(1 - y^2)(1 - k^2 y^2)}}$$

$$u = \int_0^y \frac{dy}{\sqrt{(1 - y^2)(1 - k^2 y^2)}} \tag{6.3.17}$$

This is known as the *elliptic integral of the first kind*. When $y = 1$, the pendulum has reached its positive turning point. The corresponding value of u is generally written

$$K = \int_0^1 \frac{dy}{\sqrt{(1 - y^2)(1 - k^2 y^2)}} \tag{6.3.18}$$

and called the complete *elliptic integral of the first kind*. Another form of the elliptic integrals results from making the change of variable

$$y = \sin \phi = sn\,u \tag{6.3.19}$$

We find

$$u = \int_0^\phi \frac{d\phi}{\sqrt{1 - k^2 \sin^2 \phi}} \qquad K = \int_0^{\pi/2} \frac{d\phi}{\sqrt{1 - k^2 \sin^2 \phi}} \tag{6.3.20}$$

Unfortunately, ϕ is called the *amplitude* of u in the mathematical literature. In physics, the term amplitude generally refers to the maximum displacement of a particle from equilibrium when it executes periodic motion. Also frequently used is the notation

$$u = F(k, \phi) \tag{6.3.21}$$

In order for the elliptic function $sn\,u$ to be of use, its numerical values as a function of u and k must be known.* Observe from equation (6.3.11) that

---

* Tables of the elliptic integrals (6.3.20) are available in the Chemical Rubber Company's *Standard Mathematical Tables*. From these tables the data for Fig. 6.3.3 were obtained.

$k^2 = 0.992$ means that the energy is only slightly less than the value for the sticking solution. This is quite evident in the plot of *snu* for $k^2 = 0.992$ in Fig. 6.3.3. In obtaining these figures, it is, of course, necessary only to obtain numerical data for the first quarter of a cycle $0 \le snu \le 1$. The curve for the next quarter of the cycle, corresponding to the pendulum returning to equilibrium from its positive turning point, is exactly the same shape, a fact which is due to the symmetry of the system with

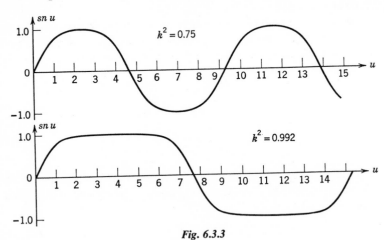

Fig. 6.3.3

respect to time inversion. The curve for the remaining half of the cycle, where the pendulum swings from equilibrium to its negative turning point and back again, is obtained simply by turning the curve for the first half of the period upside down. This is true because the potential function for the pendulum is symmetric with respect to space inversion.

The complete elliptic integral (6.3.18) is the value of u for which one quarter of a cycle is complete: $sn(K) = 1$. The period of *snu*, therefore, is $4K$. If P is the period of the pendulum,

$$\sqrt{g/s}\, P = 4K, \qquad P = \sqrt{s/g}\, 4K \qquad (6.3.22)$$

Since K depends on the modulus k which, in turn, depends on the amplitude of the motion, Eq. (6.3.11), the period of the pendulum is *amplitude dependent*. The period of a pendulum for very small amplitude is

$$P_0 = 2\pi\sqrt{s/g} \qquad (6.3.23)$$

Therefore,

$$\frac{P}{P_0} = \frac{2K}{\pi} \qquad (6.3.24)$$

Figure 6.3.4 shows P/P_0 as a function of the angular amplitude θ_2.

It is possible to obtain the period of the pendulum as a power series expansion in θ_2. Since $k^2 < 1$, the integrand of the elliptic integral (6.3.20) can be expanded by means of the binomial theorem:

$$K = \int_0^{\pi/2} (1 + \tfrac{1}{2}k^2 \sin^2 \phi + \tfrac{3}{8}k^4 \sin^4 \phi + \cdots)\, d\phi \qquad (6.3.25)$$

Integration term by term yields

$$K = \frac{\pi}{2}[1 + \tfrac{1}{4}k^2 + \tfrac{9}{64}k^4 + \cdots] \qquad (6.3.26)$$

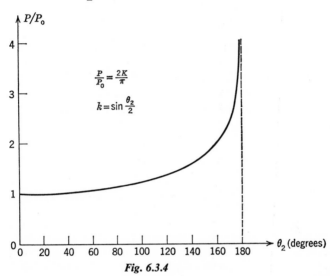

$$\frac{P}{P_0} = \frac{2K}{\pi}$$

$$k = \sin\frac{\theta_2}{2}$$

Fig. 6.3.4

The period of the plane pendulum, therefore, is

$$P = P_0\left(1 + \frac{1}{4}\sin^2\frac{\theta_2}{2} + \frac{9}{64}\sin^4\frac{\theta_2}{2} + \cdots\right) \qquad (6.3.27)$$

If the approximation $\sin \theta_2/2 = \theta_2/2$ is made and only two terms are retained,

$$P = P_0(1 + \tfrac{1}{16}\theta_2{}^2) \qquad (6.3.28)$$

The period to this approximation is also obtainable from Eq. 4.13.23.

If the modulus has the exact value $k^2 = 1$, the energy is $2mgs$ and the sticking solution is obtained. For this special case, the differential equation (6.3.15) is

$$\left(\frac{dy}{du}\right)^2 = (1 - y^2)^2 \qquad (6.3.29)$$

$$\frac{dy}{du} = \pm(1 - y^2) \qquad (6.3.30)$$

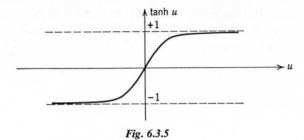

Fig. 6.3.5

If $dy/du > 0$, the solution is

$$y = \tanh u = \tanh (\sqrt{g/s}\, t) = \sin \frac{\theta}{2} \qquad (6.3.31)$$

If the pendulum passes through $\theta = 0$ with exactly the energy $2mgs$, it approaches $\theta = 180°$ $(y = 1)$ asymptotically as illustrated in Fig. 6.3.5.

If $\mathscr{E} > 2mgs$, the motion is nonperiodic. The modulus can be redefined as

$$k^2 = \frac{2mgs}{\mathscr{E}} < 1 \qquad (6.3.32)$$

Equation (6.3.10) is then

$$\dot\theta^2 = \frac{4g}{sk^2}\left[1 - k^2 \sin^2 \frac{\theta}{2}\right] \qquad (6.3.33)$$

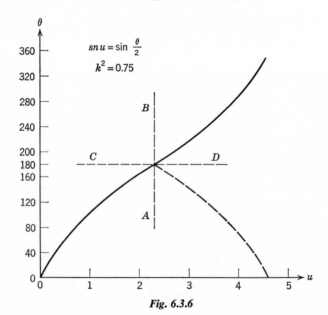

Fig. 6.3.6

The change of variable $y = \sin(\theta/2)$ results in

$$\dot{y}^2 = \frac{g}{k^2 s}(1 - y^2)(1 - k^2 y^2) \tag{6.3.34}$$

The solution is

$$y = snu = \sin\frac{\theta}{2}, \qquad u = \frac{1}{k}\sqrt{\frac{g}{s}}\,t \tag{6.3.35}$$

To interpret this result, θ as a function of u is plotted for $0 \leq \theta \leq 180°$, resulting in a portion of the graph of Fig. 6.3.6. The value of the modulus used in this example is $k^2 = 0.75$. For greater values of u, we must continue with values of θ greater than $180°$. It is evident from (6.3.35) that the same value of u results from, say, $\theta = 200°$ as from $\theta = 160°$. The motion for $180° < \theta \leq 360°$ can also be obtained by reflecting the solution for $0 \leq \theta \leq 180°$ first in the line AB, resulting in the dashed curve, and then making a second reflection in CD. It is easily verified that this double reflection procedure leads to a continuous tangent to the curve at $\theta = 180°$ meaning that no discontinuity in the velocity occurs.

6.4. THE JACOBIAN ELLIPTIC FUNCTIONS

The Jacobian elliptic function $y = snu$ has been introduced in Section 6.3 in connection with the plane pendulum. The first order differential equation (6.3.15) can be expressed

$$\frac{d}{du}(snu) = \pm\sqrt{1 - sn^2 u}\sqrt{1 - k^2 sn^2 u} \tag{6.4.1}$$

It is convenient, therefore, to introduce two new elliptic functions as

$$cnu = \pm\sqrt{1 - sn^2 u} \tag{6.4.2}$$

$$dnu = \sqrt{1 - k^2 sn^2 u} \tag{6.4.3}$$

where the plus or minus sign is used, depending on the sign of the derivative of snu. Referring to Fig. 6.3.3, and remembering that $0 < u \leq K$ is the first quarter cycle of snu, reveals that

$$\frac{d}{du}(snu) > 0 \qquad -K < u < K,\ 3K < u < 5K,\ \text{etc.}$$
$$\frac{d}{du}(snu) < 0 \qquad K < u < 3K,\ 5K < u < 7K,\ \text{etc.} \tag{6.4.4}$$

where K is the complete elliptic integral (6.3.18). The derivatives of snu at $u = K, 3K, 5K, \ldots$ are zero. These results are, of course, exactly analogous to the behavior of sin u and cos u. Equations (6.4.1), (6.4.2), and (6.4.3) are also expressible as

$$\frac{d}{du}(snu) = cnu\ dnu \tag{6.4.5}$$

$$sn^2u + cn^2u = 1 \tag{6.4.6}$$

$$dn^2u + k^2sn^2u = 1 \tag{6.4.7}$$

By differentiating (6.4.6) and (6.4.7) with respect to u it is proved that

$$\frac{d}{du}(cnu) = -snu\ dnu \tag{6.4.8}$$

$$\frac{d}{du}(dnu) = -k^2snucnu \tag{6.4.9}$$

The elimination of sn^2u between (6.4.6) and (6.4.7) yields

$$dn^2u - k^2cn^2u = 1 - k^2 \tag{6.4.10}$$

The functions cnu and dnu are plotted in Fig. 6.4.1 for $k^2 = 0.75$ and $k^2 = 0.992$. Note that cnu and dnu are *even* and snu is *odd*:

$$sn(-u) = -snu \qquad cn(-u) = cnu \qquad dn(-u) = dnu \tag{6.4.11}$$

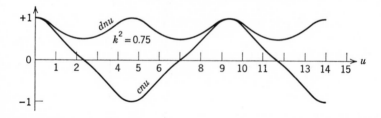

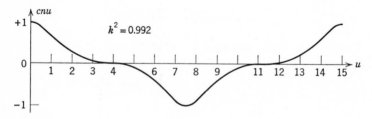

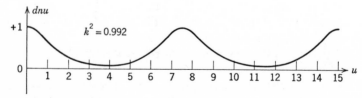

Fig. 6.4.1

The periods of both *snu* and *cnu* are $4K$ and that of *dnu* is $2K$. Note the special cases that occur when the modulus is zero or unity:

$$k = 0 \qquad snu = \sin u \qquad cnu = \cos u$$
$$dnu = 1 \tag{6.4.12}$$

$$k = 1 \qquad snu = \tanh u$$
$$cnu = dnu = \frac{1}{\cosh u} \tag{6.4.13}$$

The first order differential equation satisfied by $y = snu$ is

$$\left(\frac{dy}{du}\right)^2 = (1 - y^2)(1 - k^2 y^2) \tag{6.4.14}$$

By differentiation with respect to u,

$$2\frac{dy}{du}\frac{d^2y}{du^2} = -2y\frac{dy}{du}(1 - k^2 y^2) - 2k^2 y\frac{dy}{du}(1 - y^2) \tag{6.4.15}$$

Simplification gives

$$\frac{d^2y}{du^2} + y(1 + k^2) - 2k^2 y^3 = 0 \tag{6.4.16}$$

which is the second order differential equation obeyed by $y = snu$.
If $y = cnu$, Eq. (6.4.8) can be written

$$\left(\frac{dy}{du}\right)^2 = sn^2u\, dn^2u = (1 - y^2)(1 - k^2 + k^2 y^2) \tag{6.4.17}$$

The second order differential equation obeyed by *cnu* follows by differentiation:

$$\frac{d^2y}{du^2} + y(1 - 2k^2) + 2k^2 y^3 = 0 \tag{6.4.18}$$

Note that the differential equations (6.4.16) and (6.4.18) are basically the same type with changes in the coefficients of y and y^3. These are nonlinear second order differential equations of the same type encountered in Chapter 4 in the study of one-dimensional mechanical systems. In Appendix C will be found an extensive list of first and second order differential equations satisfied by the Jacobian elliptic functions and various combinations of them. These can be derived by the same procedure used to obtain (6.4.16) and (6.4.18).

A fourth Jacobian elliptic function can be defined by

$$tnu = \frac{snu}{cnu} \tag{6.4.19}$$

Special notations are also given to the reciprocals of *snu*, *cnu*, etc., but it does not seem worthwhile to keep track of so much terminology since the properties of all such functions may be obtained as needed from the basic three: *snu*, *cnu*, and *dnu*. As might be expected, there is a hierarchy of identities obeyed by the Jacobian elliptic functions. These and further developments of the theory of elliptic functions and integrals can be found in the references listed at the end of this chapter.

We find, for example, in Pierce (Reference 1) that the Jacobian elliptic functions obey

$$sn(u + v) = \frac{snu\,cnv\,dnv + cnu\,snv\,dnu}{1 - k^2sn^2u\,sn^2v} \qquad (6.4.20)$$

$$cn(u + v) = \frac{cnu\,cnv - snu\,snv\,dnu\,dnv}{1 - k^2sn^2u\,sn^2v} \qquad (6.4.21)$$

$$dn(u + v) = \frac{dnu\,dnv - k^2snu\,snv\,cnu\,cnv}{1 - k^2sn^2u\,sn^2v} \qquad (6.4.22)$$

These identities are quite a bit harder to prove than are the analogous trigonometric identities. (Proofs can be found in Chapter 22 of Whittaker and Watson, Reference 4.) By recalling the special values $sn(K) = 1$ and $cn(K) = 0$ it is easy to specialize the above identities to

$$sn(u + K) = \frac{cnu}{dnu} \qquad (6.4.23)$$

$$cn(u + K) = -\frac{snu}{dnu}\,dnK \qquad (6.4.24)$$

$$dn(u + K) = +\frac{dnK}{dnu} \qquad (6.4.25)$$

$$tn(u + K) = -\frac{1}{tnu\,dnK} \qquad (6.4.26)$$

$$dn(u + K)tn(u + K) = \frac{-1}{dnu\,tnu} \qquad (6.4.27)$$

Thus, for instance, the ratio *cnu/dnu* is not a fundamentally new function and would obey the same second order differential equation as *snu*. All of the fundamentally different combinations of Jacobian elliptic functions that give a second order differential equation of the form $y'' + ay + by^3 = 0$ are listed in Appendix C. Their use in solving mechanics problems is illustrated in Section 6.5.

6.5. SOLUTIONS OF DIFFERENTIAL EQUATIONS BY MEANS OF JACOBIAN ELLIPTIC FUNCTIONS

A few nonlinear second order differential equations, e.g., that of the simple pendulum, Eq. (6.3.5), can be solved exactly in terms of elliptic functions if an appropriate change of the dependent variable is made. In most cases this will not be true and it is necessary either to find some other special function or to fall back on the approximate differential equation resulting from the Taylor's series expansion of the potential function:

$$V(x) = k_0 x + \tfrac{1}{2}k_1 x^2 + \tfrac{1}{3}k_2 x^3 + \tfrac{1}{4}k_3 x^4 + \cdots \qquad (6.5.1)$$

By using the conservation of energy principle

$$\mathcal{E} = \tfrac{1}{2}m\dot{x}^2 + V(x) \qquad (6.5.2)$$

it is always possible, in principle, to obtain the motion by evaluation of

$$t = \int_{x_0}^{x} \frac{dx}{\sqrt{(2/m)(\mathcal{E} - V(x))}} \qquad (6.5.3)$$

Outside of some changes in the variables, this is essentially what was done with the plane pendulum to get the elliptic integral (6.3.17). If $V(x)$ is approximated by (6.5.1) then (6.5.3) is an integral of the form

$$t = \int_{x_0}^{x} \frac{dx}{\sqrt{a_4 x^4 + a_3 x^3 + a_2 x^2 + a_1 x + a_0}} \qquad (6.5.4)$$

It is shown in the mathematical literature that such an integral can always be reduced to an elliptic integral of the first kind. We will present here an alternative and much simpler approach to the solution of nonlinear differential equations by means of elliptic functions. If the potential function is symmetric, the odd powers can be eliminated. The equation of motion, using the same notion as in Section 4.13 is

$$\ddot{x} + \alpha x + \gamma x^3 = 0 \qquad \alpha = \frac{k_1}{m} \qquad \gamma = \frac{k_3}{m} \qquad (6.5.5)$$

If α and γ are positive, only periodic motion can occur. The substitution

$$x = A y(at) \qquad u = at \qquad (6.5.6)$$

where A and a are constants results in

$$y'' + \frac{\alpha}{a^2} y + \frac{A^2 \gamma}{a^2} y^3 = 0 \qquad (6.5.7)$$

where the prime stands for differentiation with respect to u. If the differential equation were linear, the constant A would simply cancel out! Because of the nonlinearity, A becomes entwined with the other constants, leading, as we shall see, to a dependence of the frequency on the amplitude. We require from the list in Appendix C, a differential equation that has positive coefficients. $y = cnu$ satisfies

$$y'' + (1 - 2k^2)y + 2k^2y^3 = 0 \qquad (6.5.8)$$

which will be satisfactory provided that

$$0 \leq k^2 < \tfrac{1}{2} \qquad (6.5.9)$$

since $k^2 > \tfrac{1}{2}$ means that the coefficient of the linear term is negative. Equating the coefficients of (6.5.7) and (6.5.8),

$$\frac{\alpha}{a^2} = 1 - 2k^2 \qquad \frac{A^2\gamma}{a^2} = 2k^2 \qquad (6.5.10)$$

Solving for a and k,

$$a^2 = \alpha + \gamma A^2 \qquad k^2 = \frac{\gamma A^2}{2\alpha + 2\gamma A^2} \qquad (6.5.11)$$

The condition (6.5.9) requires

$$0 \leq \gamma A^2 < \alpha + \gamma A^2 \qquad (6.5.12)$$

which means that A can take on any value. A is recognized as the amplitude of the motion and must be found from initial conditions. The exact solution of (6.5.5), containing two arbitrary constants A and t_0 is

$$x = Acn[a(t - t_0)] \qquad (6.5.13)$$

with a and the modulus k given by (6.5.11). As mentioned earlier, we ordinarily set $t_0 = 0$. The actual period of the motion is obtained from

$$aP = 4K \qquad (6.5.14)$$

where K is the complete elliptic integral (6.3.18). The period is amplitude dependent, this dependence entering both through the constant a and the modulus as given by (6.5.11).

The energy of the particle is

$$\frac{\mathscr{E}}{m} = \tfrac{1}{2}\dot{x}^2 + \tfrac{1}{2}\alpha x^2 + \tfrac{1}{4}\gamma x^4 \qquad (6.5.15)$$

$$= \tfrac{1}{2}a^2A^2(y')^2 + \tfrac{1}{2}\alpha A^2 y^2 + \tfrac{1}{4}\gamma A^4 y^4$$

when $x = A$, $\dot{x} = 0$, and $y = 1$. The energy in terms of the amplitude, therefore, is

$$\mathscr{E} = \tfrac{1}{2}m(\alpha + \tfrac{1}{2}\gamma A^2)A^2 \qquad (6.5.16)$$

The fact that $\mathscr{E}$ has been evaluated using the special value $x = A$ is of no consequence since $\mathscr{E}$ is a constant.

As a second example, consider

$$\ddot{x} + \alpha x - \gamma x^3 = 0 \qquad (6.5.17)$$

The potential function

$$\psi(x) = \tfrac{1}{2}\alpha x^2 - \tfrac{1}{4}\gamma x^4 \qquad (6.5.18)$$

is sketched in Fig. 6.5.1. The maxima of $\psi(x)$ are at $\pm x_1$ and are given by

$$x_1 = \sqrt{\alpha/\gamma} \qquad \psi_1 = \frac{\alpha^2}{4\gamma} = \tfrac{1}{4}\gamma x_1^4 \qquad (6.5.19)$$

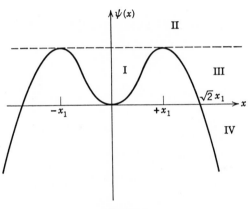

Fig. 6.5.1

The zeroes of $\psi(x)$ occur at $\pm\sqrt{2}x_1$. In terms of $x = Ay(at)$, the equation of motion is

$$y'' + \frac{\alpha}{a^2}y - A^2\frac{\gamma}{a^2}y^3 = 0 \qquad (6.5.20)$$

Comparison of (6.5.20) and (1) of Appendix C reveals that the periodic solution, i.e., the solution in region I of Fig. 6.5.1 is

$$x = Asn(at) \qquad (6.5.21)$$

$$a^2 = \alpha - \tfrac{1}{2}\gamma A^2 \qquad (6.5.22)$$

$$k^2 = \frac{\gamma A^2}{2\alpha - \gamma A^2} \qquad (6.5.23)$$

It is important to remember that the modulus is restricted to the range of values

$$0 \le k^2 < 1 \qquad (6.5.24)$$

and this can often be applied as a test to make sure that the correct solution has been chosen. From (6.5.23),

$$0 \leq \tfrac{1}{2}\gamma A^2 < \alpha - \tfrac{1}{2}\gamma A^2 \qquad (6.5.25)$$

which, by (6.5.19), is the equivalent of

$$0 \leq A < x_1 \qquad (6.5.26)$$

This is the correct range of values of the amplitude for periodic motion. The energy of the particle is

$$\frac{\mathscr{E}}{m} = \tfrac{1}{2}\alpha A^2 - \tfrac{1}{4}\gamma A^4 \qquad (6.5.27)$$

It is easily verified that (6.5.26) gives the range of energies

$$0 \leq \frac{\mathscr{E}}{m} < \psi_1 \qquad (6.5.28)$$

The sticking solution on the border between I and II is found from the limiting value of the modulus $k = 1$:

$$x = x_1 \tanh (\sqrt{\alpha/2}\, t) \qquad (6.5.29)$$

In region II of Fig. 6.5.1, the solution is

$$x = A\, dn(at)tn(at) \qquad (6.5.30)$$

$$a^2 = \tfrac{1}{2}\gamma A^2 \qquad (6.5.31)$$

$$k^2 = \frac{\alpha + \gamma A^2}{2\gamma A^2} \qquad (6.5.32)$$

At $x = 0$, all the energy is kinetic:

$$\frac{\mathscr{E}}{m} = \tfrac{1}{2}\dot{x}^2 = \tfrac{1}{2}a^2 A^2 (y')^2 \qquad (6.5.33)$$

Referring to (7) of Appendix C, $y' = 1$ when $y = 0$. Therefore,

$$\frac{\mathscr{E}}{m} = \tfrac{1}{4}\gamma A^4 \qquad (6.5.34)$$

The modulus condition (6.5.24) leads to

$$A > x_1 \qquad (6.5.35)$$

which, in terms of energy means

$$\frac{\mathscr{E}}{m} > \psi_1 \qquad (6.5.36)$$

as required for solutions in region II. Note from (6.5.32) that as $A \to \infty$, $k^2 \to \frac{1}{2}$ so that in this case the modulus is actually restricted to $\frac{1}{2} < k^2 < 1$. The function *dnutnu* is plotted in Fig. 6.5.2 for $k^2 = 0.883$. The effect on the motion due to the two humps in the potential function is clearly visible. Notice that *dnutnu* $\to \infty$ as $u \to K$ ($K = 2.50$ for $k^2 = 0.883$). In terms of the motion, this means that the particle covers an infinite

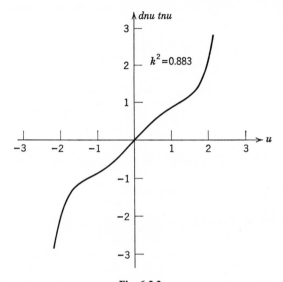

Fig. 6.5.2

distance in a finite time! It should be remembered, however, that in practical applications the potential function (6.5.18) will generally be only an approximation and will no doubt not hold over the entire range.

In region III of Fig. 6.5.1 the solution is

$$x = \frac{A}{sn(at)} \tag{6.5.37}$$

$$a^2 = \tfrac{1}{2}\gamma A^2 \tag{6.5.38}$$

$$k^2 = \frac{2\alpha - \gamma A^2}{\gamma A^2} \qquad 0 \le k^2 < 1 \tag{6.5.39}$$

$$x_1 < A < \sqrt{2}x_1 \tag{6.5.40}$$

$$\frac{\mathscr{E}}{m} = \tfrac{1}{2}\alpha A^2 - \tfrac{1}{4}\gamma A^4 \qquad 0 \le \frac{\mathscr{E}}{m} < \psi_1 \tag{6.5.41}$$

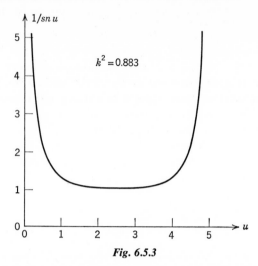

Fig. 6.5.3

In this case, A represents the single turning point of the motion. Figure 6.5.3 shows $1/snu$ for $k^2 = 0.883$. For $k = 1$, the sticking solution on the border between II and III is obtained:

$$x = \frac{x_1}{\tanh\left(\sqrt{\alpha/2}\ t\right)} \tag{6.5.42}$$

The value $k = 0$ gives the solution for zero energy on the border between III and IV:

$$x = \frac{\sqrt{2}x_1}{\sin\left(\sqrt{\alpha}\ t\right)} \tag{6.5.43}$$

Finally, in region IV the solution is found to be

$$x = \frac{A}{cn(at)} \tag{6.5.44}$$

$$a^2 = \gamma A^2 - \alpha \tag{6.5.45}$$

$$k^2 = \frac{\gamma A^2 - 2\alpha}{2\gamma A^2 - 2\alpha} \qquad 0 \le k^2 < \tfrac{1}{2} \tag{6.5.46}$$

$$\sqrt{2}\ x_1 \le A < \infty \tag{6.5.47}$$

$$\frac{\mathscr{E}}{m} = \tfrac{1}{2}\alpha A^2 - \tfrac{1}{4}\gamma A^4 \qquad -\infty < \mathscr{E} < 0 \tag{6.5.48}$$

We close this section by making some remarks about the contrast between linear and nonlinear systems. In solving the linear equation

$$\ddot{x} + \alpha x = 0 \tag{6.5.49}$$

we substitute $x = A \cos \omega t$ where ω is a parameter to be determined in terms of α from the differential equation. In solving the nonlinear equation

$$\ddot{x} + \alpha x + \gamma x^3 = 0 \tag{6.5.50}$$

we substitute $x = Ay(at)$ where y is one of the elliptic functions. There are now *two* coefficients, α and γ, in the differential equation and there are *two* parameters in the elliptic function, a and the modulus k, to be determined in terms of them. Due to the nonlinearity, these relations will also involve the amplitude. In solving (6.5.50), we *must not* try to superimpose solutions such as $x = A sn(at) + B cn(at)$ as this principle works *only* for linear differential equations.

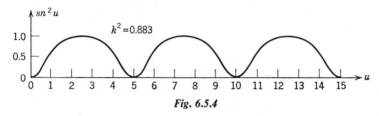

Fig. 6.5.4

If in the Taylor's series expansion of a symmetric potential one more term is included, there results the equation of motion

$$\ddot{x} + \alpha x + \gamma x^3 + \epsilon x^5 = 0 \tag{6.5.51}$$

If an exact solution is desired, we would need a *hyperelliptic function* depending on *three* independent parameters.

If $z = snu$, then z obeys the first order differential equation

$$(z')^2 = (1 - z^2)(1 - k^2 z^2) \tag{6.5.52}$$

Suppose the change of dependent variable $y = z^2$ is made. It is found that y obeys

$$(y')^2 = 4y(1 - y)(1 - k^2 y) \tag{6.5.53}$$

A second differentiation gives

$$y'' - 2 + 4(k^2 + 1)y - 6k^2 y^2 = 0 \tag{6.5.54}$$

which is the second order differential equation obeyed by $y = sn^2 u$. The differential equations satisfied by the squares of other elliptic functions are listed in Appendix C.

Figure 6.5.4 shows a plot of $sn^2 u$ for $k^2 = 0.883$.

As an application, the periodic solution of

$$\ddot{x} + \alpha x - \beta x^2 = 0 \tag{6.5.55}$$

will be found.

The potential function for (6.5.55) is

$$\psi(x) = \tfrac{1}{2}\alpha x^2 - \tfrac{1}{3}\beta x^3 \qquad (6.5.56)$$

and is sketched in Fig. 6.5.5. The location and value of the relative maximum of $\psi(x)$ are

$$x_1 = \frac{\alpha}{\beta}$$

$$\psi_1 = \frac{\alpha^3}{6\beta^2} = \tfrac{1}{6}\beta x_1^{\,3} \qquad (6.5.57)$$

The potential has the value ψ_1 also at $x = -\tfrac{1}{2}x_1$; thus as the energy approaches ψ_1 the turning points of the periodic motion approach x_1 and $-\tfrac{1}{2}x_1$. The zeroes of $\psi(x)$ are at $x = 0$ and $x = (\tfrac{3}{2})x_1$.

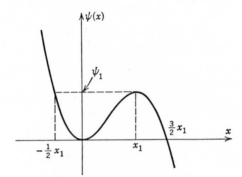

Fig. 6.5.5

The substitution

$$x = B + Ay(at) \qquad (6.5.58)$$

leads to

$$y'' + \frac{\alpha B - \beta B^2}{Aa^2} + \frac{\alpha - 2\beta B}{a^2}\, y - \frac{\beta A}{a^2}\, y^2 = 0 \qquad (6.5.59)$$

Since the potential function is not symmetric, the average position of the particle is not zero. The constant B provides a means of adjusting the average position properly depending on the energy. The comparison of (6.5.59) and (6.5.54) suggests that $y = sn^2(at)$. Equating coefficients yields

$$\frac{\alpha B - \beta B^2}{Aa^2} = -2 \qquad \frac{\alpha - 2\beta B}{a^2} = 4(k^2 + 1) \qquad \frac{\beta A}{a^2} = 6k^2 \qquad (6.5.60)$$

which are three equations relating the four parameters A, B, a, and k. One of the parameters is to be regarded as a constant of integration and the other three are then computed in terms of it. The elimination of a^2 and k^2 results in

$$B^2 + B(A - x_1) + \tfrac{1}{3}A^2 - \tfrac{1}{3}Ax_1 = 0 \qquad (6.5.61)$$

where x_1 is given by (6.5.57). The solution of (6.5.61) for B is

$$B = \tfrac{1}{2}(x_1 - A) \pm \tfrac{1}{2}\sqrt{x_1^2 - \tfrac{1}{3}A^2} \qquad (6.5.62)$$

In the limit of very small amplitudes $A \to 0$ and $B \to 0$. Then,

$$B = \tfrac{1}{2}x_1 \pm \tfrac{1}{2}x_1 \qquad (6.5.63)$$

The minus sign, therefore, is required and

$$B = \tfrac{1}{2}(x_1 - A) - \tfrac{1}{2}\sqrt{x_1^2 - \tfrac{1}{3}A^2} \qquad (6.5.64)$$

The turning points of the motion occur when $y = 1$ and $y = 0$ and are

$$x_2 = A + B \qquad x_3 = B \qquad (6.5.65)$$

As the energy approaches the limiting value ψ_1, the turning points approach $x_2 = x_1$ and $x_3 = -\tfrac{1}{2}x_1$. The limiting values of A and B, therefore, must, be

$$B = -\tfrac{1}{2}x_1 \qquad A = \tfrac{3}{2}x_1 \qquad (6.5.66)$$

Observe that (6.5.64) conforms to this limit.

From (6.5.60) the remaining parameters expressed in terms of A are found to be

$$a^2 = \beta\left[\frac{A}{12} + \tfrac{1}{4}\sqrt{x_1^2 - \tfrac{1}{3}A^2}\right] \qquad (6.5.67)$$

$$k^2 = \frac{2A}{A + 3\sqrt{x_1^2 - \tfrac{1}{3}A^2}} \qquad (6.5.68)$$

The modulus condition $0 \le k^2 < 1$ applied to (6.5.68) yields

$$0 \le A < \tfrac{3}{2}x_1 \qquad (6.5.69)$$

which gives the correct limiting value (6.5.66) for A. Since (6.5.69) includes the entire range of values of A for which periodic motion is possible, we conclude that all the periodic solutions of (6.5.55) have been found. By using the list of differential equations in Appendix C, all the nonperiodic solutions of (6.5.55) can also be found. Since by now the method should be clear we will not pursue this program in detail.

Notice that in the sketch of sn^2u in Fig. 6.5.4 the curvature is much more pronounced at $sn^2u = 0$ than at $sn^2u = 1$. This is correlated with the fact that the potential function, Fig. 6.5.5, is steeper at the negative turning point than at the positive turning point.

6.6. THE SPHERICAL PENDULUM

If a single particle of mass m is constrained to move without friction on the surface of a sphere of radius s and is acted on by a uniform gravitational field it forms a spherical pendulum. The motion can be described by using spherical polar coordinates (Section 2.13) as illustrated in Fig. 6.6.1. In order to keep the particle on the spherical surface, a nonconservative force of constraint is required. This force acts perpendicular to the surface

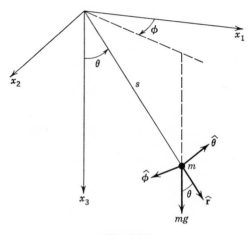

Fig. 6.6.1

of constraint and could be supplied by the surface itself, e.g., if the particle were sliding without friction on a spherical bowl, or by a string or rod to which the particle is attached. Whatever the mechanism, the equation of constraint is $r = s = $ const.

The resolution of the forces into their r, θ, and ϕ components results in

$$F_r = mg \cos \theta - F \qquad (6.6.1)$$

$$F_\theta = -mg \sin \theta \qquad (6.6.2)$$

$$F_\phi = 0 \qquad (6.6.3)$$

where F is the force of constraint. There is no force in the ϕ-direction, meaning that the system is symmetric with respect to a rotation about the x_3-axis. It is to be anticipated that there will be a constant of the motion associated with this symmetry. The kinematic expressions for acceleration have already been derived and are found in Eq. (2.21.11). Because of the constraint, $\dot{r} = 0$ and $\ddot{r} = 0$. The equations of motion are

$$mg \cos\theta - F = -m(s\dot{\theta}^2 + s \sin^2\theta\dot{\phi}^2) \qquad (6.6.4)$$

$$-mg \sin\theta = m(s\ddot{\theta} - s \sin\theta \cos\theta\dot{\phi}^2) \qquad (6.6.5)$$

$$0 = \frac{m}{s \sin\theta}\frac{d}{dt}(s^2 \sin^2\theta\dot{\phi}) \qquad (6.6.6)$$

Since r is constant, there are actually two independent coordinates θ and ϕ. The system is said to have *two degrees of freedom*. Equation (6.6.6) gives

$$ms^2 \sin^2\theta\dot{\phi} = l = \text{const.} \qquad (6.6.7)$$

which is the promised constant of the motion resulting from the rotational symmetry. It is possible to prove that (6.6.7) is identical to the x_3-component of the angular momentum of the particle:

$$l_3 = m[x_1\dot{x}_2 - \dot{x}_1 x_2] \qquad (6.6.8)$$

Writing

$$x_1 = s \sin\theta \cos\phi \qquad x_2 = s \sin\theta \sin\phi \qquad (6.6.9)$$

$$\dot{x}_1 = (s \cos\theta \cos\phi)\dot{\theta} - (s \sin\theta \sin\phi)\dot{\phi} \qquad (6.6.10)$$

$$\dot{x}_2 = (s \cos\theta \sin\phi)\dot{\theta} + (s \sin\theta \cos\phi)\dot{\phi} \qquad (6.6.11)$$

and substituting into (6.6.8) results in

$$l_3 = ms^2 \sin^2\theta\dot{\phi} = l \qquad (6.6.12)$$

This an example where a *single component only* of the angular momentum is conserved. The letter l is not to be confused with the magnitude of the total angular momentum which *is not* conserved; the subscript 3 is dropped because it is inconvenient to carry it through the subsequent calculations. By using (6.6.12), $\dot{\phi}$ can be eliminated from (6.6.5), resulting in a second order differential equation for $\theta(t)$:

$$\ddot{\theta} - \frac{l^2 \cos\theta}{m^2 s^4 \sin^3\theta} + \frac{g}{s}\sin\theta = 0 \qquad (6.6.13)$$

The equation of conservation of energy is

$$\mathscr{E} = T + V = T - mgs \cos\theta \qquad (6.6.14)$$

The potential energy is defined in such a way that it has the value $-mgs$ when the particle is at $\theta = 0$. The expression for kinetic energy in spherical coordinates is found in Eq. (2.13.10).

$$\mathscr{E} = \tfrac{1}{2}m(s^2\dot{\theta}^2 + s^2 \sin^2\theta\dot{\phi}^2) - mgs \cos\theta \qquad (6.6.15)$$

Elimination of $\dot{\phi}$ by means of (6.6.12) yields

$$\mathscr{E} = \tfrac{1}{2}ms^2\dot{\theta}^2 + \frac{l^2}{2ms^2\sin^2\theta} - mgs\cos\theta \qquad (6.6.16)$$

The equation of motion (6.6.13) and its first integral (6.6.16) involve only a single function $\theta(t)$. The problem has been reduced to the equivalent of a one-dimensional mechanical system and can be treated as such. It is convenient to define an *effective potential* as

$$V_e(\theta) = \frac{l^2}{2ms^2\sin^2\theta} - mgs\cos\theta = mgs\left[\frac{l^2}{2m^2gs^3\sin^2\theta} - \cos\theta\right] \qquad (6.6.17)$$

so that the energy equation reads

$$\mathscr{E} = \tfrac{1}{2}ms^2\dot{\theta}^2 + V_e(\theta) \qquad (6.6.18)$$

The effective potential plays exactly the same role as does the actual potential in an ordinary one-dimensional problem. In Fig. 6.6.2, V_e/mgs is plotted for three different values of the dimensionless parameter

$$\mu = \frac{l^2}{2m^2gs^3} \qquad (6.6.19)$$

As long as $l \neq 0$, the effective potential is infinite at $\theta = 0$ and $\theta = 180°$, meaning that the particle never passes through these points. The case $l = 0$ is the plane pendulum already treated in Section 6.3. Between the limiting values $0°$ and $180°$, $V_e(\theta)$ has a single minimum. Suppose that l has been given some definite value so that we can fix our attention on one of the curves of Fig. 6.6.2. If the energy of the system is exactly equal to the minimum in $V_e(\theta)$, then θ is a constant and the particle moves in a horizontal circular orbit. This is also called a *conical pendulum*. If the particle is perturbed by the addition of a little energy, $\theta(t)$ executes small oscillations about its equilibrium value. The particle still moves almost in a circular orbit, and for this reason the orbit is said to be *stable*. In general, $\theta(t)$ is a periodic function varying between two turning points that are determined from the effective potential in exactly the same way as is done with the ordinary potential in one-dimensional motion. By setting $\ddot{\theta} = 0$ in the equation of motion (6.6.5), the equilibrium value of θ can be computed in terms of the angular velocity $\dot{\phi}$:

$$\cos\theta_1 = \frac{g}{s\dot{\phi}^2} > 0 \qquad (6.6.20)$$

The value of θ_1 is always less than $90°$. As the angular velocity, and hence the angular momentum, is taken, larger and larger θ_1 approaches $90°$

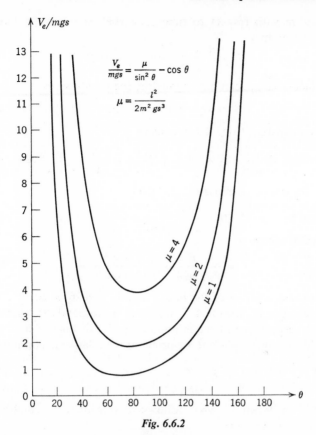

$$\frac{V_e}{mgs} = \frac{\mu}{\sin^2 \theta} - \cos \theta$$

$$\mu = \frac{l^2}{2m^2 gs^3}$$

Fig. 6.6.2

as is also evident from Fig. 6.6.2 by the fact that the minima approach $90°$ for increasing values of μ. Thus if a rock is tied to a string and whirled in a horizontal circle, the string becomes nearly horizontal for large velocities.

The energy equation can be expressed as

$$\frac{2\mathscr{E}}{mgs} = \frac{s}{g} \dot{\theta}^2 + \frac{l^2}{m^2 gs^3} \frac{1}{\sin^2 \theta} - 2 \cos \theta \qquad (6.6.21)$$

The change of dependent variable

$$z = -\cos \theta \qquad \dot{\theta}^2 = \frac{\dot{z}^2}{1 - z^2} \qquad (6.6.22)$$

yields

$$\frac{s}{g} \dot{z}^2 = \left(\frac{2\mathscr{E}}{mgs} - \frac{l^2}{m^2 gs^3} \right) - 2z - \frac{2\mathscr{E}}{mgs} z^2 + 2z^3 \qquad (6.6.23)$$

Differentiation with respect to time gives the second order differential equation obeyed by z:

$$\ddot{z} + \frac{g}{s} + \frac{2\mathscr{E}}{ms^2} z - 3\frac{g}{s} z^2 = 0 \tag{6.6.24}$$

The constant term g/s can be removed by the substitution

$$z = x - c = -\cos\theta \tag{6.6.25}$$

where c is a constant to be determined:

$$\ddot{x} + \frac{g}{s} - \frac{2\mathscr{E}}{ms^2}c - \frac{3g}{s}c^2 + \left(\frac{2\mathscr{E}}{ms^2} + \frac{6gc}{s}\right)x - \frac{3g}{s}x^2 = 0 \tag{6.6.26}$$

The constant c is to be chosen such that

$$0 = \frac{g}{s} - \frac{2\mathscr{E}}{ms^2}c - \frac{3g}{s}c^2 \tag{6.6.27}$$

The choice of the root

$$c = -\frac{\mathscr{E}}{3mgs} + \sqrt{\left(\frac{\mathscr{E}}{3mgs}\right)^2 + \frac{1}{3}} \tag{6.6.28}$$

will make the coefficient of the linear term in (6.6.26) positive. In terms of the discussion of Section 6.5, this root corresponds to choosing the origin with respect to x at the minimum of $\psi(x)$ in Fig. 6.5.5. The other root would, of course, place the x-origin at the relative maximum of $\psi(x)$.

$$\ddot{x} + \left[\frac{6g}{s}\sqrt{\left(\frac{\mathscr{E}}{3mgs}\right)^2 + \frac{1}{3}}\right]x - \frac{3g}{s}x^2 = 0 \tag{6.6.29}$$

It will be convenient in what follows to introduce the dimensionless parameter

$$\epsilon = \frac{\mathscr{E}}{3mgs} \tag{6.6.30}$$

It is seen that with

$$\alpha = \frac{6g}{s}\sqrt{\epsilon^2 + \tfrac{1}{3}} \qquad \beta = \frac{3g}{s} \tag{6.6.31}$$

Eq. (6.6.29) is identical to the differential equation in (6.5.55). The solution, therefore, is

$$\cos\theta = c - x = c - B - A\,sn^2(at) \tag{6.6.32}$$

From (6.5.64), (6.5.67), and (6.5.68)

$$B = \sqrt{\epsilon^2 + \tfrac{1}{3}} - \frac{A}{2} - \sqrt{\epsilon^2 + \tfrac{1}{3} - \tfrac{1}{12}A^2} \tag{6.6.33}$$

$$a^2 = \frac{3g}{s}\left[\frac{A}{12} + \tfrac{1}{2}\sqrt{\epsilon^2 + \tfrac{1}{3} - \tfrac{1}{12}A^2}\right] \tag{6.6.34}$$

$$k^2 = \frac{2A}{A + 6\sqrt{\epsilon^2 + \tfrac{1}{3} - \tfrac{1}{12}A^2}} \tag{6.6.35}$$

In terms of these parameters and the constant c as given by (6.6.28),

$$\cos \theta = -\epsilon + \sqrt{\epsilon^2 + \tfrac{1}{3} - \tfrac{1}{12}A^2} + A(\tfrac{1}{2} - sn^2(at)) \qquad (6.6.36)$$

Once the energy is known, only the parameter A is left undetermined. It would be possible in principle to evaluate A in terms of energy and angular momentum by inserting (6.6.36) into (6.6.23), but this leads to a cubic equation in A and is too complicated. Instead, the following method can be used. After the energy and angular momentum are specified, the effective potential can be graphed and the turning points of the motion determined as accurately as necessary by numerical methods. The turning points could also be obtained experimentally. If θ_2 is the larger turning point and θ_3 is the smaller turning point, then $sn^2(at) = 1$ at θ_2 and $sn^2(at) = 0$ at θ_3. Therefore,

$$\cos \theta_2 = -\epsilon + \sqrt{\epsilon^2 + \tfrac{1}{3} - \tfrac{1}{12}A^2} - \tfrac{1}{2}A \qquad (6.6.37)$$

$$\cos \theta_3 = -\epsilon + \sqrt{\epsilon^2 + \tfrac{1}{3} - \tfrac{1}{12}A^2} + \tfrac{1}{2}A \qquad (6.6.38)$$

Subtracting one equation from the other yields

$$A = \cos \theta_3 - \cos \theta_2 \qquad (6.6.39)$$

so that A can be found quite simply, once the turning points are known.

The determination of the azimuthal angle ϕ as a function of time is, unfortunately, not so easy. It is necessary to substitute (6.6.36) into the angular momentum equation (6.6.12) and integrate to find $\phi(t)$. This will be carried out approximately by the Fourier series method in the following discussion.

A much better qualitative and a quite satisfactory quantitative picture of the motion can be obtained in the form of a Fourier series. A Fourier series solution of (6.6.29) is given by Eq. (4.13.35) and by its use the Fourier series expansion for $\cos \theta$ is found to be

$$\cos \theta = -\epsilon + \sqrt{\epsilon^2 + \tfrac{1}{3}} - \frac{b_1^2}{4\sqrt{\epsilon^2 + \tfrac{1}{3}}} - b_1 \cos \omega t$$
$$(6.6.40)$$
$$+ \frac{b_1^2}{12\sqrt{\epsilon^2 + \tfrac{1}{3}}} \cos 2\omega t - \frac{b_1^3}{192(\epsilon^2 + \tfrac{1}{3})} \cos 3\omega t$$

$$\omega^2 = \frac{6g}{s} \sqrt{\epsilon^2 + \tfrac{1}{3}} \left[1 - \frac{5b_1^2}{24(\epsilon^2 + \tfrac{1}{3})} \right] \qquad (6.6.41)$$

The turning points are θ_2 at $\omega t = 0$ and θ_3 at $\omega t = \pi$. Inserting these two values into (6.6.40) and subtracting the two equations results in

$$b_1 = \tfrac{1}{2}(\cos \theta_3 - \cos \theta_2) - \frac{b_1^3}{192(\epsilon^2 + \tfrac{1}{3})}$$
$$(6.6.42)$$
$$\cong \tfrac{1}{2}(\cos \theta_3 - \cos \theta_2) - \frac{(\cos \theta_3 - \cos \theta_2)^3}{1536(\epsilon^2 + \tfrac{1}{3})}$$

For a numerical example, let

$$s = 98 \text{ cm} \qquad \frac{\mathscr{E}}{mgs} = 2 \qquad \epsilon = \tfrac{2}{3}$$

$$\mu = \frac{l^2}{2m^2gs^3} = 1.000 \qquad \sqrt{\epsilon^2 + \tfrac{1}{3}} = 0.882 \tag{6.6.43}$$

In Fig. 6.6.2 will be found a plot of the effective potential for $\mu = 1.000$. The turning points occur on this curve when $\mathscr{E}/(mgs) = V_e/(mgs) = 2$ and therefore, are roots of

$$\cos^3 \theta + 2 \cos^2 \theta - \cos \theta - 1 = 0 \tag{6.6.44}$$

By making a plot that is a little more accurate than Fig. 6.6.2, approximate turning points can be determined. These are then approximate roots of (6.6.44) and can be corrected, e.g., by Newton's method. The result is

$$\cos \theta_2 = -0.555 \qquad \theta_2 = 123°43'$$
$$\cos \theta_3 = 0.802 \qquad \theta_3 = 36°40' \tag{6.6.45}$$

By using (6.6.42)

$$b_1 = 0.676 \tag{6.6.46}$$

Equations (6.6.40) and (6.6.41) are then

$$\cos \theta = 0.0855 - 0.676 \cos \omega t + 0.043 \cos 2\omega t - 0.002 \cos 3\omega t \tag{6.6.47}$$

$$\omega = 6.82 \text{ sec}^{-1} = 391° \text{ sec}^{-1} \tag{6.6.48}$$

The convergence is seen to be quite good. The mean or average value of θ is given by

$$\overline{\cos \theta} = -\epsilon + \sqrt{\epsilon^2 + \tfrac{1}{3}} - \frac{b_1^2}{4\sqrt{\epsilon^2 + \tfrac{1}{3}}} = 0.0855, \qquad \bar{\theta} = 85°6' \tag{6.6.49}$$

This is not to be confused with the *equilibrium* value of θ which is the location of the minimum of $V_e(\theta)$, about 69° in this example. The average value of θ is displaced away from the equilibrium value due to the asymmetry of the effective potential.

In computing $\phi(t)$ it will be convenient to write (6.6.40) as

$$\cos \theta = a_0 - b_1 \cos \omega t + b_2 \cos 2\omega t - b_3 \cos 3\omega t \tag{6.6.50}$$

From (6.6.12) the expression for $\dot{\phi}$ is

$$\dot{\phi} = \frac{l}{ms^2} \frac{1}{1 - \cos^2 \theta} \tag{6.6.51}$$

The expansion of

$$f(z) = \frac{1}{1 - z^2} \tag{6.6.52}$$

in a Taylor's series about $z = a_0$ yields

$$\frac{1}{1 - \cos^2 \theta} = \frac{1}{1 - a_0^2} + \frac{2a_0}{(1 - a_0^2)^2} (\cos \theta - a_0)$$
$$+ \frac{1 + 3a_0^2}{(1 - a_0^2)^3} (\cos \theta - a_0)^2 + \cdots \tag{6.6.53}$$

Using the Fourier series (6.6.50) and maintaining second order accuracy gives

$$\frac{1}{1 - \cos^2 \theta} = \frac{1}{1 - a_0^2} + \frac{2a_0}{(1 - a_0^2)^2} (-b_1 \cos \omega t + b_2 \cos 2\omega t)$$
$$+ \frac{1 + 3a_0^2}{(1 - a_0^2)^3} \frac{b_1^2}{2} (1 + \cos 2\omega t) + \cdots \tag{6.6.54}$$

By combining (6.6.54) and (6.6.51) and integrating

$$\frac{ms^2}{l} \phi = \left[\frac{1}{1 - a_0^2} + \frac{b_1^2}{2} \frac{1 + 3a_0^2}{(1 - a_0^2)^3} \right] t - \frac{2b_1 a_0}{\omega(1 - a_0^2)^2} \sin \omega t$$
$$+ \frac{1}{2\omega} \left[\frac{2a_0 b_2}{(1 - a_0^2)^2} + \frac{b_1^2}{2} \frac{1 + 3a_0^2}{(1 - a_0^2)^3} \right] \sin 2\omega t + \cdots \tag{6.6.55}$$

where the initial condition $\phi = 0$ at $t = 0$ is assumed. Observe that in terms of the parameter μ given by (6.6.43)

$$\frac{l}{ms^2} = \frac{\sqrt{2\mu m^2 gs^3}}{ms^2} = \sqrt{\frac{2\mu g}{s}} \tag{6.6.56}$$

Continuing with the numerical example we find

$$\phi = 5.57t - 0.076 \sin \omega t + 0.098 \sin 2\omega t \tag{6.6.57}$$

The fact that the coefficients of the first and second harmonics are of the same order of magnitude does not indicate poor convergence. The smallness of the coefficient of $\sin \omega t$ is due to the happenstance that the mean value of θ is near $90°$ so that a_0 is small.

We could use our numerical results to plot the trajectory of the spherical pendulum on the surface of a sphere. The reader can refer to Fig. 7.4.3 which is a plot of the orbit of an electron in a radial electric field. This, and most orbit problems, is qualitatively similar to the spherical pendulum.

It is possible to obtain a differential equation which gives directly θ as a function of ϕ. This is done by introducing the dependent variable

$$x = \cot \theta \qquad (6.6.58)$$

Then,

$$\frac{dx}{dt} = \frac{dx}{d\phi}\frac{d\phi}{dt} = -\frac{1}{\sin^2 \theta}\dot\theta \qquad (6.6.59)$$

By means of the angular momentum equation $l = ms^2 \sin^2 \theta \dot\phi$,

$$\dot\theta = -\frac{dx}{d\phi}\frac{l}{ms^2} \qquad (6.6.60)$$

A second differentiation yields

$$\ddot\theta = -\frac{d^2x}{d\phi^2}\frac{l}{ms^2}\dot\phi = -\frac{d^2x}{d\phi^2}\frac{l^2}{m^2s^4 \sin^2 \theta} \qquad (6.6.61)$$

Noting that

$$\sin \theta = \frac{1}{\sqrt{1 + x^2}} \qquad (6.6.62)$$

and substituting into the second order equation (6.6.13) results in

$$\frac{d^2x}{d\phi^2} + x = \frac{gm^2s^3}{l^2}\left(\frac{1}{1 + x^2}\right)^{3/2} \qquad (6.6.63)$$

In terms of the variable x the energy equation is

$$\frac{2\mathscr{E}}{mgs} = \frac{l^2}{m^2gs^3}\left[\left(\frac{dx}{d\phi}\right)^2 + 1 + x^2\right] - \frac{2x}{\sqrt{1 + x^2}} \qquad (6.6.64)$$

6.7. LAGRANGE'S EQUATIONS

So far in our development of the mechanics of a particle, the concept of generalized force, as developed in Chapter 2, has been used as a method of calculating the kinematic expressions for acceleration in a given coordinate system. The equations of motion are then derived from the familiar $\mathbf{F} = m\mathbf{a}$. In what follows, we show how the forces can be explicitly eliminated from the formalism.

Let a particle that is moving in three dimensions be acted on by a conservative force. Assume that there are no constraints. The particle then has three independent coordinates or degrees of freedom. Recall that the generalized force components obey the covariant transformation law:

$$Q_k = \frac{\partial q_i'}{\partial q_k}Q_i' \qquad (6.7.1)$$

Suppose that the q_i' are rectangular Cartesian coordinates. Then $Q_i' = F_i$ are the rectangular components of force and

$$Q_k = \frac{\partial x_i}{\partial q_k} F_i = -\frac{\partial x_i}{\partial q_k}\frac{\partial V}{\partial x_i} = -\frac{\partial V}{\partial q_k} \qquad (6.7.2)$$

Thus Eq. (2.21.5) can be expressed

$$\frac{d}{dt}\left(\frac{\partial T}{\partial \dot{q}_k}\right) - \frac{\partial T}{\partial q_k} = -\frac{\partial V}{\partial q_k} \qquad (6.7.3)$$

Moreover, V does not depend on the velocity so that it is permissible to write (6.7.3) as

$$\frac{d}{dt}\left(\frac{\partial L}{\partial \dot{q}_k}\right) - \frac{\partial L}{\partial q_k} = 0 \qquad (6.7.4)$$

where L is the *Lagrangian* defined by

$$L = T - V \qquad (6.7.5)$$

The equations of motion can be determined directly from (6.7.4) without the necessity of actually calculating the forces.

Suppose now that there is an equation of constraint connecting the coordinates which can be expressed

$$f(q_1, q_2, q_3) = 0 \qquad (6.7.6)$$

This allows one coordinate to be expressed in terms of the other two so that the equations of transformation giving the rectangular coordinates in terms of the generalized coordinates are*

$$x_i = x_i(q_1, q_2) \qquad (6.7.7)$$

An example is provided by the spherical pendulum of Section 6.6 where the equation of constraint corresponding to (6.7.6) is $r - s = 0$ and the equations of transformation are

$$x_1 = s \sin \theta \cos \phi \qquad x_2 = s \sin \theta \sin \phi \qquad x_3 = s \cos \theta \quad (6.7.8)$$

where s is a constant and the two independent coordinates are θ and ϕ.

The equation of constraint (6.7.6) represents a surface in three-dimensional space. A definite force of constraint is necessary to keep the particle on the surface of constraint which we can think of as being exerted by the surface itself. If there is no friction, the force of constraint is necessarily perpendicular to the surface of constraint and therefore does no work on the particle.

---

* The reader will recognize that these are parametric equations of a surface in three-dimensional space as discussed in Section 2.25.

There are now only *two* generalized force components which are related to the rectangular components of force on the particle by means of

$$Q_\alpha = \frac{\partial x_i}{\partial q_\alpha} F_i \qquad \alpha = 1, 2 \qquad i = 1, 2, 3 \qquad (6.7.9)$$

where F_i are the *three* rectangular components of the net force on the particle and include the forces of constraint. As the particle is displaced an amount $d\mathbf{s}$, the work done on it is

$$dW = \mathbf{F} \cdot d\mathbf{s} = F_i \, dx_i \qquad (6.7.10)$$

The vector $d\mathbf{s}$ must of course lie in the surface of constraint. Writing

$$dx_i = \frac{\partial x_i}{\partial q_\alpha} dq_\alpha \qquad (6.7.11)$$

and substituting into (6.7.10) yields

$$F_i \, dx_i = F_i \frac{\partial x_i}{\partial q_\alpha} dq_\alpha = Q_\alpha \, dq_\alpha \qquad (6.7.12)$$

There are three terms in the sum $F_i \, dx_i$ and only two in $Q_\alpha dq_\alpha$.* Let us now divide the rectangular components of force acting on the particle into two parts one of which is the force of constraint and the other of which includes all other forces:

$$F_i = C_i + G_i \qquad (6.7.13)$$

Since the force of constraint is workless

$$C_i \, dx_i = 0 \qquad (6.7.14)$$

and (6.7.12), therefore, becomes

$$G_i \, dx_i = G_i \frac{\partial x_i}{\partial q_\alpha} dq_\alpha = Q_\alpha \, dq_\alpha \qquad (6.7.15)$$

The *two* coordinates q_1 and q_2 are independent and, therefore, it is true that

$$Q_\alpha = G_i \frac{\partial x_i}{\partial q_\alpha} \qquad (6.7.16)$$

The generalized force components do not involve the force of constraint! If all forces other than the force of constraint are derivable from a

---

* Where no confusion will result, both covariant and contravariant quantities are labeled with a subscript. Equation (6.7.12) is, of course, a generalized dot product: $dW = Q_\alpha dq^\alpha$.

potential

$$Q_\alpha = -\frac{\partial V}{\partial q_i}\frac{\partial x_i}{\partial q_\alpha} = -\frac{\partial V}{\partial q_\alpha} \tag{6.7.17}$$

Lagrange's equations, therefore, are

$$\frac{d}{dt}\left(\frac{\partial T}{\partial \dot{q}_\alpha}\right) - \frac{\partial T}{\partial q_\alpha} = Q_\alpha = -\frac{\partial V}{\partial q_\alpha}$$

$$\frac{d}{dt}\left(\frac{\partial L}{\partial \dot{q}_\alpha}\right) - \frac{\partial L}{\partial q_\alpha} = 0 \qquad L = T - V \tag{6.7.18}$$

Thus, if one of the coordinates is eliminated from the Lagrangian by means of the equation of constraint, the equations of motion for the remaining two coordinates is obtained directly from (6.7.18). The force of constraint no longer appears in the formalism! To illustrate, the Lagrangian for the spherical pendulum is

$$L = \tfrac{1}{2}m(s^2\dot{\theta}^2 + s^2\sin^2\theta\,\dot{\phi}^2) + mgs\cos\theta \tag{6.7.19}$$

The Lagrange equations are

$$\frac{d}{dt}\left(\frac{\partial L}{\partial \dot{\theta}}\right) - \frac{\partial L}{\partial \theta} = 0 \qquad \frac{d}{dt}\left(\frac{\partial L}{\partial \dot{\phi}}\right) - \frac{\partial L}{\partial \phi} = 0 \tag{6.7.20}$$

from which follows

$$ms^2\ddot{\theta} - ms^2\sin\theta\cos\theta\,\dot{\phi}^2 + mgs\sin\theta = 0 \tag{6.7.21}$$

$$\frac{\partial L}{\partial \dot{\phi}} = ms^2\sin^2\theta\,\dot{\phi} = l = \text{const.} \tag{6.7.22}$$

If there are *two* equations of constraint,

$$f_1(q_1, q_2, q_3) = 0 \qquad f_2(q_1, q_2, q_3) = 0 \tag{6.7.23}$$

there is only *one* independent coordinate. Suppose that by means of (6.7.23) q_2 and q_3 are eliminated from the Lagrangian. There is now only *one* Lagrange equation

$$\frac{d}{dt}\left(\frac{\partial L}{\partial \dot{q}_1}\right) - \frac{\partial L}{\partial q_1} = 0 \tag{6.7.24}$$

The particle is constrained to move on a curve which is the intersection of the two surfaces (6.7.23).

It is the usual practice to define

$$p_\alpha = \frac{\partial L}{\partial \dot{q}_\alpha} \tag{6.7.25}$$

as the *generalized momenta*. The terms *canonical momenta* or *conjugate momenta* are also used. Lagrange's equations are then

$$\dot{p}_\alpha - \frac{\partial L}{\partial q_\alpha} = 0 \tag{6.7.26}$$

Any time that a coordinate is absent from the Lagrangian, the conjugate momentum will be a constant of the motion. In the example of the spherical pendulum, an inspection of the Lagrangian (6.7.19) reveals that ϕ itself does not appear. Thus

$$\frac{\partial L}{\partial \phi} = 0 \qquad \dot{p}_\phi = 0 \qquad p_\phi = l = \text{const.} \tag{6.7.27}$$

A coordinate that does not appear explicitly in the Lagrangian is said to be *cyclic*. The momentum conjugate to a cyclic coordinate is a constant of the motion.

In order to bring out the transformation properties of the various quantities, let us write the Lagrangian as

$$L = \tfrac{1}{2} m g_{\alpha\beta} \dot{q}^\alpha \dot{q}^\beta - V(q_1, q_2, q_3) \tag{6.7.28}$$

The subscripts run from 1 to 3, from 1 to 2, or are restricted to the single value 1, depending on whether there are no constraints, one constraint, or two constraints. The canonical momenta are

$$p_\alpha = \frac{\partial L}{\partial \dot{q}^\alpha} = m g_{\alpha\beta} \dot{q}^\beta \tag{6.7.29}$$

thus revealing that they are related to the generalized velocity components by the process of lowering the index. The canonical momenta, therefore, transform covariantly:

$$p_\alpha{}' = \frac{\partial q^\beta}{\partial q^{\alpha'}} p_\beta \tag{6.7.30}$$

6.8. PARTICLE CONSTRAINED TO MOVE ON A CONE

Suppose that a particle is acted on by a uniform gravitational field in the negative z-direction and is constrained to move on the surface of a cone as in Fig. 6.8.1. In cylindrical coordinates the Lagrangian is

$$L = \tfrac{1}{2} m(\dot{r}^2 + r^2\dot{\theta}^2 + \dot{z}^2) - mgz \tag{6.8.1}$$

The equation of constraint can be expressed

$$z = ar \tag{6.8.2}$$

where a is a positive constant. If z is eliminated from the Lagrangian the result is

$$L = \tfrac{1}{2}m(\dot{r}^2[1 + a^2] + r^2\dot{\theta}^2) - mgar \qquad (6.8.3)$$

The Lagrange equations for the two independent coordinates r and θ are

$$\frac{d}{dt}\left(\frac{\partial L}{\partial \dot{r}}\right) - \frac{\partial L}{\partial r} = 0 \qquad \frac{d}{dt}\left(\frac{\partial L}{\partial \dot{\theta}}\right) - \frac{\partial L}{\partial \theta} = 0 \qquad (6.8.4)$$

The resulting equations of motion are

$$m(1 + a^2)\ddot{r} - mr\dot{\theta}^2 + mga = 0 \qquad (6.8.5)$$

$$mr^2\dot{\theta} = l = \text{const.} \qquad (6.8.6)$$

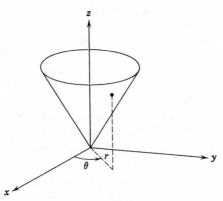

Fig. 6.8.1

Notice that θ is a cyclic coordinate. The canonically conjugate momentum is $l = \partial L / x\dot{\theta}$ and is a constant of the motion. The combination of (6.8.5) and (6.8.6) yields a differential equation for r as a function of t:

$$m(1 + a^2)\ddot{r} - \frac{l^2}{mr^3} + mga = 0 \qquad (6.8.7)$$

The energy equation is

$$\mathscr{E} = T + V = \tfrac{1}{2}m(\dot{r}^2[1 + a^2] + r^2\dot{\theta}^2) + mgar$$
$$= \tfrac{1}{2}m\dot{r}^2(1 + a^2) + \frac{l^2}{2mr^2} + mgar \qquad (6.8.8)$$

where θ has been eliminated by means of (6.8.6). A study of the effective potential

$$V_e(r) = \frac{l^2}{2mr^2} + mgar \qquad (6.8.9)$$

reveals that there is a single minimum at

$$r_1 = \left(\frac{l^2}{m^2 ga}\right)^{1/3} \qquad (6.8.10)$$

For a given angular momentum, the particle can move in a stable circular orbit of radius r_1. In general, r will be a periodic function of time and the motion will vary between two horizontal circles on the cone. The effective potential can be expanded about its minimum as

$$V_e(r) = V_1 + \frac{3mga}{2r_1}(r - r_1)^2 - \frac{2mga}{r_1^2}(r - r_1)^3 + \frac{10mga}{4r_1^3}(r - r_1)^4 + \cdots$$

$$(6.8.11)$$

The approximate equation of motion is

$$m(1 + a^2)\ddot{r} + \frac{dV_e}{dr} = 0$$

$$(6.8.12)$$

$$(1 + a^2)\ddot{r} + \frac{3ga}{r_1}(r - r_1) - \frac{6ga}{r_1^2}(r - r_1)^2 + \frac{10ga}{r_1^3}(r - r_1)^3 = 0$$

This equation can be solved by the Fourier series method.

It is possible to derive a differential equation for the orbit which does not involve the time. This is done by the change of dependent variable

$$x = \frac{1}{r} \qquad \dot{x} = \frac{dx}{d\theta}\dot{\theta} = -\frac{1}{r^2}\dot{r} \qquad (6.8.13)$$

By means of the angular momentum equation (6.8.6)

$$\dot{r} = -\frac{l}{m}\frac{dx}{d\theta} \qquad (6.8.14)$$

A second differentiation yields

$$\ddot{r} = -\frac{l}{m}\frac{d^2x}{d\theta^2}\dot{\theta} = -\frac{l^2x^2}{m^2}\frac{d^2x}{d\theta^2} \qquad (6.8.15)$$

The equation of motion (6.8.7) becomes

$$(1 + a^2)\frac{d^2x}{d\theta^2} + x - \frac{m^2ga}{l^2x^2} = 0 \qquad (6.8.16)$$

Since x is a periodic function of θ it is possible to find a Fourier series solution of the form

$$x = a_0 + b_1\cos(\lambda\theta) + b_2\cos(2\lambda\theta) + b_3\cos(3\lambda\theta) + \cdots \quad (6.8.17)$$

where λ plays the same role as does the frequency when time is the independent variable. The equilibrium value of x is

$$x_1 = \left(\frac{m^2 ga}{l^2}\right)^{1/3} \tag{6.8.18}$$

and if

$$f(x) = x - \frac{m^2 ga}{l^2 x^2} \tag{6.8.19}$$

is expanded about x_1, the resulting equation of motion is

$$(1 + a^2)\frac{d^2 x}{d\theta^2} + 3(x - x_1) - \frac{3}{x_1}(x - x_1)^2 + \frac{4}{x_1^2}(x - x_1)^3 = 0 \tag{6.8.20}$$

If the constants are identified as

$$\alpha = \frac{3}{1 + a^2} \qquad \beta = \frac{3}{x_1(1 + a^2)} \qquad \gamma = \frac{4}{x_1^2(1 + a^2)} \tag{6.8.21}$$

the solution is obtained from Eqs. (4.13.35) and (4.13.36) as

$$\frac{1}{r} = \frac{1}{r_1} + \frac{r_1 b_1^2}{2} + b_1 \cos(\lambda\theta) - \frac{r_1 b_1^2}{6}\cos(2\lambda\theta) + \frac{b_1^3 r_1^2}{16}\cos(3\lambda\theta) \tag{6.8.22}$$

$$\lambda^2 = \frac{3}{1 + a^2}[1 + \tfrac{1}{6}r_1^2 b_1^2] \tag{6.8.23}$$

where $r_1 = 1/x_1$.

The reader should observe that this example confirms the remarks made at the end of Section 4.13. The terms proportional to β and γ are of equal importance in determining the coefficient of the third harmonic in (6.8.22) and the expression of λ, Eq. (6.8.23).

Just as in the example of the spherical pendulum, the orbit is qualitatively similar to Fig. 7.4.3. We can imagine that the two dashed circles are the horizontal circles on the cone between which the motion varies. The turning points are found from $\dot{r} = 0$ and are the points where the orbit is tangent to the dashed circles. In the terminology of orbit theory, a turning point is called an *apse* (plural, *apsides*). The angle through which the particle moves in traveling from an outer to an inner turning point is found from

$$\lambda \, \Delta\theta = \pi$$

$$\Delta\theta = \pi\sqrt{(1 + a^2)/3}[1 - \tfrac{1}{12}r_1^2 b_1^2] \tag{6.8.24}$$

Thus the orbit would approximately close on itself if the slope of the cone were chosen such that

$$2 \, \Delta\theta = 2\pi, \qquad \tfrac{1}{2} \cdot 2\pi, \qquad \tfrac{1}{3} \cdot 2\pi, \dots \tag{6.8.25}$$

Also possible would be

$$2 \, \Delta\theta = 2 \cdot 2\pi, \quad 3 \cdot 2\pi \qquad (6.8.26)$$

The Lagrange formulation can be made to yield the forces of constraint, which in this example would be the force exerted on the particle by the cone. The introduction of the equation of constraint into the Lagrangian removes the force of constraint from the formalism. If the equation of constraint is *not* introduced into the Lagrangian, the three Lagrange equations are

$$\frac{d}{dt}\left(\frac{\partial T}{\partial \dot{q}_k}\right) - \frac{\partial T}{\partial q_k} = -\frac{\partial V}{\partial q_k} + C_k$$

or   (6.8.27)

$$\frac{d}{dt}\left(\frac{\partial L}{\partial \dot{q}_k}\right) - \frac{\partial L}{\partial q_k} = C_k$$

where C_k is that part of the generalized force which is not derivable from a potential. For the particle moving on a cone, the Lagrangian is

$$L = \tfrac{1}{2}m(\dot{r}^2 + r^2\dot{\theta}^2 + \dot{z}^2) - mgz \qquad (6.8.28)$$

Identifying the coordinates as $q_1 = r$, $q_2 = \theta$ and $q_3 = z$ and applying (6.8.27) yields

$$m\ddot{r} - mr\dot{\theta}^2 = C_1 \qquad (6.8.29)$$

$$\frac{d}{dt}(mr^2\dot{\theta}) = C_2 \qquad (6.8.30)$$

$$m\ddot{z} + mg = C_3 \qquad (6.8.31)$$

If there is no constraint, $C_1 = 0$ and the equations of motion of a free particle moving in a uniform gravitational field result. By using (6.8.5) and (6.8.6)

$$C_1 = -ma(a\ddot{r} + g) \qquad (6.8.32)$$

$$C_2 = 0 \qquad (6.8.33)$$

$$C_3 = m(a\ddot{r} + g) \qquad (6.8.34)$$

It is necessary, of course, to remember the connection between the covariant components of a force and its physical components. The components of the metric tensor are $g_{11} = 1$, $g_{22} = r^2$, and $g_{33} = 1$. Referring to Eq. (2.18.6) reveals that in this case C_1 and C_3 are identical to the physical components, and therefore, must be interpreted as the r- and z-components of the force which the cone exerts on the particle.

6.9. TIME-DEPENDENT CONSTRAINTS

So far we have considered simple conservative systems and constraints that can be expressed in the form $f(q_1, q_2, q_3) = 0$. The formalism can be extended to include constraints which are *time dependent*. An example is the spherical pendulum with a sphere of variable radius. The transformation equations read

$$x_1 = s(t) \sin \theta \cos \phi \qquad x_2 = s(t) \sin \theta \sin \phi \qquad x_3 = s(t) \cos \theta$$
$$(6.9.1)$$

where $s(t)$ is the variable radius of the sphere and must be regarded as a given function of time rather than a coordinate.

In general, a time-dependent constraint can be expressed in the form

$$f(q_1, q_2, q_3, t) = 0 \qquad (6.9.2)$$

The transformation equations are

$$x_i = (q_1, q_2, t) \qquad (6.9.3)$$

the coordinate q_3 having been removed by means of the equation of constraint. The rectangular components of the displacement of the particle are

$$dx_i = \frac{\partial x_i}{\partial q_\alpha} dq_\alpha + \frac{\partial x_i}{\partial t} dt \qquad (6.9.4)$$

The first term is due to the displacement of the particle within the surface of constraint and the second term is the contribution to the displacement due to the motion of the surface itself. An increment of work done on the particle is

$$F_i \, dx_i = F_i \frac{\partial x_i}{\partial q_\alpha} dq_\alpha + F_i \frac{\partial x_i}{\partial t} dt$$
$$= Q_\alpha \, dq_\alpha + F_i \frac{\partial x_i}{\partial t} dt \qquad (6.9.5)$$

Since $(\partial x_i / \partial q_\alpha) \, dq_\alpha$ is that part of the particle displacement which is parallel to the surface and since the force of constraint is perpendicular to the surface of constraint, the generalized force components Q_α still do not involve the forces of constraint. This does not mean that the forces of constraint do not contribute to the energy. The work done on the particle by the forces of constraint enters through the second term of (6.9.5). At any rate, it is still possible to use

$$\frac{d}{dt}\left(\frac{\partial T}{\partial \dot{q}_\alpha}\right) - \frac{\partial T}{\partial q_\alpha} = Q_\alpha = -\frac{\partial V}{\partial q_\alpha} \qquad (6.9.6)$$

provided that all forces other than the forces of constraint are obtainable from a potential. The effect on the motion due to the time dependence of the constraint enters (6.9.6) via the kinetic energy. The total energy, $\mathscr{E} = T + V$ is of course no longer a constant of the motion.

As an example, suppose a plane pendulum is made by tying a mass to a string and suppose that there is a mechanism whereby the string can be reeled out or pulled in. The Lagrangian is

$$L = \tfrac{1}{2}m(\dot{s}^2 + s^2\dot{\theta}^2) - mgs(1 - \cos\theta) \tag{6.9.7}$$

The Lagrange equation for the single independent coordinate θ is

$$\frac{d}{dt}\left(\frac{\partial L}{\partial \dot{\theta}}\right) - \frac{\partial L}{\partial \theta} = 0 \tag{6.9.8}$$

This gives

$$\frac{d}{dt}(ms^2\dot{\theta}) + mgs\sin\theta = 0 \tag{6.9.9}$$

It must be remembered that s is to be treated as a given function of time so that the equation of motion is

$$s\ddot{\theta} + 2\dot{s}\dot{\theta} + g\sin\theta = 0 \tag{6.9.10}$$

Once the function $s(t)$ is known, the integration can be carried out.

6.10. DISSIPATIVE FORCES

In no case so far examined has the Lagrange formulation been applied to a system where dissipative forces are present. Of common occurrence in practice are frictional forces which are directly proportional to the velocity, and we show in this section how the Lagrange formulation can be modified to include such forces. For the sake of generality, we assume that a general linear relation exists between the rectangular components of the dissipative force acting on the particle and the velocity components:

$$F_i = -b_{ij}\dot{x}_j \tag{6.10.1}$$

The *Rayleigh dissipative function* is defined to be

$$\mathscr{F} = \tfrac{1}{2}b_{ij}\dot{x}_j\dot{x}_i \tag{6.10.2}$$

If all forces other than the frictional forces (6.10.1) are conservative, the correct equations of motion of a particle result from

$$\frac{d}{dt}\left(\frac{\partial L}{\partial \dot{x}_k}\right) - \frac{\partial L}{\partial x_k} + \frac{\partial \mathscr{F}}{\partial \dot{x}_k} = 0 \tag{6.10.3}$$

where
$$L = \tfrac{1}{2}m\dot{x}_i\dot{x}_i - V(x_1, x_2, x_3) \tag{6.10.4}$$
We find

$$m\ddot{x}_k + \frac{\partial V}{\partial x_k} + b_{ki}\dot{x}_i = 0 \tag{6.10.5}$$
For instance, if
$$V = \tfrac{1}{2}k_{ij}x_ix_j \qquad b_{ki} = \delta_{ki}b \tag{6.10.6}$$
the equations of motion are

$$m\ddot{x}_k + k_{ki}x_i + b\dot{x}_k = 0 \tag{6.10.7}$$

and are seen to be the equations of motion of a three-dimensional harmonic oscillator with damping.

The conversion of the formalism to an arbitrary curvilinear coordinate system presents no problem. Since $\mathscr{F}$ is a scalar function, we may write

$$\mathscr{F} = \tfrac{1}{2}b'_{\alpha\beta}\dot{q}_\alpha\dot{q}_\beta \tag{6.10.8}$$

where the $b'_{\alpha\beta}$ are found from the b_{ij} by a second rank covariant tensor transformation:

$$b'_{\alpha\beta} = \frac{\partial x_i}{\partial q_\alpha}\frac{\partial x_j}{\partial q_\beta}b_{ij} \tag{6.10.9}$$

For example, consider the plane pendulum with damping. In rectangular coordinates, the Rayleigh dissipation function is

$$\mathscr{F} = \tfrac{1}{2}b(\dot{x}_1{}^2 + \dot{x}_2{}^2) \tag{6.10.10}$$

Without even using (6.10.9) directly, it is obvious that in plane polar coordinates

$$\mathscr{F} = \tfrac{1}{2}bs^2\dot{\theta}^2 \tag{6.10.11}$$

Thus, with

$$L = \tfrac{1}{2}ms^2\dot{\theta}^2 - mgs(1 - \cos\theta) \tag{6.10.12}$$

and

$$\frac{d}{dt}\frac{\partial L}{\partial\dot{\theta}} - \frac{\partial L}{\partial\theta} + \frac{\partial\mathscr{F}}{\partial\dot{\theta}} = 0 \tag{6.10.13}$$

we find

$$\ddot{\theta} + \frac{b}{m}\dot{\theta} + \frac{g}{s}\sin\theta = 0 \tag{6.10.14}$$

6.11. CLASSIFICATION OF CONSTRAINTS

It is the general practice to label constraints as *scleronomous* if they are independent of time and *rheonomous* if they are time dependent. Constraints are further classified as *holonomic* if they can be expressed in the form $f(q_1, q_2, q_3, t) = 0$. A particle placed inside a spherical container of radius s obeys the constraint

$$x_1^2 + x_2^2 + x_3^2 \leq s^2 \qquad (6.11.1)$$

which is *nonholonomic*. Another type of nonholonomic constraint is of the form

$$A_1 \, dq_1 + A_2 \, dq_2 + A_3 \, dq_3 = 0 \qquad (6.11.2)$$

where no scalar function f exists such that $A_i = \partial f / \partial q_i$ so that (6.11.2) is nonintegrable. Such constraints occur in certain types of rigid body motion. Nonholonomic constraints are by far the most difficult and usually require special treatment in each case. Our development of particle mechanics has so far included only holonomic constraints.

6.12. HAMILTON'S EQUATIONS

The equations of motion of a system that is described by a Lagrangian can be recast into another form known as the *Hamiltonian formulation*. The Hamiltonian is defined as

$$H = p_\alpha \dot{q}_\alpha - L \qquad (6.12.1)$$

where the sum runs over the number of degrees of freedom of the system, e.g., for a system consisting of one particle and a single holonomic constraint, $p_\alpha \dot{q}_\alpha = p_1 \dot{q}_1 + p_2 \dot{q}_2$. The total differential of H is

$$dH = p_\alpha \, d\dot{q}_\alpha + \dot{q}_\alpha \, dp_\alpha - \frac{\partial L}{\partial q_\alpha} dq_\alpha - \frac{\partial L}{\partial \dot{q}_\alpha} d\dot{q}_\alpha - \frac{\partial L}{\partial t} dt \quad (6.12.2)$$

Here, $d\dot{q}_\alpha$, dp_α, etc., are *mathematical* variations and do not necessarily correspond to the differences as would be calculated between neighboring points on the actual trajectory of a particle. Since

$$p_\alpha = \frac{\partial L}{\partial \dot{q}_\alpha} \qquad \dot{p}_\alpha = \frac{\partial L}{\partial q_\alpha} \qquad (6.12.3)$$

(6.12.2) reduces to

$$dH = \dot{q}_\alpha \, dp_\alpha - \dot{p}_\alpha \, dq_\alpha - \frac{\partial L}{\partial t} dt \qquad (6.12.4)$$

Suppose that the relations $p_\alpha = mg_{\alpha\beta}\dot{q}_\beta$ are solved for $\dot{q}_\beta$ and that $\dot{q}_\beta$ are then eliminated from the Hamiltoian so that it appears as a function of q_α, p_α, and t. The total differential of H can then be expressed

$$dH = \frac{\partial H}{\partial p_\alpha} dp_\alpha + \frac{\partial H}{\partial q_\alpha} dq_\alpha + \frac{\partial H}{\partial t} dt \qquad (6.12.5)$$

Comparison of (6.12.5) and (6.12.4) reveals that

$$\dot{q}_\alpha = \frac{\partial H}{\partial p_\alpha} \qquad \dot{p}_\alpha = -\frac{\partial H}{\partial q_\alpha} \qquad \frac{\partial H}{\partial t} = -\frac{\partial L}{\partial t} \qquad (6.12.6)$$

These are known as Hamilton's canonical equations and can be used in place of Lagrange's equations (6.12.3) to obtain the equations of motion of a system.

If L has no explicit time dependence, the Hamiltonian is a constant of the motion. The total derivative of H with respect to time is

$$\frac{dH}{dt} = \frac{\partial H}{\partial p_\alpha} \dot{p}_\alpha + \frac{\partial H}{\partial q_\alpha} \dot{q}_\alpha + \frac{\partial H}{\partial t} \qquad (6.12.7)$$

Here we are talking about the actual time rate of change of the Hamiltonian from the point of view of the particle in question. By using (6.12.6),

$$\frac{dH}{dt} = \frac{\partial H}{\partial t} = -\frac{\partial L}{\partial t} \qquad (6.12.8)$$

Thus H is a constant if $\partial L/\partial t = 0$.

It is possible to regard (6.12.6) as giving a set of first order equations in the variables q_α, p_α. For a single particle with no constraints, there would be six such first order equations. The particle can be described by its trajectory in *phase space* which is a six-dimensional space constructed out of the three coordinates and the three moments.

For simple conservative systems, it is possible to show that the Hamiltonian represents the total energy. Note that

$$p_\alpha \dot{q}_\alpha = mg_{\alpha\beta}\dot{q}_\alpha\dot{q}_\beta = 2T \qquad (6.12.9)$$

Thus

$$H = 2T - (T - V) = T + V \qquad (6.12.10)$$

The equivalence of H with the total energy has been established here only for simple conservative systems. In more general systems that can be described by the Lagrange or Hamiltonian formulations, it is not necessarily true that H represents the total energy.

As an example of the use of the Hamiltonian formulation, consider the particle constrained to move on the surface of a cone discussed in Section 6.8. The generalized coordinates are $q_1 = r, q_2 = \theta$. Reference to the Lagrangian (6.8.3) shows that the components of the metric tensor are

$$g_{11} = 1 + a^2 \qquad g_{22} = r^2 \qquad g_{12} = 0 \qquad (6.12.11)$$

The canonical momenta, therefore, are

$$p_1 = p_r = mg_{11}\dot{q}_1 = m(1 + a^2)\dot{r} \qquad (6.12.12)$$

$$p_2 = p_\theta = mg_{22}\dot{q}_2 = mr^2\dot{\theta} \qquad (6.12.13)$$

Before application of Hamilton's equations, it is necessary to express H in terms of r, θ, p_r, and p_θ. The result is

$$H = T + V = \frac{p_r^2}{2m(1 + a^2)} + \frac{p_\theta^2}{2mr^2} + mgar \qquad (6.12.14)$$

The equations of motion follow from

$$\dot{r} = \frac{\partial H}{\partial p_r} \qquad \dot{\theta} = \frac{\partial H}{\partial p_\theta} \qquad \dot{p}_r = -\frac{\partial H}{\partial r} \qquad \dot{p}_\theta = -\frac{\partial H}{\partial \theta} \qquad (6.12.15)$$

which give

$$\dot{r} = \frac{p_r}{m(1 + a^2)} \qquad \dot{\theta} = \frac{p_\theta}{mr^2}$$

$$\dot{p}_r = \frac{p_\theta^2}{mr^3} - mga \qquad \dot{p}_\theta = 0 \qquad (6.12.16)$$

Just as in the case of the Lagrange formulation, cyclic coordinates do not appear explicitly in the Hamiltonian. The combination of the equations for r and p_r yields

$$m(1 + a^2)\ddot{r} - \frac{p_\theta^2}{mr^3} + mga = 0 \qquad (6.12.17)$$

which is identical to (6.8.7)

A somewhat simpler example is that of the one-dimensional harmonic oscillator. The Hamiltonian is

$$H = \frac{p^2}{2m} + \tfrac{1}{2}kx^2 \qquad (6.12.18)$$

The equations of motion are

$$\dot{x} = \frac{\partial H}{\partial p} \qquad \dot{p} = -\frac{\partial H}{\partial x}$$

$$\dot{x} = \frac{p}{m} \qquad \dot{p} = -kx \qquad (6.12.19)$$

Since $H = \mathscr{E} = $ const., the trajectory of the particle in the two-dimensional phase space formed from p and x can be obtained directly from (6.12.18). Writing

$$1 = \frac{p^2}{2m\mathscr{E}} + \frac{x^2}{2\mathscr{E}/k} \qquad (6.12.20)$$

it is obvious that the path in phase space is an ellipse the semi-axes of which are of lengths

$$a = \sqrt{2\mathscr{E}/k} \qquad b = \sqrt{2m\mathscr{E}} \qquad (6.12.21)$$

It is evident that the Hamiltonian formulation places both the coordinates and their canonically conjugate momenta on an equal footing.

The old quantum theory (pre-1925) reached its most advanced form in the *Bohr-Sommerfeld quantization rule*. Consider a system the coordinates and conjugate momenta of which are periodic functions of the time. According to the Bohr-Sommerfeld rule, a classical system is to be quantized by means of

$$\oint p_{(\alpha)}\, dq_{(\alpha)} = n_\alpha h \quad \text{(no sum)} \tag{6.12.22}$$

where the integral is taken over one complete cycle of the motion, n_α is an integer, and h is Planck's constant:

$$h = 6.6253 \times 10^{-27} \text{ erg-sec} \tag{6.12.23}$$

The parentheses are placed around the subscripts in (6.12.22) to indicate that no sum is implied.

In the example of the harmonic oscillator $\oint p\, dx$ is the area of the ellipse in phase space. In other words, according to the quantization rule, the area enclosed by the path in phase space can have only certain definite values. The area of an ellipse is $\pi a b$ and (6.12.22) and (6.12.21) give

$$\pi ab = \pi\sqrt{2\mathscr{E}/k}\sqrt{2m\mathscr{E}} = nh$$

$$\mathscr{E} = n\frac{h}{2\pi}\sqrt{k/m} = n\hbar\omega_0 \tag{6.12.24}$$

where $\omega_0 = \sqrt{k/m}$ is the natural frequency of the classical oscillator. Since the combination $h/2\pi$ appears so often it is given special designation:

$$\hbar = \frac{h}{2\pi} = 1.05445 \times 10^{-27} \text{ erg sec} \tag{6.12.25}$$

According to (6.12.24), the oscillator can exist in only certain discrete *energy states* or *energy levels*. When the oscillator changes from one state to another, energy must be absorbed or emitted, e.g., in the form of electromagnetic radiation. Just why or how the transition is made from one energy level to another, the old quantum theory does not say. The quantum mechanics of Heisenberg and Schrödinger, introduced in about 1925, leads to a slightly different formula for the energy levels of a harmonic oscillator:

$$\mathscr{E} = (n + \tfrac{1}{2})\hbar\omega_0 \tag{6.12.26}$$

The old quantum theory is of historical interest and has some pedagogical value as a bridge between classical and quantum mechanics. It is sometimes useful as a kind of rule of thumb to tell whether quantum effects are important in a given situation.

The Hamiltonian formulation is valuable as a starting point in the further development of the theoretical structure of mechanics. It is in these developments that classical mechanics most closely resembles quantum mechanics. The Hamiltonian for a particle moving in a conservative force field can be expressed in the form

$$H = \frac{\mathbf{p} \cdot \mathbf{p}}{2m} + V \qquad (6.12.27)$$

If the formal substitutions

$$H = i\hbar \frac{\partial}{\partial t} \qquad \mathbf{p} = -i\hbar \nabla \qquad (6.12.28)$$

are made and the resulting equation allowed to operate on the wave function ψ the result is

$$i\hbar \frac{\partial \psi}{\partial t} = -\frac{\hbar^2}{2m} \nabla^2 \psi + V \psi \qquad (6.12.29)$$

which is the Schrödinger wave equation for a particle of mass m moving in a conservative force field. This should not be regarded in any sense as a derivation of the wave equation; it is merely intended to show the correspondence between classical and quantum mechanics.

PROBLEMS

6.2.1   A gun is on a cliff of height h above a level plane. Find the maximum range of the gun on the level plane and the corresponding firing angle as a function of the initial velocity. Neglect air resistance.

6.2.2   A gun at the origin of coordinates has a muzzle velocity of 1600 ft/sec. An airplane travels on a course given by $x + z = 3.5 \times 10^4$ ft, and at a constant altitude $y = 10^4$ ft. Over what portion of its course is it in danger?

6.2.3   If there is air resistance proportional to the velocity in the motion of a projectile, find an exact formula for y in terms of x. By making appropriate expansions, find an approximate expression for y in terms of x that is correct to terms linear in the damping constant b. If a cannon ball is fired at an angle of elevation of 45° with a muzzle velocity of 400 ft/sec, find the effect on the horizontal range due to damping if $b/m = 10^{-3}\ \text{sec}^{-1}$.

6.3.4   By consulting an appropriate reference, complete the numerical data in the following table. Let $k^2 = 0.992$, i.e., $k = \sin 85°$:

$\phi°$	0	10	20	30	40	50	60	70	80	90
$snu = \sin \phi$	0	0.1736	0.3420	0.5000	0.6428	0.7660	0.8660	0.9397	0.9848	1.000
u						1.009				
$\sin(\theta/2) = ksnu$										
θ										

Plot θ as a function of u for one complete cycle of the motion of a simple pendulum. If the length of the pendulum is 100 cm, how long does it take it to descend from $170°$ to $30°$?

6.4.5 Compute the first two nonzero terms of the Taylor series expansion of *snu* about $u = 0$.

6.5.6 A particle obeys the equation of motion $\ddot{x} + \gamma x^3 = 0$, where $\gamma > 0$. Show that the frequency is given by $\omega = $ const. $\sqrt{\gamma} A$ where A is the amplitude of the motion. Find the numerical value of the constant. Compare this with the answer that was obtained by the Fourier series method in Problem 4.13.28.

6.5.7 Show that, for $0 \leq k^2 \leq \frac{1}{2}$, *cnu* has one inflection point over $0 < u < 2K$, whereas if $\frac{1}{2} < k^2 < 1$, there are three.

6.5.8 A particle obeys the equation of motion $m\ddot{x} + k_1 x - k_3 x^3 = 0$ with $k_1 = 3600$ dynes/cm, $k_3 = 100$ dynes/cm^3, and $m = 10$ gm. At a certain time, the position and kinetic energy are noted to be $x_0 = 10$ cm and $T_0 = 9 \times 10^4$ ergs. Find the correct form of the solution and find the numerical values of all parameters.

6.5.9 Find all the solutions of $\ddot{x} - \alpha x + \gamma x^3 = 0$ where $\alpha > 0$ and $\gamma > 0$.

6.5.10 The equation of motion of a particle is

$$\ddot{x} - c + \alpha x - \beta x^2 = 0$$

where c, α, and β are positive constants. Under what circumstances can the constant term c be eliminated by a substitution of the form $x = y - b$ where b is a constant? Sketch the potential function for the two cases where the constant term *cannot* be eliminated and where it *can* be eliminated.

6.5.11 Find the first and second order differential equations obeyed by $y = (snu)^n$.

6.5.12 Find the first order differential equation obeyed by

$$y = \frac{A + snu}{1 + Bsnu}$$

6.6.13 Show that the equilibrium value of θ and the equilibrium energy for the spherical pendulum are related by

$$\cos \theta_1 = -\frac{\mathscr{E}_1}{3mgs} + \sqrt{\left(\frac{\mathscr{E}_1}{3mgs}\right)^2 + \frac{1}{3}}$$

Hint: In terms of the variable x of equation (6.6.29), what is the equilibrium position?

6.6.14 A spherical pendulum of length 100 cm is released at the equator ($\theta = 90°$) with a velocity of 200 cm/sec and in a direction tangent to the equator. What are the maximum and minimum values of θ (turning points) in the motion?

6.6.15 For the spherical pendulum, plot the effective potential for $\mu = 0.5$ for the range $20° \leq \theta \leq 150°$. For $\mathscr{E}/(mgs) = 0.6$, read the values of the turning points as accurately as you can from the graph. Correct the turning points numerically to obtain $\cos \theta_3$ and $\cos \theta_2$ accurate to three decimal places.

6.6.16 For the numerical values (6.6.43) and (6.6.45), compute the frequency of the θ motion to three places using the exact solution (6.6.36) and compare it with the value (6.6.48) that was obtained by the Fourier series method. *Caution:* The period of sn^2u is $2K$.

6.6.17   For the spherical pendulum, expand the effective potential about its equilibrium value θ_1. Neglect terms of order $(\theta - \theta_1)^3$ and higher. Show that for small deviations from a circular orbit the frequency of the θ motion in terms of the equilibrium value of θ is

$$\omega = \sqrt{\frac{g}{s}} \sqrt{\frac{1 + 3 \cos^2 \theta_1}{\cos \theta_1}}$$

6.7.18   A particle moving under the influence of a uniform gravitational field in the negative z-direction is constrained to move on the intersection of the two planes $x + z = b$, $z + y = c$. Find the motion by the Lagrange method if the particle starts from rest at $x = 0$.

6.7.19   A particle of mass m moves in a plane and is fixed to four springs at right angles as illustrated. The force constant of the horizontal springs is K_1 and that of the vertical springs is K_2. Neglecting gravity, find the potential

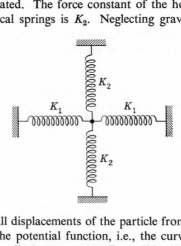

function for very small displacements of the particle from equilibrium. Sketch the *contour lines* of the potential function, i.e., the curves $V = $ const. Is the equilibrium stable for displacements in all possible directions? Write down the Lagrangian and find the equations of motion. Sketch the trajectory of the particle if $K_2 = 4K_1$ for the initial conditions $x = y = 0$ at $t = 0$. This is an example of a *Lissajous figure*.

6.7.20   In the problem of the spherical pendulum it is possible to think of the surface of the sphere over which the particle moves as a two-dimensional non-Euclidean (curved) space which is embedded in a three-dimensional space. Two possible sets of independent coordinates that can be used in this two-dimensional space are $q_1 = x, q_2 = y$ and $q_1' = \theta, q_2' = \phi$. The intrinsic geometry of the two-dimensional space is contained in the metric form which can be expressed in the two coordinate systems as

$$\alpha s^2 = \left(\frac{s^2 - y^2}{s^2 - x^2 - y^2} dx^2 + \frac{s^2 - x^2}{s^2 - x^2 - y^2} dy^2 + \frac{2xy \, dx \, dy}{s^2 - x^2 - y^2} \right)$$
$$= (s^2 \, d\theta^2 + s^2 \sin^2 \theta \, d\phi^2)$$

Verify that the metric tensor has the correct transformation property, i.e.,

$$g'_{\alpha\beta} = \frac{\partial q_\mu}{\partial q_\alpha'} \frac{\partial q_\nu}{\partial q_\beta'} g_{\mu\nu}$$

6.10.21 Find the differential equations of motion of a spherical pendulum which include the effects of damping proportional to the velocity.

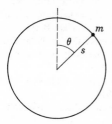

6.11.22 A particle is placed nearly at the top of a smooth sphere. As the particle slides down, at what value of θ does it leave the surface? What type of constraint is involved?

6.11.23 A particle moves on the inside surface of a smooth cylinder in the form of an ellipse of semi-axes a and b. The particle moves in a plane perpendicular to the cylinder and there is a uniform gravitational field in the

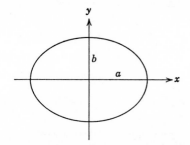

negative y-direction. What minimum speed must the particle have at the bottom of the ellipse in order to go completely around without leaving the surface?

6.12.24 The Hamiltonian of a three-dimensional anisotropic harmonic oscillator is

$$H = \frac{1}{2m}(p_x^2 + p_y^2 + p_z^2) + \tfrac{1}{2}k_1 x^2 + \tfrac{1}{2}k_2 y^2 + \tfrac{1}{2}k_3 z^2$$

where k_1, k_2, and k_3 are all different. Find the energy levels according to the old quantum theory.

6.12.25 Show that, for a conservative system, the kinetic energy can be expressed in terms of coordinates and momenta once the contravariant form of the metric tensor is known, i.e., that

$$T = \tfrac{1}{2}m g_{\alpha\beta}\dot{q}^\alpha\dot{q}^\beta = \frac{1}{2m}g^{\alpha\beta}p_\alpha p_\beta$$

Using spherical coordinates, find the contravariant components of the metric tensor for the problem of the spherical pendulum. Find the differential equations of motion by means of Hamilton's equations.

REFERENCES

1. Pierce, B. O., *A Short Table of Integrals*, Ginn, New York, 1929.
2. Byrd, Paul F., and Morris D. Friedman, *Handbook of Elliptic Integrals for Engineers and Physicists*, Springer-Verlag, Berlin, 1954.
3. Hancock, Harris, *Elliptic Integrals*, Dover, New York, 1958.
4. Whittaker, E. T., and G. N. Watson, *A Course of Modern Analysis*, Cambridge University Press, London, 1952.
5. Magnus, W., and F. Oberhettinger, *Functions of Mathematical Physics*, Chelsea Publishing Co., New York, 1954.
6. Goldstein, H., *Classical Mechanics*, Addison-Wesley, Reading, Mass., 1959.

7

Motion of a Charged Particle in an Electromagnetic Field

7.1. EQUATION OF MOTION

The basic equation of motion of a charged particle as it moves in an electromagnetic field can be stated in the Gaussian system of units as

$$m\mathbf{a} = q\mathbf{E} + \frac{q}{c}(\mathbf{u} \times \mathbf{B}) \tag{7.1.1}$$

where q is the charge on the particle (stat coulombs), $\mathbf{E}$ is the electric field (dynes per stat coulomb or stat volts per cm), c is the velocity of electromagnetic waves in vacuum (3×10^{10} cm/sec) and $\mathbf{B}$ is the magnetic field measured in gauss. Actually, the dimensions of $\mathbf{E}$ and $\mathbf{B}$ are the same, but it is the usual practice to retain the special designation "gauss." As usual, $\mathbf{a}$ and $\mathbf{u}$ are acceleration and velocity and m is the mass of the particle and is assumed to be constant. The force produced on a moving charged particle by an electromagnetic field is known as the *Lorentz force*. Equation (7.1.1) is obtained from experiment and can actually be used to define $\mathbf{E}$ and $\mathbf{B}$. By observation of the motion of a charged particle, the fields can be determined at any point. In applying the equation of motion (7.1.1), the sources which set up the fields are assumed to be undisturbed by the charged particle the motion of which is being studied.

The rate at which electromagnetic forces do work on a charged particle is

$$\begin{aligned}
\frac{dW}{dt} &= m(\mathbf{a} \cdot \mathbf{u}) = \frac{d}{dt}(\tfrac{1}{2}mu^2) \\
&= \left[q\mathbf{E} + \frac{q}{c}\mathbf{u} \times \mathbf{B}\right] \cdot \mathbf{u} = q\mathbf{E} \cdot \mathbf{u}
\end{aligned} \tag{7.1.2}$$

The force due to the magnetic field is always perpendicular to the direction of motion ($\mathbf{u} \times \mathbf{B} \cdot \mathbf{u} = 0$) and, hence, does no work on the particle. If the fields are *independent of time*, $\mathbf{E}$ is conservative and is obtainable from a scalar potential: $\mathbf{E} = -\nabla\psi$. For such fields,

$$\frac{dT}{dt} = -q \, \nabla\psi \cdot \mathbf{u} = -q \, \frac{d\psi}{dt}$$

$$\frac{d}{dt}(T + q\psi) = 0 \qquad T + q\psi = \mathscr{E} = \text{const.} \tag{7.1.3}$$

Thus in a time independent field the total energy is a constant. If the electromagnetic field is time-varying, the rate at which the fields do work on the particle is still given by (7.1.2) but the total energy of the particle is no longer a constant.

There are two restrictions on the validity of the equation of motion (7.1.1). One is that it is valid only if the velocity of the particle is much less than the velocity of light. Otherwise, it is necessary to use the generalization of (7.1.1) according to special relativity. As a rule of thumb, the relativistic correction amounts to about 0.5% if $u = 0.1c$. It is known from electromagnetic theory that a nonrelativistic accelerated charged particle radiates energy the rate of loss being given by

$$\frac{dW_R}{dt} = -\frac{2a^2q^2}{3c^3} \tag{7.1.4}$$

where a is the magnitude of the acceleration. This results in a damping effect on the motion of a charged particle which is qualitatively similar to friction. This effect is not included in the equations of motion (7.1.1). In fact, it is only in recent years that the precise form of the classical equations of motion which include the effects of radiation damping have been completely understood. The radiation rate as given by (7.1.4) will allow us to estimate the effects of radiation damping on the motion of charged particles.

Suppose that an electron is accelerated by a uniform electric field. If no radiation damping were present, the acceleration would be given by $a = qE/m$. If the damping is small, the acceleration is still nearly this value and can be used in (7.1.4) to estimate the radiation loss:

$$\frac{dW_R}{dt} = -\frac{2}{3}\frac{q^4E^2}{m^2c^3} \tag{7.1.5}$$

With

$$q = 4.80 \times 10^{-10} \text{ stat coulombs}$$

$$m = 9.11 \times 10^{-28} \text{ gm} \tag{7.1.6}$$

$$c = 3 \times 10^{10} \text{ cm/sec}$$

the radiation rate is

$$\frac{dW_R}{dt} = -1.58 \times 10^{-15} E^2 \text{ ergs/sec} \tag{7.1.7}$$

If the electric field is 1 stat volt/cm (1 stat volt = 300 mks volts) and if the electron starts from rest its velocity after moving 1 cm is

$$u = \sqrt{\frac{2q\psi}{m}} = \sqrt{\frac{2 \times 4.80 \times 10^{-10} \times 1}{9.11 \times 10^{-28}}} = 1.025 \times 10^9 \text{ cm/sec} \tag{7.1.8}$$

The time required to move this distance is

$$\Delta t = 2\frac{x}{u} = 1.852 \times 10^{-9} \text{ sec} \tag{7.1.9}$$

The energy loss by radiation, therefore, is

$$\Delta W_R = 1.58 \times 10^{-15} E^2 \Delta t = 2.94 \times 10^{-24} \text{ ergs} \tag{7.1.10}$$

By contrast, the total amount of work done on the particle by the electric field is

$$q\psi = 4.80 \times 10^{-10} \text{ ergs} \tag{7.1.11}$$

The fraction of this energy which is converted to radiation is

$$\frac{\Delta W_R}{q\psi} = 6.4 \times 10^{-15} \tag{7.1.12}$$

and is completely negligible. The remainder of this chapter concerns itself with the nonrelativistic motion of a single charged particle in a given electromagnetic field neglecting radiation damping.

7.2. MOTION IN A UNIFORM ELECTROMAGNETIC FIELD

Assume that **E** and **B** are constant and uniform throughout the region where the particle moves. Introduce a coordinate system with its x_3-axis in the same direction as **B** and its $x_1 x_3$-plane coincident with the plane determined by **E** and **B** as in Fig. 7.2.1. In this coordinate system, **E** and **B** have the components $(0, 0, B_3)$ and $(E_1, 0, E_3)$. The equation of motion (7.1.1) gives the component equations

$$m\ddot{x}_1 = qE_1 + \frac{q}{c} \dot{x}_2 B_3 \tag{7.2.1}$$

$$m\ddot{x}_2 = -\frac{q}{c} \dot{x}_1 B_3 \tag{7.2.2}$$

$$m\ddot{x}_3 = qE_3 \tag{7.2.3}$$

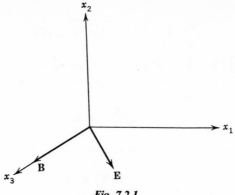

Fig. 7.2.1

The solution of (7.2.3) is easily found to be

$$x_3 = \frac{1}{2}\left(\frac{qE_3}{m}\right)t^2 + u_{03}t + x_{03} \tag{7.2.4}$$

The charge q and, hence, the acceleration $\ddot{x}_3 = qE_3/m$ can be positive or negative, depending on what type of particles are under consideration.

If (7.2.2) is multiplied by $i = \sqrt{-1}$ and added to (7.2.1) the result is

$$m(\ddot{x}_1 + i\ddot{x}_2) = qE_1 - i\frac{q}{c}B_3(\dot{x}_1 + i\dot{x}_2) \tag{7.2.5}$$

Let $z = x_1 + ix_2$. Then

$$\ddot{z} + i\omega_c\dot{z} - \frac{q}{m}E_1 = 0 \tag{7.2.6}$$

where $\omega_c = gB_3/(mc)$ is the *cyclotron frequency*. The general solution of (7.2.6) is found to be

$$z = a_1 - i\frac{E_1 c}{B_3}t - a_2 e^{-i\omega_c t} \tag{7.2.7}$$

where a_1 and a_2 are complex constants of integration. The complex velocity is

$$\dot{z} = -i\frac{E_1 c}{B_3} + a_2 i\omega_c e^{-i\omega_c t} \tag{7.2.8}$$

At $t = 0$ (7.2.7) and (7.2.8) give

$$x_{01} + ix_{02} = a_1 - a_2 \tag{7.2.9}$$

$$u_{01} + iu_{02} = -i\frac{E_1 c}{B_3} + ia_2\omega_c \tag{7.2.10}$$

The solution for a_2 is found to be

$$a_2 = \frac{u_{02}}{\omega_c} + \frac{E_1 c}{B_3 \omega_c} - i \frac{u_{01}}{\omega_c} = \rho e^{i\alpha} \tag{7.2.11}$$

where

$$\rho = \sqrt{\left(\frac{u_{02}}{\omega_c} + \frac{E_1 c}{B_3 \omega_c}\right)^2 + \left(\frac{u_{01}}{\omega_c}\right)^2} \tag{7.2.12}$$

$$\tan \alpha = - \frac{u_{01} B_3}{u_{02} B_3 + E_1 c} \tag{7.2.13}$$

The solution can now be expressed in the form

$$z = x_{01} + i x_{02} - \frac{i E_1 c}{B_3} t + \rho e^{i\alpha} - \rho e^{i(\alpha - \omega_c t)} \tag{7.2.14}$$

Suppose that the electric field is absent and that the motion is in the x_1, x_2 plane. Equations (7.2.12) and (7.2.13) reduce to

$$\rho = \sqrt{\frac{u_{02}^2 + u_{01}^2}{\omega_c^{\;2}}} = \frac{u_0}{\omega_c} \tag{7.2.15}$$

$$\tan \alpha = - \frac{u_{01}}{u_{02}} \tag{7.2.16}$$

The solution is

$$z = z_0 + \rho e^{i\alpha} - \rho e^{i(\alpha - \omega_c t)} \tag{7.2.17}$$

where $z_0 = x_{01} + i x_{02}$ is the complex initial position in the x_1, x_2 plane. The motion is circular of radius ρ and angular frequency ω_c. The complex number $z_0 + \rho e^{i\alpha}$ represents the center of the circle as shown in Fig. 7.2.2.

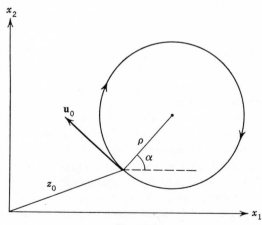

Fig. 7.2.2

If **B** is along the positive x_3-axis and if the charge on the particle is positive, the motion is *clockwise*. For a negatively charged particle, ω_c reverses sign and the particle moves counterclockwise around the circle. If the initial velocity has a component in the x_3-direction the solution for x_1 and x_2 remains unaltered. The particle spirals in the x_3-direction in the form of a helix.

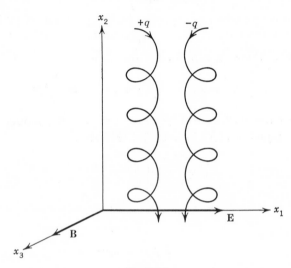

Fig. 7.2.3

Suppose now that there is an electric field in the x_1 direction. Let the motion be in the x_1x_2-plane. From (7.2.8) it is seen that this superimposes a constant velocity

$$u_d = \frac{E_1 c}{B_3}$$

in the negative x_2-direction on the circular motion. The shape of the particle trajectory is then a cycloid. The effect is to make the particle drift in the negative x_2-direction which, from Fig. 7.2.3, is seen to be in the direction of **E** × **B**. For this reason this effect is known as **E** × **B** *drift*. The direction of the drift is the same regardless of the sign of the charge on the particle as illustrated in Fig. 7.2.3. There is no net motion in the direction of the electric field but rather perpendicular to it. If the electric field has a component in the x_3-direction, there will be parabolic motion as given by Eq. (7.2.4) superimposed on the motion as pictured in Fig. 7.2.3.

7.3. FOCUSING ACTION OF UNIFORM FIELDS

If there is a uniform electric field in the negative y-direction, the motion of a charged particle is identical to that of a projectile (Section 6.2) and is given by

$$y = -\frac{1}{2}\frac{qE}{m}\frac{x^2}{u_0^2 \cos^2 \theta} + x \tan \theta \tag{7.3.1}$$

Generally, it is impossible to collimate perfectly a beam of particles and, therefore, they enter the region of interest with a certain angular spread of directions. In design problems dealing with beams of charged particles,

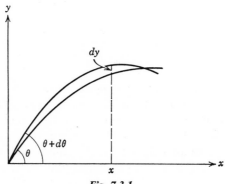

Fig. 7.3.1

it is interesting to know if the particles are brought to a focus at any point. Assume that the particles move in a plane. Let us formulate the problem in general terms, keeping (7.3.1) in mind as a specific example. Assuming that the initial speeds of the particles are all the same, the various trajectories are a one-parameter family of curves which can be expressed in the form $f(x, y, \theta) = 0$. The variation in y at a *fixed* value of x due to a variation in θ (Fig. 7.3.1) is computed by means of

$$\frac{\partial f}{\partial y} dy + \frac{\partial f}{\partial \theta} d\theta = 0 \tag{7.3.2}$$

The value of x where the two trajectories for θ and $\theta + d\theta$ intersect is the point of focus and is found by setting $dy = 0$, giving

$$\frac{\partial f}{\partial \theta} = 0 \tag{7.3.3}$$

which is identical to the envelope condition (6.2.8). The best focusing, therefore, occurs on the envelope of the one-parameter family of curves. Equation (6.2.12) gives the equation of the envelope of the parabolic trajectories given by (7.3.1).

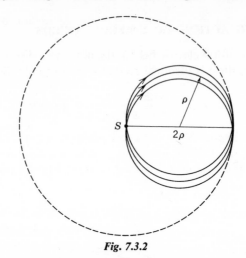

Fig. 7.3.2

If particles move in a uniform magnetic field, they move in circles of radius ρ provided that their initial direction is perpendicular to the field. Suppose that the particles come from a source S and all have the same initial speed. The various circles which are possible, depending on the initial direction, have an envelope which is a circle of radius 2ρ as illustrated in Fig. 7.3.2, and it is on this circle that the best focusing occurs.

7.4. RADIAL ELECTRIC FIELDS

If two concentric cylindrical electrodes of radii a and b are maintained at potentials ψ_a and ψ_b (Fig. 7.4.1) there will be a radial electric field between them given by

$$E = \frac{K}{r} \qquad K = \frac{\psi_a - \psi_b}{\log (b/a)} \qquad (7.4.1)$$

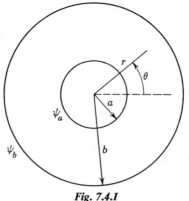

Fig. 7.4.1

The constant K can be positive or negative, depending on whether the inner electrode is at a higher or lower potential than the outer electrode. If a charged particle is placed between these electrodes, its equation of motion is $q\mathbf{E} = m\mathbf{a}$. In cylindrical coordinates, the electric field has a radial component only and the equations of motion in component form are

$$m(\ddot{r} - r\dot{\theta}^2) = \frac{qK}{r} \tag{7.4.2}$$

$$\frac{d}{dt}(mr^2\dot{\theta}) = 0 \tag{7.4.3}$$

$$\ddot{z} = 0 \tag{7.4.4}$$

where z is measured along the axis of the inner electrode. If it is assumed that the initial velocity has no component in the z-direction, the motion will take place in the r, θ-plane. The fact that the field is cylindrically symmetric with no component in the θ-direction [Eq. (7.4.3)] results in conservation of angular momentum:

$$l = mr^2\dot{\theta} = \text{const.} \tag{7.4.5}$$

It is convenient to introduce a *force constant* by means of $C = -qK$ so that the radial force on the particle is

$$F(r) = -\frac{C}{r} \tag{7.4.6}$$

and, therefore, is a force of attraction toward the inner electrode if $C > 0$. Using the angular momentum equation (7.4.5) to eliminate $\dot{\theta}$ from (7.4.2) results in

$$\ddot{r} - \frac{l^2}{m^2 r^3} + \frac{C}{mr} = 0 \tag{7.4.7}$$

The potential energy of the particle is found by integration of $dV = -F(r)\,dr$ from the inner electrode to some arbitrary value of r between a and b:

$$V = V_a + C \log\frac{r}{a} \qquad C = \frac{V_b - V_a}{\log(b/a)} \tag{7.4.8}$$

The energy equation, therefore, is

$$\mathscr{E} = \tfrac{1}{2}m\dot{r}^2 + \tfrac{1}{2}mr^2\dot{\theta}^2 + V(r)$$
$$= \tfrac{1}{2}m\dot{r}^2 + \frac{l^2}{2mr^2} + C\log\frac{r}{a} + V_a \tag{7.4.9}$$

This problem is qualitatively similar to the spherical pendulum and the particle constrained to move on the surface of a cone. The effective potential is given by

$$V_e(r) = \frac{l^2}{2mr^2} + C \log \frac{r}{a} + V_a \qquad (7.4.10)$$

and two distinctly different situations occur, depending on whether $C > 0$ (attraction toward inner electrode) or $C < 0$ (repulsion from inner electrode) as is evident from the sketches of $V_e(r)$ in Fig. 7.4.2. For attractive forces, there exist stable circular orbits of radius r_1 which are

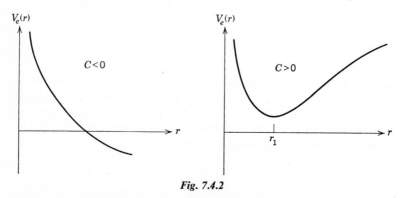

Fig. 7.4.2

found as usual by locating the minimum of $V_e(r)$:

$$r_1 = \frac{l}{\sqrt{mC}} \qquad (7.4.11)$$

More generally, for $C > 0$, r will be a periodic function of the time. For $C < 0$, periodic motion does not occur. The particle always moves continuously outward until it eventually collides with the outer electrode.

An orbit equation can be derived by exactly the same procedure that was used to obtain (6.8.16) in the case of the particle constrained to move on a cone. The result is

$$\frac{d^2x}{d\theta^2} + x - \frac{mC}{l^2x} = 0 \qquad (7.4.12)$$

where $x = 1/r$. If $C > 0$, x is a periodic function of θ. Expansion of $x - mC/(l^2x)$ about the equilibrium value $x_1 = 1/r_1$ yields the approximate equation of motion

$$\frac{d^2x}{d\theta^2} + 2(x - x_1) - \frac{1}{x_1}(x - x_1)^2 + \frac{1}{x_1^2}(x - x_1)^3 = 0 \quad (7.4.13)$$

Identification of the coefficients as

$$\alpha = 2 \qquad \beta = \frac{1}{x_1} = r_1 \qquad \gamma = \frac{1}{x_1^2} = r_1^2 \qquad (7.4.14)$$

and reference to (4.13.35) and (4.13.36) shows that the Fourier series solution, correct to third order, is

$$\frac{r_1}{r} = 1 + \frac{(b_1 r_1)^2}{4} + b_1 r_1 \cos \lambda\theta - \frac{(b_1 r_1)^2}{12} \cos 2\lambda\theta + \frac{(b_1 r_1)^3}{48} \cos 3\lambda\theta$$

(7.4.15)

$$\lambda^2 = 2[1 + \tfrac{1}{6}(r_1 b_1)^2]$$ (7.4.16)

As a numerical example, suppose $b_1 r_1 = 0.400$. Then

$$\frac{r_1}{r} = 1.040 + 0.400 \cos \lambda\theta - 0.0133 \cos 2\lambda\theta + 0.0013 \cos 3\lambda\theta$$ (7.4.17)

$$\lambda = \sqrt{2}[1 + 0.0133] = 1.434$$ (7.4.18)

The motion varies between apsides at $r/r_1 = 0.700$ and $r/r_1 = 1.600$. The angle between apsides is determined by

$$\lambda \, \Delta\theta = 180° \qquad \Delta\theta = 125.5°$$ (7.4.19)

Figure 7.4.3 is based on these values. In the construction of Fig. 7.4.3, it is only necessary to plot the portion of the orbit from its inner apse at A

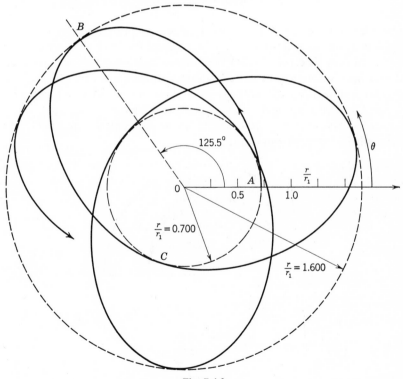

Fig. 7.4.3

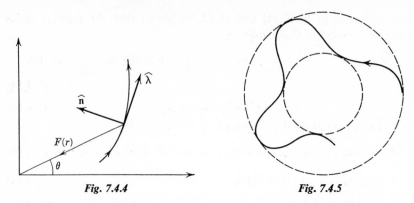

Fig. 7.4.4   Fig. 7.4.5

to its outer apse at *B*. The portion *BC* is then obtained by a simple reflection in the apsidal line *OB*. By continuing in this way, the entire figure can be produced.

According to (2.2.6), the acceleration of any particle can be expressed as

$$\mathbf{a} = \frac{u^2}{\rho}\hat{\mathbf{n}} + \frac{du}{dt}\hat{\boldsymbol{\lambda}}$$ (7.4.20)

where $\hat{\mathbf{n}}$ is the unit normal to the trajectory, ρ is the radius of curvature and $\hat{\boldsymbol{\lambda}}$ is a unit tangent vector, Fig. 7.4.4. Since the force is a *central force*, i.e., is always directed toward the origin, and since the velocity must always have a component in the positive θ-direction in order to conserve angular momentum, the unit vector $\hat{\mathbf{n}}$ must always have a component directed toward the origin. This means that the particle trajectory always curves in the *same way* in a purely attractive central force field. An impossible orbit is shown in Fig. 7.4.5.

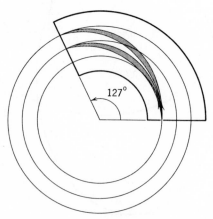

Fig. 7.4.6

A radial electric field has a focusing property. Suppose that particles are injected into the field in a tangential direction so that the starting point is a turning point. The particles are not perfectly collimated so that there is a small spread in the initial directions. A circle, the radius of which equals the value of r for the other turning point, is an envelope for the particle trajectories, and therefore, they will tend to come to a focus here. Figure 7.4.6 shows particles of two different initial velocities entering a radial electric field and being brought to a focus. If the deviation from circular orbits is not too great, focusing occurs when

$$\sqrt{2}\theta = \pi \qquad \theta = 127° \qquad (7.4.21)$$

Such a device can be used as a velocity spectrograph.

7.5. THE CYLINDRICAL MAGNETRON

Devices with radial electric fields as discussed in Section 7.4 that have in addition a magnetic field in the z-direction are generally referred to as *magnetrons*. We assume that the magnetic field is uniform. In cylindrical coordinates, the velocity of the particle and the magnetic field can be expressed at any point as

$$\mathbf{u} = u_r\hat{\mathbf{r}} + u_\theta\hat{\theta} + u_z\hat{\mathbf{z}} \qquad (7.5.1)$$

$$\mathbf{B} = B_r\hat{\mathbf{r}} + B_\theta\hat{\theta} + B_z\hat{\mathbf{z}} \qquad (7.5.2)$$

Since the coordinate system is orthogonal, the unit vectors obey

$$\hat{\mathbf{r}} \times \hat{\theta} = \hat{\mathbf{z}} \qquad \hat{\theta} \times \hat{\mathbf{z}} = \hat{\mathbf{r}} \qquad \hat{\mathbf{z}} \times \hat{\mathbf{r}} = \hat{\theta} \qquad (7.5.3)$$

The cross product of $\mathbf{u}$ and $\mathbf{B}$ is then

$$\mathbf{u} \times \mathbf{B} = (u_\theta B_z - u_z B_\theta)\hat{\mathbf{r}} + (u_z B_r - u_r B_z)\hat{\theta} + (u_r B_\theta - u_\theta B_r)\hat{\mathbf{z}} \quad (7.5.4)$$

The physical components of velocity in the $\hat{\mathbf{r}}$-, $\hat{\theta}$-, and $\hat{\mathbf{z}}$-directions are

$$u_r = \dot{r} \qquad u_\theta = r\dot{\theta} \qquad u_z = \dot{z} \qquad (7.5.5)$$

For the magnetron, $B_r = B_\theta = 0$. Therefore,

$$\mathbf{u} \times \mathbf{B} = r\dot{\theta}B_z\hat{\mathbf{r}} - \dot{r}B_z\hat{\theta} \qquad (7.5.6)$$

The equations of motion of a charged particle in a cylindrical magnetron are

$$m(\ddot{r} - r\dot{\theta}^2) = -\frac{C}{r} + \frac{q}{c}r\dot{\theta}B_z \qquad (7.5.7)$$

$$\frac{1}{r}\frac{d}{dt}(mr^2\dot{\theta}) = -\frac{q}{c}\dot{r}B_z \qquad (7.5.8)$$

$$m\ddot{z} = 0 \qquad (7.5.9)$$

The fact that the force due to the magnetic field can have a component in the θ-direction ruins the conservation of angular momentum. However, (7.5.8) can be expressed as follows:

$$\frac{d}{dt}(mr^2\dot\theta) = -\frac{q}{c} r\dot r B_z = -\frac{d}{dt}\left(\frac{q}{c}\frac{1}{2} r^2 B_z\right)$$

$$\frac{d}{dt}\left(mr^2\dot\theta + \frac{q}{2c} r^2 B_z\right) = 0$$

Thus

$$mr^2\dot\theta + \frac{q}{2c} r^2 B_z = l = \text{const.} \tag{7.5.10}$$

replaces the conservation of angular momentum. Note that (7.5.10) can be written

$$\dot\theta = -\frac{\omega_c}{2} + \frac{l}{mr^2} \tag{7.5.11}$$

showing that there is added to the θ-motion a constant angular velocity of magnitude

$$\omega_L = \frac{\omega_c}{2} = \frac{qB_z}{2mc} \tag{7.5.12}$$

which is known as the *Larmor frequency*. It should be kept in mind that q can be positive or negative. It is also possible for B_z to be negative.

The magnetron obviously has cylindrical symmetry in that neither **E** nor **B** have any dependence on the angle θ. The existence of constants of the motion is intimately connected with symmetry. As problems become more complex, it may not be possible to identify a given constant of the motion with a simple thing like ordinary linear or angular momentum. This is made evident by equation (7.5.10). The Lagrange or Hamilton formulation of mechanics allows a more systematic investigation of the existence of constants of the motion through the concept of *generalized momenta* and *cyclic coordinates*. So far, we have been able to cast the equation of motion in the form

$$\frac{d}{dt}\frac{\partial L}{\partial \dot q_k} - \frac{\partial L}{\partial q_k} = 0 \tag{7.5.13}$$

only for conservative systems but we will show in Section 7.6 that a Lagrangian can be found for a particle moving in an electromagnetic field.

The energy equation for the magnetron is

$$\mathscr{E} = \tfrac{1}{2}m\dot r^2 + \tfrac{1}{2}mr^2\dot\theta^2 + C \log\frac{r}{a} + V_a \tag{7.5.14}$$

Fig. 7.5.1

where C is given by (7.4.8). By using Eq. (7.5.10), $\dot{\theta}$ can be eliminated and an effective potential for the motion can be found. If $\dot{\theta}$ is also eliminated from (7.5.7), a differential equation for r as a function of t results. The solution can then be completed, e.g., by Fourier series. (See Problem 7.5.9.) The qualitative nature of the motion can be found by bending the cycloidal motion of Fig. 7.2.3 around in a circle. Various possible types of orbits are pictured in Fig. 7.5.1.

7.6. THE LAGRANGE FORMULATION

In the most general time-dependent electromagnetic field, it is possible to derive the fields from a scalar and a vector potential by means of

$$\mathbf{E} = -\nabla\psi - \frac{1}{c}\frac{\partial \mathbf{A}}{\partial t} \tag{7.6.1}$$

$$\mathbf{B} = \nabla \times \mathbf{A} \tag{7.6.2}$$

(see Section 5.6). If the magnetic field is absent and the electric field is static the Lagrangian for the motion of a charged particle is

$$L = T - q\psi \tag{7.6.3}$$

If an attempt is made to generalize the Lagrangian for motion in an arbitrary electromagnetic field by incorporating terms involving the vector potential, the choices are limited. The Lagrangian is an *invariant* and, hence, **A** must appear as a dot product with some other vector. Various possibilities are terms of the form

$$\mathbf{A} \cdot \mathbf{A} \qquad \mathbf{A} \cdot \mathbf{u} \qquad \mathbf{A} \cdot \mathbf{a} \tag{7.6.4}$$

The first is out because the squares of the fields do not appear in the equations of motion. The third is not possible because it would lead to a term involving the derivative of the acceleration in the equation of motion. Accordingly, we are led to assume a Lagrangian of the form

$$L = T - q\psi + \alpha(\mathbf{A} \cdot \mathbf{u}) \tag{7.6.5}$$

where α is a constant to be determined.

Assume that all quantities are expressed in rectangular coordinates. Then

$$L = \tfrac{1}{2}m\dot{x}_k\dot{x}_k - q\psi + \alpha(A_k\dot{x}_k) \tag{7.6.6}$$

The derivatives of the Lagrangian are

$$\frac{\partial L}{\partial x_i} = -q\frac{\partial \psi}{\partial x_i} + \alpha\frac{\partial A_k}{\partial x_i}\dot{x}_k \tag{7.6.7}$$

$$\frac{\partial L}{\partial \dot{x}_i} = m\dot{x}_i + \alpha A_i \tag{7.6.8}$$

The components of the vector potential are functions of position and time:

$$A_i = A_i(x_1, x_2, x_3, t) \tag{7.6.9}$$

Remember that, from the point of view of the particle, A_i changes both because the particle is moving and because the fields are time dependent:

$$\frac{dA_i}{dt} = \frac{\partial A_i}{\partial x_k}\dot{x}_k + \frac{\partial A_i}{\partial t} \tag{7.6.10}$$

Thus

$$\frac{d}{dt}\!\left(\frac{\partial L}{\partial \dot{x}_i}\right) = m\ddot{x}_i + \alpha\frac{\partial A_i}{\partial x_k}\dot{x}_k + \alpha\frac{\partial A_i}{\partial t} \tag{7.6.11}$$

and

$$\frac{d}{dt}\!\left(\frac{\partial L}{\partial \dot{x}_i}\right) - \frac{\partial L}{\partial x_i} = m\ddot{x}_i + \alpha\dot{x}_k\!\left(\frac{\partial A_i}{\partial x_k} - \frac{\partial A_k}{\partial x_i}\right) + \alpha\frac{\partial A_i}{\partial t} + q\frac{\partial \psi}{\partial x_i} \tag{7.6.12}$$

Next, note that

$$
\begin{aligned}
(\mathbf{u} \times \mathbf{B})_i &= (\mathbf{u} \times \mathbf{\nabla} \times \mathbf{A})_i = \delta_{ijk}\dot{x}_j\delta_{klm}\,\partial_l A_m \\
&= \delta_{kij}\,\delta_{klm}\dot{x}_j\partial_l A_m = (\delta_{il}\delta_{jm} - \delta_{im}\delta_{jl})\dot{x}_j\partial_l A_m \tag{7.6.13} \\
&= \dot{x}_j(\partial_i A_j - \partial_j A_i) = -\dot{x}_k(\partial_k A_i - \partial_i A_k)
\end{aligned}
$$

Hence, (7.6.12) can be written as

$$\frac{d}{dt}\left(\frac{\partial L}{\partial \dot{x}_i}\right) - \frac{\partial L}{\partial x_i} = m\ddot{x}_i - \alpha(\mathbf{u} \times \mathbf{B})_i + \alpha\frac{\partial A_i}{\partial t} + q\frac{\partial \psi}{\partial x_i} \quad (7.6.14)$$

If α is chosen as q/c and the electric field is introduced by means of (7.6.1),

$$\frac{d}{dt}\left(\frac{\partial L}{\partial \dot{x}_i}\right) - \frac{\partial L}{\partial x_i} = m\ddot{x}_i - \frac{q}{c}(\mathbf{u} \times \mathbf{B})_i - qE_i \quad (7.6.15)$$

Reference to the equation of motion (7.1.1) shows that

$$\frac{d}{dt}\left(\frac{\partial L}{\partial \dot{x}_i}\right) - \frac{\partial L}{\partial x_i} = 0 \quad (7.6.16)$$

The correct Lagrangian for a charged particle moving in an electromagnetic field, therefore, is

$$L = \tfrac{1}{2}mu^2 - q\psi + \frac{q}{c}(\mathbf{A} \cdot \mathbf{u}) \quad (7.6.17)$$

The fact that rectangular coordinates have been used to establish the correct form of the Lagrangian results in no loss of generality. Lagrange's equations satisfy the principle of general covariance meaning that if a Lagrangian is proved to be correct in one coordinate system it is automatically valid in all coordinate systems. In an arbitrary coordinate system the Lagrangian can be expressed as

$$L = \tfrac{1}{2}mg_{ij}\dot{q}^i\dot{q}^j - q\psi + \frac{q}{c}g_{ij}A^i\dot{q}^j \quad (7.6.18)$$

where A^i are the contravariant components of the vector potential. The equations of motion are to be obtained from

$$\frac{d}{dt}\frac{\partial L}{\partial \dot{q}^i} - \frac{\partial L}{\partial q^i} = 0 \quad (7.6.19)$$

The canonical momenta now have a more complicated form:

$$p_i = \frac{\partial L}{\partial \dot{q}^i} = mg_{ij}\dot{q}^j + \frac{q}{c}g_{ij}A^j \quad (7.6.20)$$

The Lagrangian (7.6.17) is not unique. As we learned in Section 5.6, new potentials which will give exactly the same electromagnetic field can be defined by means of

$$\mathbf{A} = \mathbf{A}' + \nabla\phi \quad (7.6.21)$$

$$\psi = \psi' - \frac{1}{c}\frac{\partial \phi}{\partial t} \quad (7.6.22)$$

where ϕ is an arbitrary scalar function. In terms of the new potentials the Lagrangian is

$$L = T - q\psi' + \frac{q}{c}(\mathbf{A}' \cdot \mathbf{u}) + \frac{q}{c}\left[\nabla\phi \cdot \mathbf{u} + \frac{\partial\phi}{\partial t}\right]$$

$$= L' + \frac{q}{c}\frac{d\phi}{dt} \tag{7.6.23}$$

The new Lagrangian, $L' = T - q\psi' + (q/c)(\mathbf{A}' \cdot \mathbf{u})$, will give the same equations of motion by virtue of the fact that the new potentials give the same electromagnetic field. (See Problem 7.6.11.)

As an example, consider the cylindrical magnetron discussed in Section 7.5. In cylindrical coordinates, the components of $\mathbf{B} = \nabla \times \mathbf{A}$ are

$$B_r = \frac{1}{r}\frac{\partial A_z}{\partial\theta} - \frac{\partial A_\theta}{\partial z} \tag{7.6.24}$$

$$B_\theta = \frac{\partial A_r}{\partial z} - \frac{\partial A_z}{\partial r} \tag{7.6.25}$$

$$B_z = \frac{1}{r}\frac{\partial}{\partial r}(rA_\theta) - \frac{1}{r}\frac{\partial A_r}{\partial\theta} \tag{7.6.26}$$

where all vectors are in terms of their physical components. The magnetic field is given by $B_r = 0$, $B_\theta = 0$, and $B_z = $ const., and it is easily verified that a possible vector potential is

$$A_r = 0 \qquad A_\theta = \tfrac{1}{2}rB_z \qquad A_z = 0 \tag{7.6.24}$$

Since the coordinate system is orthogonal, $\mathbf{A} \cdot \mathbf{u}$ can be calculated most conveniently by means of

$$\mathbf{A} \cdot \mathbf{u} = A_r u_r + A_\theta u_\theta + A_z u_z$$

$$= \tfrac{1}{2}r^2 B_z \dot\theta \tag{7.6.25}$$

The Lagrangian for a charged particle moving in the magnetron, therefore, is

$$L = \tfrac{1}{2}m(\dot r^2 + r^2\dot\theta^2 + \dot z^2) - C\log\frac{r}{a} + \frac{qB_z}{2c}r^2\dot\theta \tag{7.6.26}$$

It at once appears that θ is a cyclic coordinate. The canonically conjugate momentum is therefore a constant of the motion:

$$\frac{\partial L}{\partial\dot\theta} = mr^2\dot\theta + \frac{qB_z r^2}{2c} = l = \text{const.} \tag{7.6.27}$$

This is identical to Eq. (7.5.10). The remaining equations of motion follow in the usual way.

7.7. THE HAMILTON FORMULATION

According to (6.12.1), the Hamiltonian is defined as

$$H = p_i \dot{q}^i - L \qquad (7.7.1)$$

By using the expression (7.6.18) for the Lagrangian and the canonical momenta (7.6.20) we find

$$H = m g_{ij} \dot{q}^i \dot{q}^j + \frac{q}{c} g_{ij} \dot{q}^i A^j - \tfrac{1}{2} m g_{ij} \dot{q}^i \dot{q}^j + q\psi - \frac{q}{c} g_{ij} \dot{q}^i A^j \qquad (7.7.2)$$

$$= \tfrac{1}{2} m g_{ij} \dot{q}^i \dot{q}^j + q\psi = T + V$$

Thus H still represents the total energy of the particle. The Hamiltonian is a constant of the motion whenever $\partial L/\partial t = 0$. Before Hamilton's canonical equations,

$$\dot{p}_i = -\frac{\partial H}{\partial q^i} \qquad \dot{q}^i = \frac{\partial H}{\partial p_i} \qquad (7.7.3)$$

can be used, it is necessary to solve (7.6.20) for the generalized velocities and express H in terms of coordinates and canonical momenta. From the expression (7.6.20) for the canonical momenta,

$$g^{ik} p_i = m g^{ik} g_{ij} \dot{q}^j + \frac{q}{c} g^{ik} g_{ij} A^j = m \delta_j{}^k \dot{q}^j + \frac{q}{c} \delta_j{}^k A^j$$

Therefore,

$$\dot{q}^k = \frac{1}{m} \left[g^{ik} p_i - \frac{q}{c} A^k \right] \qquad (7.7.4)$$

The Hamiltonian (7.7.2) becomes

$$H = \frac{2}{2m} g_{ij} \left[g^{ri} p_r - \frac{q}{c} A^i \right] \left[g^{sj} p_s - \frac{q}{c} A^j \right] + q\psi$$

$$= \frac{1}{2m} \left[\delta_j{}^r p_r - \frac{q}{c} g_{ij} A^i \right] \left[g^{sj} p_s - \frac{q}{c} A^j \right] + q\psi \qquad (7.7.5)$$

$$= \frac{1}{2m} \left[g^{sr} p_r p_s - \frac{q}{c} p_r A^r - \frac{q}{c} A^s p_s - \frac{q^2}{c^2} g_{ij} A^i A^j \right] + q\psi$$

In conventional vector notation,

$$H = \frac{1}{2m} \left[\mathbf{p} \cdot \mathbf{p} - \frac{2q}{c} \mathbf{p} \cdot \mathbf{A} + \frac{q^2}{c^2} \mathbf{A} \cdot \mathbf{A} \right] + q\psi \qquad (7.7.6)$$

An important application of this result lies in its use to infer the correct wave equation for a particle moving in an electromagnetic field in quantum mechanics.

7.8. TWO-DIMENSIONAL MAGNETIC FIELDS

Suppose that a static magnetic field exists in a region of space and has no x_3 component. Let x_1 and x_2 components be represented by a few terms of their Taylor's series expansions:

$$B_1 = B_0 + \frac{\partial B_1}{\partial x_1} x_1 + \frac{\partial B_1}{\partial x_2} x_2 = B_0 + b_{11}x_1 + b_{12}x_2 \qquad (7.8.1)$$

$$B_2 = b_2 + \frac{\partial B_2}{\partial x_1} x_1 + \frac{\partial B_2}{\partial x_2} x_2 = b_2 + b_{21}x_1 + b_{22}x_2 \qquad (7.8.2)$$

where it is assumed that B_1 and B_2 have no functional dependence on x_3. It is not possible to choose the four derivatives b_{ij} independently. If the region of space under consideration is free of sources (electric currents), then the magnetic field obeys the field equations $\nabla \cdot \mathbf{B} = 0$ and $\nabla \times \mathbf{B} = 0$ which require that

$$b_{11} + b_{12} = 0 \qquad b_{12} = b_{21} \qquad (7.8.3)$$

The second rank tensor quantity

$$\begin{pmatrix} b_{11} & b_{12} \\ b_{21} & b_{22} \end{pmatrix} \qquad (7.8.4)$$

therefore, is *symmetric* and by simply rotating the coordinate system about the x_3-axis a new coordinate system can be found in which $b_{21} = b_{12} = 0$ (see Section 3.4). Assume further that at the origin $B_2 = 0$. The magnetic field, therefore, is

$$B_1 = B_0 + b_{11}x_1 \qquad B_2 = -b_{11}x_2 \qquad (7.8.5)$$

and represents a simple field from which we may hope to learn some of the general properties of the motion of a charged particle in a nonuniform field. The differential equation of the field lines is

$$\frac{B_2}{B_1} = \frac{dx_2}{dx_1},$$

$$b_{11}x_2 \, dx_1 + (B_0 + b_{11}x_1) \, dx_2 = 0 \qquad (7.8.6)$$

and is found to be an exact differential. So we may write

$$\frac{\partial \phi}{\partial x_1} = b_{11}x_2 \qquad \frac{\partial \phi}{\partial x_2} = B_0 + b_{11}x_1 \qquad (7.8.7)$$

Integration yields

$$\phi = B_0 x_2 + b_{11}x_1 x_2 \qquad (7.8.8)$$

The curves which represent the field lines are $\phi = C_1 = $ const. or

$$x_2 = \frac{C_1}{B_0 + b_{11}x_1} \tag{7.8.9}$$

and are seen to be hyperbolic. The field lines are sketched in Fig. 7.8.1. Since the region is source free, $\nabla \times \mathbf{B} = 0$ and it is possible to derive $\mathbf{B}$ from a scalar potential:

$$B_1 = -\frac{\partial \psi}{\partial x_1} = B_0 + b_{11}x_1 \tag{7.8.10}$$

$$B_2 = -\frac{\partial \psi}{\partial x_2} = -b_{11}x_2 \tag{7.8.11}$$

Integration yields

$$\psi = -B_0 x_1 - \tfrac{1}{2}b_{11}(x_1^2 - x_2^2) \tag{7.8.12}$$

The equipotential surfaces are given by $\psi = -C_2 = $ const. or

$$B_0 x_1 + \tfrac{1}{2}b_{11}(x_1^2 - x_2^2) = C_2 \tag{7.8.13}$$

The equipotential surfaces are hyperbolic cylindrical surfaces parallel to the x_3-axis and intersect the $x_1 x_2$-plane on the dashed curves of Fig. 7.8.1. The field lines are perpendicular to these surfaces.

Comparison of (7.8.10), (7.8.11), and (7.8.7) shows that

$$\frac{\partial \phi}{\partial x_2} = -\frac{\partial \psi}{\partial x_1} \qquad \frac{\partial \phi}{\partial x_1} = \frac{\partial \psi}{\partial x_2} \tag{7.8.14}$$

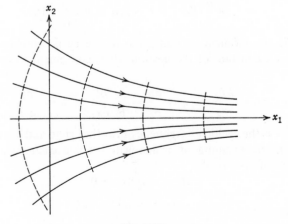

Fig. 7.8.1

Those readers familiar with complex variable theory will recognize these as the *Cauchy-Riemann equations*. The functions ϕ and ψ are called *conjugate functions*

The magnetic field is given in terms of the vector potential by means of

$$B_1 = \frac{\partial A_3}{\partial x_2} - \frac{\partial A_2}{\partial x_3} = B_0 + b_{11}x_1 \qquad (7.8.15)$$

$$B_2 = \frac{\partial A_1}{\partial x_3} - \frac{\partial A_3}{\partial x_1} = -b_{11}x_2 \qquad (7.8.16)$$

$$B_3 = \frac{\partial A_2}{\partial x_1} - \frac{\partial A_1}{\partial x_2} = 0 \qquad (7.8.17)$$

These equations are satisfied by

$$A_1 = A_2 = 0 \qquad A_3 = B_0 x_2 + b_{11}x_1 x_2 \qquad (7.8.18)$$

The Lagrangian for a charged particle moving in the field is

$$L = \tfrac{1}{2}mu^2 + \frac{q}{c}\mathbf{A}\cdot\mathbf{u}$$

$$= \tfrac{1}{2}m(\dot{x}_1{}^2 + \dot{x}_2{}^2 + \dot{x}_3{}^2) + \frac{q}{c}(B_0 x_2 + b_{11}x_1 x_2)\dot{x}_3 \qquad (7.8.19)$$

and the equations of motion are found to be

$$m\ddot{x}_1 - \frac{q}{c}b_{11}x_2\dot{x}_3 = 0 \qquad (7.8.20)$$

$$m\ddot{x}_2 - \frac{q}{c}(B_0 + b_{11}x_1)\dot{x}_3 = 0 \qquad (7.8.21)$$

$$m\dot{x}_3 + \frac{q}{c}(B_0 x_2 + b_{11}x_1 x_2) = p_3 = \text{const.} \qquad (7.8.22)$$

where p_3 is the momentum conjugate to the cyclic coordinate x_3 and, therefore, is a constant of the motion. It is convenient to introduce the parameters

$$\omega_c = \frac{qB_0}{mc} \qquad \eta = \frac{b_{11}}{B_0} = \frac{1}{B_0}\frac{\partial B_1}{\partial x_1} \qquad a = \frac{p_3}{m\omega_c} \qquad (7.8.23)$$

Note that ω_c is the cyclotron frequency near the origin and a is a characteristic length, the meaning of which will become clear presently. The equations of motion read

$$\ddot{x}_1 - \omega_c\eta x_2\dot{x}_3 = 0 \qquad (7.8.24)$$

$$\ddot{x}_2 - \omega_c(1 + \eta x_1)\dot{x}_3 = 0 \qquad (7.8.25)$$

$$\dot{x}_3 + \omega_c(1 + \eta x_1)x_2 = a\omega_c \qquad (7.8.26)$$

Since the electric field is absent, another constant of the motion is the speed of the particle:

$$\dot{x}_1^2 + \dot{x}_2^2 + \dot{x}_3^2 = u^2 = \text{const.} \tag{7.8.27}$$

Let us first consider the solutions for a *uniform* field, i.e., $\eta = 0$:

$$\ddot{x}_1 = 0 \qquad \ddot{x}_2 - \omega_c \dot{x}_3 = 0 \qquad \dot{x}_3 + \omega_c x_2 = a\omega_c \tag{7.8.28}$$

A possible set of solutions is

$$x_1 = u_{01}t \qquad x_2 = \rho_0 \cos \omega_c t + a \qquad x_3 = -\rho_0 \sin \omega_c t \tag{7.8.29}$$

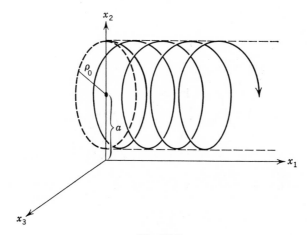

Fig. 7.8.2

where u_{01} and ρ_0 are constants of integration. Equation (7.8.27) requires that

$$u_{01}^2 + \rho_0^2 \omega_c^2 = u^2 \tag{7.8.30}$$

The particle spirals in the form of a helix on the surface of a cylinder of radius ρ_0; the axis of the cylinder is located at $x_2 = a$ as pictured in Fig. 7.8.2.

The motion in the nonuniform field is governed by a set of nonlinear coupled equations involving three unknown functions $x_1(t)$, $x_2(t)$, and $x_3(t)$. An exact solution in terms of simple functions is not possible, but a great deal of qualitative information about the motion can be obtained. The substitution of (7.8.26) into (7.8.27) yields

$$\dot{x}_1^2 + \dot{x}_2^2 + [a\omega_c - \omega_c(1 + \eta x_1)x_2]^2 = u^2 \tag{7.8.31}$$

Thus the function

$$u_3^2 = \omega_c^2[a - (1 + \eta x_1)x_2]^2 \tag{7.8.32}$$

plays the role of an effective potential as far as the x_1x_2 motion is concerned. The contours of this function are the curves determined from

$$u_3{}^2 = \omega_c{}^2 k^2, \qquad k = \text{const.}$$

$$x_2 = \frac{a \pm k}{1 + \eta x_1} \tag{7.8.33}$$

These contour lines, or equipotential lines, are seen to be identical to the lines of the magnetic field. Note also that they are the same as the curves

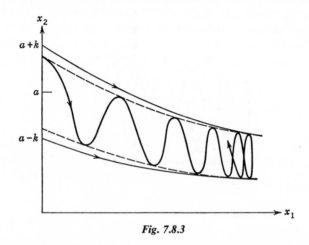

Fig. 7.8.3

obtained from $A_3 = \text{const.}$ For a given value of the constant k there are two lines of the magnetic field which cross the x_2-axis at $a + k$ and $a - k$ as illustrated in Fig. 7.8.3. For a particle of a given speed u, let the constant k be determined by $u = \omega_c k$. Equation (7.8.31) can be expressed

$$\dot{x}_1{}^2 + \dot{x}_2{}^2 = u^2 - u_3{}^2 \geq 0 \tag{7.8.34}$$

When the particle reaches one of the contour lines, $u_3 = \omega_c k$ and necessarily $\dot{x}_1{}^2 + \dot{x}_2{}^2 = 0$. The particle is confined to move between the two field lines since to go beyond would mean $\dot{x}_1{}^2 + \dot{x}_2{}^2 < 0$. If the particle is given an initial velocity in the x_1-direction it begins to spiral along as in the case where the field is uniform. At first, $\dot{x}_1$ and $\dot{x}_2$ are not simultaneously zero so that the particle never actually reaches the field lines determined by $u = \omega_c k$, but rather moves between the dashed curves as indicated in Fig. 7.8.3. As the particle progresses along in the x_1-direction, the field lines close in on it. The speed in the x_1-direction becomes less and less. Finally, the particle actually touches the limiting field lines and $\dot{x}_1{}^2 + \dot{x}_2{}^2 = 0$. This is a *turning point* or *mirror point* in the motion, and the particle now begins to spiral back in the negative x_1-direction. As it turns out, the particle does not actually remain in the vicinity of the x_1x_2-plane but drifts in the x_3-direction. (This question is taken up in Section 7.9.) We thus find that charged particles tend to spiral along the field lines and as they move into regions of greater field density their forward advance is slowed until they are finally reflected. This effect is independent of the sign of the charge.

A useful technique for obtaining approximate solutions of the equations of motion is the method of perturbations. If the curvature of the field lines is not too great, the parameter η is a small quantity and the terms involving η in the equations of motion can be regarded as perturbations on the motion. The solutions, considered as a function of η, can be represented by a Taylor's series:

$$x_1 = f_0 + f_1\eta + f_2\eta^2 + \cdots$$
$$x_2 = g_0 + g_1\eta + g_2\eta^2 + \cdots \qquad (7.8.35)$$
$$x_3 = h_0 + h_1\eta + h_2\eta^2 + \cdots$$

where $f_0, f_1, g_0, \ldots$ are functions of time to be determined. The substitution of (7.8.35) into the equations of motion (7.8.24), (7.8.25), and (7.8.26) yields

$$\ddot{f}_0 + (\ddot{f}_1 - \omega_c g_0 \dot{h}_0)\eta = 0$$
$$\ddot{g}_0 - \omega_c \dot{h}_0 + (\ddot{g}_1 - \omega_c \dot{h}_1 - \omega_c f_0 \dot{h}_0)\eta = 0 \qquad (7.8.36)$$
$$\ddot{h}_0 + \omega_c \dot{g}_0 + (\ddot{h}_1 + \omega_c \dot{g}_1 + \omega_c f_0 \dot{g}_0)\eta = a\omega_c$$

where all terms of order η^2 or higher are discarded. As long as η is kept small, the above equations must hold for any choice of η. The coefficients of like powers of η must separately vanish:

$$\ddot{f}_0 = 0 \qquad \ddot{g}_0 - \omega_c \dot{h}_0 = 0 \qquad \ddot{h}_0 + \omega_c \dot{g}_0 = a\omega_c \qquad (7.8.37)$$

$$\ddot{f}_1 - \omega_c g_0 \dot{h}_0 = 0 \qquad (7.8.38)$$

$$\ddot{g}_1 - \omega_c \dot{h}_1 - \omega_c f_0 \dot{h}_0 = 0 \qquad (7.8.39)$$

$$\ddot{h}_1 + \omega_c \dot{g}_1 + \omega_c f_0 \dot{g}_0 = 0 \qquad (7.8.40)$$

The equations involving only f_0, g_0, and h_0 are called the *zero order* equations and are seen to be identical to the equations of motion (7.8.28) where the curvature is absent. Thus

$$f_0 = u_{01}t \qquad g_0 = \rho_0 \cos \omega_c t + a \qquad h_0 = -\rho_0 \sin \omega_c t \qquad (7.8.41)$$

It is now possible to substitute (7.8.41) into the *first order* equations (7.8.38), (7.8.39), and (7.8.40) and integrate to find the functions f_1, g_1, and h_1. For instance, (7.8.38) gives

$$\ddot{f}_1 + \omega_c^2 a \rho_0 \cos \omega_c t + \rho_0^2 \omega_c^2 \cos^2 \omega_c t = 0 \qquad (7.8.42)$$

A single integration gives us an expression for the velocity in the x_1-direction:

$$\begin{aligned}
\dot{x}_1 = \dot{f}_0 + \eta \dot{f}_1 &= u_{01} - \omega_c a \rho_0 \eta \sin \omega_c t \\
&\quad - \tfrac{1}{2}\rho_0^2 \omega_c^2 \eta t - \tfrac{1}{4}\rho_0^2 \omega_c \eta \sin 2\omega_c t
\end{aligned} \qquad (7.8.43)$$

where the constant of integration is chosen so that $\dot{x}_i = u_{01}$ at $t = 0$. The terms containing $\sin \omega_c t$ and $\sin 2\omega_c t$ are rapidly oscillating. We see that the average progression of the particle in the x_1-direction is given by

$$\begin{aligned}
u_1 &= u_{01} - \tfrac{1}{2}u_\perp^2 t \eta \\
x_1 &= u_{01} t - \tfrac{1}{4}u_\perp^2 t^2 \eta
\end{aligned} \qquad (7.8.44)$$

where $u_\perp = \rho_0 \omega_c$ is the component of the initial velocity perpendicular to the field lines at $x_1 = 0$. The approximate mirror point is found by setting $u_1 = 0$:

$$t = \frac{2u_{01}}{u_\perp^2 \eta} \qquad \eta x_1 = \frac{u_{01}^2}{u_\perp^2} \qquad (7.8.45)$$

From (7.8.5) and (7.8.23) it is seen that

$$\eta x_1 = \frac{B_1 - B_0}{B_0} \qquad (7.8.46)$$

Thus

$$u_{01} = u_\perp \sqrt{(B_1/B_0) - 1} \qquad (7.8.47)$$

This result means that if the particle starts with parallel and perpendicular components of velocity u_{01} and $u_\perp$ at a point where the field strength is B_0, it will be reflected at a point where field strength has increased to B_1. As an approximate result, this criterion can be used in any non-uniform static magnetic field.

7.9. CURVATURE AND GRADIENT DRIFT

In Section 7.8, the tendency of charged particles to spiral along the magnetic field lines and to be repelled from regions of high field density was demonstrated. In this section, it is shown that charged particles also tend to drift in a direction perpendicular to the radius of curvature of the field lines.

Suppose that curved magnetic field lines exist in a region of space. Assume that in cylindrical coordinates the field can be represented as $(0, B_\theta, 0)$ over the range of motion of the particle. Such a field is shown in Fig. 7.9.1 and could be created by means of a straight wire carrying a current along the z-axis. The θ component of the field will be represented by

$$B_\theta = B_1 + \left(\frac{dB_\theta}{dr}\right)_{r_1} (r - r_1) = B_1[1 - \eta(r - r_1)] \qquad (7.9.1)$$

where

$$\eta = -\frac{1}{B_1}\left(\frac{dB_\theta}{dr}\right)_{r_1} \qquad (7.9.2)$$

For example, if the field is due to a current J flowing on the z-axis,

$$B_\theta = \frac{2J}{rc} \qquad \left(\frac{dB_\theta}{dr}\right)_{r_1} = -\frac{2J}{r_1{}^2 c} \qquad (7.9.3)$$

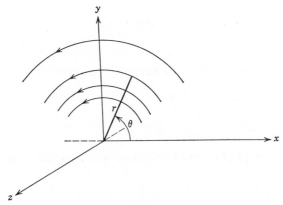

Fig. 7.9.1

so that η as defined by (7.9.2) is positive. We are therefore dealing with a field that has *curvature* and *gradient*. By using the expressions for the components of $\mathbf{B} = \nabla \times \mathbf{A}$ in cylindrical coordinates, Eqs. (7.6.24), (7.6.25), and (7.6.26), it is verified that a possible vector potential is

$$A_r = A_\theta = 0 \qquad A_z = -B_1 r + \tfrac{1}{2}\eta B_1 (r - r_1)^2 \qquad (7.9.4)$$

The Lagrangian is

$$L = \tfrac{1}{2}m(\dot{x}_1{}^2 + \dot{x}_2{}^2 + \dot{x}_3{}^2) - \frac{q}{c}\dot{z}B_1 r + \frac{q}{2c}\eta B_1 \dot{z}(r - r_1)^2 \qquad (7.9.5)$$

The equations of motion are found to be

$$\ddot{r} - r\dot{\theta}^2 + \omega_c \dot{z} - \eta\omega_c \dot{z}(r - r_1) = 0 \qquad (7.9.6)$$

$$\frac{\partial L}{\partial \dot{\theta}} = mr^2\dot{\theta} = l = \text{const.} \qquad (7.9.7)$$

$$\frac{\partial L}{\partial \dot{z}} = m[\dot{z} - \omega_c r + \tfrac{1}{2}\eta\omega_c(r - r_1)^2] = p_z = \text{const.} \qquad (7.9.8)$$

There are two canonical momenta given by (7.9.7) and (7.9.8) which are constants of the motion. $\omega_c = qB_1/(mc)$ is the cyclotron frequency evaluated at $r = r_1$. It is convenient to introduce a constant a which has the dimensions of length by means of

$$p_z = -m\omega_c a$$

Equation (7.9.8) can then be written

$$\dot{z} = \omega_c(r - a) - \tfrac{1}{2}\eta\omega_c(r - r_1)^2 \tag{7.9.9}$$

The elimination of $\dot{\theta}$ and $\dot{z}$ from (7.9.6) by means of (7.9.7) and (7.9.9) leads to

$$\ddot{r} - \frac{l^2}{m^2 r^3} + \omega_c^2[r - a - \eta(r - a)(r - r_1) \\ - \tfrac{1}{2}\eta(r - r_1)^2 + \tfrac{1}{2}\eta^2(r - r_1)^3] = 0 \tag{7.9.10}$$

Since there is no electric field, all the energy is kinetic:

$$\mathscr{E} = \tfrac{1}{2}m(\dot{r}^2 + r^2\dot{\theta}^2 + \dot{z}^2) \\ = \tfrac{1}{2}m\dot{r}^2 + \frac{l^2}{2mr^2} + \tfrac{1}{2}m\omega_c^2[r - a - \tfrac{1}{2}\eta(r - r_1)^2]^2 \tag{7.9.11}$$

The effective potential for motion in the r-direction, therefore, is

$$V_e(r) = \frac{l^2}{2mr^2} + \tfrac{1}{2}m\omega_c^2[r - a - \tfrac{1}{2}\eta(r - r_1)^2]^2 \tag{7.9.12}$$

A little consideration shows that the general form of this function in the vicinity of r_1 is as illustrated in Fig. 7.9.2. Originally, the field component B_θ was expanded in a Taylor's series about an arbitrary point $r = r_1$ and it is now convenient to choose this point to coincide with the minimum in

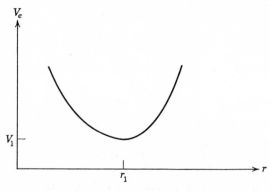

Fig. 7.9.2

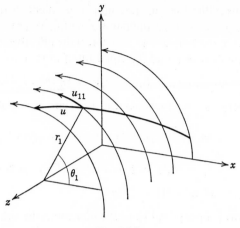

Fig. 7.9.3

$V_e(r)$. The field line at $r = r_1$ is therefore a position of *stable equilibrium*, at least insofar as the r-motion is concerned.

Suppose that the initial conditions are such that $\mathscr{E} = V_1$ exactly. Then $r = r_1 = $ const. and from the equation of motion (7.9.10),

$$ -\frac{l^2}{m^2 r_1^3} + \omega_c^2(r_1 - a) = 0 \tag{7.9.13} $$

The θ and z motions are then

$$ \dot{\theta}_1 = \frac{l}{mr_1^2} = \text{const.} \tag{7.9.14} $$

$$ \dot{z}_1 = \omega_c(r_1 - a) = \frac{l^2}{m^2 r_1^3 \omega_c} \tag{7.9.15} $$

The component of the velocity parallel to the field lines is

$$ u_{11} = r_1 \dot{\theta}_1 \tag{7.9.16} $$

and (7.9.14) and (7.9.15) can be combined into the expression

$$ \dot{z} = \frac{u_{11}^2}{r_1 \omega_c} \tag{7.9.17} $$

The particle, therefore, moves over the cylindrical surface $r = r_1 = $ const. without spiralling, as shown in Fig. 7.9.3. Since the field strength is the same at all points of the particle trajectory, the effects of the gradient of the field, dB_θ/dr, are not felt. The velocity (7.9.17) is perpendicular to the radius of curvature r_1 of the field lines and is generally known as *curvature drift*.

In order to examine the phenomenon of gradient drift in the simplest circumstance, assume that the angular momentum is zero so that the motion takes place in a plane $\theta = $ const. For $l = 0$ the effective potential (7.9.12) has its minimum at $r_1 = a$ and the equation of motion for r is found to be

$$\ddot{r} + \omega_c^2[r - r_1 - \tfrac{3}{2}\eta(r - r_1)^2 + \tfrac{1}{2}\eta^2(r - r_1)^3] = 0 \qquad (7.9.18)$$

Since $r(t)$ is periodic, we may use the Fourier series solution (4.13.35). With

$$\alpha = \omega_c^2 \qquad \beta = \tfrac{3}{2}\eta\omega_c^2 \qquad \gamma = \tfrac{1}{2}\eta^2\omega_c^2 \qquad (7.9.19)$$

We find

$$r - r_1 = \tfrac{3}{4}b_1^2\eta + b_1 \cos \omega t - \tfrac{1}{4}b_1^2\eta \cos 2\omega t$$

$$\omega^2 = \omega_c^2[1 - \tfrac{3}{2}\eta^2 b_1^2] \qquad (7.9.20)$$

Neglecting terms proportional to η^2 the motion in the z-direction is found from (7.9.9) to be

$$\dot{z} = \omega_c[\tfrac{3}{4}b_1^2\eta + b_1 \cos \omega t - \tfrac{1}{4}b_1^2\eta \cos 2\omega t - \tfrac{1}{2}\eta b_1^2 \cos^2 \omega t + \cdots]$$

$$\dot{z} = \omega_c[\tfrac{1}{2}\eta b_1^2 + b_1 \cos \omega t - \tfrac{1}{2}\eta b_1^2 \cos 2\omega t + \cdots] \qquad (7.9.21)$$

where $\cos^2 \omega t = \tfrac{1}{2}(1 + \cos 2\omega t)$ is used. Integration yields

$$z = \tfrac{1}{2}\omega_c\eta b_1^2 t + \frac{\omega_c}{\omega} b_1 \sin \omega t - \frac{1}{4}\frac{\omega_c}{\omega} \eta b_1^2 \sin 2\omega t \qquad (7.9.22)$$

The magnetic field (7.9.1) should, of course, satisfy the field equation $\nabla \times \mathbf{B} = 0$. This condition can be met only approximately. It is found that the r and θ components of $\nabla \times \mathbf{B}$ are identically zero whereas

$$(\nabla \times \mathbf{B})_z = \frac{1}{r}\frac{\partial}{\partial r}(rB_\theta) = B_1\left(\frac{1}{r} - 2\eta + \frac{\eta r_1}{r}\right) \qquad (7.9.23)$$

Thus, if

$$\eta = \frac{1}{r_1} = -\frac{1}{B_1}\left(\frac{dB_\theta}{dr}\right)_{r_1} \qquad (7.9.24)$$

then $\nabla \times \mathbf{B} = 0$ at $r = r_1$. A better approximation to both the field equations and the equation of motion can be obtained by taking another term in the Taylor's series expansion of B_θ.

As a numerical example, suppose the particle is a proton:

$$m_p = 1.67 \times 10^{-24} \text{ gm} \qquad q = 4.80 \times 10^{-10} \text{ stat coul} \qquad (7.9.25)$$

Furthermore, let

$$r_1 = 10 \text{ cm} \qquad \eta = \frac{1}{r_1} = 0.1 \text{ cm}^{-1}$$

$$B_1 = 10^4 \text{ gauss} \qquad b_1 = 2 \text{ cm} \qquad (7.9.26)$$

Then

$$\omega_c = \frac{qB_1}{mc} = 1.0 \times 10^6 \text{ sec}^{-1} \qquad (7.9.27)$$

If t is measured in *microseconds*,

$$r = 10.3 + 2.0 \cos t - 0.1 \cos 2t + \cdots$$
$$z = 0.2t + 2.0 \sin t - 0.1 \sin 2t + \cdots \qquad (7.9.28)$$

Figure 7.9.4 is based on these equations. The term proportional to t gives rise to a constant drift in the positive z-direction at a rate 0.2 cm/microsecond. Qualitatively, this is because the field is stronger at the bottom

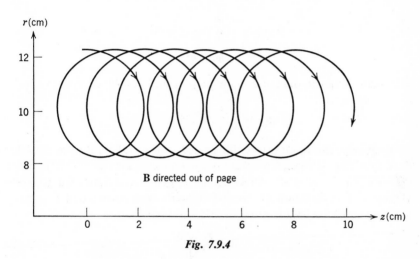

Fig. 7.9.4

of the spirals than at the top, causing a greater curvature of the path and also a slight modification of the frequency as given by (7.9.20).

The *gradient drift velocity* is the constant term in Eq. (7.9.21):

$$u_G = \tfrac{1}{2}\eta\omega_c b_1{}^2 \qquad (7.9.29)$$

As the particles are injected into the field, their velocity perpendicular to the field lines is essentially

$$u_\perp = \omega_c b_1 \qquad (7.9.30)$$

and (7.9.29), therefore, is

$$u_G = \tfrac{1}{2}\eta \frac{u_\perp{}^2}{\omega_c} = \frac{u_\perp{}^2}{2r_1\omega_c} \qquad (7.9.31)$$

where (7.9.24) is used. When the angular momentum is not zero, so that the particles are progressing along the field lines, the effects of curvature

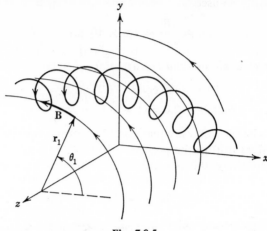

Fig. 7.9.5

drift and gradient drift are both present and can be combined into a single expression:

$$u_D = \frac{1}{r_1 \omega_c} (u_{11}^2 + \tfrac{1}{2} u_\perp^2) \tag{7.9.32}$$

Qualitatively, the motion is as illustrated in Fig. 7.9.5 with the particles drifting in the z-direction at a rate given approximately by (7.9.32). Since $\omega_c = qB_1/(mc)$, the drift will be in the negative z-direction for negative particles. The direction of the drift velocity at a given point in a non-uniform magnetic field is given by the unit vector

$$\hat{n} = \frac{\mathbf{r}_1 \times \mathbf{B}}{r_1 B} \tag{7.9.33}$$

where r_1 is the radius of curvature of the field lines at the point in question. Hence, the drift velocity in vector form is

$$\mathbf{u}_D = \frac{\mathbf{r}_1 \times \mathbf{B}}{B r_1{}^2 \omega_c} (u_{11}^2 + \tfrac{1}{2} u_\perp^2) \tag{7.9.34}$$

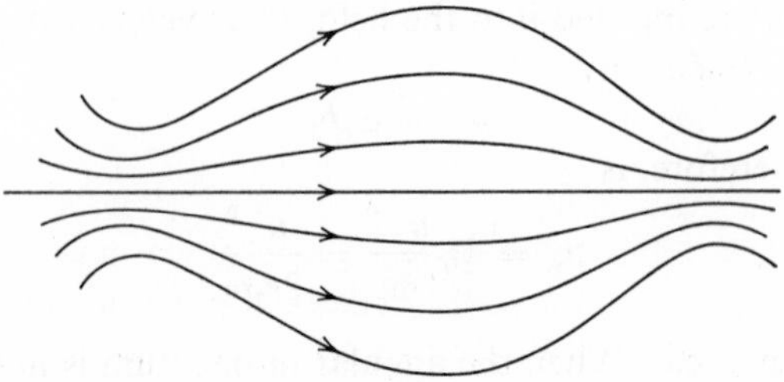

Fig. 7.9.6

and can actually be used to estimate the rate of drift of particles in any non-uniform magnetic field.

The reflection of charged particles from regions of high field density is a valuable property of magnetic fields in the design of devices to contain charged particles, e.g., in the construction of a fusion reactor. Figure 7.9.6 shows a possible field configuration that might be used as a "magnetic bottle." On the other hand, curvature drift and gradient drift are troublesome in this regard.

7.10. VAN ALLEN RADIATION

The Van Allen radiation belts surrounding the earth consist essentially of charged particles trapped in the magnetic field of the earth. The trapping of the charged particles can be qualitatively understood on the basis of the results obtained in Sections 7.8 and 7.9. The magnetic field of the earth is approximately a dipole field. In spherical coordinates

$$B_r = -2\mu \frac{\cos\theta}{r^3} \qquad B_\theta = -\frac{\mu\sin\theta}{r^3} \qquad B_\phi = 0 \qquad (7.10.1)$$

where μ is the strength of the dipole and can be found by measuring the magnetic field at a known point on the surface of the earth. The differential equation of the magnetic field lines is

$$\frac{dr}{r\,d\theta} = \frac{B_r}{B_\theta} \qquad (7.10.2)$$

or

$$\frac{1}{r}\,dr - 2\frac{\cos\theta}{\sin\theta}\,d\theta = 0 \qquad (7.10.3)$$

Integration yields

$$r = k\sin^2\theta \qquad (7.10.4)$$

where k is a constant. The field lines for various values of the constant k are sketched in Fig. 7.10.1.

It was shown in Section 7.8 that the particles spiral along the field lines and, therefore, we may expect the range of motion of a given particle to be roughly between two field lines as indicated by the shaded region of Fig. 7.10.1. Since the particle will be repelled from regions of high field density, they can become trapped in the field for long periods of time. Because the field lines converge in the polar regions, it is here that the particles are likely ultimately to enter the earth's atmosphere, accounting for the aurora borealis. Due to the curvature and gradient of the field, the particles will drift in the ϕ direction so that an original concentration of particles will eventually diffuse out and encircle the entire earth.

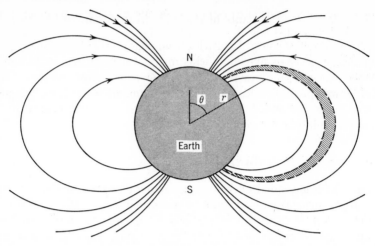

Fig. 7.10.1

Reference to Eq. (2.24.5), (2.24.6), and (2.24.7) shows that the components of $\mathbf{B} = \nabla \times \mathbf{A}$ in spherical coordinates are

$$(\nabla \times \mathbf{A})_r = \frac{1}{r \sin \theta}\left(\frac{\partial}{\partial \theta}[\sin \theta A_\phi] - \frac{\partial A_\theta}{\partial \phi}\right) = -2\mu \frac{\cos \theta}{r^3} \quad (7.10.5)$$

$$(\nabla \times \mathbf{A})_\theta = \frac{1}{r \sin \theta}\left(\frac{\partial A_r}{\partial \phi} - \frac{\partial}{\partial r}[r \sin \theta A_\phi]\right) = -\mu \frac{\sin \theta}{r^3} \quad (7.10.6)$$

$$(\nabla \times \mathbf{A})_\phi = \frac{1}{r}\left(\frac{\partial}{\partial r}[rA_\theta] - \frac{\partial A_r}{\partial \theta}\right) = 0 \quad (7.10.7)$$

A vector potential which satisfies these relations is

$$A_r = 0, \qquad A_\theta = 0, \qquad A_\phi = -\frac{\mu \sin \theta}{r^2} \quad (7.10.8)$$

The Lagrangian for a charged particle moving in a magnetic dipole field, therefore, is

$$L = \tfrac{1}{2}m(\dot{r}^2 + r^2\dot{\theta}^2 + r^2 \sin^2 \theta \dot{\phi}^2) - \frac{q\mu}{c}\frac{\sin^2 \theta}{r}\dot{\phi} \quad (7.10.9)$$

The coordinate ϕ is cyclic so that

$$\frac{\partial L}{\partial \dot{\phi}} = l = mr^2 \sin^2 \theta \dot{\phi} - \frac{q\mu}{rc}\sin^2 \theta \quad (7.10.10)$$

is a constant of the motion. A second constant of the motion is the magnitude of the velocity:

$$u^2 = \dot{r}^2 + r^2\dot{\theta}^2 + r^2 \sin^2 \theta \dot{\phi}^2 \quad (7.10.11)$$

By using these two constants of the motion, the approximate range of motion of a particle with given initial conditions can be determined and the regions from which the particle is definitely excluded can be found.

The solution of (7.10.10) for $\dot{\phi}$ is

$$\dot{\phi} = \frac{l}{mr^2 \sin^2 \theta} + \frac{q\mu}{mr^3c} \qquad (7.10.12)$$

The ϕ component of the velocity of the particle is

$$u_\phi = r \sin \theta \dot{\phi} \qquad (7.10.13)$$

The motion of any particle is obviously restricted by the condition

$$|u_\phi| \leq u \qquad (7.10.14)$$

The equality can hold only when $\dot{r}$ and $\dot{\theta}$ are both zero. If $|u_\phi| > u$ we would have $\dot{r}^2 + r^2\dot{\theta}^2 < 0$ which is obviously not possible. The combination of (7.10.12), (7.10.13), and (7.10.14) gives

$$\left| \frac{lc}{q\mu r} \frac{1}{\sin \theta} + \frac{\sin \theta}{r^2} \right| \leq \frac{mcu}{q\mu} \qquad (7.10.15)$$

The quantity

$$a_0 = \sqrt{\frac{q\mu}{mcu}} \qquad (7.10.16)$$

has the dimensions of length and is a convenient parameter to work with. For concreteness, suppose that we are dealing with protons moving at $0.1c$. The approximate value of the magnetic dipole moment of the earth is

$$\mu = 8.1 \times 10^{25} \text{ stat coulomb cm} \qquad (7.10.17)$$

$$a_0 = \sqrt{\frac{4.80 \times 10^{-10} \times 8.1 \times 10^{25}}{1.67 \times 10^{-24} \times 9 \times 10^{20} \times 0.1}} = 1.61 \times 10^{10} \text{ cm} \quad (7.10.18)$$

(See Appendix D for a selected list of physical constants.) By contrast the mean value of the earth's radius is

$$R = 6.37 \times 10^8 \text{ cm} \qquad (7.10.19)$$

so that $a_0 \simeq 25$ earth radii. Suppose that the protons start at the equator ($\sin \theta_0 = 1$) at a distance r_0 from the earth's center and that the initial velocity has no component in the ϕ-direction. Then

$$l = - \frac{q\mu}{r_0 c} \qquad (7.10.20)$$

and (7.10.15) can be expressed

$$\left| \frac{\sin \theta}{r^2} - \frac{1}{rr_0 \sin \theta} \right| \le \frac{1}{a_0^2} \qquad (7.10.21)$$

For our protons moving at a speed of 0.1c, $a_0^{-2} = 0.0016R^{-2}$. In the limit of very low speed particles,

$$\frac{\sin \theta}{r^2} - \frac{1}{rr_0 \sin \theta} \sim 0$$

$$r \sim r_0 \sin^2 \theta \qquad (7.10.22)$$

This, of course, is identical to the equation of a field line and means, as far as the motion with respect to r is concerned, that a particle spirals along a field line—a fact that we already knew.

Suppose that the protons are released at an elevation of 1 earth radius above the earth's surface and at the equatorial plane. The magnetic field here is

$$B_0 = \frac{\mu}{r_0^3} = \frac{8.1 \times 10^{25}}{(2 \times 6.37 \times 10^8)^3} = 0.04 \text{ gauss} \qquad (7.10.23)$$

In the case where the protons moving at a speed of 0.1c are released perpendicular to the field lines the radius of the spirals, sometimes called the *gyration radius*, is

$$\rho_0 = \frac{mcu}{qB_0} = \frac{0.1mc^2}{qB_0} = 7.8 \times 10^6 \text{ cm} = 78 \text{ km}. \qquad (7.10.24)$$

Thus ρ_0 is very small compared to the elevation of the particles above the earth.

By using (7.10.21), the so-called *forbidden regions* for the particle motion can be studied. These are the regions of space which a particle with given initial conditions never reaches. The boundary separating the allowed from the forbidden regions results if the equality is used in (7.10.21):

$$\frac{\sin \theta}{r^2} - \frac{1}{rr_0 \sin \theta} = \pm \frac{1}{a_0^2} \qquad (7.10.25)$$

Both the choice of the plus sign and the minus sign gives a quadratic equation for r. One of these has one positive root, the other has two. These are

$$\frac{r}{a_0} = \frac{-1 + \sqrt{1 + 4(r_0/a_0)^2 \sin^3 \theta}}{2(r_0/a_0) \sin \theta} \qquad (7.10.26)$$

$$\frac{r}{a_0} = \frac{1 \pm \sqrt{1 - 4(r_0/a_0)^2 \sin^3 \theta}}{2(r_0/a_0) \sin \theta} \qquad (7.10.27)$$

Figure 7.10.2 shows a sketch of the three curves given by (7.10.26) and (7.10.27) for the case where $r_0/a_0 < \frac{1}{2}$. The dashed curve represents the field line $r = r_0 \sin^2 \theta$. The shaded regions are the forbidden regions. Since the particles are actually released at r_0, trapping of the particles occurs. Other cases of interest are $r_0/a_0 = \frac{1}{2}$ and $r_0/a_0 > \frac{1}{2}$; the reader can sketch the qualitative form of (7.10.26) and (7.10.27) for these cases.

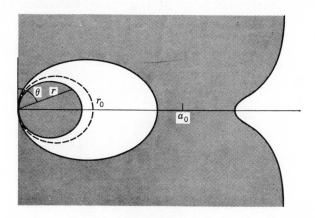

Fig. 7.10.2

A given particle does not reach all parts of the allowed region. For instance, a particle which is released in a direction perpendicular to the field lines, i.e., with its initial velocity vector in the equatorial plane, will remain in the equatorial plane. The study of the motion for this special case is left as a problem. If a particle is released with a component of velocity both parallel and perpendicular to the field line at r_0, its point of reflection, i.e., the greatest distance along the field line measured from the equatorial plane that it reaches, can be estimated. The dipole field strength at the equator and at the point of reflection is

$$B_0 = \frac{\mu}{r_0^{\,3}} \qquad B = \sqrt{B_r^{\,2} + B_\theta^{\,2}} = \frac{\mu}{r^3}\sqrt{1 + 3\cos^2\theta} \qquad (7.10.28)$$

Since the particle follows the field lines we may write $r = r_0 \sin^2 \theta$. Hence,

$$\frac{B}{B_0} = \frac{r_0^{\,3}}{r^3}\sqrt{1 + 3\cos^2\theta} = \frac{\sqrt{1 + 3\cos^2\theta}}{\sin^6\theta} \qquad (7.10.29)$$

By means of (7.8.47),

$$\frac{u_\parallel^{\,2}}{u_\perp^{\,2}} = \frac{\sqrt{1 + 3\cos^2\theta}}{\sin^6\theta} - 1 \qquad (7.10.30)$$

7.11. AXIALLY SYMMETRIC FIELDS: ELECTRON AND ION OPTICS

An important class of static electromagnetic fields are those which have symmetry about some axis. In cylindrical coordinates r, θ, z the fields have no components in the θ-direction and the remaining components have no functional dependence on θ. Simple magnetostatic and electrostatic lenses used in the focusing of beams of charged particles, e.g., in an electron microscope or in an ordinary television picture tube, are this type.

Let us try to find a vector potential with $A_r = A = 0$. Since $B_\theta = 0$, $\mathbf{B} = \nabla \times \mathbf{A}$ reduces to

$$B_r = -\frac{\partial A_\theta}{\partial z} \qquad B_z = \frac{1}{r}\frac{\partial}{\partial r}(rA_\theta) \tag{7.11.1}$$

Integration of the equation for B_z results in

$$A_\theta = \frac{1}{r}\int_0^r rB_z\,dr \tag{7.11.2}$$

It must be true that

$$B_r = -\frac{\partial A_\theta}{\partial z} = -\frac{1}{r}\int_0^r r\frac{\partial B_z}{\partial z}\,dr \tag{7.11.3}$$

The field equation $\nabla \cdot \mathbf{B} = 0$ reads

$$\frac{1}{r}\frac{\partial}{\partial r}(rB_r) + \frac{\partial B_z}{\partial z} = 0 \tag{7.11.4}$$

Hence,

$$\frac{1}{r}\int_0^r r\frac{\partial B_z}{\partial z}\,dr = -\frac{1}{r}\int_0^r \frac{\partial}{\partial r}(rB_r)\,dr = -B_r \tag{7.11.5}$$

and, therefore, (7.11.3) is satisfied. The Lagrangian for a charged particle moving in this field is

$$L = \tfrac{1}{2}m(\dot{r}^2 + r^2\dot{\theta}^2 + \dot{z}^2) - q\psi + \frac{q}{c}\dot{\theta}\int_0^r rB_z\,dr \tag{7.11.6}$$

where $\psi(r, z)$ is the scalar potential from which the electric field is to be obtained by means of $\mathbf{E} = -\nabla\psi$. Since θ is a cyclic coordinate we have the important result that

$$\frac{\partial L}{\partial\dot{\theta}} = l = mr^2\dot{\theta} + \frac{q}{c}\int_0^r rB_z\,dr \tag{7.11.7}$$

is a constant of the motion. Experts on the subject of electron optics refer to this result as *Busch's theorem*. Notice that

$$\int_0^r B_z \cdot 2\pi r\,dr = \Phi \tag{7.11.8}$$

is the flux of the magnetic field which passes through a circle of radius r about the z-axis. Busch's theorem may then be written,

$$l = mr^2\dot{\theta} + \frac{q}{2\pi c}\,\Phi \tag{7.11.9}$$

The remaining equations of motion are obtained from the Lagrangian in the usual way:

$$m\ddot{z} + q\frac{\partial\psi}{\partial z} - \frac{q}{c}\dot{\theta}\int_0^r r\frac{\partial B_z}{\partial z}\,dr = 0 \tag{7.11.10}$$

$$m\ddot{r} - mr\dot{\theta}^2 + q\frac{\partial\psi}{\partial r} - \frac{q}{c}\dot{\theta}\frac{\partial}{\partial r}\int_0^r (rB_z)\,dr = 0 \tag{7.11.11}$$

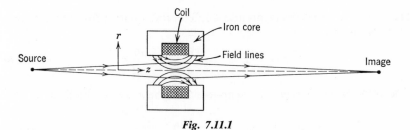

Fig. 7.11.1

By means of $E_r = -\partial\psi/\partial r$, $E_z = -\partial\psi/\partial z$, and (7.11.3),

$$m\ddot{z} = qE_z - \frac{q}{c}r\dot{\theta}B_r \tag{7.11.12}$$

$$m\ddot{r} - mr\dot{\theta}^2 = qE_r + \frac{q}{c}r\dot{\theta}B_z \tag{7.11.13}$$

There is one more constant of the motion, namely, the total energy

$$\mathscr{E} = \tfrac{1}{2}m(\dot{r}^2 + r^2\dot{\theta}^2 + \dot{z}^2) + q\psi \tag{7.11.14}$$

Equations (7.11.7), (7.11.12), (7.11.13), and (7.11.14) are a fundamental set of equations of motion on which the study of axially symmetric electric and magnetic focusing devices depends.

Figure 7.11.1 shows a simple magnetic lens. In application, the particles move close to the axis and the magnetic field exists only over a portion of the path. The actual deflection of the particles is small. The situation is similar to the geometrical optics of thin lenses. Since the particles never deviate very far from the axis, the dependence of B_z on r can be neglected as a first approximation. Busch's theorem gives approximately

$$l = mr^2\dot{\theta} + \frac{q}{c}\tfrac{1}{2}r^2B_z \tag{7.11.15}$$

At the source (Fig. 7.11.1) the particles have no component of velocity in the θ direction. Moreover, the magnetic field is zero here. Thus $l = 0$ and (7.11.15) gives for points interior to the lens

$$\dot{\theta} = - \frac{qB_z}{2mc} \tag{7.11.16}$$

Note that this is just the Larmor frequency and is the angular rate at which the particles precess about the z-axis as they pass through the lens. Since the electric field is absent, the substitution of (7.11.16) into (7.11.13) yields

$$\ddot{r} + \left(\frac{qB_z}{2mc}\right)^2 r = 0 \tag{7.11.17}$$

It is convenient to convert this into a differential equation for r as a function of z. Note that

$$\dot{r} = \frac{dr}{dz} \dot{z} \qquad \ddot{r} = \frac{d^2r}{dz^2} \dot{z}^2 + \frac{dr}{dz} \ddot{z} \tag{7.11.18}$$

The velocity has barely any components in the r and θ directions so that

$$\dot{z}^2 = u^2 = \text{const.} \qquad \ddot{z} = 0 \qquad \ddot{r} = \frac{d^2r}{dz^2} u^2 \tag{7.11.19}$$

approximately. Thus

$$\frac{d^2r}{dz^2} + \left(\frac{qB_z}{2mcu}\right)^2 r = 0 \tag{7.11.20}$$

This is the *paraxial ray* equation for a magnetic lens and is valid as long as the particles move near the axis of the lens. An important conclusion to be drawn from this result is that $d^2r/dz^2 < 0$ so that the trajectories are curved toward the axis. The lens is thus *positive* or *converging* independent of the sign of the charge of the particles.

Suppose that the particles originally move parallel to the axis. Let z_1 and z_2 be points just to the left and right of the lens. From (7.11.20),

$$\left(\frac{dr}{dz}\right)_2 - \left(\frac{dr}{dz}\right)_1 = -r \int_{z_1}^{z_2} \left(\frac{qB_z}{2mcu}\right)^2 dz \tag{7.11.21}$$

But $(dr/dz)_1 = 0$ and we note from the geometry of Fig. 7.11.2 that

$$\left(\frac{dr}{dz}\right)_2 = - \frac{r}{f} \tag{7.11.22}$$

where f is the *focal length* of the lens. It is assumed that r does not change much over the range $z_1 < z < z_2$, i.e., we make the *thin lens* approximation.

Thus

$$\frac{1}{f} = \int_{z_1}^{z_2} \left(\frac{qB_z}{2mcu} \right)^2 dz \tag{7.11.23}$$

Once the focusing property has been established and the focal length determined, the procedure for the location of images is exactly the same as in ordinary optics. There is one further complication. As the particles pass through the lense, they precess about the z-axis with the Larmor frequency (7.11.16). Any image formed by the lens, therefore, will be

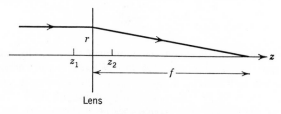

Fig. 7.11.2

rotated (in addition to being inverted as in geometrical optics) with respect to the source.

A very readable account of electric and magnetic lenses is to be found in Pierce, Reference 1 at the end of this chapter.

PROBLEMS

7.1.1 An electron moves at right angles to a uniform magnetic field of strength 100 gauss. For what value of the radius of the orbit does the speed equal $0.1c$?

7.1.2 Show that if $x_i(t)$ is a solution of

$$m\mathbf{a} = q\mathbf{E} + (q/c)(\mathbf{u} \times \mathbf{B})$$

and the fields are scaled up according to

$$\mathbf{E}' = b^2\mathbf{E} \qquad \mathbf{B}' = b\mathbf{E}$$

then $x_i(bt)$ is the new solution. The particles follow the same trajectories but at different rates.

7.2.3 A charged particle moves in the x_1x_2-plane in a region where there is a uniform electric field in the x_1-direction and a uniform magnetic field in the x_3-direction. Particles enter with a speed u_0 and a direction θ as indicated. Find θ in terms of u_0, E, and B such that the motion has cusps. For what initial conditions does the particle move in a straight line in the negative x_2-direction?

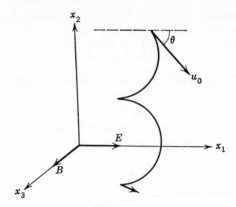

7.3.4 One type of mass spectrometer utilizes the focusing properties of a uniform magnetic field. Suppose that ions of mass m and charge q enter the field with an angular spread of directions α. After traveling through a half circle, the ions are brought to a focus on a photographic plate. Find the width of the spectral line produced. Assume that all the ions have the same speed.

7.4.5 The two cylindrical electrodes of a radii a and b have a potential difference $\psi_a - \psi_b > 0$. Electrons are emitted from the inner electrode in the plane of the paper at a velocity u_0 and a direction α as illustrated. Find a formula

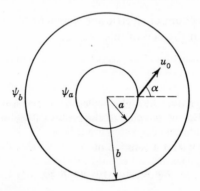

for the minimum kinetic energy that an electron must have in order to just reach the outer electrode in terms of a, b, $\psi_a - \psi_b$ and α. How should the electrodes be designed in order to minimize the dependence on α?

7.4.6 A charged particle moving in a circular orbit in a radial electric field is perturbed by the addition of a very small amount of energy $\Delta\mathscr{E}$. Show that

$$\Delta\mathscr{E} = \frac{l^2}{m}\,b_1{}^2$$

approximately. Assume that the perturbation does not alter the angular momentum.

7.5.7   Find the effective potential for a charged particle moving in a cylindrical magnetron. Show that it always has a single minimum and approaches $+\infty$ both as $r \to 0$ and $r \to \infty$ regardless of whether the electric field is attractive or repulsive. Find the equilibrium value of r.

7.5.8   Show that the velocity that an ion must have when released at the inner electrode of a magnetron in order to just reach the outer electrode before turning back is given by

$$u_0^2 \left(1 - \frac{a^2}{b^2} \sin^2 \alpha \right) + u_0 a\omega \sin \alpha \left(1 - \frac{a^2}{b^2} \right) - \frac{\omega^2 b^2}{4} \left(1 - \frac{a^2}{b^2} \right)^2 + \frac{2q}{m}(\psi_a - \psi_b) = 0$$

where ω is the cyclotron frequency. (Refer to the figure of Problem 7.4.5.) If the inner electrode is a fine wire, $a \to 0$ and

$$u_0^2 \cong \tfrac{1}{4}\omega^2 b^2 - \frac{2q}{m}(\psi_a - \psi_b)$$

7.5.9   For a charged particle moving in a magnetron, find the second order differential equation for r as a function of t. Show that the approximate differential equation is

$$\ddot{r} + \left[\omega_c^2 + \frac{2C}{mr_1^2} \right] (r - r_1) - \left[\frac{3\omega_c^2}{2r_1} + \frac{5C}{mr_1^3} \right] (r - r_1)^2$$
$$+ \left[\frac{5\omega_c^2}{2r_1^2} + \frac{9C}{mr_1^4} \right] (r - r_1)^3 = 0$$

7.5.10   An electron moves in a magnetron in which there is only a radial electric field. The radii of the electrodes are $a = 5$ cm and $b = 15$ cm. The potential difference between electrodes is 1 stat volt. Assuming the orbits to be circular, quantize the motion according to the Bohr-Sommerfield rule. If the radius of the orbit is 10 cm, what is the value of the quantum number? What is the fractional change in the radius of the orbit if the quantum number changes by one unit? Are quantum effects important here?

7.6.11   Prove that if L is a Lagrangian for a particle, then $L' = L + d\phi/dt$ where $\phi = \phi(q_1, q_2, q_3, t)$ is also a Lagrangian which gives the same equations of motion. Do this by proving directly that $d\phi/dt$ satisfies Lagrange's equations identically, i.e.,

$$\frac{d}{dt} \frac{\partial}{\partial \dot{q}_k} \left(\frac{d\phi}{dt} \right) - \frac{\partial}{\partial q_k} \left(\frac{d\phi}{dt} \right) \equiv 0$$

7.8.12   Derive Eqs. (7.8.20), (7.8.21), and (7.8.22) by the Hamilton method.

7.9.13   Prove that no magnetic field with curvature only and no gradient can exist in source-free space. Do this by showing that in cylindrical coordinates no magnetic field with components $(0, B_\theta, 0)$, where B_θ is strictly constant, exists.

7.10.14   Qualitatively, what do the boundaries between forbidden and allowed regions as given by (7.10.27) and (7.10.26) look like for $r_0/a_0 = \tfrac{1}{2}$ and $r_0/a_0 > \tfrac{1}{2}$?

7.10.15 Show that the radius of curvature of a dipole field line at the point where it crosses the equatorial plane ($\theta = \pi/2$, $r = r_1$) is $\rho = \frac{1}{3}r_1$.

7.10.16 Suppose that a particle moves in the equatorial plane of the earth's dipole field ($\theta = \pi/2$). Find the equations of motion and the effective potential. Show that the effective potential has a maximum and a minimum, and sketch the form of the function. Show that the minimum corresponds to a particle at rest in stable equilibrium and that the maximum corresponds to a particle moving clear around the earth in an *unstable* circular orbit.

7.10.17 For a charged particle moving in the equatorial plane of the earth's dipole field, show that the approximate differential equation for $r(t)$ is

$$\ddot{r} + \omega_c^2 \left[(r - r_1) - \frac{6}{r_1}(r - r_1)^2 + \frac{20}{r_1^2}(r - r_1)^3 \right] = 0$$

where r_1 is the position of stable equilibrium and ω_c is the cyclotron frequency at $r = r_1$. Compute the Fourier series solution up to the second harmonic. Compute the drift velocity of the particles in the ϕ direction and compare it with Eq. (7.9.31).

7.11.18 From the Biot-Savart law

$$\mathbf{B} = \frac{1}{c} \oint \frac{J \, d\mathbf{s} \times \mathbf{r}}{r^3}$$

compute the magnetic field on the axis of a current loop of radius a. Calculate the focal length of the current loop if it is used as a "thin" magnetic lens. *Hint:* The assumption that the lens is thin means that it acts only over a small portion of the path of the particle. Since the field falls off rapidly with distance away from the loop, the limits in Eq. (7.11.23) can be extended to $\pm\infty$ without appreciable error.

7.11.19 In treating the motion of charged particles near the axis of an axially symmetric *electrostatic* lens, it is necessary to know the field only at points slightly removed from the axis. Show that if $\phi(z)$ represents the electrostatic potential *on* the axis, then the potential off the axis is approximately

$$\psi(r, z) = \phi(z) - \tfrac{1}{4}\phi''(z)r^2$$

where the primes indicate derivatives. *Hint:* Write

$$\psi(r,z) = \phi(z) + \phi_1(z)r + \phi_2(z)r^2 + \cdots$$

and use the fact that ψ obeys Laplace's equation to evaluate $\phi_1, \phi_2, \ldots$.

7.11.20 If the total energy of a particle is chosen to be zero so that $\frac{1}{2}mu^2 = -q\psi$, show that the paraxial ray equation for a purely electrostatic lens is

$$r'' + r'\frac{\phi'}{2\phi} + r\frac{\phi''}{4\phi} = 0$$

(See Problem 7.11.19 for the meaning of ϕ.)

REFERENCES

1. Pierce, J. R., *Theory And Design Of Electron Beams*. Van Nostrand, New York, 1954.
2. Alfven, H., *Cosmical Electrodynamics*, Oxford University Press, London, 1963.
3. Zworykin, V. K., et al., *Electron Optics And The Electron Microscope*, John Wiley and Sons, New York, 1945.
4. Lehnert, Bo., *Dynamics Of Charged Particles*, John Wiley and Sons (Interscience), New York, 1964.
5. Vallarta, M. S., *Handbuck der Physik*, S. Flugge Ed., **46**, Springer-Verlag, Berlin, 1961.

8

Many-Particle Systems

Previous chapters deal with the motion of a single particle in a prescribed externally generated force field. In applications, such as the design of apparatus to contain charged particles, the interactions of the particles with one another is of great importance. In this chapter, we begin the study of many-particle systems.

8.1. GALILEAN TRANSFORMATIONS

In Chapters 1 and 2, the transformations from one *stationary* coordinate system to another were extensively studied. These transformations had no time dependence. Considered here will be the simplest example of a time-dependent transformation.

Pictured in Fig. 8.1.1 are two rectangular Cartesian coordinate systems. Suppose that a particle has the position vector **r** relative to the origin O of frame F and position vector **r**′ relative to the origin O' of F'. Let **R** be the vector from O to O'. The three vectors are related by

$$\mathbf{r} = \mathbf{r}' + \mathbf{R} \qquad (8.1.1)$$

Each of the three vectors **r**, **r**′, and **R** has a set of three components in each of the coordinate systems. The components of a given vector in the two coordinate systems are related to one another by an orthogonal transformation the components of which are denoted S_{ij}. In order to keep track of the components of the three displacement vectors in the two

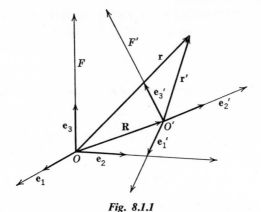

Fig. 8.1.1

frames of reference, we construct the following table

	r	**r′**	**R**
F	x_i	$S_{ji}x_j'$	X_i
F'	$S_{ij}x_j$	x_i'	$X_i' = S_{ij}X_j$

Equation (8.1.1) can be written in component form in either frame F or F':

$$x_i = S_{ji}x_j' + X_i \tag{8.1.2}$$

$$S_{ij}x_j = x_i' + X_i' \tag{8.1.3}$$

The unit coordinate vectors of F and F' are related by

$$\hat{e}_i' = S_{ij}\hat{e}_j \tag{8.1.4}$$

$$\hat{e}_j = S_{ij}\hat{e}_i' \tag{8.1.5}$$

Thus (8.1.1) can also be expressed in the form

$$x_i\hat{e}_i = x_i'\hat{e}_i' + X_i\hat{e}_i \tag{8.1.6}$$

In the most general circumstance, the coordinate systems are moving relative to one another. The motion will involve both rotation and translation of F' relative to F so that the components of the orthogonal transformation as well as the vector **R** are functions of the time. If the transformation involves uniform translation only, without acceleration and without rotation, it is called *Galilean*. The remainder of this section is devoted to the study of Galilean transformations. It is useful to think of F as being a fixed frame and of F' as moving relative to it. The fixed frame is usually called the laboratory frame of reference.

Suppose that observers in F and F' each measure the velocity of a particle. The results of their measurements can be related by differentiating (8.1.2):

$$u_i = S_{ji}u_j' + v_i \qquad (8.1.7)$$

where

$$v_i = \frac{dX_i}{dt} \qquad (8.1.8)$$

are the components in frame F of the velocity of $0'$ relative to 0 and S_{ij} is regarded as constant. In vector form, (8.1.7) can be stated as

$$\mathbf{u} = \mathbf{u}' + \mathbf{v} \qquad (8.1.9)$$

which is known as the Galilean addition theorem for velocities. To the observers stationed in F and F', the trajectory of the particle might appear

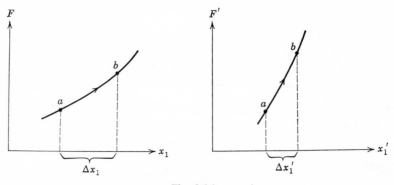

Fig. 8.1.2

as depicted in Fig. 8.1.2. Let a and b refer to two events which occur on the trajectory of the particle. For concreteness, suppose that the particle flashes a light at these two points. If our observers in F and F' wish to calculate the velocity of the particle, they each observe the *time interval* and the *distance interval* between a and b *in their own frame of reference*. The two observers then compute velocity components by means of

$$u_i = \frac{\Delta x_i}{\Delta t} \qquad u_i' = \frac{\Delta x_i'}{\Delta t'} \qquad (8.1.10)$$

It is trivial in this case that $\Delta t = \Delta t'$. It is important to emphasize that both observers use the same two events (flashes of light) to perform the calculation indicated in (8.1.10). The results of these measurements are related by means of (8.1.7).

For a Galilean transformation, v_i is to be regarded as a constant. The transformation law of particle acceleration is obtained by differentiation of (8.1.7):

$$a_i = S_{ji}a_j' \tag{8.1.11}$$

which is precisely the same as for stationary coordinate systems. This means that Newton's second law retains its covariance under Galilean

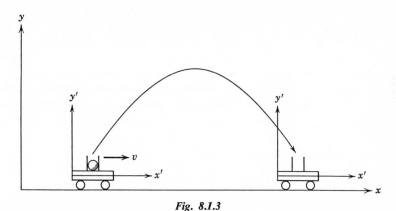

Fig. 8.1.3

transformations. The equations of motion of a particle can be expressed in component form in either frame as

$$F_i = ma_i \qquad F_i' = ma_i' \tag{8.1.12}$$

where

$$F_i' = S_{ij}F_j \tag{8.1.13}$$

If the coordinate axis of F' are parallel to those of F, then $S_{ij} = \delta_{ij}$ and the transformation equations of position velocity and acceleration reduce to

$$x_i = x_i' + X_i \tag{8.1.14}$$

$$u_i = u_i' + v_i \tag{8.1.15}$$

$$a_i = a_i' \tag{8.1.16}$$

These equations contain all of the significant features of a Galilean transformation and will be adequate for the applications of Galilean transformation theory that we shall want to make.

A simple example illustrating the covariance of Newtonian mechanics is the following demonstration experiment frequently done in elementary courses. A spring gun is mounted vertically on a cart which moves without friction on a level surface as illustrated in Fig. 8.1.3. As the car moves along, the gun fires a ball which eventually drops back into the gun. Let a Galilean frame of reference be rigidly attached to the car. Newton's second law is valid both in the moving frame of reference and the frame which is stationary in the laboratory. In each, there is a uniform gravitational field in the negative y-direction. According to an observer in the moving frame, the ball is fired straight up and so naturally it drops back into the barrel of the gun. According to an observer in the laboratory frame, the initial velocity of the ball in the x-direction is identical to the velocity of the car, and since the x-component of the velocity of both car and ball remains constant, the ball remains right over the gun and eventually drops back into it. The point of view is different in the two frames and different initial conditions are used, but the *basic principle*, Newton's second law, is the same for both observers and they both predict the same outcome for the experiment. This is the essence of a covariant theory.

The subject matter of this section is sometimes referred to as *Newtonian relativity*.

8.2. SYSTEMS OF INTERACTING PARTICLES

This section discusses some general theorems which pertain to many particle systems. As examples, the hydrogen atom is a system consisting of two interacting particles, namely, an electron and a proton; a helium atom is composed of an α-particle and two electrons; the earth and the moon can, as a first approximation, be treated as a two-particle system although at first sight it seems highly inappropriate to classify the earth as a particle! We treat the earth as a particle here because we are interested only in its motion as a whole, not in the relative motion of its individual parts.

The total linear momentum of a system of particles at any instant is the vector sum of the momenta of the individual particles:

$$\mathbf{p} = \sum_{\alpha=1}^{N} \mathbf{p}_\alpha = \sum_{\alpha=1}^{N} m_\alpha \mathbf{u}_\alpha \tag{8.2.1}$$

where N is the total number of particles present. The masses m_α of the individual particles are assumed to be strictly constant. The total force on a system of particles is the vector sum of the forces on the individual particles:

$$\mathbf{F} = \sum_{\alpha=1}^{N} \mathbf{F}_\alpha \tag{8.2.2}$$

where $\mathbf{F}_\alpha$ is the *net force* on particle α and includes both the forces which other particles in the group exert (internal forces) and forces which are exerted by external agents (external forces). By Newton's second law,

$\mathbf{F}_\alpha = \dot{\mathbf{p}}_\alpha$ so that the total force is the time rate of change of the total momentum:

$$\mathbf{F} = \sum_{\alpha=1}^{N} \dot{\mathbf{p}}_\alpha = \frac{d}{dt} \sum_{\alpha=1}^{N} \mathbf{p}_\alpha = \dot{\mathbf{p}} \qquad (8.2.3)$$

Let O be any fixed point in space and let $\mathbf{r}_\alpha$ be the position vector of particle α with respect to this point (Fig. 8.2.1). The total angular momentum of the system about this point is the vector sum of the angular momenta of the individual particles:

$$\mathbf{l} = \sum_{\alpha=1}^{N} \mathbf{l}_\alpha = \sum_{\alpha=1}^{N} \mathbf{r}_\alpha \times \mathbf{p}_\alpha \qquad (8.2.4)$$

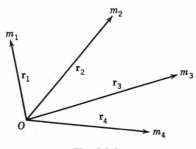

Fig. 8.2.1

The total torque on the system of particles is the vector sum of the torques on the individual particles, including torques produced by both internal and external forces. It is easily shown that the total torque on the system is the time rate of change of the total angular momentum:

$$\mathbf{N} = \sum_{\alpha=1}^{N} \mathbf{N}_\alpha = \sum_{\alpha=1}^{N} \mathbf{r}_\alpha \times \mathbf{F}_\alpha = \frac{d}{dt} \sum_{\alpha=1}^{N} \mathbf{r}_\alpha \times \dot{\mathbf{p}}_\alpha = \dot{\mathbf{l}} \qquad (8.2.5)$$

The net force on a given particle can always be divided into an internal force and an external force:

$$\mathbf{F}_\alpha = \mathbf{F}_\alpha^{in} + \mathbf{F}_\alpha^{ex} \qquad (8.2.6)$$

The net torque on a given particle can be similarly divided:

$$\mathbf{N}_\alpha = \mathbf{r}_\alpha \times \mathbf{F}_\alpha = \mathbf{r}_\alpha \times (\mathbf{F}_\alpha^{in} + \mathbf{F}_\alpha^{ex}) = \mathbf{N}_\alpha^{in} + \mathbf{N}_\alpha^{ex} \qquad (8.2.7)$$

For some simple types of forces it is possible to prove directly that

$$\sum_{\alpha=1}^{N} \mathbf{F}_\alpha^{in} = 0 \qquad \sum_{\alpha=1}^{N} \mathbf{N}_\alpha^{in} = 0 \qquad (8.2.8)$$

meaning that a system is not capable of exerting a net force or torque on itself. Suppose, for example, that the forces which the particles in a group exert on one another are directed along the lines joining their centers such as would be the case if the particles were charged and the electrostatic forces between the particles were being considered. For simplicity, suppose the system consists of three particles as in Fig. 8.2.2. According to Newton's third law, the force which particle 1 exerts on particle 2 is equal

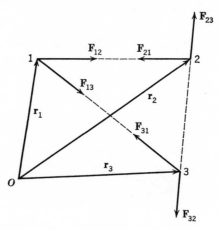

Fig. 8.2.2

in magnitude but opposite in direction to the force that particle 2 exerts on particle 1:

$$\mathbf{F}_{12} = -\mathbf{F}_{21} \qquad (8.2.9)$$

Similarly,

$$\mathbf{F}_{13} = -\mathbf{F}_{31} \qquad \mathbf{F}_{23} = -\mathbf{F}_{32} \qquad (8.2.10)$$

Therefore,

$$\sum_{\alpha=1}^{3} \mathbf{F}_{\alpha}^{in} = \mathbf{F}_{1}^{in} + \mathbf{F}_{2}^{in} + \mathbf{F}_{3}^{in}$$
$$= \mathbf{F}_{12} + \mathbf{F}_{13} + \mathbf{F}_{21} + \mathbf{F}_{23} + \mathbf{F}_{31} + \mathbf{F}_{32} = 0 \qquad (8.2.11)$$

The net internal torque can be written

$$\mathbf{N}^{in} = \sum_{\alpha=1}^{3} \mathbf{r}_{\alpha} \times \mathbf{F}_{\alpha}^{in}$$
$$= \mathbf{r}_1 \times (\mathbf{F}_{12} + \mathbf{F}_{13}) + \mathbf{r}_2 \times (\mathbf{F}_{21} + \mathbf{F}_{23}) + \mathbf{r}_3 \times (\mathbf{F}_{31} + \mathbf{F}_{32})$$
$$= \mathbf{r}_1 \times (\mathbf{F}_{12} - \mathbf{F}_{31}) + \mathbf{r}_2 \times (-\mathbf{F}_{12} + \mathbf{F}_{23}) + \mathbf{r}_3 \times (\mathbf{F}_{31} - \mathbf{F}_{23}) \qquad (8.2.12)$$
$$= \mathbf{F}_{31} \times (\mathbf{r}_1 - \mathbf{r}_3) + \mathbf{F}_{12} \times (\mathbf{r}_2 - \mathbf{r}_1) + \mathbf{F}_{23} \times (\mathbf{r}_3 - \mathbf{r}_2)$$

But $\mathbf{r}_1 - \mathbf{r}_3$ is parallel to $\mathbf{F}_{31}$, $\mathbf{r}_2 - \mathbf{r}_1$ is parallel to $\mathbf{F}_{12}$, and $\mathbf{r}_3 - \mathbf{r}_2$ is parallel to $\mathbf{F}_{23}$. Hence,

$$\mathbf{N}_\alpha^{\text{in}} = 0 \tag{8.2.13}$$

A somewhat more complicated situation is provided by two interacting current loops. The currents J_1 and J_2 are assumed to be constant and the current loops are fixed in space. According to Ampere's law, the force on the current element $J_1 \, ds_1$ due to the magnetic field of current element $J_2 \, ds_2$ (Fig. 8.2.3) is

$$d\mathbf{F}_1 = \frac{J_1 J_2}{c^2} \frac{ds_1 \times (ds_2 \times \mathbf{r})}{r^3} \tag{8.2.14}$$

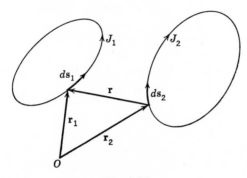

Fig. 8.2.3

The force on element $J_2 \, ds_2$ due to the magnetic field of element $J_1 \, ds_1$ is

$$d\mathbf{F}_2 = \frac{J_1 J_2}{c^2} \frac{ds_2 \times (ds_1 \times [-\mathbf{r}])}{r^3} \tag{8.2.15}$$

The double cross products can be expanded by means of (1.10.7):

$$d\mathbf{F}_1 = \frac{J_1 J_2}{r^3 c^2} [ds_2(ds_1 \cdot \mathbf{r}) - \mathbf{r}(ds_1 \cdot ds_2)] \tag{8.2.16}$$

$$d\mathbf{F}_2 = \frac{J_1 J_2}{r^3 c^2} [-ds_1(ds_2 \cdot \mathbf{r}) + \mathbf{r}(ds_1 \cdot ds_2)] \tag{8.2.17}$$

Hence $d\mathbf{F}_1 \neq -d\mathbf{F}_2$. The magnetostatic force between two current elements considered as particles does not obey Newton's third law!

To compute the net force which $J_2 \, ds_2$ exerts on the first current loop, it is necessary to integrate (8.2.16) around loop number one:

$$d\mathbf{F}_1(\text{net}) = \frac{J_1 J_2}{c^2} \left[ds_2 \oint_1 \frac{ds_1 \cdot \mathbf{r}}{r^3} - \oint_1 \frac{\mathbf{r}(ds_1 \cdot ds_2)}{r^3} \right] \tag{8.2.18}$$

Note that

$$\oint_1 \frac{ds_1 \cdot \mathbf{r}}{r^3} = -\oint_1 \nabla_1\left(\frac{1}{r}\right) \cdot ds_1 = -\oint_1 d\left(\frac{1}{r}\right) = 0 \qquad (8.2.19)$$

where ∇_1 is the gradient with respect to the coordinates of circuit element $J_1\, ds_1$. The integral (8.2.19) is zero because the integral of an exact differential around a closed path is zero. Hence,

$$d\mathbf{F}_1(\text{net}) = -\frac{J_1 J_2}{c^2} \oint_1 \frac{\mathbf{r}(ds_1 \cdot ds_2)}{r^3} \qquad (8.2.20)$$

This represents the force on the first loop due to the element $J_2\, ds_2$ of the second loop. In order to find the total force exerted on the first loop, it is necessary to integrate (8.2.20) around the second loop:

$$\mathbf{F}_1 = -\frac{J_1 J_2}{c^2} \oint_2 \oint_1 \frac{\mathbf{r}(ds_1 \cdot ds_2)}{r^3} \qquad (8.2.21)$$

Similarly, it is found that the force exerted on the second loop by the first is

$$\mathbf{F}_2 = \frac{J_1 J_2}{c^2} \oint_1 \oint_2 \frac{\mathbf{r}(ds_1 \cdot ds_2)}{r^3} \qquad (8.2.22)$$

Thus $\mathbf{F}_1 + \mathbf{F}_2 = 0$ and the *net* force acting between the two current loops obeys Newton's third law. Actually, we have not proved that the sum of the internal forces is zero. To do this properly, *all* forces must be considered. This would include, for example, the force which is exerted on a current element by the magnetic field produced by the other parts of that same circuit. The fact that the current loops are stationary in space presupposes the existence of certain mechanical forces to support them, which should be included in any proof that the sum of all internal forces vanishes. The proof of (8.2.13) for the current loops is much harder; we do not attempt to prove this here.

A still more complex situation occurs when moving charged particles interact. Part of the momentum of the system is actually in the electromagnetic field. Accelerated charges lose momentum through radiation, and this effect must be taken into account.

An *isolated system* shall be defined to be a system on which there are no external forces of any kind. It seems preferable to adopt the principle of conservation of linear and angular momentum as fundamental hypotheses rather than Newton's third law. Consequently, the following hypothesis is assumed: *The total linear and the total angular momentum of an isolated system are constants.* As a consequence, it follows from (8.2.3) and (8.2.5) that

$$\sum_{\alpha=1}^{N} \mathbf{F}_\alpha^{\text{in}} = 0 \qquad \sum_{\alpha=1}^{N} \mathbf{N}_\alpha^{\text{in}} = 0 \qquad (8.2.23)$$

for an isolated system. For continuous systems, the sums must be replaced by integrals. Having adopted the conservation of momentum as fundamental, it becomes unnecessary to re-establish (8.2.23) for each kind of force and each special situation.

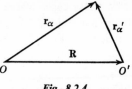

Fig. 8.2.4

If the net force on a system is zero and if the net torque about any one point is known to be zero, then the net torque about any other point is necessarily zero. Suppose it has been established that about point O in Fig. 8.2.4

$$\sum_{\alpha=1}^{N} \mathbf{r}_\alpha \times \mathbf{F}_\alpha = 0 \qquad (8.2.24)$$

If O' is any other fixed point, then

$$\mathbf{r}_\alpha = \mathbf{R} + \mathbf{r}_\alpha'$$

and

$$0 = \sum_{\alpha=1}^{N} (\mathbf{R} + \mathbf{r}_\alpha') \times \mathbf{F}_\alpha = \mathbf{R} \times \sum_{\alpha=1}^{N} \mathbf{F}_\alpha + \sum_{\alpha=1}^{N} \mathbf{r}_\alpha' \times \mathbf{F}_\alpha$$

But the first term vanishes by virtue of the assumption that the net force on the system is zero. Hence,

$$\sum_{\alpha=1}^{N} \mathbf{r}_\alpha' \times \mathbf{F}_\alpha = 0 \qquad (8.2.25)$$

This theorem is important in the study of the statics of rigid bodies. Necessary and sufficient conditions for a rigid body to be in equilibrium are that the net force and the net torque *about any one point* vanish.

If the configuration of a system of particles undergoes a differential change, the displacements of the individual particles will be ds_α. The work done on the system by all forces, *including both external and internal forces* is

$$dW = \sum_{\alpha=1}^{N} \mathbf{F}_\alpha \cdot d\mathbf{s}_\alpha = \sum_{\alpha=1}^{N} m_\alpha \frac{d\mathbf{u}_\alpha}{dt} \cdot \mathbf{u}_\alpha \, dt$$

$$= \sum_{\alpha=1}^{N} d(\tfrac{1}{2} m_\alpha u_\alpha{}^2) = dT \qquad (8.2.26)$$

where

$$T = \sum_{\alpha=1}^{N} \tfrac{1}{2} m_\alpha u_\alpha{}^2 \qquad (8.2.27)$$

represents the kinetic energy of the system. Thus the work done on a system of particles by the *net force* equals the change in the kinetic energy. The internal forces contribute! The fact that $\sum \mathbf{F}_\alpha^{\text{in}} = 0$ does not imply that $\sum \mathbf{F}_\alpha^{\text{in}} \cdot d\mathbf{s}_\alpha = 0$!

In those cases where the forces are conservative, they may be derived from a potential function that depends on the $3N$ coordinates of the particles making up the system:

$$V = V(x_1, x_2, x_3, x_4, \ldots, x_{3N}) \qquad (8.2.28)$$

where x_1, x_2, and x_3 are the rectangular coordinates of the first particle, x_4, x_5, and x_6 are the rectangular coordinates of the second particle, and so on. The force on a given particle is

$$\mathbf{F}_\alpha = -\boldsymbol{\nabla}_\alpha V \qquad (8.2.29)$$

where $\boldsymbol{\nabla}_\alpha$ is the gradient with respect to the coordinates of the particle under consideration. In terms of the potential function, the increment of work done by the forces is

$$
\begin{aligned}
dW &= -\sum_{\alpha=1}^{N} \boldsymbol{\nabla}_\alpha V \cdot d\mathbf{s}_\alpha \\
&= -\left(\frac{\partial V}{\partial x_1} dx_1 + \frac{\partial V}{\partial x_2} dx_2 + \cdots + \frac{\partial V}{\partial x_{3N}} dx_{3N}\right) = -dV
\end{aligned}
\qquad (8.2.30)
$$

By combining (8.2.26) and (8.2.30)

$$d(T + V) = 0 \qquad (8.2.31)$$

Thus the total energy

$$\mathscr{E} = T + V \qquad (8.2.32)$$

is a constant of the motion.

8.3. CENTER OF MASS

The location of the *center of mass* of a system of particles is defined by means of

$$\mathbf{R}_c = \frac{1}{M} \sum_{\alpha=1}^{N} m_\alpha \mathbf{r}_\alpha \qquad M = \sum_{\alpha=1}^{N} m_\alpha \qquad (8.3.1)$$

where M is the total mass of the system and $\mathbf{r}_\alpha$ are the position vectors of the particles with respect to an arbitrary fixed point. The properties of the center of mass make it a convenient reference point in discussing the dynamics of a system of particles as the following developments will show.

By differentiation of (8.3.1)

$$M\mathbf{v}_c = \sum_{\alpha=1}^{N} m_\alpha \mathbf{u}_\alpha = \mathbf{p} \qquad \mathbf{v}_c = \frac{d\mathbf{R}_c}{dt} \qquad (8.3.2)$$

The total linear momentum of a system of particles is therefore the product of the velocity of the center of mass and the total mass.

If there are forces present, the net force on the system is

$$\mathbf{F} = \frac{d\mathbf{p}}{dt} = M\mathbf{a}_c \qquad \mathbf{a}_c = \frac{d\mathbf{v}_c}{dt} \tag{8.3.3}$$

By the hypothesis of conservation of momentum $\mathbf{a}_c = 0$ if external forces are absent. If external forces are present, the internal forces still cancel and (8.3.3) shows that the center of mass moves like a single particle of mass M.

Let $\mathbf{r}_\alpha'$ be the position vectors of the particles making up the system measured from the center of mass. Then

$$\mathbf{r}_\alpha = \mathbf{R}_c + \mathbf{r}_\alpha' \tag{8.3.4}$$

and (8.3.1) gives

$$\mathbf{R}_c = \frac{1}{M} \sum_{\alpha=1}^{N} (\mathbf{R}_c + \mathbf{r}_\alpha')m_\alpha = \mathbf{R}_c + \frac{1}{M} \sum_{\alpha=1}^{N} m_\alpha \mathbf{r}_\alpha'$$

$$\therefore \quad \sum_{\alpha=1}^{N} m_\alpha \mathbf{r}_\alpha' = 0 \tag{8.3.5}$$

By differentiating (8.3.4) and (8.3.5)

$$\mathbf{u}_\alpha = \mathbf{v}_c + \mathbf{w}_\alpha \tag{8.3.6}$$

$$\sum_{\alpha=1}^{N} m_\alpha \mathbf{w}_\alpha = 0 \tag{8.3.7}$$

where $\mathbf{w}_\alpha = \dot{\mathbf{r}}_\alpha'$ is the velocity of the particle under consideration relative to the center of mass as would be measured by observers in the laboratory frame of reference. The use of $\mathbf{u}_\alpha'$ to represent this velocity is avoided here because later we shall want to introduce a moving coordinate system with its origin at the center of mass. The moving coordinate system will, in general, be rotating and the particle velocity $\mathbf{u}_\alpha'$ as measured by observers in the moving frame will not be the same as $\mathbf{w}_\alpha$. The components of $\mathbf{w}_\alpha$ are found by first expressing the components of $\mathbf{r}_\alpha'$ in the laboratory frame and then differentiating them.

The total angular momentum of the system with respect to a fixed point in the laboratory frame of reference is

$$\begin{aligned}
\mathbf{l} &= \sum_{\alpha=1}^{N} \mathbf{r}_\alpha \times m_\alpha \mathbf{u}_\alpha = \sum_{\alpha=1}^{N} (\mathbf{R}_c + \mathbf{r}_\alpha') \times m_\alpha (\mathbf{v}_c + \mathbf{w}_\alpha) \\
&= (\mathbf{R}_c \times \mathbf{v}_c) \sum_{\alpha=1}^{N} m_\alpha + \mathbf{R}_c \times \sum_{\alpha=1}^{N} m_\alpha \mathbf{w}_\alpha \\
&\quad + \sum_{\alpha=1}^{N} m_\alpha \mathbf{r}_\alpha' \times \mathbf{v}_c + \sum_{\alpha=1}^{N} (\mathbf{r}_\alpha' \times m_\alpha \mathbf{w}_\alpha)
\end{aligned} \tag{8.3.8}$$

By means of (8.3.5) and (8.3.7)

$$l = \mathbf{R}_c \times M\mathbf{v}_c + \sum_{\alpha=1}^{N} (\mathbf{r}_\alpha{}' \times m_\alpha \mathbf{w}_\alpha) \tag{8.3.9}$$

Therefore, it is possible to separate the total angular momentum into a term which is equivalent to the angular momentum of a single particle of mass M located at the center of mass and a term which represents the angular momentum of the system about the center of mass. The kinetic energy of the system can be similarly separated:

$$\begin{aligned} T &= \sum_{\alpha=1}^{N} \tfrac{1}{2} m_\alpha u_\alpha{}^2 = \sum_{\alpha=1}^{N} \tfrac{1}{2} m_\alpha (\mathbf{v}_c + \mathbf{w}_\alpha) \cdot (\mathbf{v}_c + \mathbf{w}_\alpha) \\ &= \sum_{\alpha=1}^{N} \tfrac{1}{2} m_\alpha v_c{}^2 + \mathbf{v}_c \cdot \sum_{\alpha=1}^{N} m_\alpha \mathbf{w}_\alpha + \sum_{\alpha=1}^{N} \tfrac{1}{2} m_\alpha w_\alpha{}^2 \\ &= \tfrac{1}{2} M v_c{}^2 + \sum_{\alpha=1}^{N} \tfrac{1}{2} m_\alpha w_\alpha{}^2 \end{aligned} \tag{8.3.10}$$

Suppose that a system of particles is acted on by a uniform gravitational field. The net force is

$$\mathbf{F} = \sum_{\alpha=1}^{N} m_\alpha \mathbf{g} = M\mathbf{g} \tag{8.3.11}$$

and the net torque about an arbitary fixed point is

$$\begin{aligned} \mathbf{N} &= \sum_{\alpha=1}^{N} \mathbf{r}_\alpha \times m_\alpha \mathbf{g} = \left(\sum_{\alpha=1}^{N} \mathbf{r}_\alpha m_\alpha \right) \times \mathbf{g} \\ &= M\mathbf{R}_c \times \mathbf{g} = \mathbf{R}_c \times M\mathbf{g} \end{aligned} \tag{8.3.12}$$

For example, when treating the statics or dynamics of a rigid body in the uniform gravitational field of the earth, the effect of the gravitational field is entirely equivalent to that of a single force equal to the weight of the body and acting at the center of mass.

8.4. COLLISIONS BETWEEN PARTICLES

In the most general type of collision process, a number of particles with momenta $\mathbf{p}_1, \mathbf{p}_2, \ldots$ come together and interact after which particles with momenta $\mathbf{q}_1, \mathbf{q}_2, \ldots$ emerge as in Fig. 8.4.1. In the most general case, the masses of the individual particles need not be the same after the collision; there may even be a different number of particles. The total mass remains the same, however. If there are no forces present other than the forces of interaction between the particles, the system is isolated and the total linear momentum remains constant:

$$\mathbf{p}_1 + \mathbf{p}_2 + \cdots = \mathbf{q}_1 + \mathbf{q}_2 + \cdots \tag{8.4.1}$$

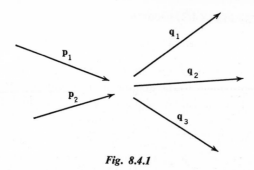

Fig. 8.4.1

The nature of the interaction is not important as far as momentum conservation is concerned. If charged particles are involved and large accelerations occur when the particles interact, momentum loss through electromagnetic radiation may become important requiring that a term that takes this into account be included in (8.4.1).

For a two-particle collision in which there is no exchange of mass so that the same two particles emerge after the collision, the momentum conservation principle can be expressed

$$m_1\mathbf{u}_1 + m_2\mathbf{u}_2 = m_1\mathbf{w}_1 + m_2\mathbf{w}_2 \qquad (8.4.2)$$

Here, $\mathbf{w}_1$ and $\mathbf{w}_2$ are used to represent the particle velocities after collision. In general, the initial and final kinetic energies are not the same, the difference being accounted for by the conversion of energy into other forms such as heat or internal energy:

$$\tfrac{1}{2}m_1u_1{}^2 + \tfrac{1}{2}m_2u_2{}^2 = \tfrac{1}{2}m_1w_1{}^2 + \tfrac{1}{2}m_2w_2{}^2 + H \qquad (8.4.3)$$

where H can be either positive or negative. As an example, let a spring of force constant k be compressed an amount s between two carts of mass m_1 and m_2 as in Fig. 8.4.2. The carts are temporarily held together by a string. When the string is cut, the potential energy of the spring, which is to be regarded as a type of internal energy, is converted into the final kinetic energy of the carts:

$$H = -\tfrac{1}{2}ks^2 \qquad (8.4.4)$$

H is negative because a decrease in internal energy is involved. The initial velocities are zero. The conservation of energy and momentum

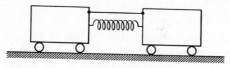

Fig. 8.4.2

equations (8.4.2) and (8.4.3) become

$$0 = m_1 w_1 + m_2 w_2 \qquad (8.4.5)$$

$$0 = \tfrac{1}{2} m_1 w_1{}^2 + \tfrac{1}{2} m_2 w_2{}^2 - \tfrac{1}{2} k s^2 \qquad (8.4.6)$$

and can be solved for the final velocities w_1 and w_2. This example is analogous to an unstable atomic nucleus which decays into two fragments. Collisions for which $H = 0$ are called *elastic*.

Of frequent occurrence in practice is a collision process in which one of the particles is initially stationary, e.g., in a Rutherford scattering experiment in which atomic nuclei are bombarded by α particles. After

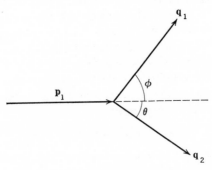

Fig. 8.4.3

collision, the two particles will in general go off at angles θ and ϕ with respect to the direction of the momentum of the bombarding particle (Fig. 8.4.3). The final momenta, $\mathbf{q}_1$ and $\mathbf{q}_2$, determine a plane and since the total momentum of the system perpendicular to this plane is zero, the initial momentum vector $\mathbf{p}_1$ must also lie in this plane. Assume that the particles themselves are unaltered by the collision process. Conservation of linear momentum gives two component equations

$$q_1 \cos \phi + q_2 \cos \theta = p_1 \qquad (8.4.7)$$

$$q_1 \sin \phi - q_2 \sin \theta = 0 \qquad (8.4.8)$$

where p_1 and q_1 are the magnitudes of the initial and final momenta of the bombarding particle, and q_2 is the magnitude of the final momentum of the originally stationary target particle. Simultaneous solution of (8.4.7) and (8.4.8) yields

$$q_1 = \frac{p_1 \sin \theta}{\sin (\theta + \phi)} \qquad q_2 = \frac{p_1 \sin \phi}{\sin (\theta + \phi)} \qquad (8.4.9)$$

The conservation of energy equation can be expressed in the form

$$\frac{p_1^2}{2m_1} = \frac{q_1^2}{2m_1} + \frac{q_2^2}{2m_2} + H \tag{8.4.10}$$

If the collision is elastic, $H = 0$ and the substitution of (8.4.9) into (8.4.10) yields

$$\frac{m_1}{m_2} = \frac{\sin^2(\theta + \phi) - \sin^2\theta}{\sin^2\phi} \tag{8.4.11}$$

This result is transformed by means of various trigonometric identities as follows:

$$\frac{m_1}{m_2} = \frac{[\sin(\theta + \phi) + \sin\theta][\sin(\theta + \phi) - \sin\theta]}{\sin^2\phi}$$

$$= \frac{\left[2\sin\dfrac{2\theta + \phi}{2}\cos\dfrac{\phi}{2}\right]\left[2\sin\dfrac{\phi}{2}\cos\dfrac{2\theta + \phi}{2}\right]}{\sin^2\phi}$$

$$= \frac{2\sin\dfrac{2\theta + \phi}{2}\cos\dfrac{2\theta + \phi}{2}\cdot 2\cos\dfrac{\phi}{2}\sin\dfrac{\phi}{2}}{\sin^2\phi} = \frac{\sin(2\theta + \phi)\sin\phi}{\sin^2\phi}$$

$$\frac{m_1}{m_2} = \frac{\sin(2\theta + \phi)}{\sin\phi} \tag{8.4.12}$$

Certain general conclusions about the collision can be drawn from (8.4.12). If the incident particle is very light,

$$\frac{m_1}{m_2} \to 0 \qquad \sin(2\theta + \phi) \to 0 \tag{8.4.13}$$

One possibility is

$$2\theta + \phi = 0 \qquad \theta = -\frac{\phi}{2} \tag{8.4.14}$$

This gives a collision such as is pictured in Fig. 8.4.4., which is clearly not possible. The reader can analyze the special case of a head-on elastic

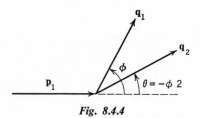

Fig. 8.4.4

collision and prove that if $m_1 < m_2$, the bombarding particle recoils straight back, i.e., $\phi = 180°$. Thus ϕ and θ cannot both be zero. The remaining possibility which will satisfy (8.4.13) is

$$2\theta + \phi = \pi \qquad (8.4.15)$$

which means that the direction of recoil of the heavy particle bisects the angle between the initial and final directions of the incident particle.

If the two masses are equal, $\sin (2\theta + \phi) = \sin \phi$. The two possibilities are

$$2\theta + \phi = \phi \qquad 2\theta + \phi = \pi - \phi \qquad (8.4.16)$$

The first gives $\theta = 0$, which means that the struck particle goes straight ahead. This is the special case of a head-on collision between particles of equal mass, and the reader can prove that all of the energy is transferred to the target particle so that the incident particle remains stationary after the collision. If the collision is not head-on, the second condition applies and can be expressed

$$\theta + \phi = \frac{\pi}{2} \qquad (8.4.17)$$

meaning that the angle between the directions of the final momenta of the two particles is $90°$.

Finally, let us find the maximum possible angle of scatter of the incident particle. We write (8.4.12) in the form

$$\frac{m_1}{m_2} \sin \phi = \sin (2\theta + \phi) \qquad (8.4.18)$$

and regard ϕ as a function of θ. Differentiation yields

$$\frac{m_1}{m_2} \cos \phi \frac{d\phi}{d\theta} = \cos (2\theta + \phi) \cdot \left(2 + \frac{d\phi}{d\theta}\right) \qquad (8.4.19)$$

The condition for ϕ to be a maximum is $d\phi/d\theta = 0$. Hence, $\cos (2\theta + \phi) = 0$ and $\sin (2\theta + \phi) = 1$. Equation (8.4.18) then gives

$$\sin \phi_{\max} = \frac{m_2}{m_1} \qquad (8.4.20)$$

Suppose that m_1 is very large. The incident particle is barely deviated so that (8.4.20) gives a value of ϕ_m near zero. If m_1 is gradually decreased, the maximum angle through which it can scatter increases, finally reaching $90°$ if $m_1 = m_2$. If $m_1 < m_2$, (8.4.20) gives no solution, so all angles of scatter are possible.

8.5. MOMENTUM CONSERVATION AND NEWTON'S THIRD LAW

We have adopted the momentum conservation principle as being more fundamental than Newton's third law. Nevertheless, Newton's third law is a useful concept and we investigate here its connection with momentum conservation in simple collision problems. When two bodies collide, they exert forces on one another for a short period of time. If, as in Fig. 8.5.1, these forces are $\mathbf{F}_1$ and $\mathbf{F}_2$, then according to Newton's third law

$$\mathbf{F}_1 = -\mathbf{F}_2 \tag{8.5.1}$$

The *impulse* of a force over a period of time is defined to be

$$\mathbf{J} = \int_{t_1}^{t_2} \mathbf{F} \, dt \tag{8.5.2}$$

To find the total impulses of the forces $\mathbf{F}_1$ and $\mathbf{F}_2$ it is necessary to integrate over the entire time that the bodies are in contact:

$$\mathbf{J}_1 = \int_{t_1}^{t_2} \mathbf{F}_1 \, dt \qquad \mathbf{J}_2 = \int_{t_1}^{t_2} \mathbf{F}_2 \, dt \tag{8.5.3}$$

Since $\mathbf{F}_1$ is *the only* force acting on the first body $\mathbf{F}_1 = m_1 \, d\mathbf{u}_1/dt$ and

$$\mathbf{J}_1 = \int_{t_1}^{t_2} m_1 \frac{d\mathbf{u}_1}{dt} \, dt = m_1 \mathbf{w}_1 - m_1 \mathbf{u}_1 \tag{8.5.4}$$

The total impulse received by the body during the collision is therefore equal to the change in the momentum of the body. Similarly, for the second body

$$\mathbf{J}_2 = m_2 \mathbf{w}_2 - m_2 \mathbf{u}_2 \tag{8.5.5}$$

From (8.5.1) it follows that $\mathbf{J}_1 = -\mathbf{J}_2$. Therefore,

$$m_1 \mathbf{w}_1 + m_2 \mathbf{w}_2 = m_1 \mathbf{u}_1 + m_2 \mathbf{u}_2 \tag{8.5.6}$$

As has been previously mentioned, there is an intimate connection between constants of the motion and symmetry. To establish a relation between conservation of linear momentum and symmetry, suppose two identical particles travel toward one another at equal speeds and make a head-on collision. The assumption of symmetry requires that one particle not be given preference over the other. Therefore, it is necessary to assume that the particles

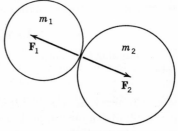

Fig. 8.5.1

move away from each other at equal speeds after the collision, i.e., momentum is conserved.

8.6. GALILEAN INVARIANCE OF MOMENTUM AND ENERGY CONSERVATION

For simplicity, the discussion is limited to a two-particle system. It will be shown that if two particles interact and if linear momentum, angular momentum, and energy are conserved in one Galilean frame of reference, then these quantities are conserved in all such frames. Assume that in one Galilean frame it has been established that

$$m_1\mathbf{u}_1 + m_2\mathbf{u}_2 = m_1\mathbf{w}_1 + m_2\mathbf{w}_2 \tag{8.6.1}$$

The transformation of velocities to a second frame moving at a constant velocity of translation $\mathbf{v}$ relative to the first is

$$\mathbf{u}_1 = \mathbf{u}_1' + \mathbf{v}, \qquad \mathbf{w}_1 = \mathbf{w}_1' + \mathbf{v}, \qquad \text{etc.} \tag{8.6.2}$$

Hence,

$$m_1(\mathbf{u}_1' + \mathbf{v}) + m_2(\mathbf{u}_2' + \mathbf{v}) = m_1(\mathbf{w}_1' + \mathbf{v}) + m_2(\mathbf{w}_2' + \mathbf{v}) \tag{8.6.3}$$

The terms involving $\mathbf{v}$ cancel leaving

$$m_1\mathbf{u}_1' + m_2\mathbf{u}_2' = m_1\mathbf{w}_1' + m_2\mathbf{w}_2' \tag{8.6.4}$$

Thus the principle of linear momentum conservation is preserved by a Galilean transformation. Note that the *actual values* of the momenta are altered by the transformation, but that momentum conservation as a basic principle remains intact.

Assume that the energy conservation principle

$$\tfrac{1}{2}m_1u_1^2 + \tfrac{1}{2}m_2u_2^2 = \tfrac{1}{2}m_1w_1^2 + \tfrac{1}{2}m_2w_2^2 + H \tag{8.6.5}$$

has been established in one Galilean frame of reference. The substitution of (8.6.2) into (8.6.5) yields

$$\begin{aligned} \tfrac{1}{2}m_1(\mathbf{u}_1' + \mathbf{v}) \cdot (\mathbf{u}_1' + \mathbf{v}) + \tfrac{1}{2}m_2(\mathbf{u}_2' + \mathbf{v}) \cdot (\mathbf{u}_2' + \mathbf{v}) \\ = \tfrac{1}{2}m_1(\mathbf{w}_1' + \mathbf{v}) \cdot (\mathbf{w}_1' + \mathbf{v}) + \tfrac{1}{2}m_2(\mathbf{w}_2' + \mathbf{v}) \cdot (\mathbf{w}_2' + \mathbf{v}) + H \end{aligned} \tag{8.6.6}$$

If the dot products are multiplied out and (8.6.4) is used, the result is

$$\tfrac{1}{2}m_1u_1'^2 + \tfrac{1}{2}m_2u_2'^2 = \tfrac{1}{2}m_1w_1'^2 + \tfrac{1}{2}m_2w_2'^2 + H \tag{8.6.7}$$

The angular momentum of a system of particles can be expressed in the form

$$\mathbf{l} = \mathbf{R}_c \times M\mathbf{v}_c + \mathbf{l}_{cm} \tag{8.6.8}$$

where $\mathbf{R}_c$ and $\mathbf{v}_c$ are the position and velocity of the center of mass and $\mathbf{l}_{cm}$ is the angular momentum measured relative to the center of mass and is unaltered by a Galilean transformation. If linear momentum is conserved, the center of mass moves in a straight line so that $\mathbf{R}_c \times M\mathbf{v}_c = \text{const.}$ in any Galilean frame of reference. Thus, if $\mathbf{l}$ is constant in one Galilean frame of reference, it must be constant in all such frames.

It is frequently more convenient to solve collision problems in a Galilean frame of reference other than the laboratory frame from which the process is actually observed. Since the total momentum of the system is a constant, it is possible to introduce a Galilean frame of reference in which the center of mass is stationary. This is called the *center of mass* coordinate

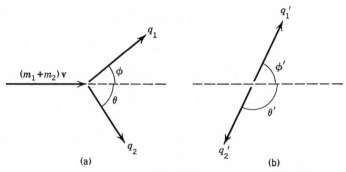

(a)   (b)

Fig. 8.6.1

system and is especially convenient for the solution of many collision and scattering problems. In a two-particle collision problem, the velocity of the center of mass is

$$\mathbf{v} = \frac{m_1\mathbf{u}_1 + m_2\mathbf{u}_2}{m_1 + m_2} = \frac{m_1\mathbf{w}_1 + m_2\mathbf{w}_2}{m_1 + m_2} \tag{8.6.9}$$

The total momentum of the system with respect to the center of mass frame of reference is zero.

As an example, suppose a particle initially moving at a velocity $\mathbf{v}$ decays into two fragments. In the laboratory, the decay appears as in Fig. 8.6.1a. In the center of mass coordinate system the particle is originally stationary and when it decays, fragments move in opposite directions with equal momenta as in Fig. 8.6.1b. In this example, the velocity of the center of mass and the velocity of the original particle are identical.

Suppose that it is required to find q_1 as a function of the angle ϕ in the laboratory coordinate system. If H is the amount of internal energy released in the decay process, the equations of conservation of momentum and energy in the center of mass coordinate system are

$$q_1' = q_2' \qquad \frac{q_1'^2}{2m_1} + \frac{q_2'^2}{2m_2} = H \tag{8.6.10}$$

Notice that H is defined here in such a way that it is a positive number. Solution for q_1' yields

$$q_1'^2 = \frac{2Hm_1m_2}{m_1 + m_2} \tag{8.6.11}$$

The relation between the components of the momentum of the particle of mass m_1 in the laboratory and center of mass coordinate system is

$$q_1' \sin \phi' = q_1 \sin \phi \tag{8.6.12}$$

$$q_1' \cos \phi' = q_1 \cos \phi - m_1 v \tag{8.6.13}$$

By squaring and adding the above equations we find

$$q_1'^2 = q_1^2 - 2m_1 q_1 v \cos \phi + m_1^2 v^2. \tag{8.6.14}$$

which is a quadratic equation for q_1. The solutions are

$$q_1 = m_1 v \cos \phi \pm \sqrt{q_1'^2 - m_1^2 v^2 \sin^2 \phi} \tag{8.6.15}$$

Obviously, since $q_1 = q_1'$, if $v = 0$ it is necessary to use the positive sign in (8.6.15). Hence,

$$q_1 = m_1 v \cos \phi + \sqrt{\frac{2Hm_1m_2}{m_1 + m_2} - m_1^2 v^2 \sin^2 \phi} \tag{8.6.16}$$

8.7. PARTICLES WITH VARIABLE MASS

The concepts of impulse and momentum developed in the preceding sections lead quite naturally to a theory of particles with variable mass.

Suppose that a fluid flows through a pipe with a right angle bend in it as in Fig. 8.7.1. Let the rate of flow of mass be dm/dt. As the particles of the fluid impinge on the bend in the pipe, they receive an impulse and a consequent change in momentum. If the velocity of the fluid is u, the impulse in the x-direction received by an increment of mass dm as it makes the 90° turn is

$$dJ_x = u\, dm \tag{8.7.1}$$

The walls of the pipe, therefore, must exert a force on the fluid in the x-direction given by

$$F_x\, dt = dJ_x = u\, dm, \qquad F_x = u\frac{dm}{dt} \tag{8.7.2}$$

By Newton's third law, there is a reaction force of equal magnitude *exerted on the pipe* in the *opposite* direction.

The same principle applies to a particle which is expelling or picking up mass. For concreteness, suppose we are talking about a rocket ship which is expelling fuel at a rate dm_f/dt. If $\mathbf{u}_f'$ is the exhaust velocity *relative to the ship*, then the thrust produced by the rocket motor is $-\mathbf{u}_f' \, dm_f/dt$. The basic equation of motion, force equals mass times acceleration, holds:

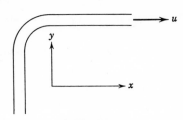

$$\mathbf{F} - \mathbf{u}_f' \frac{dm_f}{dt} = m \frac{d\mathbf{u}}{dt} \quad (8.7.3)$$

Fig. 8.7.1

where $\mathbf{F}$ is the applied external force, e.g., the gravitational force produced on the rocket by the earth. The velocity $\mathbf{u}_f'$ is measured by an observer on the accelerated reference frame of the ship and is related to the exhaust velocity as measured by an observer in the Galilean frame from which the motion is being observed, e.g., the earth, by

$$\mathbf{u}_f = \mathbf{u}_f' + \mathbf{u} \quad (8.7.4)$$

The rate of change of the mass of the ship is

$$\frac{dm}{dt} = -\frac{dm_f}{dt} \quad (8.7.5)$$

The equation of motion (8.7.3) can therefore be expressed in the alternative form

$$\mathbf{F} - \mathbf{u}_f \frac{dm_f}{dt} = \mathbf{u} \frac{dm}{dt} + m \frac{d\mathbf{u}}{dt} = \frac{d}{dt}(m\mathbf{u}) \quad (8.7.6)$$

In rocketry, Eq. (8.7.3) is generally the most convenient form of the equation of motion since it is $\mathbf{u}_f'$ which is known.

Consider the vertical take off of a rocket from the earth. The portion of the flight which is under power is so short that the gravitational field is approximately uniform. Air resistance is neglected. If the rate of ejection of fuel is a constant, the mass of the ship as a function of time is

$$m = -kt + m_0 \quad (8.7.7)$$

where m_0 is the initial mass. Let y be the vertical displacement above the earth's surface. The exhaust velocity is a constant h directed in the *negative y*-direction with respect to an observer on the ship. In Eq. (8.7.3), $\mathbf{F} = -mg\hat{\mathbf{j}}$, $u_f' = -h\hat{\mathbf{j}}$, and $dm_f/dt = -dm/dt = +k$. Hence,

$$-mg - (-h)(k) = m\ddot{y}$$

or

$$(kt - m_0)g + hk = (m_0 - kt)\ddot{y} \quad (8.7.8)$$

Writing (8.7.8) in the form

$$dy = \frac{kh\,dt}{m_0 - kt} - g\,dt \qquad (8.7.9)$$

and integrating from 0 to t yields

$$\dot{y} = -h \log \left(\frac{m_0 - kt}{m_0} \right) - gt \qquad (8.7.10)$$

where the velocity at $t = 0$ is assumed to be zero.

Suppose that nine-tenths of the original mass of the rocket is fuel. The time at which powered flight ends is therefore determined by $kt = 0.9m_0$. At this time, the velocity is

$$\dot{y} = h \log 10 - \frac{0.9m_0 g}{k} \qquad (8.7.11)$$

The greater the exhaust velocity, the greater will be the final velocity of the ship. If the ship is in free space so that $g = 0$, the final velocity depends *only* on the exhaust velocity and on the fraction of the original mass of the ship that is fuel. An ion propulsion engine which ejects ions at a very high velocity, even though the rate of ejection of mass may be small, is very efficient in the long run for propelling a rocket in free space. Of course, if the rate of ejection of mass is small, it will take the ship a long time to attain its final velocity.

The integration of (8.7.10) from 0 to t gives the position of the rocket as a function of time:

$$y = \frac{hm_0}{k} \left[\frac{m_0 - kt}{m_0} \log \left(\frac{m_0 - kt}{m_0} \right) + \frac{kt}{m_0} \right] - \tfrac{1}{2}gt^2 \qquad (8.7.12)$$

where the initial condition $y = 0$ at $t = 0$ is assumed.

Let us make one further remark about the basic equation of motion. If the ship is in free space so that $\mathbf{F} = 0$, Eq. (8.7.6) can be expressed as

$$0 = \mathbf{u}_f \, dm_f + d(m\mathbf{u}) \qquad (8.7.13)$$

The system, rocket plus ejected fuel, is *isolated* and (8.7.13) is simply a statement of the fact that the total momentum of the system does not change. If the ship starts from rest at a certain point in free space, its center of mass remains there forever! The passengers are propelled to the far reaches of space by virtue of the fact that ejected matter in the form of fuel has suffered a similar fate but in the opposite direction.

As another example, suppose that a particle traveling in the positive x-direction encounters a medium at $x = 0$ from which it accumulates matter in proportion to the distance it has traveled. The mass of the particle is then given by

$$m = m_0 = \text{const.} \qquad x < 0$$
$$m = kx + m_0 \qquad x > 0 \qquad (8.7.14)$$

Since the medium is stationary, the relative velocity of particle and medium is the same as the velocity of the particle. The particle, therefore, experiences a force $-u\,dm/dt$. Hence,

$$-u\frac{dm}{dt} = m\frac{du}{dt} \qquad (8.7.15)$$

In a sense, this is the converse of the rocket problem with the particle accumulating mass rather than ejecting it and being slowed down as a result. Alternatively, (8.7.15) can be expressed

$$0 = \frac{d}{dt}(mu) \qquad (8.7.16)$$

which says that the total momentum of the system (particle plus medium) does not change. If (8.7.16) is integrated and combined with (8.7.14), the result is

$$(kx + m_0)\dot{x} = m_0 u_0 \qquad (8.7.17)$$

where u_0 is the initial velocity at $x = 0$. Writing

$$(kx + m_0)\,dx = m_0 u_0\,dt$$

and integrating from 0 to t yields

$$\tfrac{1}{2}kx^2 + m_0 x = m_0 u_0 t$$

This is solved for x to get

$$x = -\frac{m_0}{k} + \sqrt{\frac{m_0^{\,2}}{k^2} + \frac{2m_0 u_0 t}{k}} \qquad (8.7.18)$$

We note further that dissipation of energy is involved. The loss in kinetic energy is

$$\Delta T = \tfrac{1}{2}m_0 u_0^2 - \tfrac{1}{2}m\dot{x}^2 \qquad (8.7.19)$$

Equation (8.7.17) can be expressed as $m\dot{x} = m_0 u_0$ so that (8.7.19) becomes

$$\Delta T = \tfrac{1}{2}m_0 u_0^{\,2}\left(1 - \frac{m_0}{m}\right) \qquad (8.7.20)$$

The process is actually a continuous inelastic collision. Conversely, the rocket is a continuous decay process in which internal energy is released by the burning of fuel and converted into kinetic energy.

8.8. THE LAGRANGE FORMULATION FOR MANY-PARTICLE SYSTEMS

In a general type of mechanical system, there are N-particles described by $3N$ rectangular Cartesian coordinates. The particles individually obey Newtonian mechanics and it is convenient to write Newton's second law as

$$F_i = m_{ij}\ddot{x}_j \qquad (8.8.1)$$

where the repeated subscript is now summed from 1 to N and the tensor m_{ij} is diagonal and has the components

$$
\begin{matrix}
m_1 & 0 & 0 & 0 & 0 & 0 & \cdots \\
0 & m_1 & 0 & 0 & 0 & 0 & \cdots \\
0 & 0 & m_1 & 0 & 0 & 0 & \cdots \\
0 & 0 & 0 & m_2 & 0 & 0 & \cdots \\
0 & 0 & 0 & 0 & m_2 & 0 & \cdots \\
0 & 0 & 0 & 0 & 0 & m_2 & \cdots \\
\cdot & \cdot & \cdot & \cdot & \cdot & \cdot & \cdots
\end{matrix}
\qquad (8.8.2)
$$

Here, m_1 is the mass of the first particle in the system, m_2 is the mass of the second particle, and so on. Of course, only the term for which $j = i$ is nonzero in (8.8.1). It will be assumed that there are n time-dependent holonomic constraints which can be expressed in the form

$$
\begin{aligned}
f_1(x_1, \ldots, x_{3N}, t) &= 0 \\
f_2(x_1, \ldots, x_{3N}, t) &= 0 \\
f_n(x_1, \ldots, x_{3N}, t) &= 0
\end{aligned}
\qquad (8.8.3)
$$

meaning that the actual number of degrees of freedom or independent coordinates is

$$f = 3N - n \qquad (8.8.4)$$

It is sometimes useful to think of the system as being represented by a single point in a rectangular Cartesian space of $3N$ dimensions. Each of the equations of constraint (8.8.3) represents a hypersurface in this $3N$-dimensional space and the actual motion of the representative point is restricted to the f-dimensional subspace which is the common intersection of these hypersurfaces. Explicit time dependence is included in (8.8.3) to allow for the fact that the hypersurfaces are changing in time. The conceptual f-dimensional space, in which the representative point moves, is called *configuration space*.

We will show that the same form of Lagrange's equations with which we are already familiar is still valid. Since there are f-degrees of freedom, it is possible to introduce f generalized coordinates q_α. The $3N$-rectangular coordinates are related to the generalized coordinates by transformation equations of the form

$$x_i = x_i(q_\alpha, t) \qquad i = 1, \ldots, 3N \qquad \alpha = 1, \ldots, f \qquad (8.8.5)$$

The explicit time dependence in (8.8.5) results from the time dependence of the equations of constraint. Differentiation of the transformation equation with respect to the time yields

$$\dot{x}_i = \frac{\partial x_i}{\partial q_\beta} \dot{q}_\beta + \frac{\partial x_i}{\partial t} \qquad (8.8.6)*$$

The kinetic energy of the system can be expressed as

$$T = \tfrac{1}{2} m_{ij} \dot{x}_i \dot{x}_j \qquad (8.8.7)$$

Differentiation with respect to q_α gives

$$\frac{\partial T}{\partial q_\alpha} = m_{ij} \frac{\partial \dot{x}_i}{\partial q_\alpha} \dot{x}_j \qquad (8.8.8)$$

The functional dependence of $\dot{x}_i$ on q_α is to be found from (8.8.6). By differentiating (8.8.6) and substituting the result in (8.8.8) we get

$$\frac{\partial T}{\partial q_\alpha} = m_{ij} \left[\frac{\partial^2 x_i}{\partial q_\alpha \partial q_\beta} \dot{q}_\beta + \frac{\partial^2 x_i}{\partial q_\alpha \partial t} \right] \dot{x}_j \qquad (8.8.9)$$

Again using (8.8.6), we note that

$$\frac{\partial \dot{x}_i}{\partial \dot{q}_\alpha} = \frac{\partial x_i}{\partial q_\alpha} \qquad (8.8.10)$$

Therefore, differentiation of T with respect to $\dot{q}_\alpha$ yields

$$\frac{\partial T}{\partial \dot{q}_\alpha} = m_{ij} \frac{\partial \dot{x}_i}{\partial \dot{q}_\alpha} \dot{x}_j = m_{ij} \frac{\partial x_i}{\partial q_\alpha} \dot{x}_j \qquad (8.8.11)$$

Differentiating totally with respect to time gives

$$\frac{d}{dt} \left(\frac{\partial T}{\partial \dot{q}_\alpha} \right) = m_{ij} \left[\frac{\partial^2 x_i}{\partial q_\beta \partial q_\alpha} \dot{q}_\beta + \frac{\partial^2 x_i}{\partial t \, \partial q_\alpha} \right] \dot{x}_j + m_{ij} \frac{\partial x_i}{\partial q_\alpha} \ddot{x}_j \qquad (8.8.12)$$

---

* The $\dot{q}_\beta$ are, of course, contravariant. The notation is a little less cumbersome if we label everything with subscripts.

Combining (8.8.9) and (8.8.12) results in

$$\frac{d}{dt}\left(\frac{\partial T}{\partial \dot{q}_\alpha}\right) - \frac{\partial T}{\partial q_\alpha} = m_{ij}\ddot{x}_j \frac{\partial x_i}{\partial q_\alpha} = F_i \frac{\partial x_i}{\partial q_\alpha} = Q_\alpha \qquad (8.8.13)$$

which are Lagrange's equations in exactly the same form as were developed in Chapter 6 for a single particle.

If there are no dissipative forces present and if all forces other than the forces of constraint are obtainable from a potential function, it is necessary merely to extend the argument of Section 6.9 to show that the forces of constraint cancel out. Let i run from 1 to N and α run from 1 to f in Eqs. (6.9.4) and (6.9.5). Thus,

$$Q_\alpha = -\frac{\partial V}{\partial q_\alpha} \qquad (8.8.14)$$

and

$$\frac{d}{dt}\left(\frac{\partial L}{\partial \dot{q}_\alpha}\right) - \frac{\partial L}{\partial q_\alpha} = 0, \qquad L = T - V \qquad (8.8.15)$$

If the constraints are time dependent, the formal expression for the kinetic energy in terms of the generalized velocities becomes somewhat complicated. Combining (8.8.6) and (8.8.7) gives

$$\begin{aligned}
T &= \tfrac{1}{2}m_{ij}\left[\frac{\partial x_i}{\partial q_\beta}\dot{q}_\beta + \frac{\partial x_i}{\partial t}\right]\left[\frac{\partial x_j}{\partial q_\alpha}\dot{q}_\alpha + \frac{\partial x_j}{\partial t}\right] \\
&= \tfrac{1}{2}m_{ij}\frac{\partial x_i}{\partial q_\alpha}\frac{\partial x_j}{\partial q_\beta}\dot{q}_\alpha\dot{q}_\beta + m_{ij}\frac{\partial x_i}{\partial q_\beta}\frac{\partial x_j}{\partial t}\dot{q}_\beta + \tfrac{1}{2}m_{ij}\frac{\partial x_i}{\partial t}\frac{\partial x_j}{\partial t}
\end{aligned} \qquad (8.8.16)$$

The terms involving $\partial x_i/\partial t$ are contributions to the kinetic energy from the motion of the surfaces of constraint. The metric tensor in the subspace which is the common intersection of the surfaces of constraint is

$$g_{\alpha\beta} = m_{ij}\frac{\partial x_i}{\partial q_\alpha}\frac{\partial x_j}{\partial q_\beta} \qquad (8.8.17)$$

and has been defined in such a way that it incorporates the masses of the individual particles. If the constraints are not time dependent, the kinetic energy is simply

$$T = \tfrac{1}{2}g_{\alpha\beta}\dot{q}_\alpha\dot{q}_\beta \qquad (8.8.18)$$

In simple problems involving time-dependent constraints it is seldom necessary to calculate the kinetic energy by using (8.8.16) directly. As an example, suppose a wheel with wire spokes mounted on a verticle shaft is turning at a constant angular velocity ω. There is a bead on one of the spokes which can slide without friction. The bead is released at a distance r_0 from the shaft with zero initial velocity in the r-direction and it is desired to find its motion. The *only* forces are forces of constraint exerted on the bead by the spokes of the wheel. The kinetic energy can be written in the usual way in polar coordinates as

$$T = \tfrac{1}{2}m(\dot{r}^2 + r^2\dot{\theta}^2) \tag{8.8.19}$$

The time-dependent holonomic constraint is

$$\theta = \omega t \tag{8.8.20}$$

Hence, the Lagrangian involving the single independent coordinate r is

$$L = T = \tfrac{1}{2}m(\dot{r}^2 + r^2\omega^2) \tag{8.8.21}$$

The equation of motion is found to be

$$\ddot{r} - r\omega^2 = 0 \tag{8.8.22}$$

The solution satisfying the initial conditions is

$$r = r_0 \cosh \omega t \tag{8.8.23}$$

8.9. SOME APPLICATIONS OF LAGRANGE'S EQUATIONS

A pulley system is set up as illustrated in Fig. 8.9.1 with equal masses on the ends of the cords. The pulleys are massless and frictionless. The position coordinates of the two masses are x_1 and x_2, but these coordinates are *not* independent. If x_1 increases by dx_1, x_2 decreases by twice this amount:

$$dx_2 = -2\,dx_1 \tag{8.9.1}$$

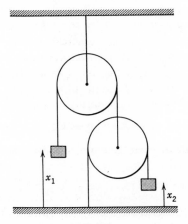

Fig. 8.9.1

Integration yields the equation of constraint

$$x_2 = -2x_1 + C \tag{8.9.2}$$

The value of the constant C is not important. The Lagrangian is

$$L = \tfrac{1}{2}m\dot{x}_1{}^2 + \tfrac{1}{2}m\dot{x}_2{}^2 - mgx_1 - mgx_2 \tag{8.9.3}$$

Introduction of the constraint results in

$$L = \tfrac{5}{2}m\dot{x}_1{}^2 + mgx_1 - mgC \tag{8.9.4}$$

The equation of motion is

$$\frac{d}{dt}\left(\frac{\partial L}{\partial \dot{x}_1}\right) - \frac{\partial L}{\partial x_1} = 0 \tag{8.9.5}$$

and yields

$$\ddot{x}_1 = \tfrac{1}{5}g \tag{8.9.6}$$

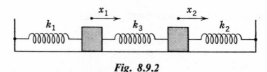

Fig. 8.9.2

The forces of constraint are the tensions in the cords. Let F_1 be the tension in the cord supporting the first mass. Then directly from Newton's second law

$$F_1 - mg = m\ddot{x}_1 \tag{8.9.7}$$

By combining (8.9.6) and (8.9.7),

$$F_1 = mg + \tfrac{1}{5}mg = \tfrac{6}{5}mg \tag{8.9.8}$$

Suppose that two masses are constrained to move without friction in one dimension and are connected by springs of force constants k_1, k_2, and k_3. The springs k_1 and k_2 are attached to supports as illustrated in Fig. 8.9.2. This is an example of *coupled oscillators*. Let the displacements x_1 and x_2 of the two masses be measured from the equilibrium positions and let s_1, s_2, and s_3 be the extensions of the springs when $x_1 = x_2 = 0$. At equilibrium, the tensions in the springs balance:

$$k_1 s_1 = k_2 s_2 = k_3 s_3 \tag{8.9.9}$$

When the particles are displaced from equilibrium, the potential energy of the system is

$$V = \tfrac{1}{2}k_1(x_1 + s_1)^2 + \tfrac{1}{2}k_3(x_2 - x_1 + s_3)^2 + \tfrac{1}{2}k_2(s_2 - x_2)^2 \tag{8.9.10}$$

When the squares in (8.9.10) are multiplied out, the cross terms involving s_1, s_2, and s_3 drop out by virtue of (8.9.9). Constant terms such as $\tfrac{1}{2}k_1 s_1{}^2$

can be discarded. Therefore, a suitable potential function which can be used in place of (8.9.10) is

$$V = \tfrac{1}{2}k_1x_1{}^2 + \tfrac{1}{2}k_3(x_2 - x_1)^2 + \tfrac{1}{2}k_2x_2{}^2 \qquad (8.9.11)$$

The potential (8.9.11) has the value zero when the particles are in their equilibrium position and is *positive definite* for any displacement of the system from equilibrium indicating that the equilibrium is stable.

It is useful to represent the system by a single point in a two-dimensional Cartesian space, the coordinate axes of which are x_1 and x_2. The equipotential curves in this space are given by $V = $ const. Since V is a positive

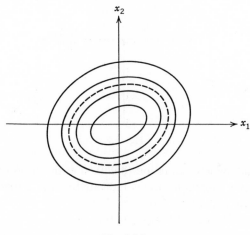

Fig. 8.9.3

definite quadratic form, the equipotentials are ellipses as illustrated in Fig. 8.9.3. Suppose that the system has been given an energy $\mathscr{E}$. Let the dashed curve of Fig. 8.9.3 represent the equipotential $\mathscr{E} = V$. Since $\mathscr{E} = T + V$ and $T \geq 0$, the allowed region for the motion is *inside* the dashed curve. If the energy is small, the maximum displacement of the representative point from the origin is also small. This is essentially the criterion for a stable equilibrium. The coordinate system of Fig. 8.9.3 is the two-dimensional configuration space of the system.

The Lagrangian for the system is

$$L = \tfrac{1}{2}m_1\dot{x}_1{}^2 + \tfrac{1}{2}m_2\dot{x}_2{}^2 - \tfrac{1}{2}k_1x_1{}^2 - \tfrac{1}{2}k_3(x_2 - x_1)^2 - \tfrac{1}{2}k_2x_2{}^2 \quad (8.9.12)$$

By means of

$$\frac{d}{dt}\frac{\partial L}{\partial \dot{x}_1} - \frac{\partial L}{\partial x_1} = 0 \qquad \frac{d}{dt}\frac{\partial L}{\partial \dot{x}_2} - \frac{\partial L}{\partial x_2} = 0 \qquad (8.9.13)$$

the equations of motion are found to be

$$m_1\ddot{x}_1 + k_1 x_1 - k_3(x_2 - x_1) = 0 \tag{8.9.14}$$

$$m_2\ddot{x}_2 + k_2 x_2 + k_3(x_2 - x_1) = 0 \tag{8.9.15}$$

and are coupled linear equations involving two unknowns. A detailed solution of (8.9.14) and (8.9.15) is postponed until Section 8.10.

An electrical circuit the dynamical behavior of which is identical to the mechanical system just considered is illustrated in Fig. 8.9.4. When the circuit is quiescent, the voltages on the capacitors are equal:

$$\frac{q_{10}}{C_1} = \frac{q_{30}}{C_3} = \frac{q_{20}}{C_2} \tag{8.9.16}$$

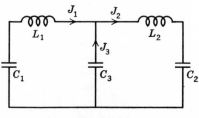

Fig. 8.9.4

which is the analogue of equation (8.9.9). Let the currents be defined as positive when the directions are as indicated by the arrows in Fig. 8.9.4. The currents are related by

$$J_2 = J_1 + J_3$$

and if q_1, q_2, and q_3 represent displacements of charge in the system,

$$q_2 = q_1 + q_3$$

The potential energy is

$$V = \frac{1}{2C_1}(q_{10} - q_1)^2 + \frac{1}{2C_3}(q_{30} - [q_2 - q_1])^2 + \frac{1}{2C_3}(q_{20} + q_2)^2 \tag{8.9.17}$$

Just as in the mechanical system, cross terms involving q_{10}, q_{20}, and q_{30} cancel out. Discarding constant terms, the Lagrangian is

$$L = \tfrac{1}{2}L_1\dot{q}_1^2 + \tfrac{1}{2}L_2\dot{q}_2^2 - \frac{q_1^2}{2C_1} - \frac{(q_2 - q_1)^2}{2C_3} - \frac{q_2^2}{2C_2} \tag{8.9.18}$$

and is to be compared to Eq. (8.9.12).

An acoustical system which has the same dynamical behavior as the mechanical and electrical systems just discussed can be constructed from three Helmholtz resonators (Section 4.9) as shown in Fig. 8.9.5.

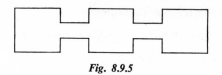

Fig. 8.9.5

A double plane pendulum consists of two masses suspended by strings or massless rigid rods as in Fig. 8.9.6. The system has two degrees of freedom and the angles θ_1 and θ_2 can be used as generalized coordinates. The potential energy of the system can be expressed

$$V = -m_1 g s_1 \cos \theta_1 - m_2 g (s_1 \cos \theta_1 + s_2 \cos \theta_2) \qquad (8.9.19)$$

The rectangular coordinates of m_1 and m_2 are (x_1, x_2) and (x_3, x_4). The transformation equations connecting rectangular and generalized co-ordinates are

$$x_1 = s_1 \sin \theta_1 \qquad x_2 = s_1 \cos \theta_1$$
$$x_3 = s_1 \sin \theta_1 + s_2 \sin \theta_2 \qquad (8.9.20)$$
$$x_4 = s_1 \cos \theta_1 + s_2 \cos \theta_2$$

The kinetic energy is written

$$T = \tfrac{1}{2} m_1 (\dot{x}_1^2 + \dot{x}_2^2) + \tfrac{1}{2} m_2 (\dot{x}_3^2 + \dot{x}_4^2) \qquad (8.9.21)$$

and computed in terms of θ_1 and θ_2 by means of (8.9.20). The resulting Lagrangian for the system is

$$L = \tfrac{1}{2}(m_1 + m_2) s_1^2 \dot{\theta}_1^2 + m_2 s_1 s_2 \dot{\theta}_1 \dot{\theta}_2 \cos(\theta_2 - \theta_1) + \tfrac{1}{2} m_2 s_2^2 \dot{\theta}_2^2$$
$$+ (m_1 + m_2) g s_1 \cos \theta_1 + m_2 g s_2 \cos \theta_2 \qquad (8.9.22)$$

It is interesting to view the motion as being given by a single representative point in a space the coordinates of which are θ_1 and θ_2. The metric tensor

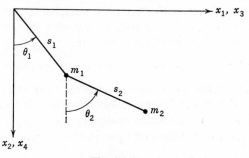

Fig. 8.9.6

of this coordinate system has components

$$\begin{pmatrix} (m_1 + m_2)s_1{}^2 & m_2 s_1 s_2 \cos(\theta_2 - \theta_1) \\ m_2 s_1 s_2 \cos(\theta_2 - \theta_1) & m_2 s_2{}^2 \end{pmatrix} \tag{8.9.23}$$

The coordinate system is nonorthogonal.

The general motion of the system is described by complicated nonlinear differential equations. If the angles are small, the cosine terms can be expanded in Taylor's series. All terms such as $\theta_1 \theta_2 \theta_2{}^2$, $\theta_2{}^4$, etc., are discarded. The resulting approximate Lagrangian is

$$\begin{aligned} L = {}&\tfrac{1}{2}(m_1 + m_2)s_1{}^2\dot{\theta}_1{}^2 + m_2 s_1 s_2 \dot{\theta}_1 \dot{\theta}_2 + \tfrac{1}{2}m_2 s_2{}^2 \dot{\theta}_2{}^2 \\ &- \tfrac{1}{2}(m_1 + m_2)g s_1 \theta_1{}^2 - \tfrac{1}{2}m_2 g s_2 \theta_2{}^2 \end{aligned} \tag{8.9.24}$$

The equations of motion are found to be

$$(m_1 + m_2)s_1{}^2\ddot{\theta}_1 + m_2 s_1 s_2 \ddot{\theta}_2 + (m_1 + m_2)g s_1 \theta_1 = 0 \tag{8.9.25}$$

$$m_2 s_2{}^2 \ddot{\theta}_2 + m_2 s_1 s_2 \ddot{\theta}_1 + m_2 g s_2 \theta_2 = 0 \tag{8.9.26}$$

and are coupled linear equations.

If holonomic constraints are present and the Lagrangian is expressed in terms of the correct number of degrees of freedom, the forces of constraint do not appear in the formalism. Let us compute the generalized force components for the double pendulum and see in detail how the forces of constraint cancel out. By means of

$$Q_1 = \frac{\partial x_i}{\partial \theta_1} F_i \qquad Q_2 = \frac{\partial x_i}{\partial \theta_2} F_i \tag{8.9.27}$$

and the transformation equations (8.9.20) we find

$$Q_1 = s_1 \cos \theta_1 (F_1 + F_3) - s_1 \sin \theta_1 (F_2 + F_4) \tag{8.9.28}$$

$$Q_2 = F_3 s_2 \cos \theta_2 - F_4 s_2 \sin \theta_2 \tag{8.9.29}$$

The rectangular force components F_3 and F_4 on the second particle include the components of the force of constraint exerted on the particle by the string or rod to which it is attached, and also the externally applied forces, in this case the weight of the particle:

$$Q_2 = -C_3 s_2 \cos \theta_2 - (-C_4 + m_2 g)s_2 \sin \theta_2 \tag{8.9.30}$$

where the components of the force of constraint are written $(-C_3, -C_4)$. By Newton's third law, the forces exerted on the rod by the particle are equal in magnitude but opposite in direction to $-C_3$ and $-C_4$. From Fig. 8.9.7, it is evident that the torque on the rod about its upper end is

$$N = C_3 s_2 \cos \theta_2 - C_4 s_2 \sin \theta_2 \tag{8.9.31}$$

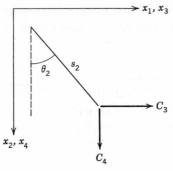

Fig. 8.9.7

Since the rod has no mass, it can support no torque and $N = 0$. Thus (8.9.30) becomes

$$Q_2 = -m_2 g s_2 \sin \theta_2 \qquad (8.9.32)$$

and does not involve forces of constraint. A similar proof holds for Q_1.

8.10. LINEAR VIBRATIONS

Equations (8.9.14) and (8.9.15) provide an example of a linear system with two degrees of freedom. The term *linear* as used here refers to the fact that the differential equations are linear, and not to the fact that the masses move along a straight line. A clue as to how to procede to solve (8.9.14) and (8.9.15) is provided by the discussion of the potential function. The ellipses which represent equipotentials in Fig. 8.9.3 are tilted at an angle to the x_1-axis. It seems reasonable to suppose that a simplification could be accomplished by introducing a new set of coordinate axes in the configuration space that coincide with the principle axes of the ellipses; in short a principle axis transformation. The potential function (8.9.11) can be expressed

$$
\begin{aligned}
V &= \tfrac{1}{2}(k + k_3)x_1{}^2 - k_3 x_1 x_2 + \tfrac{1}{2}(k + k_3)x_2{}^2 \\
&= \tfrac{1}{2}(x_1 \quad x_2)\begin{pmatrix} k + k_3 & -k_3 \\ -k_3 & k + k_3 \end{pmatrix}\begin{pmatrix} x_1 \\ x_2 \end{pmatrix}
\end{aligned}
\qquad (8.10.1)
$$

where for simplicity we have set $k_1 = k_2 = k$. (See Chapter 3 for a discussion of matrix notation.) The problem of finding the principle axis transformation is discussed in detail in Section 3.4. Let y_1 and y_2 refer to the rotated coordinate axes. The transformation which diagonalizes the 2×2 symmetric matrix in (8.10.1) is found to be

$$x_1 = \frac{1}{\sqrt{2}}(y_1 - y_2) \qquad x_2 = \frac{1}{\sqrt{2}}(y_1 + y_2) \qquad (8.10.2)$$

and is seen to represent a 45° rotation in the configuration space. Direct substitution of the transformation equations (8.10.2) into equations of motion (8.9.14) and (8.9.15) gives

$$m_1\ddot{y}_1 - m_1\ddot{y}_2 + ky_1 - (k + 2k_3)y_2 = 0 \tag{8.10.3}$$

$$m_2\ddot{y}_1 + m_2\ddot{y}_2 + ky_1 + (k + 2k_3)y_2 = 0 \tag{8.10.4}$$

One further simplification will be made by considering the special case $m_1 = m_2 = m$. Under this circumstance, addition of (8.10.3) and (8.10.4) results in

$$\ddot{y}_1 + \frac{k}{m}\, y_1 = 0 \tag{8.10.5}$$

Similarly, by subtraction,

$$\ddot{y}_2 + \frac{k + 2k_3}{m}\, y_2 = 0 \tag{8.10.6}$$

The solution of these equations is trivial:

$$y_1 = \sqrt{2}A_1 \cos(\omega_1 t - \alpha_1) \qquad \omega_1 = \sqrt{k/m} \tag{8.10.7}$$

$$y_2 = \sqrt{2}A_2 \cos(\omega_2 t - \alpha_2) \qquad \omega_2 = \sqrt{(k + 2k_3)/m} \tag{8.10.8}$$

where A_1, A_2, α_1, and α_2 are the four constants of integration needed to fit two initial positions and two initial velocities. The factors of $\sqrt{2}$ are included for convenience. There are two natural frequencies, each belonging to a *normal mode* of vibration of the system. In general, the solution (8.10.2) is a *mixture or superposition* of these two normal modes of vibration. We can choose the initial conditions so that the system oscillates in a pure mode. Thus if $A_2 = 0$ in (8.10.8), only the mode y_1 is present:

$$x_1 = A_1 \cos(\omega_1 t - \alpha_1) \qquad x_2 = A_1 \cos(\omega_1 t - \alpha_1) \tag{8.10.9}$$

The two masses vibrate exactly in phase and with the same amplitudes. It is just as though the connecting spring k_3 were absent and the two masses were vibrating separately. On the other hand, choosing $A_1 = 0$ in (8.10.7) results in

$$x_1 = -A_2 \cos(\omega_2 t - \alpha_2) \qquad x_2 = A_2 \cos(\omega_2 t - \alpha_2) \tag{8.10.10}$$

The two masses now vibrate exactly out of phase with one another and at a modified frequency.

One further set of initial conditions is of interest. By choosing $A_1 = A_2 = A$ and the phase angles α_1 and α_2 to be zero,

$$x_1 = A \cos \omega_1 t - A \cos \omega_2 t \tag{8.10.11}$$

$$x_2 = A \cos \omega_1 t + A \cos \omega_2 t \tag{8.10.12}$$

A little consideration shows that this corresponds to the initial conditions

$$x_{10} = 0 \qquad x_{20} = 2A \qquad u_{10} = 0 \qquad u_{20} = 0 \qquad (8.10.13)$$

By means of trigonometric identities, the solutions (8.10.11) and (8.10.12) are expressed in the form

$$x_1 = 2A \sin \left(\frac{\omega_2 + \omega_1}{2} t \right) \sin \left(\frac{\omega_2 - \omega_1}{2} t \right) \qquad (8.10.14)$$

$$x_2 = 2A \cos \left(\frac{\omega_2 + \omega_1}{2} t \right) \cos \left(\frac{\omega_2 - \omega_1}{2} t \right) \qquad (8.10.15)$$

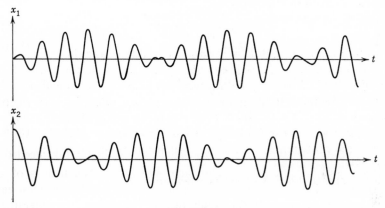

Fig. 8.10.1

The spring k_3 can be thought of as a coupling spring between the two oscillators. If this spring is very weak, then

$$\omega_2 = \sqrt{\frac{k}{m}} \sqrt{1 + 2\frac{k_3}{k}} \cong \sqrt{\frac{k}{m}} \left(1 + \frac{k_3}{k} \right) = \omega_1 + \delta \qquad (8.10.16)$$

$$\omega_2 + \omega_1 \cong 2\omega_1 \qquad \omega_2 - \omega_1 \cong \delta \qquad (8.10.17)$$

The motion is then given approximately by

$$x_1 = 2A \sin (\omega_1 t) \sin (\tfrac{1}{2}\delta t) \qquad (8.10.18)$$

$$x_2 = 2A \cos (\omega_1 t) \cos (\tfrac{1}{2}\delta t) \qquad (8.10.19)$$

These are periodic functions of frequency ω_1, modulated by a term of low frequency. Since the low frequency factors in x_1 and x_2 are exactly out of phase, the total energy transfers slowly back and forth between the two weakly coupled masses. The solutions are illustrated in Fig. 8.10.1 and are seen to be similar to the beats which occur when sound waves of slightly different frequency are added together.

The coordinates y_1 and y_2 which result in the simple differential equations (8.10.5) and (8.10.6) are known as *normal coordinates*. If in (8.10.3) and (8.10.4), the two masses *are not* equal, a transformation of a more general type than a simple rotation in the configuration space is needed to obtain the normal coordinates. We now take up the general problem.

In the general problem of linear vibrations of a system of many degrees of freedom, the kinetic and potential energies are expressed

$$T = \tfrac{1}{2} g_{ij} \dot{q}_i \dot{q}_j \qquad V = \tfrac{1}{2} k_{ij} q_i q_j \tag{8.10.20}$$

where g_{ij} and k_{ij} are components of symmetric tensors and are constants. The equations of motion are

$$g_{ij} \ddot{q}_j + k_{ij} q_j = 0 \tag{8.10.21}$$

It will be assumed that V is a positive definite quadratic form so that $q_i = 0$ is a position of stable equilibrium. In many problems the linear equations (10.21) with g_{ij} and k_{ij} regarded as constants represent only a first approximation to the motion. For a specific example, reference can be made to Eqs. (8.9.25) and (8.9.26) for the double pendulum. Since the metric tensor is constant, but not necessarily diagonal, the coordinate axis in the configuration space are straight lines but are not necessarily orthogonal.

It is useful to write (8.10.21) in matrix form as

$$G\ddot{Q} + KQ = 0 \tag{8.10.22}$$

where G and K are $f \times f$ square symmetric matrices and Q is a column matrix with components $q_1, q_2, \ldots, q_f$. Equation (8.10.22) is satisfied by

$$Q = X \cos (\omega t - \alpha) \tag{8.10.23}$$

where X is a column matrix the components of which are constants. Substituting (8.10.23) into (8.10.22) results in

$$(K - \lambda G)X = 0 \qquad \lambda = \omega^2 \tag{8.10.24}$$

It is evident that this is an eigenvalue problem of a somewhat more general nature than was treated in Section 3.4. In fact, our earlier treatment of the eigenvalue problem can be regarded as the special case $G = I$. The eigenvalues λ are the squares of the frequencies of the normal modes of vibration of the system. The secular equation is

$$\det (K - \lambda G) = 0 \tag{8.10.25}$$

and is a polynomial of degree f in λ. Thus there are f normal modes of vibration. For each eigenvalue, there is a corresponding eigenvector X to be determined from (8.10.24). The general solution of (8.10.22) is then a superposition of the normal modes of vibration:

$$Q = a_1 X_1 \cos(\omega_1 t - \alpha_1) + \cdots + a_f X_f \cos(\omega_f t - \alpha_f) \qquad (8.10.26)$$

Note that there are f multiplicative constants a_i and f-phase angles α_i, giving a total of $2f$ arbitrary constants.

The eigenvalues are real and positive. Write (8.10.24) as $KX = \lambda GX$, multiply by $\tilde{X}^*$ transpose the resulting equation, and take the complex conjugate:

$$\tilde{X}^* KX = \lambda \tilde{X}^* GX, \qquad \tilde{X} KX^* = \lambda \tilde{X} GX^*$$
$$\tilde{X}^* KX = \lambda^* \tilde{X}^* GX. \qquad (8.10.27)$$

Thus $\lambda^* = \lambda$ and λ is real. The eigenvector X is then also real and

$$\tilde{X} KX = \lambda \tilde{X} GX \qquad (8.10.28)$$

But $\tilde{X} KX$ and $\tilde{X} GX$ are positive definite quadratic forms. Hence, $\lambda > 0$.

The eigenvectors corresponding to distinct eigenvalues are orthogonal. Let λ and μ be two distinct eigenvalues with eigenvectors X_1 and X_2. Then $KX_1 = \lambda GX_1$ and $KX_2 = \mu GX_2$. Multiply the first equation by $\tilde{X}_2$ and the second by $\tilde{X}_1$ to get

$$\tilde{X}_2 KX_1 = \lambda \tilde{X}_2 GX_1 \qquad \tilde{X}_1 KX_2 = \mu \tilde{X}_1 GX_2 \qquad (8.10.29)$$

By transposing the first equation and combining it with the second we find

$$(\lambda - \mu)\tilde{X}_1 GX_2 = 0$$

or

$$\tilde{X}_1 GX_2 = 0 \qquad (8.10.30)$$

This is just the expression for the dot product of two vectors in a coordinate system with metric G. Hence, the eigenvectors are orthogonal. It will be convenient to choose the eigenvectors as unit vectors so that

$$\tilde{X}_i GX_j = \delta_{ij} \qquad (8.10.31)$$

Let the *rows* of a matrix S be constructed out of the eigenvectors. Then from (8.10.31),

$$SG\tilde{S} = I \qquad (8.10.32)$$

The inverse of S is $G\tilde{S}$. If a diagonal matrix Λ is defined as

$$\Lambda = \begin{pmatrix} \lambda(1) & 0 & 0 & \cdots \\ 0 & \lambda(2) & 0 & \cdots \\ 0 & 0 & \lambda(3) & \cdots \\ \cdot & \cdot & \cdot & \cdots \end{pmatrix} \qquad (8.10.33)$$

then the eigenvalue equations

$$KX_1 = \lambda(1)GX_1, \qquad KX_2 = \lambda(2)GX_2, \ldots \qquad (8.10.34)$$

can all be combined into the single equation

$$K\tilde{S} = \Lambda G\tilde{S} \qquad (8.10.35)$$

We introduce a new set of coordinates by means of

$$Q = \tilde{S}Q' \qquad (8.10.36)$$

The equation of motion (8.10.22) becomes

$$G\tilde{S}\ddot{Q}' + K\tilde{S}Q' = 0 \qquad (8.10.37)$$

Multiplying by S and using (8.10.32) and (8.10.35),

$$\ddot{Q}' + \Lambda Q' = 0 \qquad (8.10.38)$$

In component form

$$\ddot{q}_i' + \lambda(i)q_i' = 0 \qquad (8.10.39)$$

Thus q_i' are the normal coordinates.

Let us consider once more the equations of motion (8.9.14) and (8.9.15) with $k_1 = k_2 = k$ and $m_1 = m_2 = m$. According to the general scheme just developed, we should substitute a solution of the form

$$x_1 = A_1 \cos(\omega t - \alpha) \qquad x_2 = A_2 \cos(\omega t - \alpha) \qquad (8.10.40)$$

We find

$$(k + k_3 - m\omega^2)A_1 - k_3 A_2 = 0$$
$$-k_3 A_1 + (k + k_3 - m\omega^2)A_2 = 0 \qquad (8.10.41)$$

To obtain a nonzero solution for A_1 and A_2, it must be true that

$$\begin{vmatrix} k + k_3 - m\omega^2 & -k_3 \\ -k_3 & k + k_3 - m\omega^2 \end{vmatrix} = 0 \qquad (8.10.42)$$

which is recognized as the secular equation. The roots are

$$\omega_1^2 = \frac{k}{m} \qquad \omega_2^2 = \frac{k + 2k_3}{m} \qquad (8.10.43)$$

For $\omega = \omega_1$ (8.10.41) gives $A_1 = A_2$ and for $\omega = \omega_2$ the results is $A_1 = -A_2$. The general solution is a superposition of these two solutions:

$$\begin{pmatrix} x_1 \\ x_2 \end{pmatrix} = \begin{pmatrix} A_1 \\ A_1 \end{pmatrix} \cos(\omega_1 t - \alpha_1) + \begin{pmatrix} -A_2 \\ A_2 \end{pmatrix} \cos(\omega_2 t - \alpha_2) \qquad (8.10.44)$$

This is identical to the solution as given by (8.10.2), (8.10.7), and (8.10.8).

8.11. MOTION OF A RIGID BODY ABOUT A FIXED AXIS

A rigid body can be regarded as a system of a very large number of particles, but the number of constraints is also large so that the number of degrees of freedom are few. The nature of the constraints is such that the particles making up the body are held fixed relative to one another. If a body is in the form of a thin lamina and is mounted on a shaft which is perpendicular to the plane of the lamina, there is only one degree of freedom which we may choose to be the angle of rotation about the shaft. Let O represent the fixed axis of rotation in Fig. 8.11.1 and let $\mathbf{r}_\alpha$ be the

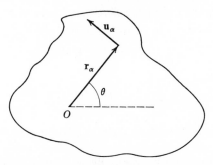

Fig. 8.11.1

position vector of a typical particle of the body relative to this axis. $\mathbf{r}_\alpha$ is perpendicular to the axis of rotation and the particle velocity $\mathbf{u}_\alpha$ is perpendicular to $\mathbf{r}_\alpha$. The angular momentum of the body about O is in the direction of the shaft and can be expressed

$$\mathbf{l} = \sum_{\alpha=1}^{N} \mathbf{r}_\alpha \times m_\alpha \mathbf{u}_\alpha = \hat{\mathbf{n}} \sum_{\alpha=1}^{N} m_\alpha r_\alpha u_\alpha = \hat{\mathbf{n}} \sum_{\alpha=1}^{N} (m_\alpha r_\alpha^2)\dot{\theta} \qquad (8.11.1)$$

where $\hat{n}$ is a unit vector along the axis of rotation, i.e., perpendicular to the plane of the page in Fig. 8.11.1, and the magnitude of the velocity has been expressed

$$u_\alpha = r_\alpha \dot{\theta} \qquad (8.11.2)$$

The quantity

$$I = \sum_{\alpha=1}^{N} (m_\alpha r_\alpha^2) = \int r^2 \, dm \qquad (8.11.3)$$

is known as the *moment of inertia* of the body about the axis through O. As indicated, the sum is replaced by an integral if the body is continuous. In terms of the moment of inertia, the magnitude of the angular momentum is

$$l = I\dot{\theta} \qquad (8.11.4)$$

The kinetic energy of the body is also expressible in a simple way in terms of the moment of inertia:

$$T = \sum_{\alpha=1}^{N} \tfrac{1}{2} m_\alpha u_\alpha{}^2 = \tfrac{1}{2} \left(\sum_{\alpha=1}^{N} m_\alpha r_\alpha{}^2 \right) \dot{\theta}^2 = \tfrac{1}{2} I \dot{\theta}^2 \qquad (8.11.5)$$

The body has been assumed to be in the shape of a thin lamina but a body of more general shape which rotates about a fixed axis can be thought of as being built up out of a series of such thin lamina.

An interesting theorem, known as the *law of parallel axes*, allows the moment of inertia about any axis to be calculated once the moment of inertia about a parallel axis through the center of mass is known. If $\mathbf{r}_\alpha{}'$ is the position vector of a typical particle relative to an axis through the center of mass and $\mathbf{R}$ is the vector from 0 to the center of mass, then

$$\mathbf{r}_\alpha = \mathbf{r}_\alpha{}' + \mathbf{R} \qquad (8.11.6)$$

and

$$
\begin{aligned}
I &= \sum_{\alpha=1}^{N} m_\alpha r_\alpha{}^2 = \sum_{\alpha=1}^{N} m_\alpha (\mathbf{r}_\alpha{}' + \mathbf{R}) \cdot (\mathbf{r}_\alpha{}' + \mathbf{R}) \\
&= \sum_{\alpha=1}^{N} m_\alpha r_\alpha{}'^2 + 2\mathbf{R} \cdot \sum_{\alpha=1}^{N} m_\alpha \mathbf{r}_\alpha{}' + \sum_{\alpha=1}^{N} m_\alpha R^2 = I_c + MR^2
\end{aligned}
\qquad (8.11.7)
$$

where I_c is the moment of inertia of the body about an axis through the center of mass and parallel to the original axis through O.

As an example, the moment of inertia of a uniform disc of radius a about an axis through its center and perpendicular to the plane of the disc is found by noting that the mass of a ring of radius r and thickness dr is

$$dm = \lambda 2\pi r \, dr \qquad (8.11.8)$$

where $\lambda = M/(\pi a^2)$ is the mass per unit area of the disc. Then

$$I_c = \int_{r=0}^{a} r^2 \, dm = \frac{2M}{a^2} \int_0^a r^3 \, dr = \tfrac{1}{2} Ma^2 \qquad (8.11.9)$$

The moment of inertia about a parallel axis passing through a point on the rim of the disc is found by using (8.11.7):

$$I = I_c + Ma^2 = \tfrac{1}{2} Ma^2 + Ma^2 = \tfrac{3}{2} Ma^2 \qquad (8.11.10)$$

It is usual to define the *radius of gyration* of a body about an axis by means of

$$I = Mk^2 \qquad (8.11.11)$$

For a uniform disc of radius a, the radius of gyration about an axis through the center is

$$k_c = \frac{1}{\sqrt{2}} a \qquad (8.11.12)$$

By means of (8.11.4) it is found that a torque on a body about the axis of rotation produces an angular acceleration given by

$$N = \frac{dl}{dt} = I\ddot{\theta} \qquad (8.11.13)$$

which is to be regarded as a statement of Newton's second law.

A simple example is provided by the physical or compound pendulum which consists of a rigid body pivoted about an axis through O (Fig. 8.11.2) and acted on by a uniform gravitational field. It was shown in Section 8.3 Eqs. (8.3.11) and (8.3.12), that all the effects of a uniform gravitational field can be accounted for by assuming that a force equal to the weight acts at the center of mass of the body. Let R be the distance from the pivot O to the center of mass. The torque exerted about point O by the gravitational field is therefore $N = -MgR \sin \theta$ and the equation of motion is

$$I\ddot{\theta} + MgR \sin \theta = 0 \qquad (8.11.14)$$

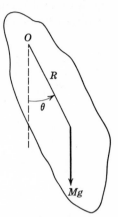

Fig. 8.11.2

The equation of motion of a compound pendulum is seen to be identical in form to that of a simple plane pendulum. For oscillations of small angular amplitude, the frequency is given by

$$\omega = \sqrt{MgR/I} = \sqrt{gR/k^2} \qquad (8.11.15)$$

where k is the radius of gyration about O. By means of the parallel axis theorem

$$k^2 = k_c{}^2 + R^2 \qquad (8.11.16)$$

where k_c is the radius of gyration about the center of mass. Hence,

$$\omega = \sqrt{gR/(k_c{}^2 + R^2)} \qquad (8.11.17)$$

which expresses the frequency as a function of the distance between the pivot and the center of mass.

Since a rigid body is a large number of particles subject to nondissipative constraints, the Lagrange formulation of mechanics applies. The Lagrangian for the compound pendulum is

$$L = \tfrac{1}{2}I\dot{\theta}^2 - MgR(1 - \cos \theta) \qquad (8.11.18)$$

and the equation of motion (8.11.14) follows.

An *Atwood machine* consists of a pulley of moment of inertia I over which passes a cord from which are suspended two masses m_1 and m_2.

Let the positions of the two masses be measured relative to the same level as in Fig. 8.11.3. The equations of constraint connecting the three coordinates x_1, x_2, and θ are

$$x_1 = x_2 = a\theta \tag{8.11.19}$$

so that the system actually has only one degree of freedom. Friction in the bearing of the pulley will be neglected. Let us adopt the Lagrange method with its obvious advantages. The potential and kinetic energies

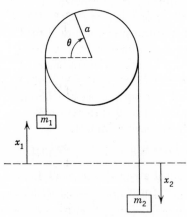

Fig. 8.11.3

of the system are

$$V = m_1 g x_1 - m_2 g x_2 = (m_1 - m_2)ga\theta \tag{8.11.20}$$

$$T = \tfrac{1}{2}I\dot{\theta}^2 + \tfrac{1}{2}m_1\dot{x}_1{}^2 + \tfrac{1}{2}m_2\dot{x}_2{}^2$$
$$= \tfrac{1}{2}(I + m_1 a^2 + m_2 a^2)\dot{\theta}^2 \tag{8.11.21}$$

where x_1 and x_2 have been eliminated by using (8.11.19). The Lagrangian and the resulting equation of motion are

$$L = \tfrac{1}{2}(I + m_1 a^2 + m_2 a^2)\dot{\theta}^2 - (m_1 - m_2)ga\theta \tag{8.11.22}$$

$$\ddot{\theta} = \frac{(m_2 - m_1)ga}{I + m_1 a^2 + m_2 a^2} \tag{8.11.23}$$

It is interesting to note that a frictional torque in the bearing can be included by writing the Lagrange equation as

$$\frac{d}{dt}\frac{\partial L}{\partial \dot{\theta}} - \frac{\partial L}{\partial \theta} = N_f \tag{8.11.24}$$

There results, in place of (8.11.23),

$$\ddot{\theta} = \frac{(m_2 - m_1)ga + N_f}{I + m_1 a^2 + m_2 a^2} \qquad (8.11.25)$$

Generally, $N_f < 0$ when $\dot{\theta} > 0$ and $N_f > 0$ when $\dot{\theta} < 0$.

8.12. MOTION OF A RIGID BODY PARALLEL TO A PLANE

In this section, we consider a rigid body which is constrained to move parallel to a plane but not about a fixed axis, as was the case in Section 8.11. Examples are a wheel rolling in a straight line or an object sliding across a horizontal surface.

According to (8.3.3), the motion of the center of mass of a body when external forces are applied is the same as that of a single particle of the same mass as the body independently of where the external forces are actually applied:

$$\mathbf{F} = M\mathbf{a}_c \qquad (8.12.1)$$

The most general displacement of a rigid body can be constructed from a translation of the center of mass plus a rotation about the center of mass. According to equation (8.3.9), the total angular momentum of a body calculated about an arbitrary fixed reference point in space separates into a part which is equivalent to the angular momentum of a single point particle located at the center of mass and a part which is the angular momentum about the center of mass:

$$\mathbf{l} = \mathbf{R} \times M\mathbf{v}_c + \mathbf{l}_c \qquad (8.12.2)$$

The total torque on the body calculated about the fixed reference point therefore also divides into two parts:

$$\mathbf{N} = \frac{d\mathbf{l}}{dt} = \mathbf{R} \times \mathbf{F} + \frac{d\mathbf{l}_c}{dt} \qquad (8.12.3)$$

Since the motion of the center of mass is already taken care of by (8.12.1) it is necessary only to consider the torques on the body about the center of mass:

$$\mathbf{N}_c = \frac{d\mathbf{l}_c}{dt} \qquad (8.12.4)$$

Since the rigid body is constrained to move in two dimensions, $\mathbf{N}_c$ and $\mathbf{l}_c$ are always in the same direction and, therefore, it is possible to replace (8.12.4) by the scalar equation

$$N_c = I_c \ddot{\theta} \qquad (8.12.5)$$

where θ is the angle of rotation of the rigid body, and I_c is the moment of inertia about an axis through the center of mass. According to (8.3.10), the kinetic energy of the body can be written

$$T = \tfrac{1}{2}Mv_c^2 + \tfrac{1}{2}I_c\dot{\theta}^2 \tag{8.12.6}$$

Equations (8.12.1), (8.12.5), and (8.12.6) can be regarded as three fundamental equations for the description of the motion of a rigid body that is constrained to move in a plane. Of course, the Lagrange formulation presents an alternative approach, and this method will be illustrated by examples.

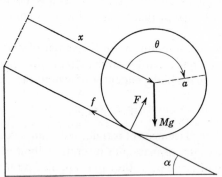

Fig. 8.12.1

It is essential in applying (8.12.5) to use the center of mass as the point about which torques are computed except, of course, in the special case where the body is constrained to rotate about a fixed axis. Consider, for instance, a uniform rod which is falling freely in the earth's gravitational field. The gravitational field produces no net torque about the center of mass and it follows from (8.12.5) that $\ddot{\theta} = 0$ meaning that the rod experiences no angular acceleration. However, if one end of the rod is used as a reference point it is found that the torque about this point due to the gravitational field is not zero! If $N = I\ddot{\theta}$ is applied about this point an erroneous conclusion is drawn.

Many of the principles are illustrated by the simple example of a wheel rolling down an inclined plane. The three forces which act on the body are its weight, a force F exerted by the plane in a direction perpendicular to the surface, and a friction force f (Fig. 8.12.1). The forces f and F are forces of constraint. The condition of rolling is actually a constraint and implies the existence of the friction force f. Without a sufficiently high coefficient of friction between wheel and plane, the wheel would slip and the more complicated situation of a force of constraint which involves dissipation of energy would result.

If x represents the linear displacement of the center of mass and θ is the angle of rotation of the wheel, then the equation of constraint connecting these two coordinates is

$$x = a\theta \tag{8.12.7}$$

where a is the radius of the wheel. By application of (8.12.1), the equation expressing the motion of the center of the wheel is

$$Mg \sin \alpha - f = M\ddot{x} = Ma\ddot{\theta} \tag{8.12.8}$$

where α is the angle of the incline. Equation (8.12.5) yields

$$fa = I_c\ddot{\theta} \tag{8.12.9}$$

The elimination of f between (8.12.8) and (8.12.9) results in

$$\ddot{\theta} = \frac{Mga \sin \alpha}{Ma^2 + I_c} = \frac{ga \sin \alpha}{a^2 + k_c^2} \tag{8.12.10}$$

where k_c is the radius of gyration of the wheel about its center of mass.

The Lagrange formulation has the advantage that it allows us to ignore the forces of constraint. The kinetic and potential energies are

$$T = \tfrac{1}{2}I_c\dot{\theta}^2 + \tfrac{1}{2}M\dot{x}^2 = \tfrac{1}{2}(I_c + Ma^2)\dot{\theta}^2 \tag{8.12.11}$$

$$V = -Mgx \sin \alpha = -Mga\theta \sin \alpha \tag{8.12.12}$$

where use has been made of (8.12.6). The Lagrangian is

$$L = \tfrac{1}{2}(I_c + Ma^2)\dot{\theta}^2 + Mga\theta \sin \alpha \tag{8.12.13}$$

and (8.12.10) follows from

$$\frac{d}{dt}\frac{\partial L}{\partial \dot{\theta}} - \frac{\partial L}{\partial \theta} = 0 \tag{8.12.14}$$

A third approach is the direct use of the conservation of energy principle. If the wheel starts from rest at $x = 0$, its angular velocity after it has rolled a distance $x = a\theta$ is found from

$$\tfrac{1}{2}(I_c + Ma^2)\dot{\theta}^2 = Mga\theta \sin \alpha \tag{8.12.15}$$

Suppose that a ball bearing of radius a rolls back and forth on a circular track of radius s as in Fig. 8.12.2. The angle ϕ with vertex at the center of the circle of radius s measures the position of the center of mass of the sphere. The sphere rotates through an angle θ when the center of mass moves through an angle ϕ. During this motion, the point P moves to P'. Since the arc lengths PC and $P'C$ are equal, the equation of constraint connecting the coordinates θ and ϕ is

$$s\phi = a(\theta + \phi) \tag{8.12.16}$$

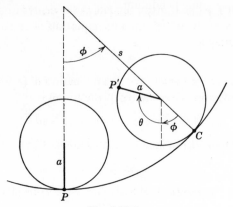

Fig. 8.12.2

We note that the velocity of the center of the sphere is $(s - a)\dot{\phi}$ and that the kinetic energy, therefore, is

$$T = \tfrac{1}{2}M(s - a)^2\dot{\phi}^2 + \tfrac{1}{2}I_c\dot{\theta}^2$$

$$= \tfrac{1}{2}M\left[1 + \frac{k_c^2}{a^2}\right](s - a)^2\dot{\phi}^2$$

(8.12.17)

where $I_c = Mk_c^2$ and $\dot{\theta}^2$ has been eliminated by using (8.12.16). The potential energy of the sphere is

$$V = Mg(s - a)(1 - \cos\phi)$$

(8.12.18)

By means of

$$L = T - V, \quad \frac{d}{dt}\frac{\partial L}{\partial\dot{\phi}} - \frac{\partial L}{\partial\phi} = 0$$

(8.12.19)

the equation of motion is found to be

$$\left(1 + \frac{k_c^2}{a^2}\right)(s - a)\ddot{\phi} + g\sin\phi = 0$$

(8.12.20)

For small amplitudes, the frequency of the motion is

$$\omega = \sqrt{\frac{g}{(s - a)\left(1 + \frac{k_c^2}{a^2}\right)}}$$

(8.12.21)

For a sphere $k_c^2 = (2/5)a^2$ and

$$\omega = \sqrt{\frac{5}{7}\frac{g}{s - a}}$$

(8.12.22)

In the limit $a \to 0$, the sphere presumably becomes a "particle" with zero moment of inertia. It may seem that the frequency should become the same as that of a plane pendulum. Actually, the limit is

$$\lim_{a \to 0} \omega = \sqrt{\frac{5}{7} \frac{g}{s}} \tag{8.12.23}$$

The constraint of rolling is responsible for the factor of $(5/7)$. If the track is frictionless so that the ball bearing slides rather than rolls, the frequency becomes identical to that of a plane pendulum, $\sqrt{g/s}$, in the limit $a \to 0$.

8.13. KINEMATICS OF RIGID-BODY MOTION

Let us suppose that our rigid body which is constrained to move in a plane is in the form of a thin lamina and has a rectangular Cartesian

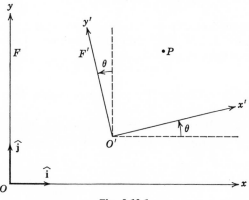

Fig. 8.13.1

coordinate system F' painted on it as in Fig. 8.13.1. The frame F is stationary in the laboratory. It is usual to refer to F as *the space set of axes* and to F' as *the body set of axes*. Since the z- and z'-axes always remain parallel, it is possible to specialize the transformation (8.1.2) to

$$x = x' \cos \theta - y' \sin \theta + X \tag{8.13.1}$$

$$y = x' \sin \theta + y' \cos \theta + Y \tag{8.13.2}$$

where (x, y) and (x', y') are the coordinates of the point P relative to F and F', respectively, and (X, Y) are the coordinates of the origin O' relative to F. The point P is to be regarded as fixed within the body. For the moment, the origin O' is arbitrary and need not coincide with some special point of the body, such as the center of mass. The position of the rigid body is completely specified once the translation of O' and the rotation of F', i.e., X, Y, and θ, are known. The rigid body is, in general, both rotating and translating so that X, Y, and θ are functions of time. Thus (8.13.1) and (8.13.2) provide a simple example of a time-dependent orthogonal transformation.

The velocity of point P is zero with respect to F'. Its velocity with respect to F is obtained by differentiation of (8.13.1) and (8.13.2):

$$\dot{x} = (-x' \sin\theta - y' \cos\theta)\dot{\theta} + v_x = -(y - Y)\dot{\theta} + v_x \quad (8.13.3)$$

$$\dot{y} = (x' \cos\theta - y' \sin\theta)\dot{\theta} + v_y = (x - X)\dot{\theta} + v_y \quad (8.13.4)$$

where $\dot{\theta}$ is the instantaneous angular velocity of rotation of F' relative to F. At a given instant, the value of $\dot{\theta}$ is the same for all frames of reference rigidly attached to the body. Noting that

$$\mathbf{r}' = (x - X)\hat{\mathbf{i}} + (y - Y)\hat{\mathbf{j}} \quad (8.13.5)$$

is the displacement vector from O' to P and writing the angular velocity as $\boldsymbol{\omega} = \hat{\mathbf{k}}\dot{\theta}$, it is seen that (8.13.3) and (8.13.4) can be expressed by

$$\mathbf{u} = (\boldsymbol{\omega} \times \mathbf{r}') + \mathbf{v} \quad (8.13.6)$$

As will be shown in Chapter 10, this result is actually valid for three-dimensional motion.

An interesting result follows from (8.13.3) and (8.13.4). There is always a point of the body frame of reference which is momentarily stationary. Setting $\dot{x} = \dot{y} = 0$ gives

$$x = X - \frac{v_y}{\dot{\theta}} \qquad y = Y + \frac{v_x}{\dot{\theta}} \quad (8.13.7)$$

as the coordinates of this point. This point is at ∞ when $\dot{\theta} = 0$, i.e., when the body is in pure translational motion. As time progresses, the coordinates given by (8.13.7) trace out a curve in the laboratory frame of reference which is known as the *space centrode*. There is always a point of the body frame of reference which is in contact with the space centrode and is instantaneously stationary. The motion of the rigid body can be described as a rotation about this point considered as an *instantaneous fixed axis* (point P of Fig. 8.13.2). As time goes on, there will be a continuum of points *in the body frame of reference* which come into contact with the space centrode. This curve is known as the *body centrode* and the motion of the rigid body can be described by saying that the body centrode rolls with slipping on the space centrode.

The magnitude of the distance from the origin O' of frame F' to the instantaneous point of contact P of the body and space centrodes is

$$b = \sqrt{(x - X)^2 + (y - Y)^2} \quad (8.13.8)$$

By means of (8.13.7),

$$v_x = (y - Y)\dot\theta, \qquad v_y = -(x - X)\dot\theta \qquad v = \sqrt{v_x{}^2 + v_y{}^2} = b\dot\theta$$

$$(8.13.9)$$

Since O' can be chosen arbitrarily, (8.13.9) shows that all points of the rigid body are in rotation with the angular velocity $\dot\theta$ about P as an instantaneous fixed axis. In particular, let O' be the center of mass of the body. Then writing $v_c = b_c\dot\theta$, the kinetic energy (8.12.6) of the body can be expressed by

$$T = \tfrac{1}{2}(I_c + Mb_c{}^2)\dot\theta^2 \qquad (8.13.10)$$

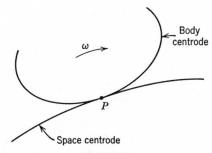

Fig. 8.13.2

By means of the parallel axis theorem (8.11.7) we recognize that $I_c + Mb_c{}^2$ is the moment of inertia of the body about point P. Thus (8.13.10) shows that we may calculate the kinetic energy of the body about P considered as an instantaneous fixed axis, and this fact simplifies the analysis of some problems.

As an example, let us construct a lop-sided wheel by placing a point mass m on its rim. The wheel rolls in a straight line along a level surface; θ is the angle of rotation defined in such a way that $\theta = 0$ when the point mass is at the bottom. Here, the body and space centrodes are known before-hand and are simply the rim of the wheel and the straight line along which it rolls. It is not necessary actually to locate the center of mass of the particle-wheel system. In order to calculate the instantaneous moment of inertia about P (Fig. 8.13.3) we write

$$I_p = I_{\text{wheel}} + I_{\text{particle}} \qquad (8.13.11)$$

If a is the radius of the wheel, the distance from P to the particle is $2a \sin(\theta/2)$. Hence,

$$I_{\text{particle}} = m \cdot \left(2a \sin\frac{\theta}{2}\right)^2 \qquad (8.13.12)$$

and

$$I_p = I_c + Ma^2 + 4ma^2 \sin^2 \frac{\theta}{2} \qquad (8.13.13)$$

where I_c is the moment of inertia of the wheel alone about its center of mass and M is the mass of the wheel alone. We have here a situation in which the moment of inertia is a function of the coordinate θ. The kinetic energy of the system is

$$T = \tfrac{1}{2}\left(I_c + Ma^2 + 4ma^2 \sin^2 \frac{\theta}{2}\right)\dot{\theta}^2 \qquad (8.13.14)$$

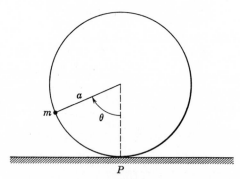

Fig. 8.13.3

Since the potential energy of the wheel does not change, a suitable potential function for the system is

$$V = mga(1 - \cos \theta) \qquad (8.13.15)$$

Forming the Lagrangian and the Lagrangian equations of motion in the usual way yields

$$[I_c + Ma^2 + 2ma^2(1 - \cos \theta)]\ddot{\theta} + ma^2\dot{\theta}^2 \sin \theta + mga \sin \theta = 0$$
$$(8.13.16)$$

It is incorrect to apply the Newtonian equation $N = I\ddot{\theta}$ about point P considered as a fixed axis! The torque about P is $N = -mga \sin \theta$. Using (8.13.13), $N = I\ddot{\theta}$ gives an equation of motion in obvious disagreement with (8.13.16). We can, of course, locate the center of mass of the system and apply (8.12.1) and (8.12.5), but this is tedious in this case.

Equation (8.13.16) is a nonlinear differential equation of a different type than has previously been encountered. Writing

$$I = I_c + Ma^2 \qquad (8.13.17)$$

the total energy is

$$\mathscr{E} = \tfrac{1}{2}\left(I + 4ma^2 \sin^2 \frac{\theta}{2}\right)\dot{\theta}^2 + mga(1 - \cos \theta) \qquad (8.13.18)$$

In spite of the functional dependence of the coefficient of $\dot{\theta}^2$ on θ it is still possible to graph the potential function and obtain a qualitative picture of the motion in much the same way as was done for the plane pendulum. For a given energy, the turning points are found as usual when $\dot{\theta} = 0$, i.e., $\mathscr{E} = V$. The motion is thus qualitatively similar to that of a plane pendulum with periodic solutions occurring when $\mathscr{E} < 2mga$ and nonperiodic solutions occurring when $\mathscr{E} > 2mga$. If the solutions are periodic and of very small amplitude the approximations $\cos \theta = 1$ and $\sin \theta = \theta$ can be made in (8.13.16) yielding

$$I\ddot{\theta} + ma^2\dot{\theta}^2\theta + mga\theta = 0 \tag{8.13.19}$$

The first term in the Fourier series solution is

$$\theta = b_1 \cos \omega t \tag{8.13.20}$$

and if this is substituted into (8.13.19) and all but the terms which are linear in b_1 discarded the result is

$$\omega = \sqrt{mga/I} \qquad I = I_c + ma^2 \tag{8.13.21}$$

8.14. IMPULSE AND MOMENTUM IN RIGID-BODY PROBLEMS

The use of the principle of conservation of angular momentum in solving problems involving rigid bodies is illustrated by an example. Suppose that a disc of radius a is mounted on a shaft of radius b. The shaft rolls on horizontal rails that are just high enough so that the disc clears the floor as in Fig. 8.14.1. The shaft rolls with no slippage on the rails. Eventually, the end of the rails is reached and the disc drops onto the floor at point P. There is friction and after all slippage between floor and disc ceases, what is the final velocity of the disc?

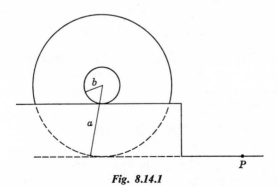

Fig. 8.14.1

When the disc engages the floor at point P, the external forces are those exerted by the floor and the gravitational field. None of these forces exerts a torque about point P. Angular momentum about this point is therefore conserved. Angular momentum about any point above the floor *would not* be conserved because the frictional force between floor and disc would produce a torque about such a point. Initially, the angular momentum about point P is

$$l_p = Mv_ca + I_c\dot{\theta} \tag{8.14.1}$$

where $v_c = b\dot{\theta}$ is the linear velocity of the center of mass and use has been made of (8.3.9). After the disc is rolling on the floor, the angular momentum about P is

$$l_p = Mw_ca + I_c\dot{\phi} \tag{8.14.2}$$

where $w_c = a\dot{\phi}$ is the final linear velocity of the center of mass and $\dot{\phi}$ is the final angular velocity. By equating (8.14.1) and (8.14.2),

$$(Mab + I_c)\dot{\theta} = (Ma^2 + I_c)\dot{\phi} \tag{8.14.3}$$

which determines $\dot{\phi}$ when $\dot{\theta}$ is known.

When the disc first engages the floor, heat is produced until slippage ceases. The amount of energy converted to heat is

$$
\begin{aligned}
H &= \tfrac{1}{2}Mv_c{}^2 + \tfrac{1}{2}I_c\dot{\theta}^2 - \tfrac{1}{2}Mw_c{}^2 - \tfrac{1}{2}I_c\dot{\phi}^2 \\
&= \tfrac{1}{2}(Mb^2 + I_c)\dot{\theta}^2 - \tfrac{1}{2}(Ma^2 + I_c)\dot{\phi}^2
\end{aligned}
\tag{8.14.4}
$$

By using (8.14.3) to eliminate $\dot{\phi}$ this can be expressed as

$$H = \tfrac{1}{2}\frac{MI_c(a - b)^2}{Ma^2 + I_c}\dot{\theta}^2 \tag{8.14.5}$$

If N represents the net torque on a body, either around a fixed axis or around the center of mass, then the *angular impulse* is

$$K = \int N\, dt = \int I\ddot{\theta}\, dt = I\dot{\phi} - I\dot{\theta} \tag{8.14.6}$$

where $\dot{\phi}$ is the final angular velocity and $\dot{\theta}$ is the initial angular velocity.

Suppose that a body resting on a frictionless surface, such as a plank resting on a frozen pond, is struck a sharp blow as in Fig. 8.14.2. The *linear* impulse is then

$$J = \int F\, dt = Mv_c \tag{8.14.7}$$

where v_c is the velocity of the center of mass after the blow and is in the same direction as the impulse. There is also an *angular* impulse around the center of mass given by

$$K_c = \int N_c \, dt = \int FR \, dt = JR \qquad (8.14.8)$$

where R is the perpendicular distance from the center of mass to the line of action of the force. The angular velocity of rotation of the body after the blow is therefore given by

$$JR = Mv_c R = I_c \dot{\phi} \qquad (8.14.9)$$

As we know, there is a point in the body frame of reference, possibly outside the body itself, which is momentarily stationary. The distance b

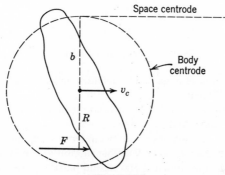

Fig. 8.14.2

of this point from the center of mass is given by $v_c = b\dot{\phi}$ and is measured in a direction perpendicular to v_c as in Fig. 8.14.2. This locates the first point of the space centrode. Since the center of mass moves at a constant velocity v_c and since $\dot{\phi}$ is a constant, the space centrode is a straight line a perpendicular distance $R + b$ from the point where the blow was struck. The motion of the body is the same as the rolling of a disc of radius b along the space centrode. The actual physical body is to be thought of as rigidly attached to this disc. The rim of the disc is, of course, the body centrode. It is evident that the force-free motion of *any* rigid body in two dimensions is equivalent to the rolling of an appropriate disc along a straight line.

PROBLEMS

8.1.1 Obtain Eq. (8.1.3) by transforming Eq. (8.1.2).

8.2.2 If a system of particles is isolated, then necessarily

$$\sum_{\alpha=1}^{N} \mathbf{F}_\alpha = 0 \quad \text{and} \quad \sum_{\alpha=1}^{N} \mathbf{N}_\alpha = 0$$

Is the converse necessarily true?

8.2.3   If the net force on a system of particles is zero, i.e.,

$$\sum_{\alpha=1}^{N} \mathbf{F}_\alpha = 0$$

it is not necessarily true that the net torque is zero. Show, however, that under this circumstance the net torque has the *same value* when taken about any point whatsoever.

8.2.4   Let V_{12} be the potential energy between two particles defined in such a way that $V_{12} = 0$ when the particles are an infinite distance apart. Note that $V_{12} = V_{21}$. If there are N such particles present, show that the potential energy of the system is

$$V = \tfrac{1}{2} \sum_{\alpha=1}^{N} \sum_{\beta=1}^{N} V_{\alpha\beta}$$

provided that

$$V_{11} = V_{22} = \cdots = 0$$

8.3.5   Find the center of mass of a thin hemispherical shell over which a mass M is uniformly distributed.

8.4.6   For two particles which make a head-on *elastic* collision, show that the relative velocity is conserved, i.e., that $u_1 - u_2 = w_2 - w_1$. Find the general expressions for the final velocities w_1 and w_2 in terms of u_1 and u_2. If the second particle is initially at rest ($u_2 = 0$), find the fraction of the initial energy of the first particle which is transmitted to the second particle in the collision process. Show that the greatest transfer of energy occurs when $m_1 = m_2$ and that, in fact, all of the energy is transferred under this circumstance.

8.4.7   A block of wood of mass M rests on a frictionless surface. A bullet of momentum mu_1 traveling parallel to the surface strikes and embeds itself in the block. Show that the fraction of the original kinetic energy which is converted to heat is $M/(m + M)$.

8.4.8   By elimination of θ from (8.4.7) and (8.4.8) show that

$$\frac{q_1}{p_1} = \frac{m_1 \cos\phi + \sqrt{m_2{}^2 - m_1{}^2 \sin^2\phi}}{m_1 + m_2}$$

Neutrons enter a block of graphite and make elastic collisions with the carbon atoms. Show that for one of these collisions

$$\frac{q_1}{p_1} \cong \frac{\cos\phi + 12}{13}$$

What is the order of magnitude of the fractional energy loss of a neutron per collision? This is the principle whereby energetic neutrons are moderated. Give some good reasons why graphite is used for this purpose rather than some other material.

8.6.9   A Galilean frame of reference F' moves with velocity v in the x-direction of frame F. The x- and x'-axes are parallel. If a particle moves at a

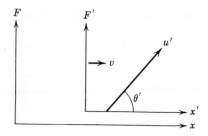

speed u and at an angle θ with the x-axis in frame F, show that its speed and direction in Frame F' are given by

$$\tan \theta' = \frac{\sin \theta}{\cos \theta - (v/u)} \qquad u'^2 = u^2 + v^2 - 2uv \cos \theta$$

8.6.10 Derive Eq. (8.6.16) without using the center of mass coordinate system. If $m_1 = m_2$, what is the ratio of H to the initial kinetic energy in the laboratory coordinate system such that the angle ϕ is never greater than $60°$? Does the fact that $m_1 = m_2$ imply $\phi = \theta$?

8.7.11 An open-topped freight car weighing 10 tons is coasting without friction along a level track. It is raining, with the rain falling vertically. The initial velocity of the car is 2 ft/sec. There is a drain pipe in the floor so that the water runs out as fast as it comes in. What is the velocity of the car after it has traveled long enough for 2 tons of rain water to run through it?

8.7.12 How is the equation of motion (8.7.15) to be modified if the medium which the particle encounters at $x = 0$ is moving in the positive x-direction with velocity v? If the rate of accumulation of mass is proportional to the distance the particle travels *relative to the medium*, show that the solution is

$$x = vt - \frac{m_0}{k} + \sqrt{\frac{m_0^2}{k^2} + \frac{2m_0(u_0 - v)t}{k}}$$

8.7.13 A jet engine moving through still air at 500 ft/sec scoops in air at a rate of 32 lb/sec. Burned fuel and air are expelled at a rate of 40 lb/sec and at an exhaust velocity of 1000 ft/sec measured relative to the engine. What is the thrust of the engine?

8.8.14 In the figure, the two masses are connected by a cord that passes over a massless and frictionless pulley. There is no friction between m_1 and the

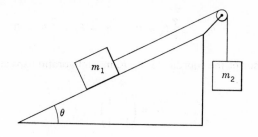

inclined plane. Find the acceleration of the system by the Lagrange method. What is the tension in the cord?

8.8.15   In the figure, $m_2 > m_1$ and k is the force constant of the spring. The pulley is massless and frictionless. Initially, the system is maintained at rest by holding m_1 at $x_1 = 0$. The initial value of x_2 is given by $kx_{20} = m_2 g$. Find the motion by the Lagrange method when the system is released.

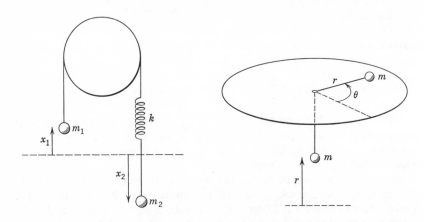

8.8.16   Two particles of equal mass are connected by a string. One particle moves on a horizontal frictionless surface. The string passes through a small hole in the surface so that the other particle hangs below as illustrated. Find the equations of motion. Reduce the problem to an equivalent one-dimensional problem and discuss the qualitative nature of the solution by using an appropriate effective potential. Show in particular that there exists a stable circular orbit for the particle moving on the horizontal surface. By writing $x = 1/r$, find a differential equation for x in terms of θ.

8.8.17   For the problem of the bead sliding without friction on the spoke of a rotating wheel discussed at the end of Section 8.8, answer the following questions. (a) What is the Hamiltonian? (b) Is the Hamiltonian identical to the total energy? (c) Is the Hamiltonian a constant of the motion? (d) Is the total energy a constant of the motion?

8.10.18   For the special case $m_1 = m_2 = m$ and $s_1 = s_2 = s$, the linearized equations (8.9.25) and (8.9.26) for the double pendulum can be expressed

$$2\ddot{\theta}_1 + \ddot{\theta}_2 + 2\frac{g}{s}\theta_1 = 0 \qquad \ddot{\theta}_1 + \ddot{\theta}_2 + \frac{g}{s}\theta_2 = 0$$

The metric tensor of the coordinates θ_1, θ_2 in configuration space can therefore be expressed

$$G = \begin{pmatrix} 2 & 1 \\ 1 & 1 \end{pmatrix}$$

(a) In the configuration space, what is the angle between the θ_1- and θ_2-axes? Refer to Eq. (2.16.20). (b) Find the general solution of the equations of motion. Describe qualitatively what the motion looks like for the two cases where only one mode of vibration is present. (c) Sketch the coordinate axes θ_1 and θ_2 in the configuration space and indicate the directions of the two eigenvectors. **Caution:** The components of the eigenvectors are *contravariant*. Recall the difference between the contravariant components of a vector and its physical components. See Eq. (2.18.6). (d) Normalize the eigenvectors according to (8.10.31) and find the normal coordinates. Sketch the coordinate axes of the normal coordinate system and show that they coincide with the contravariant components of the eigenvectors. *Hint:* Consider a point with coordinates $(q_1', 0)$ and use (8.10.36).

8.10.19 Two plane pendulums of length s are connected by a coupling spring of force constant k_3. There is no tension in the spring when $\theta_1 = \theta_2 = 0$.

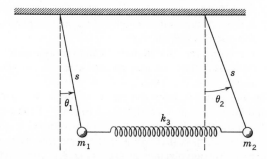

Find the Lagrangian and the linearized differential equations of motion for very small amplitudes. Note that the equations of motion are identical in form to (8.9.14) and (8.9.15).

8.10.20 Three equal masses rest on a frictionless track and are joined together by identical springs. The end springs are rigidly attached to supports. Find the normal modes of vibration and write the expression for the general

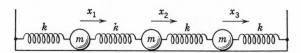

solution. Describe the motion qualitatively when the system is vibrating in each of its normal modes. Can you guess what the normal modes are going to look like if there are four equal masses and five identical springs? Verify your guess by making a calculation.

8.10.21 A thin hollow tube of mass M and length a slides without friction on a horizontal wire. On each end hang simple plane pendulums of mass m and length s. The position of the end of the tube is measured by a coordinate z; θ, and θ_2 are the angular displacements of the pendulums. Find a Lagrangian for the system. Find the differential equations of motion in the linear approximation.

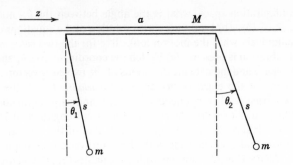

8.11.22 Find the moments of inertia of:

(a) A uniform disc of radius a about a diameter.

(b) A uniform sphere of radius a about an axis through its center.

(c) A thin uniform rod of length s about an axis perpendicular to the rod and through one end.

(d) A uniform thin disc of mass M and radius a with a point mass m on its rim about an axis perpendicular to the plane of the disc and through the center.

8.11.23 Find the distance from the center of mass that a plane physical pendulum should be pivoted in order that the frequency of the oscillations shall be a maximum. Show that, in general, there are two possible distances of the pivot from the center of mass that will give the same frequency of oscillation.

8.11.24 Solve the Atwood machine problem (Fig. 8.11.3) entirely by Newtonian methods. What are the tensions in the cord on the two sides of the pulley? Neglect friction in the bearing of the pulley.

8.11.25 A uniform rod of moment of inertia I about its center is pivoted at its center. A bead of mass m is placed on the rod and slides without friction.

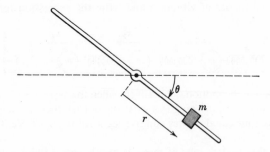

Find the differential equations of motion of the system. Do this problem by both the Lagrangian and the Newtonian methods. The system moves in a vertical plane.

8.11.26 A wire stretched tight between two supports has two uniform rods of length s and mass m mounted on it at equal intervals. The rods are perpendicular to the wire. When the wires are twisted they exert restoring torques proportional to the angle of twist, e.g., the wire between the support and the first rod exerts at torque $N_1 = -\tau\theta_1$ where τ is a constant. Find the potential energy of the system when the rods are at angles θ_1 and θ_2. Find the Lagrangian and the differential equations of motion of the system. The equations of motion should be identical in form to (8.9.14) and (8.9.15) if $m_1 = m_2$ and $k_1 = k_2 = k_3$.

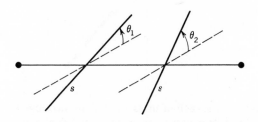

8.12.27   Two cylinders have the same mass and the same radius, but one is hollow and has all its mass concentrated in a thin layer near the surface. Both cylinders are allowed to roll down a plane, starting from rest. Find the ratio of the linear velocities of the two cylinders after each has rolled the same distance.

8.12.28   A cart of total mass m_1 (including wheels) with four identical wheels each of moment of inertia I and radius a moves on a horizontal surface. Attached

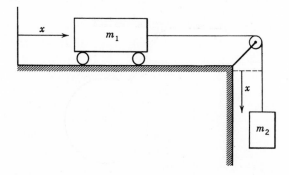

to the cart is a cord that passes over a massless and frictionless pulley to a hanging weight of mass m_2. Compute the acceleration of the system. Neglect friction in the bearings of the wheels, but do assume that there is friction between wheels and level surface so that the wheels do not slide.

8.12.29   A uniform thin rod of length s and mass M is hung from one end by a string of length r. The motion is confined to the plane of the paper. Find the linearized differential equations of motion for small amplitudes of vibration. The equations of motion should be of the same general form as (8.9.25) and (8.9.26).

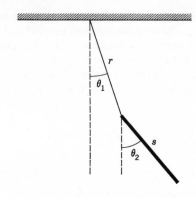

8.12.30   Two particles, each of mass 1 gram, are tied together by a string of length 1 cm. The system moves on a horizontal frictionless surface and is spinning about its center of mass with an angular velocity of 40 sec^{-1}. The center of mass is moving at a linear speed of 20 cm/sec. The string breaks and one of the particles is observed to move at an angle of 60° with the original direction of motion of the center of mass. Find the magnitudes of the final momenta of both particles. Also find the direction of motion of the second particle.

8.12.31   The center of mass of a wheel is constrained to move on the arc of a circle by means of a weightless connecting rod as in the diagram. The bearings at both ends of the connecting rod are frictionless and the system moves in a

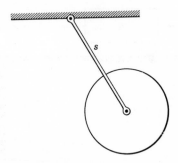

vertical plane. Show that the motion of the center of mass of the wheel is exactly the same as that of a simple plane pendulum of the same length as the connecting rod.

8.12.32   A sphere of radius a is placed almost at the top of a fixed sphere of radius b. As the sphere of radius a rolls down, at what angular displacement will the spheres separate? Assume that there is sufficient friction so that rolling takes place as long as the spheres are in contact.

8.13.33   Carry the approximate solution of (8.13.16) another step by retaining all terms up to third order, i.e., terms proportional to θ^3, θ^2, $\ddot{\theta}$, $\dot{\theta}^2\,\theta$, etc. Substitute a solution of the form

$$\theta = b_1 \cos \omega t + b_3 \cos 3\omega t$$

and compute b_3 and ω in terms of b_1.

8.14.34 Two particles of mass m are at rest and connected by a rigid massless rod of length s. A third particle also of mass m, and moving with a velocity u

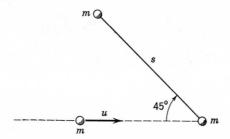

at an angle of 45° with the rigid rod, strikes one of the particles and sticks to it as in the figure. Find the linear velocity of the center of mass and the angular velocity of rotation of the system after the collision. Locate the space centrode.

9

The Two-Body Central Force Problem and Scattering Theory

In the discussion of a system of a large number of particles, a simplification results if the number of constraints is large so that the number of degrees of freedom is small, such as in the example of a rigid body. We now pass to the other extreme where the system consists of only two interacting particles and find that such a system is again mathematically tractable even in the absence of constraints. It is interesting that the general problem of three interacting particles has never been solved in general except by numerical methods.

The motion of a planet around the sun can be treated to a first approximation as a two-body central force problem. The effects of other planets in the solar system are then introduced as perturbations. The fact that the planets are not point particles introduces no major effect since the mass distribution of a planet is very nearly spherically symmetric and, therefore, the gravitational field is almost the same as that of a true point particle. Moreover, any effects of the nonspherical distribution of mass is mitigated by the circumstance that interplanetary distances are generally very large compared to the dimensions of the planets themselves. The nonspherical distribution of mass in the earth, however, does introduce measurable perturbations in the orbits of artificial satellites. The study of these perturbations is useful in determining the distribution of matter in the earth.

Historically, the problem of the scattering of α-particles by atomic nuclei was first treated as a two-body problem by the methods of classical mechanics.

9.1. EQUATIONS OF MOTION OF TWO INTERACTING POINT PARTICLES

If there are no forces present other than those which the two particles exert on one another, the center of mass moves at a constant velocity. It is convenient to work in the Galilean frame of reference in which the center of mass is stationary, i.e., in the center of mass coordinate system. The total angular momentum and the total kinetic energy with respect to

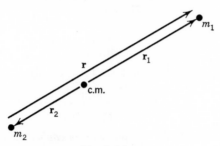

Fig. 9.1.1

the center of mass coordinate system are

$$l = r_1 \times m_1 u_1 + r_2 \times m_2 u_2 \tag{9.1.1}$$

$$T = \tfrac{1}{2}m_1 u_1^2 + \tfrac{1}{2}m_2 u_2^2 \tag{9.1.2}$$

where r_1 and r_2 are the position vectors of the two particles measured from the center of mass (Fig. 9.1.1) and u_1 and u_2 are the particle velocities with respect to the center of mass:

$$u_1 = \frac{dr_1}{dt} \qquad u_2 = \frac{dr_2}{dt} \tag{9.1.3}$$

A simplification results if the position vector r of m_1 with respect to m_2 is introduced:

$$r = r_1 - r_2 \tag{9.1.4}$$

Since r_1 and r_2 are position vectors with respect to the center of mass,

$$m_1 r_1 + m_2 r_2 = 0 \tag{9.1.5}$$

If (9.1.4) and (9.1.5) are solved for r_1 and r_2, the result is

$$r_1 = \frac{m_2}{m_1 + m_2}\, r \qquad r_2 = -\frac{m_1}{m_1 + m_2}\, r \tag{9.1.6}$$

By differentiation with respect to the time

$$\mathbf{u}_1 = \frac{m_2}{m_1 + m_2}\,\mathbf{u} \qquad \mathbf{u}_2 = -\frac{m_1}{m_1 + m_2}\,\mathbf{u} \qquad (9.1.7)$$

where $\mathbf{u} = d\mathbf{r}/dt$ is the velocity of m_1 with respect to m_2. By substituting (9.1.6) and (9.1.7) into (9.1.1) and (9.1.2) it is found that the angular momentum and kinetic energy can be expressed by

$$\mathbf{l} = \mathbf{r} \times \frac{m_1 m_2}{m_1 + m_2}\,\mathbf{u} = \mathbf{r} \times m\mathbf{u} \qquad (9.1.8)$$

$$T = \frac{1}{2}\frac{m_1 m_2}{m_1 + m_2}\,u^2 = \tfrac{1}{2}mu^2 \qquad (9.1.9)$$

where m is called the *reduced mass* and is given by

$$m = \frac{m_1 m_2}{m_1 + m_2} \qquad (9.1.10)$$

The motion of a system of two particles is *equivalent* to the motion of a single particle of mass m about one of them considered as a fixed center of force. Of course, neither of the particles is fixed. The only true fixed point is the center of mass, and it must be remembered that (9.1.8) and (9.1.9) give the angular momentum and kinetic energy measured in a Galilean frame of reference in which the center of mass is stationary. If one of the particles is much more massive than the other, $m_2 \gg m_1$, then the massive particle is almost coincident with the center of mass and the reduced mass is nearly the same as the mass of the lighter particle:

$$m = m_1 \frac{1}{1 + (m_1/m_2)} \cong m_1\left(1 - \frac{m_1}{m_2}\right) \cong m_1 \qquad (9.1.11)$$

Since there are no external forces, the angular momentum vector $\mathbf{l}$ is a constant. Therefore, in the center of mass coordinate system, the motion takes place in a plane. Only two coordinates are needed for the description of the motion and these are chosen to be the plane polar coordinates (r, θ) of the mass m_1 referred to m_2 as the origin.

A force which depends only on the magnitude of the distance between the two particles and is directed along the line joining them is called a *central force*. It can be written

$$\mathbf{F} = F(r)\hat{\mathbf{r}} \qquad (9.1.12)$$

For example, the gravitational force between two point particles is $F(r) = -Gm_1 m_1/r^2$. The force of attraction between the earth and an artificial satellite is approximately the same, but there are correction terms depending on direction due to the nonspherically symmetric distribution of matter in the earth. The force as given by (9.1.12) is derivable from a scalar potential and the Lagrangian for the two interacting particles, therefore, is

$$L = \tfrac{1}{2}m(\dot{r}^2 + r^2\dot{\theta}^2) - V(r) \tag{9.1.13}$$

Remember that m is the *reduced mass*. The equations of motion are found to be

$$m\ddot{r} - mr\dot{\theta}^2 = -\frac{dV}{dr} = F(r) \tag{9.1.14}$$

$$mr^2\dot{\theta} = l = \text{const.} \tag{9.1.15}$$

where l is the magnitude of the angular momentum of the system. The elimination of $\dot{\theta}$ between (9.1.14) and (9.1.15) yields a differential equation for r as a function of t:

$$m\ddot{r} - \frac{l^2}{mr^3} = F(r) \tag{9.1.16}$$

Introducing the change of dependent variable

$$x = \frac{1}{r} \tag{9.1.17}$$

and using Eqs. (6.8.14) and (6.8.15), a differential equation for the orbit can be derived:

$$\left(\frac{d^2x}{d\theta^2} + x\right)x^2 = -\frac{m}{l^2}F(r) \tag{9.1.18}$$

The total energy of the particle can be expressed in various ways as follows:

$$\mathcal{E} = \tfrac{1}{2}m(\dot{r}^2 + r^2\dot{\theta}^2) + V(r) \tag{9.1.19}$$

$$\mathcal{E} = \tfrac{1}{2}m\dot{r}^2 + \frac{l^2}{2mr^2} + V(r) \tag{9.1.20}$$

$$\mathcal{E} = \frac{1}{2}\frac{l^2}{m}\left(\frac{dx}{d\theta}\right)^2 + \frac{l^2x^2}{2m} + V(r) \tag{9.1.21}$$

9.2. THE GENERAL POWER LAW OF FORCE

Of frequent occurrence in practice is a central force given by

$$F(r) = -\frac{C}{r^n} \tag{9.2.1}$$

where C and n are constants. If $C > 0$, the force is one of attraction between the two particles, and if $C < 0$, the force is repulsive. We will refer to C as the *force constant*. Except for the case $n = 1$, the potential function is

$$V(r) = -\frac{C}{n-1}\frac{1}{r^{n-1}} = \frac{-C}{n-1}x^{n-1} \qquad (9.2.2)$$

The equation for the orbit is

$$\frac{d^2x}{d\theta^2} + x = \frac{mC}{l^2}x^{n-2} \qquad (9.2.3)$$

A charged particle moving in a radial electric field is an example of the case $n = 1$ and is treated in Section 7.4. With some minor changes in the

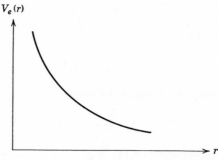

$V_e(r)$

r

Fig. 9.2.1

constants, the orbit equation for a particle constrained to move on a cone, Eq. (6.8.16), results if $n = 0$. Examples of $n = 2$ are the gravitational force between point masses and the electrostatic force between point charges.

From (9.1.20) and (9.2.2) it is seen that the effective potential for the radial motion is

$$V_e(r) = \frac{l^2}{2mr^2} - \frac{C}{n-1}\frac{1}{r^{n-1}} \qquad (9.2.4)$$

For $C < 0$, the force is repulsive. $V_e(r)$ has no relative maxima or minima for any value of n ($n = 1$ must be treated as a special case). Figure 9.2.1 shows a sketch of $V_e(r)$ for $C < 0$ and a typical value of n. There is only one turning point in the motion and the orbits are unbounded.

For $C > 0$, the force is attractive. There exists a single point of $V_e(r)$ for which $dV_e/dr = 0$ given by

$$r_1 = \left(\frac{l^2}{mC}\right)^{1/(3-n)} \qquad V_1 = \frac{C\,(n-3)}{2\,(n-1)}\left(\frac{mC}{l^2}\right)^{(n-1)/(3-n)} \qquad (9.2.5)$$

These formulas break down for both $n = 1$ and $n = 3$ which must be treated as special cases. Sketches of the effective potential are shown in Fig. 9.2.2.

It is seen that the form of the function is qualitatively different for the three ranges of the exponent $n < 1$, $1 < n < 3$, and $n > 3$. For $n < 1$, there exists bound orbits only. An important example of $n < 1$ is the harmonic oscillator potential which results if $n = -1$. For $1 < n < 3$, there exists bound orbits if $V_1 \leq \mathscr{E} < 0$ and unbound orbits if $\mathscr{E} \geq 0$. Planetary motion falls in this category. For $n > 3$, there exists bound orbits passing through the center of force if $-\infty < \mathscr{E} < V_1$; unbound

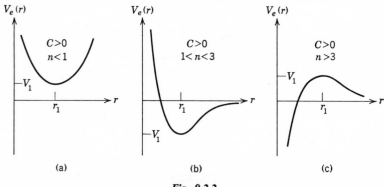

Fig. 9.2.2

orbits can occur if $0 < \mathscr{E} < V_1$ and only unbound orbits occur if $\mathscr{E} > V_1$. If the energy is exactly V_1, the sticking solution results; $r = r_1$ gives an unstable circular orbit in this case. Note that when $n < 3$, the particle never passes through the center of force.

For the range of values $1 < n < 3$, a particle with $\mathscr{E} > 0$ is able to escape the center of force, a result which is independent of the initial direction. It is well known, for instance, that for a particle starting from the surface of the earth there is a well-defined escape velocity which does not depend on the initial direction. For $n > 3$, we note from Fig. 9.2.2c that a particle starting from $r < r_1$ must have an energy greater than V_1 in order to escape. Since the value of V_1 depends on the angular momentum, there *is* a dependence of the escape velocity on initial direction for this type of force.

For the special case $n = 3$, the effective potential can be written

$$V_e(r) = \frac{l^2}{2mr^2}\left(1 - \frac{mC}{l^2}\right) \qquad (9.2.6)$$

Three cases can be distinguished, depending on the value of the parameter $1 - mC/l^2$. These are illustrated in Fig. 9.2.3. It is evident that if the initial conditions are such that the angular momentum is $l = \sqrt{mC}$ so that $1 - mC/l^2 = 0$, a circular orbit will result for *any* value of r. Such circular orbits should be regarded as unstable since a slight change in the value of l will result in an effective potential corresponding to either $1 - mC/l^2 > 0$ or $1 - mC/l^2 < 0$ and a consequent spiraling of the particle either away from the center of force or toward it.

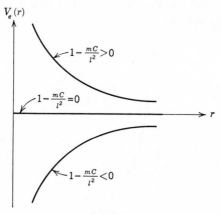

Fig. 9.2.3

The differential equation for the orbit is exactly soluble for the following values of n:

$n = -1$ (harmonic oscillator)

$$\frac{d^2x}{d\theta^2} + x - \frac{mC}{l^2}\frac{1}{x^3} = 0 \tag{9.2.7}$$

$n = 2$ (gravitational or electrostatic force)

$$\frac{d^2x}{d\theta^2} + x = \frac{mC}{l^2} \tag{9.2.8}$$

$n = 3$

$$\frac{d^2x}{d\theta^2} + x\left(1 - \frac{mC}{l^2}\right) = 0 \tag{9.2.9}$$

$n = 4$

$$\frac{d^2x}{d\theta^2} + x - \frac{mC}{l^2}x^2 = 0 \tag{9.2.10}$$

$n = 5$

$$\frac{d^2x}{d\theta^2} + x - \frac{mC}{l^2}x^3 = 0 \tag{9.2.11}$$

Other values of n may give differential equations that are reducible to exactly soluble forms by an appropriate change of variable.

The harmonic oscillator is decidedly easier to solve in rectangular coordinates. The equations of motion are

$$m\ddot{x}_1 + Cx_1 = 0 \qquad m\ddot{x}_2 + Cx_2 = 0 \qquad (9.2.12)$$

The solutions for $C > 0$ can be expressed by

$$x_1 = a \cos \omega t \qquad x_2 = b \sin \omega t \qquad (9.2.13)$$

The orbit is an ellipse given by

$$\left(\frac{x_1}{a}\right)^2 + \left(\frac{x_2}{b}\right)^2 = 1 \qquad (9.2.14)$$

Converting to polar coordinates

$$x_1 = \frac{1}{x} \cos \theta \qquad x_2 = \frac{1}{x} \sin \theta \qquad (9.2.15)$$

results in the orbit equation

$$\frac{\cos^2 \theta}{a^2} + \frac{\sin^2 \theta}{b^2} = x^2 \qquad (9.2.16)$$

which should be a solution of the nonlinear Eq. (9.2.7). Verification is left as a problem. Notice that the center of the ellipse (9.2.14) coincides with the center of force. Solution of the orbit equation for the inverse r^2 force, $n = 2$, is taken up in detail in Section 9.3 where it is shown that the orbit for this force law is also an ellipse but that the center of force is at one of the foci.

The detailed solution of the case $n = 3$, Eq. (9.2.9), is left as a problem.

The solutions of differential equations of the form (9.2.10) and (9.2.11) come out in terms of Jacobian elliptic functions and are given in Section 6.5.

9.3. THE INVERSE r^2 FORCE

One of the basic force laws of classical physics is

$$F(r) = -\frac{C}{r^2} \qquad (9.3.1)$$

For the gravitational attraction between two bodies with spherical distributions of mass the force constant is

$$C = Gm_1m_2$$
$$G = 6.670 \times 10^{-8} \frac{\text{cm}^3}{\text{gm sec}^2} = 3.42 \times 10^{-8} \frac{\text{ft}^3}{\text{slug sec}^2} \qquad (9.3.2)$$

For the electrostatic force of attraction between an electron and an atomic nucleus

$$C = Zq^2 \qquad q = 4.803 \times 10^{-10} \text{ stat coulomb} \qquad (9.3.3)$$

where Z is the atomic number of the nucleus. If the force is the electrostatic force of repulsion between two atomic nuclei,

$$C = -Z_1 Z_2 q^2 \qquad (9.3.4)$$

From (9.2.4) the effective potential is

$$V_e(r) = \frac{l^2}{2mr^2} - \frac{C}{r} \qquad (9.3.5)$$

Recall that m is the *reduced mass* given by (9.1.10). For attractive forces $C > 0$ and $V_e(r)$ resembles Fig. 9.2.2b with a single minimum given by

$$r_1 = \frac{l^2}{mC} \qquad V_1 = -\frac{mC^2}{2l^2} \qquad (9.3.6)$$

The total energy is

$$\mathcal{E} = \tfrac{1}{2}m\dot{r}^2 + V_e(r) \qquad (9.3.7)$$

For $V_1 \le \mathcal{E} < 0$, the orbits are bound and there are two turning points found from (9.3.7) by setting $\dot{r} = 0$:

$$r_2 = \frac{C}{2|\mathcal{E}|} + \sqrt{\frac{C^2}{4\mathcal{E}^2} - \frac{l^2}{2m|\mathcal{E}|}} \qquad r_3 = \frac{C}{2|\mathcal{E}|} - \sqrt{\frac{C^2}{4\mathcal{E}^2} - \frac{l^2}{2m|\mathcal{E}|}} \qquad (9.3.8)$$

where $|\mathcal{E}| = -\mathcal{E}$. If $\mathcal{E} = V_1$ exactly, a stable circular orbit of radius r_1 results. For $\mathcal{E} > 0$, the orbits are not bound and there is only one turning point given by

$$r_2 = -\frac{C}{2\mathcal{E}} + \sqrt{\frac{C^2}{4\mathcal{E}^2} + \frac{l^2}{2m\mathcal{E}}} \qquad (9.3.9)$$

Finally, setting both $\mathcal{E}$ and $\dot{r}$ equal to zero in (9.3.7) gives the single turning point

$$r_2 = \frac{l^2}{2mC} \qquad (9.3.10)$$

If $C < 0$, corresponding to a force of repulsion between the two particles, $V_e(r)$ resembles Fig. 9.2.1 and no bound orbits exist. The energy is always positive and the single turning point is given by

$$r_2 = \frac{|C|}{2\mathcal{E}} + \sqrt{\frac{C^2}{4\mathcal{E}^2} + \frac{l^2}{2m\mathcal{E}}} \qquad (9.3.11)$$

The differential equation of the orbit is given by (9.2.8) and its solution is

$$x = \frac{1}{r} = \frac{mC}{l^2} + b_1 \cos (\theta - \alpha) \qquad (9.3.12)$$

where b_1 and α are constants of integration. Without loss of generality, we can set $\alpha = 0$ The general nature of a bound orbit is discussed in Section 7.4.

The orbit given by (9.3.12) is obviously a very special situation in that the turning points, or apsides, are located exactly 180° apart. The orbit,

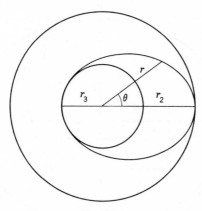

Fig. 9.3.1

therefore, closes on itself after 360° and retraces its path exactly, as in Fig. 9.3.1. It is interesting that the most important force law that actually occurs in nature is of this especially simple type. It can be concluded that any sort of perturbation or slight deviation from the inverse r^2 law of force would cause the orbit not to close on itself and give a precession or rosette pattern. Such an effect is actually observed in the orbit of the planet Mercury. Mercury is closest to the sun of any of the planets and moves in such a strong gravitational field that relativistic effects produce a slight perturbation on the motion.

In discussing the motion of the moon or artificial satellites around the earth the terms *perigee,* for the lesser turning point, and *apogee,* for the greater turning point, are used. For the motion of planets around the sun, these terms are replaced by *perihelion* and *aphelion.*

The constant b_1 in (9.3.12) can be evaluated in terms of the turning points or in terms of the constants of the motion l and $\mathscr{E}$. Setting $\alpha = 0$, the particle is at the larger turning point when $\theta = 0$:

$$\frac{1}{r_2} = \frac{mC}{l^2} + b_1 \qquad (9.3.13)$$

For *bound orbits*, r_2 is given by (9.3.8). Hence,

$$\frac{1}{r} = \frac{mC}{l^2} - \sqrt{\left(\frac{mC}{l^2}\right)^2 - \frac{2m\,|\mathscr{E}|}{l^2}} \cos\theta \qquad (9.3.14)$$

The force constant C, of course, must be positive.

The actual shape of the bound orbit is that of an ellipse. An ellipse is obtained if the point P of Fig. 9.3.2 moves in such a way that

$$r + r' = \text{const.} \qquad (9.3.15)$$

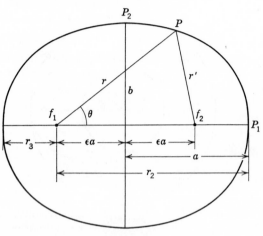

Fig. 9.3.2

The points f_1 and f_2 are the *foci*. The center of force is at the focus f_1 in Fig. 9.3.2. It should be born in mind that the particle of mass m_2 is arbitrarily designated as the "center of force" and the particle of mass m_1 is thought of as moving around it. Actually, both particles move around the common center of mass as a fixed point. The distances a and b are the major and minor axes of the ellipse. By considering the special case where P coincides with P_1 it is found that

$$r + r' = r_2 + r_3 = 2a \qquad (9.3.16)$$

The eccentricity of the ellipse is introduced by writing the distance between the center of the ellipse and either focus as ϵa. The eccentricity can have the range of values

$$0 \le \epsilon < 1 \qquad (9.3.17)$$

The value $\epsilon = 0$ gives a circle. By considering the special case in Fig. 9.3.2, where P coincides with P_2, the relation

$$a^2 = b^2 + \epsilon^2 a^2 \qquad (9.3.18)$$

is obtained. Now, by means of the cosine law,

$$r'^2 = r^2 + (2\epsilon a)^2 - 4\epsilon ar \cos\theta \qquad (9.3.19)$$

Combining (9.3.16) and (9.3.19) yields

$$\frac{1}{r} = \frac{1 - \epsilon\cos\theta}{a(1 - \epsilon^2)} \qquad (9.3.20)$$

The eccentricity can be eliminated by means of (9.3.18):

$$\frac{1}{r} = \frac{a}{b^2} - \sqrt{\left(\frac{a}{b^2}\right)^2 - \frac{1}{b^2}} \cos\theta \qquad (9.3.21)$$

Direct comparison of (9.3.21) and (9.3.14) proves that the orbit is an ellipse and also gives the following relations between the parameters of the ellipse and the total energy and angular momentum:

$$\frac{a}{b^2} = \frac{mC}{l^2} \qquad \frac{1}{b^2} = \frac{2m|\mathscr{E}|}{l^2} \qquad (9.3.22)$$

By solving for a and b

$$b = \frac{l}{\sqrt{2m|\mathscr{E}|}} \qquad a = \frac{C}{2|\mathscr{E}|} \qquad (9.3.23)$$

That the orbits of the planets are ellipses with the sun located at one of the foci is known as *Kepler's first law*. It was discovered by him as a result of his analysis of planetary observations and was published by him in 1610.

From the expression for the angular momentum we can write

$$dt = \frac{mr^2\,d\theta}{l} = \frac{2m}{l} \cdot \tfrac{1}{2}r^2\,d\theta \qquad (9.3.24)$$

Note that

$$d\sigma = \tfrac{1}{2}r^2\,d\theta \qquad (9.3.25)$$

is the element of area swept over by r as θ advances by $d\theta$, Fig. 9.3.3. Thus,

$$dt = \frac{2m}{l}\,d\sigma \qquad (9.3.26)$$

showing that the time required for the particle to travel over a portion of the elliptical orbit is directly proportional to the area swept out by the displacement vector from the center of force to the particle. This is *Kepler's second law* and is sometimes referred to as "the conservation of areal velocity." Kepler deduced his second law from observations on planetary motion, but since it depends for its validity only on the conservation of angular momentum it is valid for any two-body central force problem. The two shaded portions of Fig. 9.3.3 represent equal areas and, therefore, it is qualitatively evident that the closer the particle is to the center of force the faster it will move.

From (9.3.26) it follows that the period of the orbit is given by

$$P = \frac{2m}{l}\sigma = \frac{2m\pi ab}{l} \tag{9.3.27}$$

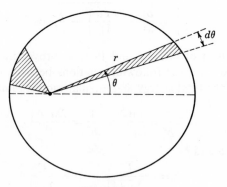

Fig. 9.3.3

where $\sigma = \pi ab$ is the area of the ellipse. By means of (9.3.23)

$$P = 2\pi a^{3/2}\sqrt{m/C} \tag{9.3.28}$$

Recalling the formula for the reduced mass (9.1.10) and that $C = Gm_1 m_2$ for planetary motion, we find for the period

$$P = \frac{2\pi a^{3/2}}{\sqrt{G(m_1 + m_2)}} \tag{9.3.29}$$

This result is *Kepler's third law*.

If the total energy has the value zero, there is a single turning point in the motion given by equation (9.3.10). Let us choose the coordinate system in such a way that r has this value when $\theta = \pi$ as in Fig. 9.3.4. From the general solution (9.3.12),

$$\frac{2mC}{l^2} = \frac{mC}{l^2} - b_1 \tag{9.3.30}$$

Hence,

$$\frac{1}{r} = \frac{mC}{l^2}(1 - \cos\theta) \tag{9.3.31}$$

or

$$2r_2 = r(1 - \cos \theta) \qquad (9.3.32)$$

By converting to rectangular coordinates it is easy to see that (9.3.32) represents a parabola:

$$2r_2 = \sqrt{x_1{}^2 + x_2{}^2} - x_1$$
$$x_2 = \pm\sqrt{4r_2{}^2 + 4r_2 x_1} \qquad (9.3.33)$$

If a particle at the surface of the earth is given exactly the right kinetic energy to make the total energy zero, it will escape along a parabolic

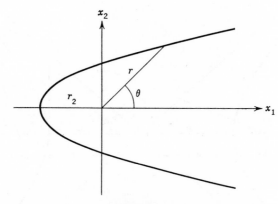

Fig. 9.3.4

trajectory. Accordingly, the *escape velocity* is given by

$$\mathscr{E} = 0 = \tfrac{1}{2}m_1 u_0{}^2 - \frac{Gm_1 m_2}{r_e} \qquad u_0 = \sqrt{\frac{2Gm_2}{r_e}} \qquad (9.3.34)$$

where m_2 is earth's mass and r_e is its radius. The reduced mass is replaced by m_1 in this case. As mentioned previously, the value of the escape velocity (9.3.34) does not depend on the initial direction—a result that is not true of all force laws.

Continuing with attractive forces, the orbits for $\mathscr{E} > 0$ are not bound. The single turning point is given by (9.3.9). If, as in Fig. 9.3.5, $r = r_2$ when $\theta = \pi$, we find from (9.3.12) that

$$\frac{1}{r_2} = \frac{mC}{l^2} - b_1 \qquad (9.3.35)$$

Hence,

$$\frac{1}{r} = \frac{mC}{l^2} - \sqrt{\left(\frac{mC}{l^2}\right)^2 + \frac{2m\mathscr{E}}{l^2}} \cos \theta \qquad (9.3.36)$$

This represents the equation of a hyperbola as we now show. Referring to Fig. 9.3.5, the hyperbola can be defined by the relation

$$r' - r = 2a \qquad (9.3.37)$$

where $2a$ is the distance between the vertices of the two branches. The center of force is at the focus f_2. The distance between the two foci is written as $2\epsilon a$ where ϵ is the eccentricity and has the range of values $\epsilon > 1$. By means of the cosine law,

$$r'^2 = r^2 + 4\epsilon^2 a^2 + 4\epsilon a r \cos \theta \qquad (9.3.38)$$

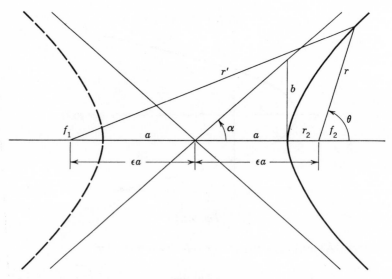

Fig. 9.3.5

The elimination of r' between (9.3.37) and (9.3.38) results in

$$\frac{1}{r} = \frac{1 - \epsilon \cos \theta}{a(\epsilon^2 - 1)} \qquad (9.3.39)$$

Let α be the slope of the asymptote to the hyperbola. As $r \to \infty$, $\theta \to \alpha$, and from (9.3.39), $\cos \alpha \to 1/\epsilon$. Then, from Fig. 9.3.5,

$$\cos \alpha = \frac{1}{\epsilon} = \frac{a}{\sqrt{a^2 + b^2}} \qquad (9.3.40)$$

Using (9.3.40) to eliminate the eccentricity from (9.3.39) results in

$$\frac{1}{r} = \frac{a}{b^2} - \sqrt{\left(\frac{a}{b^2}\right)^2 + \frac{1}{b^2}} \cos \theta \qquad (9.3.41)$$

Direct comparison of (9.3.36) and (9.3.41) yields

$$\frac{a}{b^2} = \frac{mC}{l^2} \qquad \frac{2m\mathcal{E}}{l^2} = \frac{1}{b^2}$$

$$a = \frac{C}{2\mathcal{E}} \qquad b = \frac{l}{\sqrt{2m\mathcal{E}}}$$

(9.3.42)

These results hold for C > 0 and $\mathcal{E} > 0$.

There remains the case where the force is repulsive, i.e., $C < 0$. *Only positive energies and unbound orbits can occur.* The orbits are hyperbolic

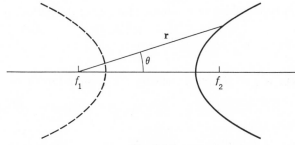

Fig. 9.3.6

and it is found that the only essential change is the center of force is at f_1 rather than f_2 in Fig. 9.3.5. The turning point is given by equation (9.3.11). Referring to Fig. 9.3.6, the following equations for the orbit and the orbit parameters can be determined:

$$\frac{1}{r} = -\frac{a}{b^2} + \sqrt{\left(\frac{a}{b^2}\right)^2 + \frac{1}{b^2}} \cos \theta$$

$$= -\frac{m\,|C|}{l^2} + \sqrt{\left(\frac{mC}{l^2}\right)^2 + \frac{2m\mathcal{E}}{l^2}} \cos \theta$$

(9.3.43)

$$a = \frac{|C|}{2\mathcal{E}} \qquad b = \frac{l}{\sqrt{2m\mathcal{E}}}$$

(9.3.44)

These results hold for C < 0 and $\mathcal{E} > 0$.

9.4. THE VIRIAL THEOREM

Consider an isolated system of N interacting particles. Let a scalar function χ be defined by means of

$$\chi = \sum_{\alpha=1}^{N} \mathbf{p}_\alpha \cdot \mathbf{r}_\alpha$$

(9.4.1)

where $\mathbf{p}_\alpha$ and $\mathbf{r}_\alpha$ are the momentum and position vectors of the particles. It will be assumed that the system remains bounded for all time, i.e., that it can be enclosed by a limiting fixed surface. The time derivative of χ is

$$\frac{d\chi}{dt} = \sum_{\alpha=1}^{N} (\dot{\mathbf{p}}_\alpha \cdot \mathbf{r}_\alpha + \mathbf{p}_\alpha \cdot \dot{\mathbf{r}}_\alpha) \tag{9.4.2}$$

The average value of $d\chi/dt$ over a period of time τ is

$$\overline{\frac{d\chi}{dt}} = \frac{1}{\tau} \int_0^\tau \frac{d\chi}{dt} \, dt = \frac{\chi(\tau) - \chi(0)}{\tau} \tag{9.4.3}$$

If the system is periodic and τ is some multiple of the period, then

$$\overline{\frac{d\chi}{dt}} = 0 \tag{9.4.4}$$

If the system is not periodic, then by the assumption of boundedness

$$\lim_{\tau \to \infty} \overline{\frac{d\chi}{dt}} = 0 \tag{9.4.5}$$

In either case, (9.4.2) gives

$$\overline{\sum_{\alpha=1}^{N} \dot{\mathbf{p}}_\alpha \cdot \mathbf{r}_\alpha} = -\overline{\sum_{\alpha=1}^{N} \mathbf{p}_\alpha \cdot \dot{\mathbf{r}}_\alpha} \tag{9.4.6}$$

Note that

$$\sum_{\alpha=1}^{N} \mathbf{p}_\alpha \cdot \dot{\mathbf{r}}_\alpha = \sum_{\alpha=1}^{N} m_\alpha u_\alpha{}^2 = 2T \tag{9.4.7}$$

where T is the total kinetic energy of the system. Since the system is isolated, $\dot{\mathbf{p}}_\alpha = \mathbf{F}_\alpha$ is the force on the αth particle due to all the other particles in the system. Thus (9.4.6) becomes

$$\bar{T} = -\frac{1}{2} \overline{\sum_{\alpha=1}^{N} \mathbf{F}_\alpha \cdot \mathbf{r}_\alpha} \tag{9.4.8}$$

This result is known as the *virial theorem*. The right-hand side of (9.4.8) is called *virial* of the system. The virial theorem has application in the kinetic theory of gases.

For two particles which interact by means of a central power law of force

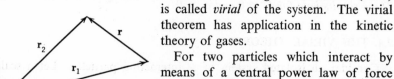

Fig. 9.4.1

$$\mathbf{F}_1 = -\mathbf{F}_2 = -\frac{C}{r^n}\left(\frac{\mathbf{r}}{r}\right) \tag{9.4.9}$$

where, as in Fig. 9.4.1, $\mathbf{r}$ is the relative displacement of the two particles and $\mathbf{r}_1$ and $\mathbf{r}_2$ are displacements relative to a fixed point O in space:

$$\mathbf{r} = \mathbf{r}_2 - \mathbf{r}_1 \tag{9.4.10}$$

Thus,

$$\sum_{\alpha=1}^{N} \mathbf{F}_\alpha \cdot \mathbf{r}_\alpha = \mathbf{F}_1 \cdot \mathbf{r}_1 + \mathbf{F}_2 \cdot \mathbf{r}_2$$

$$= -\frac{C}{r^{n+1}}(\mathbf{r} \cdot \mathbf{r}_1) + \frac{C}{r^{n+1}}(\mathbf{r} \cdot \mathbf{r}_2) \tag{9.4.11}$$

$$= -\frac{C}{r^{n+1}}\mathbf{r} \cdot (\mathbf{r}_1 - \mathbf{r}_2) = -\frac{C}{r^{n-1}}$$

The potential function is

$$V(r) = -\frac{C}{(n-1)r^{n-1}} \tag{9.4.12}$$

and by combining (9.4.8), (9.4.11), and (9.4.12) the virial theorem becomes

$$\bar{T} = \frac{1-n}{2}\bar{V} \tag{9.4.13}$$

For example, $n = 2$ for planetary motion and

$$\bar{T} = -\tfrac{1}{2}\bar{V} \tag{9.4.14}$$

9.5. SOLUTION BY FOURIER SERIES

If the motion is bound and varies periodically between two turning points, an approximate solution is readily obtainable in terms of a Fourier series. Such a circumstance arises with attractive power laws of force, $F(r) = -C/r^n$, when $1 < n < 3$ and the energy is negative and for all energies when $n < 1$. By means of (9.2.5),

$$\frac{mC}{l^2} = r_1^{n-3} \tag{9.5.1}$$

where r_1 is the equilibrium value of r. The orbit equation (9.2.3), therefore, can be written

$$\frac{d^2x}{d\theta^2} + x - r_1^{n-3}x^{n-2} = 0 \tag{9.5.2}$$

The expansion of

$$f(x) = x - r_1^{n-3}x^{n-2} \tag{9.5.3}$$

about the equilibrium value $x_1 = 1/r_1$ yields

$$f(x) = (3-n)(x-x_1) - \tfrac{1}{2}r_1(2-n)(3-n)(x-x_1)^2$$
$$+ \tfrac{1}{6}r_1^2(2-n)(3-n)(4-n)(x-x_1)^3 + \cdots \tag{9.5.4}$$

Identifying the coefficients as

$$\alpha = 3 - n \qquad \beta = \tfrac{1}{2}r_1(2 - n)(3 - n) \qquad \gamma = \tfrac{1}{6}r_1^2(2 - n)(3 - n)(4 - n)$$
$$(9.5.5)$$

and comparing with (4.13.35) and (4.13.36) yields the Fourier series solution

$$x - x_1 = \tfrac{1}{4}(2 - n)r_1b_1^2 + b_1 \cos (\lambda\theta) - \tfrac{1}{12}(2 - n)r_1b_1^2 \cos (2\lambda\theta)$$
$$+ \tfrac{1}{96}(2 - n)(3 - n)r_1^2b_1^3 \cos (3\lambda\theta) + \cdots \qquad (9.5.6)$$

$$\lambda^2 = (3 - n)[1 - \tfrac{1}{12}(n - 2)(n + 1)b_1^2r_1^2] \qquad (9.5.7)$$

Notice that if $n = 2$, the Fourier series solution has only the one term

$$x - x_1 = b_1 \cos \theta \qquad \lambda = 1 \qquad (9.5.8)$$

which, since $x_1 = mC/l^2$, is identical to the exact solution (9.3.12) already obtained. If n is only slightly different from 2, $n = 2 + \delta$, say, the Fourier series converges very rapidly. The value of λ is approximately

$$\lambda \cong \sqrt{3 - n} \cong \sqrt{1 - \delta} \qquad (9.5.9)$$

The angle through which the particle moves in order to complete one cycle of its motion is given by

$$\lambda\theta = 2\pi \qquad \theta \cong \frac{2\pi}{\sqrt{1 - \delta}} \cong 2\pi(1 + \tfrac{1}{2}\delta) \qquad (9.5.10)$$

The orbit is still almost an ellipse, but it does not quite close on itself so that the orbit can be visualized as an ellipse, the axes of which turn slowly in time. This phenomenon is called *precession*.

For $\delta > 0$, the perihelion of the orbit *advances* by

$$\Delta\theta = \pi\delta \qquad (9.5.11)$$

per revolution. Using the formula (9.3.28) for the period of the orbit gives

$$\omega_P = \frac{\Delta\theta}{P} = \frac{\delta}{2a^{3/2}}\sqrt{\frac{C(m_1 + m_2)}{m_1m_2}} \qquad (9.5.12)$$

as the approximate angular velocity at which the orbit precesses.

A very interesting fact has become apparent. There is nothing wrong with the solution (9.5.6) when $\mathscr{E} > 0$ and $1 < n < 3$. We have therefore discovered that the *nonperiodic* solutions of nonlinear differential equations can, at least in some cases, be represented by Fourier series. The secret is in the use of the variable $x = 1/r$. Let us orient the coordinate system in such a way that $\theta = 0$ when the particle is at its turning point as in Fig.

in the incident beam there are N particles per second per unit area. All those particles that scatter into the angular range ϕ to $\phi + d\phi$ pass through the annular ring of radius s and area $2\pi s\, ds$. If dN_ϕ is the number of particles per second which are counted in the angular range ϕ to $\phi + d\phi$, then

$$dN_\phi = N \cdot 2\pi s\, ds \tag{9.6.1}$$

The *differential cross section* is defined by means of the relation

$$d\sigma(\phi) = \frac{dN_\phi}{N} \tag{9.6.2}$$

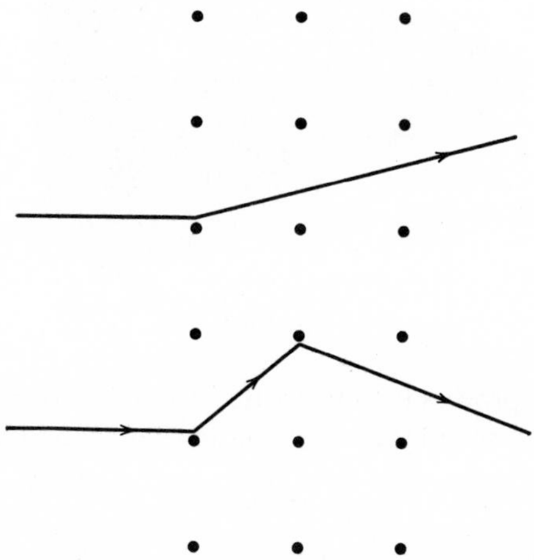

Fig. 9.6.2

This is an operational definition in that dN_ϕ and N are both measurable quantities. The definition is therefore valid whether the interaction between incident and target particles is described by quantum or classical mechanics, provided that the interaction is a function of r only and also provided that processes such as the creation of new particles are not involved. In any actual experiment there is more than one target particle, and in order to find the differential cross section of a single particle it is necessary to divide the actual number of scattered particles counted by the number of target particles involved.

Another assumption is that the density of target particles is so sparse that any one incident particle is scattered by at most one target particle. Figure 9.6.2 shows a single scattering event and the much rarer double scattering. In Rutherford's α particle experiments, the metal foils used as targets were very thin and we know the nuclei of the atoms are separated by distances that are very great compared to the nuclear size. If Σ is the combined area presented by all the nuclei in an area A of target material, then Σ/A is the probability of a single scattering and $(\Sigma/A)^2$ is essentially the probability of a double scattering. Since $\Sigma \ll A$, most α-particles pass through the foils undeviated, a few encounter a single nucleus and scatter, and virtually none go through a double scattering. Actually it was Rutherford's experiments that confirmed this view.

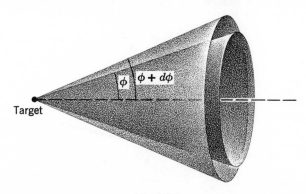

Fig. 9.6.3

If the scattering process is to be treated theoretically by the methods of classical mechanics, then by comparison of (9.6.1) and (9.6.2),

$$d\sigma = 2\pi s \, ds \tag{9.6.3}$$

The differential cross section, then, is actually an area. In quantum mechanics the basic definition (9.6.2) of the differential cross section holds, but the concept of impact parameter has no meaning. It is usual to introduce the element of solid angle

$$d\Omega = 2\pi \sin\phi \, d\phi \tag{9.6.4}$$

which is the solid angle into which the particles with impact parameters between s and $s + ds$ scatter. It is also the solid angle between the cones defined by $\phi = $ const. and $\phi + d\phi = $ const. and illustrated in Fig. 9.6.3. If (9.6.3) is divided by (9.6.4),

$$\frac{d\sigma}{d\Omega} = -\frac{s}{\sin\phi}\frac{ds}{d\phi} \tag{9.6.5}$$

which is the differential cross section per unit solid angle. The minus sign is introduced because the derivative $ds/d\phi$ is negative, i.e., as the impact parameter increases the angle of scatter ϕ decreases.

For the treatment of scattering by a repulsive inverse r^2 force field, the formula for the hyperbolic orbit can be used. As $r \to \infty$, $\theta \to \alpha$, Fig. 9.6.1. From Eq. (9.3.43)

$$\cos \alpha = \frac{1}{\sqrt{1 + (b^2/a^2)}} \tag{9.6.6}$$

and from (9.3.44)

$$\frac{b^2}{a^2} = \frac{2l^2 \mathscr{E}}{mC^2} \tag{9.6.7}$$

The angular momentum is related to the impact parameter by means of

$$l = mu_0 s \tag{9.6.8}$$

where u_0 is the speed of the incident particle relative to the target particle before scattering and when the particles are still far apart. From the geometry of Fig. 9.6.1

$$2\alpha + \phi = \pi \tag{9.6.9}$$

Therefore,

$$\cos \alpha = \cos \left(\frac{\pi}{2} - \frac{\phi}{2} \right) = \sin \frac{\phi}{2} = 1/\sqrt{1 + (b^2/a^2)}$$

$$1 + \frac{2mu_0^2 s^2 \mathscr{E}}{C^2} = \frac{1}{\sin^2 (\phi/2)} \tag{9.6.10}$$

which gives the functional relationship between the impact parameter s and the angle of scatter ϕ. By differentiation,

$$\frac{4mu_0^2 \mathscr{E}}{C^2} \frac{s \, ds}{d\phi} = -\frac{\cos (\phi/2)}{\sin^3 (\phi/2)} \tag{9.6.11}$$

Introducing the total energy by means of

$$\mathscr{E} = \tfrac{1}{2} mu_0^2 \tag{9.6.12}$$

and substituting (9.6.11) into (9.6.5) yields

$$\frac{d\sigma}{d\Omega} = \frac{C^2}{2m^2 u_0^4} \frac{\cos (\phi/2)}{\sin \phi \sin^3 (\phi/2)} \tag{9.6.13}$$

By means of trigonometric identities

$$\frac{d\sigma}{d\Omega} = \frac{C^2}{4m^2 u_0^4} \frac{1}{\sin^4 (\phi/2)} = \frac{C^2}{16 \mathscr{E}^2} \frac{1}{\sin^4 (\phi/2)} \tag{9.6.14}$$

For the scattering of α-particles of atomic nuclei, the force constant is

$$|C| = Zq^2 \tag{9.6.15}$$

There is no difference in the scattering formula (9.6.14) if the force is one of attraction instead of repulsion as would be the case, for example, if electrons were scattered from atomic nuclei.

In an actual experiment a counter is used, the aperture of which subtends a solid angle $d\Omega_c$ at the target. Let dN_t be the total number of scattered particles counted and N_t be the total number of incident particles. Let n be the number of target particles per unit volume, A the area of the target being utilized, and Δx be its thickness. The differential cross section per target particle is then

$$d\sigma = \frac{dN_t}{(N_t/A) \cdot n\Delta xA} \cdot \frac{2\pi \sin \phi \, d\phi}{d\Omega_c} \qquad (9.6.16)$$

or

$$\frac{d\sigma}{d\Omega} = \frac{1}{N_t n\Delta x} \frac{dN_t}{d\Omega_c} \qquad (9.6.17)$$

Here $dN_t/d\Omega_c$ is the total number of particles received by the counter divided by the solid angle subtended at the target by the counter and $d\sigma/d\Omega$ has the same significance as is given by Eq. (9.6.14).

The *total* cross section is defined by

$$\sigma = \int_{\text{all solid angle}} d\sigma \qquad (9.6.18)$$

For Rutherford scattering this is

$$\sigma = \int_{\phi=0}^{\pi} \frac{C^2}{4m^2u_0^4} \frac{1}{\sin^4(\phi/2)} 2\pi \sin \phi \, d\phi \qquad (9.6.19)$$

The result is infinite because *some* scattering takes place, no matter how large the impact parameter is. A force which cuts off at a finite distance from the target particle will give a finite result for the total cross section.

It is possible to express the scattering angle as an integral for an arbitrary central force. The expression for the total energy (9.1.21),

$$\mathscr{E} = \frac{l^2}{2m}\left[\left(\frac{dx}{d\theta}\right)^2 + x^2\right] + V(r) \qquad (9.6.20)$$

can be solved for $d\theta$ to yield

$$d\theta = \pm \frac{dx}{\sqrt{(2m/l^2)(\mathscr{E} - V) - x^2}} \qquad (9.6.21)$$

The impact parameter s can be introduced by means of

$$l = mu_0 s = \sqrt{2m\mathscr{E}} \, s \qquad (9.6.22)$$

This yields

$$d\theta = \pm \frac{s \, dx}{\sqrt{1 - (V/\mathscr{E}) - s^2 x^2}} \tag{9.6.23}$$

Referring to Fig. 9.6.1, we note that as r increases from r_2 to ∞, x *decreases* from $1/r_2$ to 0. At the same time, θ *increases* from 0 to α. Over this portion of the trajectory, the minus sign must be used in (9.6.23). Integration yields

$$\alpha = -\int_{x_2}^{0} \frac{s \, dx}{\sqrt{1 - (V/\mathscr{E}) - s^2 x^2}} = \int_{0}^{x_2} \frac{s \, dx}{\sqrt{1 - (V/\mathscr{E}) - s^2 x^2}} \tag{9.6.24}$$

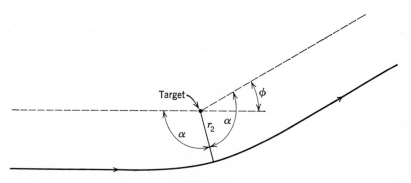

Fig. 9.6.4

The limit x_2 is the turning point determined from

$$1 - (V/\mathscr{E}) - s^2 x^2 = 0 \tag{9.6.25}$$

Since $2\alpha + \phi = \pi$, the angle of scatter is

$$\phi = \pi - 2\int_{0}^{x_2} \frac{s \, dx}{\sqrt{1 - (V/\mathscr{E}) - s^2 x^2}} \tag{9.6.26}$$

which gives the general functional relationship between ϕ and s that is to be used in the formula for the differential cross section Eq. (9.6.5).

If the force between the two particles is attractive instead of repulsive, the geometry of Fig. 9.6.1 must be replaced by that of Fig. 9.6.4. Here $2\alpha - \phi = \pi$ so that in place of (9.6.26),

$$\phi = -\pi + 2\int_{0}^{x_2} \frac{s \, dx}{\sqrt{1 - (V/\mathscr{E}) - s^2 x^2}} \tag{9.6.27}$$

9.7. TRANSFORMATION OF SCATTERING CROSS SECTION FROM CENTER OF MASS TO LABORATORY COORDINATES

So far in this chapter all calculations are done in the center of mass coordinate system. In a typical scattering experiment, the target is originally stationary in the laboratory. If the target particles are very heavy compared to the bombarding particles, the laboratory and the center of mass coordinate system are practically the same, but in those cases where this is not true it is necessary to transform from the center

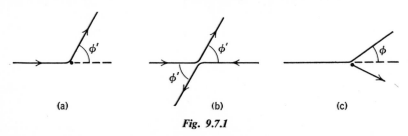

(a)   (b)   (c)

Fig. 9.7.1

of mass to the laboratory system. The appearance of the scattering process from various points of view is illustrated in Fig. 9.7.1.

In Fig. 9.7.1a, the target particle is regarded as stationary and the incident particle scatters at an angle ϕ'. This is the point of view from which all calculations have been done so far in this chapter. A person whose frame of reference is the center of mass coordinate system will first see the two particles move toward one another, and after scattering away from one another, each at the angle ϕ' with respect to the original direction as in Fig. 9.7.1b. All quantities that refer to the center of mass coordinate system are now labeled with primes. Consequently, when using formulas from previous sections of this chapter it is necessary to relabel all quantities accordingly. Finally, Fig. 9.7.1c shows the scattering as it would be observed in the laboratory.

The total energy in the center of mass coordinate system is

$$\mathcal{E}' = \tfrac{1}{2}m_1 u_1'^2 + \tfrac{1}{2}m_2 u_2'^2 = \tfrac{1}{2}\frac{m_1 m_2}{m_1 + m_2}u^2 \tag{9.7.1}$$

where u_1' and u_2' are the velocities of the particles with respect to the center of mass before the scattering and u is the *relative* velocity of the two particles. No potential energy terms appear because the velocities are all measured while the particles are still far apart. The energy of the system *as measured in the laboratory* is by Eq. (8.3.10),

$$\mathcal{E} = \tfrac{1}{2}(m_1 + m_2)v^2 + \mathcal{E}' \tag{9.7.2}$$

where v is the velocity of the center of mass given by

$$v = \frac{m_1}{m_1 + m_2} u_1 \tag{9.7.3}$$

Since the target particle is initially stationary, an alternative expression for the laboratory energy is

$$\mathcal{E} = \tfrac{1}{2} m_1 u_1^2 \tag{9.7.4}$$

which is just the initial kinetic energy of the incident particles and would be identical to the energy imparted to the incident particles by an accelerator

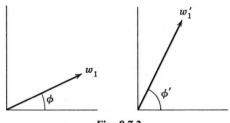

Fig. 9.7.2

if this happened to be the particle source. By combining (9.7.2), (9.7.3), and (9.7.4),

$$\mathcal{E}' = \frac{m_2}{m_1 + m_2} \mathcal{E} \tag{9.7.5}$$

which gives the relation between the laboratory energy and the center of mass energy.

The components of the velocity of the incident particle in the laboratory and center of mass coordinates after scattering are related by means of a Galilean transformation (Fig. 9.7.2):

$$w_1 \cos \phi = w_1' \cos \phi' + v \tag{9.7.6}$$

$$w_1 \sin \phi = w_1' \sin \phi' \tag{9.7.7}$$

Solution for $\tan \phi$ yields

$$\tan \phi = \frac{\sin \phi'}{\cos \phi' + (v/w_1')} \tag{9.7.8}$$

This is the basic relation connecting the angle of scatter as measured in the laboratory and the center of mass coordinate systems. It is convenient to define the parameter γ by means of

$$\gamma = \frac{v}{w_1'} = \frac{u_1}{w_1'} \frac{m_1}{m_1 + m_2} \tag{9.7.9}$$

where (9.7.3) is used.

The velocities u_1 and w_1' can be eliminated from γ as given by (9.7.9). Application of the principle of conservation of linear momentum in the center of mass coordinate system yields

$$m_1 u_1' = m_2 u_2' \tag{9.7.10}$$

$$m_1 w_1' = m_2 w_2' \tag{9.7.11}$$

The center of mass energy can be written in terms of the velocities w_1' and w_2' after scattering as

$$\mathscr{E}' = \tfrac{1}{2} m_1 w_1'^2 + \tfrac{1}{2} m_2 w_2'^2 + H \tag{9.7.12}$$

where allowance is made for the possibility that the scattering is not perfectly elastic. Though it is negligible, H is not strictly zero even in Rutherford scattering since the interaction of two charged particles produces electromagnetic radiation. The elimination of w_2' between (9.7.11) and (9.7.12) and the use of (9.7.5) yields

$$\mathscr{E} \frac{m_2}{m_1 + m_2} - H = \tfrac{1}{2} m_1 w_1'^2 \left[\frac{m_1 + m_2}{m_2} \right] \tag{9.7.13}$$

Dividing by $\mathscr{E}$ and using (9.7.4) result in

$$\frac{m_2}{m_1 + m_2} - \frac{H}{\mathscr{E}} = \frac{m_1 w_1'^2}{m u_1^2} \left[\frac{m_1 + m_2}{m_2} \right] \tag{9.7.14}$$

This relation is solved for the ratio u_1/w_1' and γ is computed as

$$\gamma = \frac{m_1}{m_2} \frac{1}{\sqrt{1 - \dfrac{m_1 + m_2}{m_2} \dfrac{H}{\mathscr{E}}}} \tag{9.7.15}$$

If the collision is elastic, then $H = 0$ and

$$\gamma = \frac{m_1}{m_2} \tag{9.7.16}$$

It is now necessary to use (9.7.8) to calculate the relation between the differential cross section per unit solid angle as measured in the laboratory and center of mass coordinate systems. Since the impact parameter is the same in either coordinate system,

$$\left(\frac{d\sigma}{d\Omega} \right)' = - \frac{s\, ds}{\sin \phi'\, d\phi'} \qquad \frac{d\sigma}{d\Omega} = - \frac{s\, ds}{\sin \phi\, d\phi} \tag{9.7.17}$$

Hence,

$$\frac{d\sigma}{d\Omega} \sin \phi\, d\phi = \left(\frac{d\sigma}{d\Omega} \right)' \sin \phi'\, d\phi' \tag{9.7.18}$$

If (9.7.8) is written as

$$\frac{\sin^2 \phi}{1 - \sin^2 \phi} = \frac{\sin^2 \phi'}{(\cos \phi' + \gamma)^2} \tag{9.7.19}$$

and solved for $\sin \phi$, the result is

$$\sin \phi = \sin \phi' \frac{1}{\sqrt{1 + \gamma^2 + 2\gamma \cos \phi'}} \tag{9.7.20}$$

By differentiation of (9.7.8),

$$\frac{1}{\cos^2 \phi} \frac{d\phi}{d\phi'} = \frac{1 + \gamma \cos \phi'}{(\gamma + \cos \phi')^2} \tag{9.7.21}$$

Equation (9.7.20) can be converted into the form

$$\cos^2 \phi = \frac{(\gamma + \cos \phi')^2}{1 + \gamma^2 + 2\gamma \cos \phi'} \tag{9.7.22}$$

Now by using (9.7.20), (9.7.21), and (9.7.22) we find

$$\frac{\sin \phi \, d\phi}{\sin \phi' \, d\phi'} = \frac{1 + \gamma \cos \phi'}{(1 + \gamma^2 + 2\gamma \cos \phi')^{3/2}} \tag{9.7.23}$$

Combining this with (9.7.18) yields

$$\frac{d\sigma}{d\Omega} = \left(\frac{d\sigma}{d\Omega}\right)' \frac{(1 + \gamma^2 + 2\gamma \cos \phi')^{3/2}}{1 + \gamma \cos \phi'} \tag{9.7.24}$$

Equation (9.7.24) is valid for quantum mechanical calculations since it depends on only the validity of a Galilean transformation and in no way on Newtonian mechanics.

For example, if Rutherford scattering is involved, we find from Eq. (9.6.14) that

$$\frac{d\sigma}{d\Omega} = \frac{Z^2 q^4}{16\mathscr{E}'^2} \frac{1}{\sin^4 (\phi'/2)} \frac{(1 + \gamma^2 + 2\gamma \cos \phi')^{3/2}}{1 + \gamma \cos \phi'} \tag{9.7.25}$$

It is not practical to try to express $d\sigma/d\Omega$ in terms of ϕ directly. Therefore, if $d\sigma/d\Omega$ as a function of ϕ is desired for the purposes of comparison with an experiment, $d\sigma/d\Omega$ is first calculated for a given ϕ' from (9.7.24). The corresponding value of ϕ is then computed from (9.7.8).

The number of $dN_{\phi'}$ of particles that scatter into the angular range ϕ' to $\phi' + d\phi'$ appear in the angular range ϕ to $\phi + d\phi$ in the laboratory. Thus $dN_\phi = dN_{\phi'}$ and from the basic definition (9.6.2) of the differential cross section

$$d\sigma(\phi') = \frac{dN_{\phi'}}{N} = \frac{dN_\phi}{N} = d\sigma(\phi) \tag{9.7.26}$$

For classical scattering, this result also follows from (9.6.1) since the impact parameter is the same in either coordinate system. Thus the differential cross section has the same value in both the laboratory and the center of mass coordinate systems. The same is true of the total cross section:

$$\sigma' = \int d\sigma' = \int d\sigma = \sigma \tag{9.7.27}$$

The element of solid angle into which a given bunch of particles goes is different in the two coordinate systems: $d\Omega \neq d\Omega'$, and this is what is responsible for Eq. (7.24).

9.8. THE BOHR-SOMMERFELD THEORY OF THE HYDROGEN ATOM

According to the Bohr-Sommerfeld rule (6.12.22) for the quantization of a classical system, the elliptic orbits of an electron in a hydrogen atom are to be quantized by means of

$$\oint p_r \, dr = n_r h \qquad \oint p_\theta \, d\theta = n_\theta h \tag{9.8.1}$$

where p_r and p_θ are the canonical momenta determined by

$$p_r = \frac{\partial L}{\partial \dot{r}} = m\dot{r} \qquad p_\theta = \frac{\partial L}{\partial \dot{\theta}} = mr^2\dot{\theta} \tag{9.8.2}$$

and h is Planck's constant. Since $p_\theta = l = $ const.,

$$\oint p_\theta \, d\theta = 2\pi l = n_\theta h, \qquad n_\theta = 1, 2, 3, \dots \tag{9.8.3}$$

The value $n_\theta = 0$ is excluded because, classically speaking, this would result in a "pendulum orbit" with the electron moving back and forth on a straight line through the nucleus of the atom. By contrast, zero angular momentum is quite possible in the wave-mechanical treatment.

To evaluate the remaining integral in (9.8.1), note that the kinetic energy of the electron can be expressed

$$T = \tfrac{1}{2}m(\dot{r}^2 + r^2\dot{\theta}^2) = \tfrac{1}{2}(\dot{r}p_r + \dot{\theta}p_\theta) \tag{9.8.4}$$

Thus

$$p_r \, dr + p_\theta \, d\theta = 2T \, dt \tag{9.8.5}$$

Integration over one cycle of the motion results in

$$(n_r + n_\theta)h = 2\bar{T}P \tag{9.8.6}$$

where P is the period and T is the average kinetic energy. By using the virial theorem (9.4.14) and $\mathscr{E} = T + V$ we find

$$\mathscr{E} = -\bar{T} \tag{9.8.7}$$

By using Eq. (9.3.28) for the period and (9.3.23) for the major and minor axes of the elliptical orbit, the following results are obtained:

$$\mathscr{E} = -\frac{2\pi^2 m Z^2 q^4}{n^2 h^2} \tag{9.8.8}$$

$$a = \frac{n^2 h^2}{4\pi^2 Z q^2 m} \tag{9.8.9}$$

$$\frac{a}{b} = \frac{n}{n_\theta} \tag{9.8.10}$$

where $n = n_r + n_\theta$ is called *the principal quantum number* and the force constant C is replaced by Zq^2. Of course, $Z = 1$ for a hydrogen atom, but our results apply to any one-electron atom, e.g., singly ionized helium. For a given value of n, the angular momentum quantum number n_θ can take on the values $1, 2, \ldots, n$. For example, if $n = 1$ the hydrogen atom is in its lowest energy state and from (9.8.9) and (9.8.10)

$$a = b = \frac{h^2}{4\pi^2 m q^2} = 5.29 \times 10^{-9} \, \text{cm} \tag{9.8.11}$$

which is known as *the first Bohr radius*. The orbit is circular as pictured in Fig. 9.8.1a. If $n = 2$, then n_θ can be either 1 or 2, resulting in the two possibilities $a = b$ and $a = 2b$. These orbits are shown in Fig. 9.8.1b. Similarly, if $n = 3$, three different angular momenta are possible, resulting in the three orbits sketched in Fig. 9.8.1c. For a given n, the different angular momenta all correspond to the same energy, a state

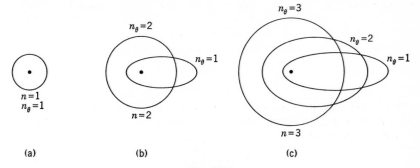

(a)   (b)   (c)

Fig. 9.8.1

of affairs which is known as *degeneracy*. This degeneracy is removed when such refinements as spin-orbit interaction and relativistic effects are taken into account. Neglecting such effects, wave mechanics also gives the formula (9.8.8) for the possible energy levels of the hydrogen atom.

PROBLEMS

9.2.1   Solve the orbit equation (9.2.9) for the inverse r^3 central force. Plot graphs of the orbit for the following numerical values:

$$\sqrt{1 - \frac{mC}{l^2}} = 3, 0.15, 0; \qquad \sqrt{\frac{mC}{l^2} - 1} = 0.15$$

9.2.2   For the general power law of force between two particles, $F(r) = -C/r^n$, make the change of dependent variable $x = y^{-p}$ in the energy equation (9.1.21). By systematically investigating values of n and p, find all the values of n that will lead to an equation of the form

$$\left(\frac{dy}{d\theta}\right)^2 = C_0 + C_1 y + C_2 y^2 + C_3 y^3 + C_4 y^4$$

with an appropriate choice of p.

9.2.3   Verify that (9.2.16) is a solution of (9.2.7). Show that

$$ab = \sqrt{\frac{l^2}{mC}} \qquad a^2 + b^2 = \frac{2\mathscr{E}}{C}$$

9.3.4   A particle is released from rest at a distance r_0 from the center of the earth. Find the time required for it to fall to a new position r.

9.3.5   Show that the elliptic orbit of a planet can be expressed

$$\frac{1}{r} = \frac{1}{2}\left(\frac{1}{r_3} + \frac{1}{r_2}\right) - \frac{1}{2}\left(\frac{1}{r_3} - \frac{1}{r_2}\right)\cos\theta$$

where r_2 and r_3 are the apsidal distances from the center of force.

9.3.6   Two gravitating masses m_1 and m_2 are separated by a distance r_0 and released from rest. Show that when the separation is r, the speeds are

$$u_1 = m_2 \sqrt{\frac{2G}{m_1 + m_2}\left(\frac{1}{r} - \frac{1}{r_0}\right)}$$

$$u_2 = m_1 \sqrt{\frac{2G}{m_1 + m_2}\left(\frac{1}{r} - \frac{1}{r_0}\right)}$$

9.3.7   Let λ be the ratio of the greatest to the least angular velocity of a particle moving in an elliptic orbit in an inverse r^2 force field. Show that the eccentricity of the orbit is

$$\epsilon = \frac{\sqrt{\lambda} - 1}{\sqrt{\lambda} + 1}$$

9.3.8 A particle moves in an attractive inverse r^5 force-field: $F(r) = -C/r^5$. If the particle is given a velocity u_0 at a distance R from the center of force and in a direction α, find the dependence of u_0 on α if u_0 is the minimum velocity for escape.

9.3.9 A projectile is fired at an angle of 60° above the horizontal and at a velocity of 5×10^5 cm/sec. Neglecting air resistance, what would be the range of the projectile measured on the earth's surface? To what height above the earth's surface does the projectile rise? Neglect any effect of the earth's rotation.

9.3.10 Suppose that it is possible to measure the location of the center of mass and the period of rotation of a double star system. Find formulas for the masses of the stars in terms of these data. The orbits may be considered to be approximately circular.

9.3.11 Find a formula for the major axis of an artificial satellite moving around the earth in terms of the period. Choose the units so that the major axis comes out in kilometers if the period is expressed in minutes. The period of the Sputnik I rocket carrier was quoted as 96.0 min and the eccentricity as 0.05. Find the greatest and least elevation of the rocket carrier above the surface of the earth. Assume that the earth is a sphere of radius 6.371×10^6 meters.

9.3.12 A comet moves on a parabolic orbit in the same plane as the orbit of the earth. Assume that the earth's orbit is a circle of radius r_e. Show that the time which the comet spends within the orbit of the earth is given by

$$t = \frac{P}{3\pi}\sqrt{2\left(1 - \frac{r_2}{r_e}\right)\left(1 + 2\frac{r_2}{r_e}\right)}$$

where P is the period of the earth's motion around the sun and r_2 is the distance of closest approach of the comet to the sun.

9.4.13 For two particles interacting by means of a harmonic oscillator, force $n = -1$ and the virial theorem (9.4.13) give $\bar{T} = \bar{V}$. Verify this result directly by using the solutions (9.2.13).

9.5.14 The spherical pendulum, treated in Section 6.6, has many features in common with the two-body central force problem. In particular, the differential equation for the orbit is given by (6.6.63). If x_1 is the equilibrium position, show that in the linear approximation

$$\frac{d^2x}{d\phi^2} + \left[1 + \frac{3x_1^2}{1 + x_1^2}\right](x - x_1) = 0$$

where x_1 is given approximately by

$$x_1^2 = \sqrt{gm^2s^3/l^2}$$

if θ is small. Show that for each complete orbit of the particle the advance of the apse is approximately

$$\Delta\phi = \frac{3\sigma}{4s^2}$$

where σ is the area of the ellipse and that the precessional velocity, i.e., the rate at which the apse advances, is approximately

$$\omega_p = \frac{3\sigma}{8\pi s^2}\sqrt{\frac{g}{s}}$$

Hint: Since $x = \cot\theta$, x is *large* when θ is small.

9.5.15 Show that the amplitude b_1 which appears in the Fourier series (9.5.6) is approximately

$$b_1 = \sqrt{\frac{1}{n-1}\left(\frac{mC}{l^2}\right)^{2/(3-n)} + \frac{2m\mathscr{E}}{(3-n)l^2}}$$

when computed in terms of the energy and the angular momentum. In making the computation, retain second order terms in (9.5.6). Notice that $b_1 > 0$ for a bound orbit, provided that the coordinate system is chosen such that $\theta = 0$ at the turning point closest to the center of force.

9.6.16 Hard spheres of radius b are elastically scattered by hard spheres of radius a. Find the relation between the scattering angle and the impact parameter in the center of mass coordinate system. Find the differential scattering cross section per unit solid angle, $d\sigma/d\Omega$. Find the total cross section.

9.6.17 A particle of energy $\mathscr{E}$ crosses a boundary where the potential energy changes abruptly from 0 to $-V_1$. As a consequence, the direction of its motion changes as indicated in the figure. If the "index of refraction" is defined as

$$n = \sin\theta_1/\sin\theta_2$$

show that

$$n = \sqrt{1 + (V_1/\mathscr{E})}$$

What is the obvious qualitative difference between this and the refraction of light?

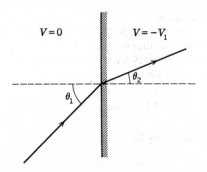

9.6.18 Frequently used in nuclear physics is the "potential hole" or "square well potential" to give an approximate description of the interaction between nucleons. It is defined by

$$V(r) = 0 \quad r > a, \qquad V(r) = -V_0 \quad 0 \le r \le a$$

Sketch the effective potential for such an interaction between two particles. Can bound orbits exist? Find the differential and total cross sections for scattering by such a potential. The results are conveniently expressed in terms of the index of refraction introduced in Problem 9.6.17. What happens to the total cross section as V_0 becomes very large compared to the energy of the incident particle?

9.6.19 Let N_t be the total number of particles which are incident on a target over a period of time and let dN be the number of particles which are scattered from the incident beam over a thickness dx of target material. Let n be the number of target particles per unit volume. Assume that the total cross section of a single target particle has the finite value σ. Find a formula for the number of incident particles which survive a distance x into the target material without being scattered. How is the total cross section related to the probability that a given particle will scatter in a thickness dx of the target material?

9.7.20 Show that the magnitude of the angular momentum in the laboratory and center of mass coordinates are related by

$$l' = \frac{m_2}{m_1 + m_2} l$$

9.7.21 For an elastic collision, plot the scattering angle ϕ as measured in the laboratory as a function of the scattering angle ϕ' as measured in the center of mass coordinates (Eq. 9.7.8) for $\gamma = \frac{1}{2}$, 1, and 2. These values correspond to $m_1 < m_2$, $m_1 = m_2$, and $m_1 > m_2$. For $\gamma > 1$, find the maximum angle of scatter in the laboratory. Check your answer against Eq. (8.4.20). When $\phi = \phi_{max}$, what is the value of ϕ'? Does this mean that there is a restriction on the possible angles of scatter in the center of mass coordinate system? What modification of Eq. (9.7.24) for the differential cross section per unit solid angle is needed when $\gamma > 1$?

9.7.22 Solve $\tan \phi = \sin \phi'/(\gamma + \cos \phi')$ formally for $\cos \phi'$. Consider the three cases $\gamma < 1$, $\gamma = 1$, and $\gamma > 1$.

9.7.23 If $\gamma < 1$, show that

$$\frac{(1 + \gamma^2 + 2\gamma \cos \phi')^{3/2}}{1 + \gamma \cos \phi'} \cong 1 + 2\gamma \cos \phi + \tfrac{1}{2}\gamma^2(3 \cos^2 \phi - 1)$$

correct to terms of order γ^2. Show that the Rutherford scattering cross section correct to order γ^2 is of the form

$$\frac{d\sigma}{d\Omega} = \frac{Z^2 q^4}{16 \mathscr{E}^2} \frac{1}{\sin^4 (\phi/2)} [1 + \gamma^2 f(\phi)]$$

where all quantities are expressed in the laboratory coordinates. [It is not necessary actually to compute $f(\phi)$; the significant feature is that no term of order γ appears.]

10

Relative Motion and Rigid Body Dynamics

10.1. ACCELERATED FRAMES OF REFERENCE

Section 8.1 introduced the idea of a time-dependent transformation between two rectangular coordinate systems. However, in that section only constant relative velocity between coordinate systems, i.e., a Galilean transformation, was considered. In Section 8.13, the transformation theory was extended to include the arbitrary relative motion of two coordinate systems in two dimensions.

It is now useful to consider the simplest example of an accelerated frame of reference. If a frame F' moves at a velocity v relative to a frame F in such a way that the x- and x'-axes always remain parallel (Fig. 10.1.1), then the components of the velocity of a particle as measured in F and F' are related by the Galilean addition theorem for velocities:

$$u_x' = u_x - v \qquad (10.1.1)$$

$$u_y' = u_y \qquad (10.1.2)$$

Since F' is accelerated, v will not be a constant. Differentiation of (10.1.1) and (10.1.2) yields the relation between the components of particle acceleration as measured in the two coordinate systems:

$$a_x' = a_x - a_0 \qquad (10.1.3)$$

$$a_y' = a_y \qquad (10.1.4)$$

where $a_0 = dv/dt$ is the acceleration of F' relative to F.

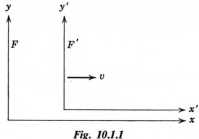

Fig. 10.1.1

In the nonaccelerated reference frame F, Newton's second law in component form is

$$F_x = ma_x \qquad F_y = ma_y \tag{10.1.5}$$

Although the practice is not universal in textbooks on intermediate mechanics, the point of view will be adopted here that Newton's second law is *generally covariant*. The force as measured by observers in the accelerated frame of reference is operationally *defined* by means of

$$F_x' = ma_x' \qquad F_y' = ma_y' \tag{10.1.6}$$

The mass of a particle is an *invariant scalar* and has the same value in all coordinate systems. The observers stationed in frame F' release a test particle of known mass and measure its acceleration in their own frame of reference. The force is then computed by means of (10.1.6).

The transformation law for force is deduced by combining (10.1.3), (10.1.4), (10.1.5), and (10.1.6):

$$F_x' = m(a_x - a_0) = F_x - ma_0 \tag{10.1.7}$$

$$F_y' = ma_y = F_y \tag{10.1.8}$$

Thus an observer stationed in frame F' interprets the term $-ma_0$ as a force. If there is no force on the particle in the nonaccelerated frame F, then in the accelerated frame F' the only force in the x-direction is $-ma_0$. If an observer in F' releases a test particle it will "fall" in the negative x-direction with acceleration a_0. It is impossible for him to tell any difference between a_0 and a uniform gravitational field of the same strength produced by a gravitating body. This fact is often referred to as *the principle of equivalence* and is one of the basic assumptions in advanced theories of gravitation. It is the practice in many texts to refer to $-ma_0$ as a "fictitious force." This force is very real to an observer in F'; therefore, we will avoid this point of view.

A frame of reference in which a neutral test body undergoes no acceleration when it is released is called an *inertial frame of reference*. The test body is specified to be neutral in order to eliminate the possibility of acceleration by electric or magnetic fields. Inertial frames can move relative to one another at constant relative velocity. The transformations among all possible inertial frames are therefore Galilean transformations, provided that the velocities involved are much less than the velocity of light. For higher velocities, the Galilean transformations must be replaced by the Lorentz transformations of special relativity.

A laboratory located on the surface of the earth is *not* an inertial frame of reference. An earth-bound laboratory is equivalent in first approximation to a laboratory on a space ship accelerating at 980 cm/sec². The qualification in first approximation is used because there are minor, but measurable, departures from uniformity in the earth's gravitational field even over the dimensions of a small laboratory. Also, there are effects due to the rotation of the earth which would not be found on a uniformly accelerated space ship. In treating problems involving the motion of charged particles in electromagnetic fields in earth-bound laboratories, the gravitational force on the particles is generally so much weaker than the electromagnetic force that it may be neglected and the laboratory considered as an inertial frame.

It is possible for a reference frame to be *locally inertial*. Suppose an elevator is released at the top of a tall building and that a Cartesian coordinate system is rigidly attached to it. Let the elevator be the origin. If an experimentor in the elevator releases a test particle, it will not fall to the floor. However, if he goes out along one of the coordinate axes, the effects of the nonuniformity of the earth's gravitational field will be found. The coordinate system is everywhere accelerated at 980 cm/sec², but at a great distance from the origin of our coordinate system the gravitational field of the earth is different from 980 cm/sec². If an experimenter at such a point releases a test particle, he will observe it to accelerate.

An artificial satellite is a local system of inertia. A particle that becomes detached from it follows it in orbit. The apparent dependence of the orbit equation (9.3.14) on the mass of the particle is artificial since C, l, and $\mathscr{E}$ all have factors of m in them so that the mass cancels out. This of course is not true if the reduced mass of the earth-satellite system is substantially different from the mass of the satellite.

10.2. THE MOST GENERAL PROPER TRANSFORMATION BETWEEN THREE-DIMENSIONAL CARTESIAN COORDINATE SYSTEMS

For the purpose of application to problems in classical mechanics, improper transformations, i.e., transformations from a right- to a left-hand coordinate system, can for the most part be left out of consideration. For example, since it is impossible to obtain a left-hand system from a right-hand system by a rotation, the motion of a rigid body is necessarily described by a time-dependent proper orthogonal transformation. Figure 8.1.1 illustrates the geometry of the transformation. We can think of **r** and **r**′ as representing the position vectors of a particle relative to the origins O and $O′$ of frames F and $F′$. Our purpose is to calculate the general relations connecting the velocity and acceleration of a particle as measured in $F′$ with these quantities as measured in F. Reference frame F is an *inertial frame* and $F′$ is moving relative to it in a completely arbitrary way.

The components of the three vectors **r**, **r**′, and **R** in frames F and $F′$ are displayed in the table on page 317 of Chapter 8. Equations (8.1.2) through (8.1.6) are all valid, provided, of course, that the transformation coefficients S_{ij} are regarded as functions of time. Since F is an *inertial frame*, the unit coordinate vectors $\hat{\mathbf{e}}_i$ are not time-dependent. Since $F′$ is in general undergoing a rotation as well as a translation, the unit vectors $\hat{\mathbf{e}}_i′$ must be treated as time dependent. By differentiation of (8.1.6),

$$\dot{x}_i\hat{\mathbf{e}}_i = \dot{x}_i′\hat{\mathbf{e}}_i′ + x_i′\frac{d\hat{\mathbf{e}}_i′}{dt} + v_i\hat{\mathbf{e}}_i \qquad (10.2.1)$$

We first give a geometrical interpretation of the derivative $d\hat{\mathbf{e}}_i′/dt$. Figure 10.2.1 shows the moving coordinate system $F′$ at two slightly different times. Actually, the origin of $F′$ is moving too, but only the change in the orientation of the unit vectors $\hat{\mathbf{e}}_i′$ contributes to the derivatives

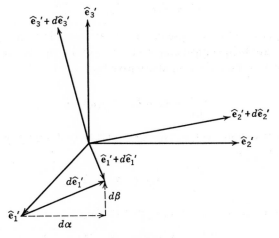

Fig. 10.2.1

$d\hat{e}_i'/dt$. The change $d\hat{e}_i'$ in the unit coordinate vector $\hat{e}_1'$ caused by the movement of the coordinate system has a component of magnitude $d\alpha$ in the $\hat{e}_2'$ direction and a component of magnitude $d\beta$ in the $\hat{e}_3'$ direction. Since $\hat{e}_1'$ has magnitude unity, $d\alpha$ and $d\beta$ represent angles of rotation. Thus,

$$d\hat{e}_1' = d\alpha\hat{e}_2' + d\beta\hat{e}_3' \tag{10.2.2}$$

The components of the angular velocity of rotation of F' can be introduced by means of

$$\omega_3' = \frac{d\alpha}{dt} \qquad \omega_2' = -\frac{d\beta}{dt} \tag{10.2.3}$$

As indicated by the primes, these are components of the angular velocity vector along the F' coordinate axes. The negative sign appears in the equation for ω_2' because a right-hand screw twisted in the sense indicated by $d\beta$ in Fig. 10.2.1 advances in the direction of $-\hat{e}_2'$. The division of (10.2.2) by dt yields

$$\frac{d\hat{e}_1'}{dt} = \omega_3'\hat{e}_2' - \omega_2'\hat{e}_3' \tag{10.2.4}$$

Treated as a vector field, the angular velocity vector $\boldsymbol{\omega}$ is a function of time only; at a given instant of time it has the same value at all points of the coordinate system F'. The angular velocity can be represented in either frame F or F':

$$\boldsymbol{\omega} = \omega_i\hat{e}_i = \omega_i'\hat{e}_i' \tag{10.2.5}$$

The connection between the components of $\boldsymbol{\omega}$ as measured in F and F' is

$$\omega_i' = S_{ij}\omega_j \qquad \omega_j = S_{ij}\omega_i' \tag{10.2.6}$$

Of course, $\boldsymbol{\omega}$ is actually a pseudo- or axial vector but only proper transformations are being considered here. It is easily shown that (10.2.4) can be expressed in the following alternative forms:

$$\frac{d\hat{e}_1'}{dt} = \boldsymbol{\omega} \times \hat{e}_1' = \delta_{1jk}\hat{e}_j'\omega_k' \tag{10.2.7}$$

The other two unit vectors can be differentiated similarly. The results are summarized as

$$\frac{d\hat{e}_i'}{dt} = \boldsymbol{\omega} \times \hat{e}_i' = \delta_{ijk}\hat{e}_j'\omega_k' \tag{10.2.8}$$

The substitution of (10.2.8) into (10.2.1) yields

$$\dot{x}_i\hat{e}_i = \dot{x}_i'\hat{e}_i' + \boldsymbol{\omega} \times x_i'\hat{e}_i' + v_i\hat{e}_i \tag{10.2.9}$$

The velocity of a particle as measured in F and F' is

$$\mathbf{u} = \dot{x}_i \hat{\mathbf{e}}_i \qquad \mathbf{u}' = \dot{x}_i' \hat{\mathbf{e}}_i' \tag{10.2.10}$$

Hence, (10.2.9) can be written

$$\mathbf{u} = \mathbf{u}' + \boldsymbol{\omega} \times \mathbf{r}' + \mathbf{v} \tag{10.2.11}$$

This is the transformation law for particle velocity. Equation (8.13.6) is a special case of Eq. (10.2.11); $u' = 0$ because there we were considering a point which was stationary in frame F'.

It is instructive to interpret (10.2.11) from the point of view of an observer in the inertial frame F. At a given instant the particle under consideration will be coincident with a point P' of F'. According to the observer in F, the origin O' of F' is moving at a velocity $\mathbf{v}$ with respect to him. Due to the rotation of F', the point P' moves with respect to O' at a velocity

$$\mathbf{w} = \boldsymbol{\omega} \times \mathbf{r}' \tag{10.2.12}$$

Simultaneously, the particle in question is moving at the velocity $\mathbf{u}'$ with respect to P'. Thus, if our experimenter in F makes a vector addition of these velocities, he gets

$$\mathbf{u} = \mathbf{v} + \mathbf{w} + \mathbf{u}' \tag{10.2.13}$$

which he interprets as the velocity of the particle with respect to himself.

To find the transformation law of particle acceleration, it is necessary to differentiate (10.2.11). We first consider the derivatives of the various vectors which appear in (10.2.11). The differentiation of (10.2.10) yields

$$\frac{d\mathbf{u}}{dt} = \ddot{x}_i \hat{\mathbf{e}}_i = \mathbf{a} \tag{10.2.14}$$

$$\frac{d\mathbf{u}'}{dt} = \ddot{x}_i' \hat{\mathbf{e}}_i' + \dot{x}_i'(\boldsymbol{\omega} \times \hat{\mathbf{e}}_i') = \mathbf{a}' + \boldsymbol{\omega} \times \mathbf{u}' \tag{10.2.15}$$

Now using (10.2.5) and (10.2.8),

$$\begin{aligned}
\dot{\boldsymbol{\omega}} = \dot{\omega}_i \hat{\mathbf{e}}_i &= \dot{\omega}_i' \hat{\mathbf{e}}_i' + \omega_i'(\boldsymbol{\omega} \times \hat{\mathbf{e}}_i') \\
&= \dot{\omega}_i' \hat{\mathbf{e}}_i' + \boldsymbol{\omega} \times \boldsymbol{\omega} = \dot{\omega}_i' \hat{\mathbf{e}}_i'
\end{aligned} \tag{10.2.16}$$

The differentiation of $\mathbf{r}'$ results in

$$\frac{d\mathbf{r}'}{dt} = \mathbf{u}' + \boldsymbol{\omega} \times \mathbf{r}' \tag{10.2.17}$$

Finally,

$$\mathbf{a}_0 = \frac{d\mathbf{v}}{dt} = \dot{v}_i \hat{\mathbf{e}}_i \tag{10.2.18}$$

is the acceleration of the origin O' of F' with respect to F. Now, by differentiating (10.2.11) and using (10.2.14) through (10.2.18),

$$\mathbf{a} = \frac{d\mathbf{u}'}{dt} + \dot{\boldsymbol{\omega}} \times \mathbf{r}' + \boldsymbol{\omega} \times \frac{d\mathbf{r}'}{dt} + \mathbf{a}_0$$

$$= \mathbf{a}' + \mathbf{a}_0 + 2\boldsymbol{\omega} \times \mathbf{u}' + \boldsymbol{\omega} \times (\boldsymbol{\omega} \times \mathbf{r}') + \dot{\boldsymbol{\omega}} \times \mathbf{r}' \qquad (10.2.19)$$

where $\mathbf{a}'$ is the acceleration of a particle as measured by observers in F' and $\mathbf{a}$ is the acceleration of the same particle as measured with respect to F. As a special case, suppose that a test particle is released from rest at O' by observers in F'. At the instant of release, $\mathbf{r}' = 0$ and $\mathbf{u}' = 0$ and (10.2.19) specializes to

$$\mathbf{a} = \mathbf{a}' + \mathbf{a}_0 \qquad (10.2.20)$$

Suppose further that there are no forces on the test particle in the inertial frame F. Then

$$\mathbf{a}' = -\mathbf{a}_0 \qquad (10.2.21)$$

Thus the instantaneous acceleration of the test particle at the instant of release is just $-\mathbf{a}_0$. To put it another way, if experimenters at O' make a static determination of the weight of a particle at O', e.g., by means of a spring balance, they will get the value $-m\mathbf{a}_0$.

In component form, (10.2.21) is

$$a_i'\hat{\mathbf{e}}_i' = -a_{0i}\hat{\mathbf{e}}_i \qquad (10.2.22)$$

By using the transformation property of the unit vectors (8.1.4) it is readily proved that

$$a_i' = -S_{ij}a_{0j} \qquad (10.2.23)$$

In Eq. (10.2.16) it is shown that the angular acceleration vector of F' can be expressed in either F or F' as

$$\dot{\boldsymbol{\omega}} = \dot{\omega}_i\hat{\mathbf{e}}_i = \dot{\omega}_i'\hat{\mathbf{e}}_i' \qquad (10.2.24)$$

Again employing (8.1.4) for the transformation of the unit vectors, we discover that the components of $\dot{\boldsymbol{\omega}}$ transform according to the simple rule

$$\dot{\omega}_i' = S_{ij}\dot{\omega}_j \qquad (10.2.25)$$

The reason why the angular velocity and the angular acceleration of F' retain their simple vector transformation property is that at a given instant they have the *same value* at all points of F' independent of the coordinates. On the other hand, the *linear* velocity and acceleration of a point in F' are functions of position and this results in a much more complicated law of transformation.

10.3. THE ANGULAR VELOCITY TENSOR

The expression (10.2.8) for the time rate of change of the unit vectors $\hat{\mathbf{e}}_i'$ of F' was obtained by a geometrical method. An alternative procedure is to differentiate the transformation equation,

$$\hat{\mathbf{e}}_i' = S_{ik}\hat{\mathbf{e}}_k \qquad (10.3.1)$$

Remembering that the unit vectors $\hat{\mathbf{e}}_k$ of the inertial frame F are constant,

$$\frac{d\hat{\mathbf{e}}_i'}{dt} = \dot{S}_{ik}\hat{\mathbf{e}}_k \qquad (10.3.2)$$

Now using the inverse of (10.3.1),

$$\hat{\mathbf{e}}_k = S_{jk}\hat{\mathbf{e}}_j' \qquad (10.3.3)$$

we arrive at

$$\frac{d\hat{\mathbf{e}}_i'}{dt} = \dot{S}_{ik}S_{jk}\hat{\mathbf{e}}_j' \qquad (10.3.4)$$

It is convenient to define

$$\Omega_{ij}' = \dot{S}_{ik}S_{jk} \qquad (10.3.5)$$

which will be shown presently to be the components of a second rank tensor.

Let us establish that Ω_{ij}' is antisymmetric. The differentiation of the orthogonality relation

$$S_{ik}S_{jk} = \delta_{ij} \qquad (10.3.6)$$

yields

$$\dot{S}_{ik}S_{jk} + S_{ik}\dot{S}_{jk} = 0$$

which is recognized to be equivalent to

$$\Omega_{ij}' + \Omega_{ji}' = 0 \qquad (10.3.7)$$

The direct comparison of (10.3.4) and (10.2.8) shows that

$$\Omega_{ij}' = \delta_{ijk}\omega_k' \qquad (10.3.8)$$

Since it is already known that the permutation symbol is a third rank pseudo-tensor and that ω_k' are the components of a pseudo-vector, (10.3.8) establishes that Ω_{ij}' are the components of an ordinary second rank antisymmetric tensor which we denote as $\boldsymbol{\Omega}$ and call the angular velocity tensor. The components of $\boldsymbol{\Omega}$ in the inertial frame F are to be found from

$$\Omega_{kl} = S_{ik}S_{jl}\Omega_{ij}' \qquad (10.3.9)$$

Using (10.3.5) and the orthogonality conditions for the transformation coefficients yields

$$\Omega_{kl} = S_{ik}S_{jl}\dot{S}_{im}S_{jm}$$
$$= S_{ik}\dot{S}_{im}\delta_{lm} = S_{ik}\dot{S}_{il} \tag{10.3.10}$$

It is revealing to work out the components Ω'_{ij}. By using (10.3.8) we find, for example, that

$$\Omega'_{11} = \delta_{11k}\omega_k' = 0$$
$$\Omega'_{12} = \delta_{12k}\omega_k' = \delta_{123}\omega_3' = \omega_3'$$
$$\Omega'_{21} = \delta_{21k}\omega_k' = \delta_{213}\omega_3' = -\omega_3'$$

Proceeding in this way, it is established that the components of $\boldsymbol{\Omega}$ in F' can be displayed in the matrix

$$\Omega' = \begin{pmatrix} 0 & \omega_3' & -\omega_2' \\ -\omega_3' & 0 & \omega_1' \\ \omega_2' & -\omega_1' & 0 \end{pmatrix} \tag{10.3.11}$$

Finally, let us solve (10.3.8) for the components of angular velocity. Multiplication by δ_{lij} yields

$$\delta_{lij}\Omega'_{ij} = \delta_{lij}\delta_{ijk}\omega_k'$$
$$= \delta_{ijl}\delta_{ijk}\omega_k'$$
$$= (\delta_{jj}\delta_{lk} - \delta_{jk}\delta_{lj})\omega_k'$$
$$= (3\delta_{lk} - \delta_{lk})\omega_k' = 2\omega_l';$$
$$\omega_l' = \tfrac{1}{2}\delta_{lij}\Omega'_{ij} = \tfrac{1}{2}\delta_{lij}\dot{S}_{ik}S_{jk} \tag{10.3.12}$$

Of course, the same relation exists between the components of $\boldsymbol{\omega}$ and $\boldsymbol{\Omega}$ in the inertial frame:

$$\omega_l = \tfrac{1}{2}\delta_{lij}\Omega_{ij} = \tfrac{1}{2}\delta_{lij}S_{ki}\dot{S}_{kj} \tag{10.3.13}$$

10.4. MECHANICS OF A PARTICLE IN AN ACCELERATED REFERENCE FRAME

Following the philosophy laid down in Section 10.1, the force on a particle in an accelerated frame of reference is defined by

$$\mathbf{F}' = m\mathbf{a}' \tag{10.4.1}$$

The force on the same particle as measured in an inertial frame is

$$\mathbf{F} = m\mathbf{a} \tag{10.4.2}$$

The transformation law of force is deduced from (10.2.19):

$$\mathbf{F}' = \mathbf{F} - m\mathbf{a}_0 + 2m(\mathbf{u}' \times \boldsymbol{\omega}) + m\boldsymbol{\omega} \times (\mathbf{r}' \times \boldsymbol{\omega}) + m(\mathbf{r}' \times \dot{\boldsymbol{\omega}}) \tag{10.4.3}$$

The force $\mathbf{F}$ which is present in the inertial frame is due to electromagnetic forces or forces of constraint. Forces due to the gravitational field produced by gravitating bodies can also be included in F, even though this makes the unprimed frame noninertial in the strict sense. The term $2m(\mathbf{u}' \times \boldsymbol{\omega})$ is known as the *Coriolis force*. Note its resemblance to the magnetic force on a moving charged particle: $(q/c)(\mathbf{u} \times \mathbf{B})$. The term $m\boldsymbol{\omega} \times (\mathbf{r}' \times \boldsymbol{\omega})$ is the *centrifugal force*. Note that

$$m\boldsymbol{\omega} \times (\mathbf{r}' \times \boldsymbol{\omega}) = m\mathbf{r}'\omega^2 - m\boldsymbol{\omega}(\mathbf{r}' \cdot \boldsymbol{\omega}) \qquad (10.4.4)$$

As a simple example, suppose that frames F and F' have their origins coincident so that $\mathbf{r} = \mathbf{r}'$ and $\mathbf{a}_0 = 0$. Let the z- and z'-axes be coincident

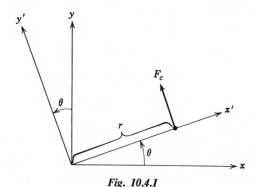

Fig. 10.4.1

and let F' rotate at a constant angular velocity about the common z- and z'-axes. The angular velocity vector is pointed out of the plane of thy paper of Fig. 10.4.1 and has magnitude $\omega = \dot{\theta}$. Suppose there is a bead on the x'-axis which slides without friction. [This problem has alreade been treated by the Lagrangian method in Section 8.8; see equations (8.8.19) through (8.8.23).] Using the Newtonian approach and taking the point of view of an observer stationed in the inertial frame, we would say that there is a single force acting on the particle, namely, the force of constraint F_c acting perpendicular to the x'-axis, Fig. 10.4.1. The equations of motion, therefore, are

$$F_r = ma_r = m(\ddot{r} - r\dot{\theta}^2) = 0 \qquad (10.4.5)$$

$$F_\theta = ma_\theta = m(r\ddot{\theta} + 2\dot{r}\dot{\theta}) = F_c \qquad (10.4.6)$$

Since $\dot{\theta} = \omega = \text{const.}$,

$$\ddot{r} - r\omega^2 = 0 \qquad 2m\dot{r}\omega = F_c \qquad (10.4.7)$$

Now taking the point of view of an observer in the accelerated frame, Fig. 10.4.2, we would say that there is a centrifugal force of magnitude $mr\omega^2$ in the x'-direction; note that $\mathbf{r} \cdot \boldsymbol{\omega} = 0$ in Eq. (10.4.4) since $\mathbf{r}$ and $\boldsymbol{\omega}$ are perpendicular. Thus,

$$mr\omega^2 = m\ddot{r} \tag{10.4.8}$$

In addition, there is a Coriolis force $2m(\mathbf{u} \times \boldsymbol{\omega}) = -2m\dot{r}\omega\hat{\mathbf{j}}'$ acting on the particle, which, since there is no motion perpendicular to the x'-axis, is

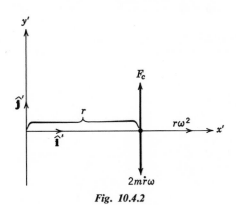

Fig. 10.4.2

exactly balanced by the force of constraint exerted on the particle by the x'-axis:

$$F_c = 2m\dot{r}\omega \tag{10.4.9}$$

Notice that (10.4.8) and (10.4.9) are identical to (10.4.7).

10.5. THE MECHANICS OF A PARTICLE NEAR THE SURFACE OF THE EARTH

An important example of an accelerated reference frame is a coordinate system fixed to the surface of the earth. Let us first inquire as to the effect on the motion of particles relative to such a coordinate system due to the motion of the earth about the sun. For this purpose, consider an earth which does not turn on its own axis as it moves around the sun as illustrated in Fig. 10.5.1. A Cartesian coordinate system fixed to the earth does not turn on its axis. Hence, $\boldsymbol{\omega} = 0$ and there is no Coriolis or centrifugal force effect due to the motion of the earth around the sun. The center of the earth is a local system of inertia in that a test particle released here would experience no acceleration relative to the earth. Even if the earth's gravitational field could be turned off, a test particle released at the surface of the earth would experience a slight acceleration relative to the surface of the earth due to the sun's gravitational field at the earth's surface being slightly different from what it is at the earth's center. This is substantially the same effect as is illustrated by Problem 10.1.3. There is a similar variation in the gravitational field over the surface of the earth due to the moon; see Problem 10.5.8. It can be concluded that any centrifugal or Coriolis force effect observed on the surface of the earth is

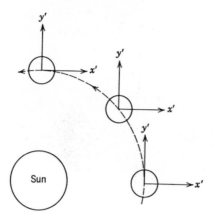

Fig. 10.5.1

due to the spin of the earth on its own axis. It will be assumed in what follows that the spin angular velocity of the earth is a constant, both in magnitude and direction.

When treating the mechanics of a particle over a limited region near the earth's surface, e.g., over the dimensions of an average laboratory, it is convenient to introduce a coordinate system fixed to the surface of the earth as illustrated in Fig. 10.5.2. The z-axis is in the direction of a plumb line, the x-axis points south, and the y-axis points east. Since we are going to work entirely in the accelerated frame of reference, primes are dropped. For an observer located at the origin O, the xy-plane is sometimes called the *horizon plane* and the xz-plane the *meridian plane*.

The earth is not a perfect sphere but, except for local variations, it is well approximated by an ellipsoid of revolution. This is due essentially to the centrifugal force produced by the earth's rotation. If the earth were made of a plastic material, the surface would conform to an equipotential surface of the combined gravitational and centrifugal forces. A plumb line would be perpendicular to this hypothetical surface. The equatorial radius of the earth is $a = 6{,}378.4$ Km and the polar radius is $b = 6{,}356.9$ Km. The departure of the earth from a spherical shape is not large.

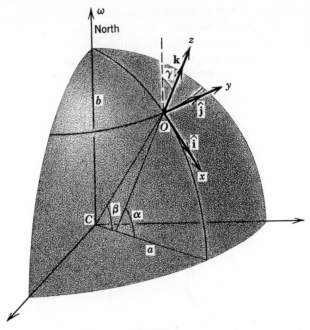

Fig. 10.5.2

If the direction of a plumb line were continued into the earth, its would not pass through the center of the earth but would eventually intersect the equatorial plane at an angle α as indicated in Fig. 10.5.2. If the direction of the plumb line is corrected for "station error," i.e., local variations in the gravitational field due to the nonuniformity of the earth's crust, then α is called the *geographical latitude* of the observer at O. If a line is drawn from O to the center C of the earth, it intersects the equatorial plane at an angle β called the *geocentric latitude*. The difference between α and β is not large, amounting at most to about $0.2°$. Finally, γ is the angle between the z-axis and the direction of the angular velocity of rotation of the earth. Obviously $\gamma = 90° - \alpha$.

We are now ready to apply Eq. (10.4.3) to determine the motion of a particle relative to a coordinate system fixed to the surface of the earth as pictured in Fig. 10.5.2. In Eq. (10.4.3) it is convenient to set

$$\mathbf{F} - m\mathbf{a}_0 = -mg\mathbf{\hat{k}} \qquad (10.5.1)$$

where mg is the measured value of the weight of a particle at O. Actually, it is the force $\mathbf{F} - m\mathbf{a}_0$ that determines the direction of a plumb line and hence the z-axis. The force $\mathbf{F}$ is produced by the gravitational field of the matter in the earth (and a very small correction due to the gravitational fields of the sun and moon) and $\mathbf{a}_0$ is the acceleration of the origin O. If the earth were a perfect sphere of radius r_e, it would be true that

$$\mathbf{a}_0 = -r_e \sin \gamma \omega^2 (\hat{\mathbf{i}} \cos \gamma + \hat{\mathbf{k}} \sin \gamma) \qquad (10.5.2)$$

Note from Fig. 10.5.2 that

$$\boldsymbol{\omega} = -\omega \sin \gamma \hat{\mathbf{i}} + \omega \cos \gamma \hat{\mathbf{k}} \qquad (10.5.3)$$

Since $\dot{\boldsymbol{\omega}} = 0$, Eq. (10.4.3) gives

$$\mathbf{F} = -mg\hat{\mathbf{k}} + 2m(\mathbf{u} \times \boldsymbol{\omega}) + m\boldsymbol{\omega} \times (\mathbf{r} \times \boldsymbol{\omega}) \qquad (10.5.4)$$

where $\mathbf{u}$ is the velocity of the particle with respect to an observer on earth and $\mathbf{r}$ is the position vector of the particle with respect to the origin O of the coordinate system pictured in Fig. 10.5.2. All quantities are now measured with respect to the accelerated frame and primes are dropped. The angular velocity of rotation of the earth is

$$\omega = 7.27 \times 10^{-5} \sec^{-1} \qquad (10.5.5)$$

The centrifugal force term $m\boldsymbol{\omega} \times (\mathbf{r} \times \boldsymbol{\omega})$ in (10.5.4) is proportional to $\omega^2 \cong 53 \times 10^{-10}$; therefore, it can be neglected, provided that the distance r of the particle from the origin is not great. The remaining terms in (10.5.4) are then exactly analogous to the force which a charged particle experiences in an electromagnetic field, namely

$$\mathbf{F} = q\mathbf{E} + \frac{q}{c}(\mathbf{u} \times \mathbf{B}) \qquad (10.5.6)$$

The important difference here is that, due to the smallness of ω, the Coriolis force term $2m\mathbf{u} \times \boldsymbol{\omega}$ is to be regarded as a very small perturbation or correction on the motion produced by the gravitational attraction $-mg$ Setting

$$\mathbf{F} = m\mathbf{a} = m(\hat{\mathbf{i}}\ddot{x} + \hat{\mathbf{j}}\ddot{y} + \hat{\mathbf{k}}\ddot{z})$$

in (10.5.4) gives the component equations

$$\ddot{x} = 2\dot{y}\omega \cos \gamma$$
$$\ddot{y} = -2\dot{z}\omega \sin \gamma - 2\dot{x}\omega \cos \gamma \qquad (10.5.7)$$
$$\ddot{z} = -g + 2\dot{y}\omega \sin \gamma$$

The solutions of these equations will be functions of the parameter ω. Since ω is small, it is convenient to express the solutions as Taylor's series expansions with respect to ω:

$$x = x_1 + x_2\omega + \cdots$$
$$y = y_1 + y_2\omega + \cdots \qquad (10.5.8)$$
$$z = z_1 + z_2\omega + \cdots$$

There is no point in including higher terms since terms of order ω^2 have already been neglected in discarding the centrifugal force terms. The coefficients x_1, x_2, y_1, etc. are functions of time to be determined. By substituting (10.5.8) into (10.5.7) and equating the coefficients of the like powers of ω the following system of equations results:

$$\ddot{x}_1 = 0 \qquad \ddot{y}_1 = 0 \qquad \ddot{z}_1 = -g \qquad (10.5.9)$$

$$\ddot{x}_2 = 2\dot{y}_1 \cos \gamma$$

$$\ddot{y}_2 = -2\dot{z}_1 \sin \gamma - 2\dot{x}_1 \cos \gamma \qquad (10.5.10)$$

$$\ddot{z}_2 = 2\dot{y}_1 \sin \gamma$$

Equations (10.5.9) give the first approximation to the solutions and are recognized to be the equations of motion of a particle moving in a uniform gravitational field. These equations would be exact if $\omega = 0$. Equations (10.5.10) provide a correction to the motion due to the Coriolis effect. As an example, suppose that a particle is dropped with zero initial velocity from a height z_0 above the surface of the earth. The solutions of (10.5.9) are then

$$x_1 = 0 \qquad y_1 = 0 \qquad z_1 = z_0 - \tfrac{1}{2}gt^2 \qquad (10.5.11)$$

Substituting (10.5.11) into (10.5.10) results in

$$\ddot{x}_2 = 0 \qquad \ddot{y}_2 = 2gt \sin \gamma \qquad \ddot{z}_2 = 0 \qquad (10.5.12)$$

Integration yields

$$x_2 = 0 \qquad y_2 = \tfrac{1}{3}gt^3 \sin \gamma \qquad z_2 = 0 \qquad (10.5.13)$$

Combining (10.5.8), (10.5.11), and (10.5.13) gives as the final result:

$$x = 0 \qquad y = \tfrac{1}{3}gt^3\omega \sin \gamma \qquad z = z_0 - \tfrac{1}{2}gt^2 \qquad (10.5.14)$$

It is seen that there is a small deflection of the particle in an easterly direction. Reference to Fig. 10.5.2 shows that if a particle is moving in the negative z-direction, then the direction of the Coriolis force $2m\mathbf{u} \times \boldsymbol{\omega}$ as determined by the right-hand rule is indeed in the y-direction. As a numerical example, if a particle is dropped from a height of 100 ft at the equator, $\gamma = \pi/2$, the horizontal deflection is about 0.15 inch.

It is not necessarily true that Eq. (10.4.3) provides the easiest approach to the solution of mechanics problems with respect to an accelerated frame of reference. For instance, in determining the motion of an artificial satellite, it may be easier to visualize the orbit of the satellite as determining a fixed plane (neglecting perturbations due to the nonspherical distribution of mass in the earth) relative to an inertial frame located somewhere in free space. The earth rotates with respect to the inertial frame and the location of the satellite with respect to the earth at a given time is found from the combined motions.

10.6. THE LAGRANGE AND HAMILTON FORMULATIONS

In this section we develop a Lagrangian which gives the correct equation of motion of a particle with respect to an accelerated reference frame. This puts at our disposal all the advantages of the Lagrange formulation such as the easy treatment of forces of constraint and the derivation of constants of the motion. The law of force with which we are dealing is conveniently expressed as

$$\mathbf{F} = m\mathbf{g} + 2m(\mathbf{u} \times \boldsymbol{\omega}) + m\mathbf{r} \times \dot{\boldsymbol{\omega}} \qquad (10.6.1)$$

where

$$\mathbf{g} = \mathbf{r}\omega^2 - \boldsymbol{\omega}(\mathbf{r} \cdot \boldsymbol{\omega}) - \mathbf{a_0} \qquad (10.6.2)$$

In the present discussion, forces due to gravitating bodies or electromagnetic forces are not included. Since we already know how to treat such forces, they can be added as needed.

In component notation, (10.6.2) can be expressed by

$$g_i = x_j(\delta_{ij}\omega^2 - \omega_i\omega_j) - a_{0i}$$

$$= -\frac{\partial\psi}{\partial x_i} \qquad (10.6.3)$$

where the potential function ψ is given by

$$\psi = -\tfrac{1}{2}\delta_{ij}x_i x_j \omega^2 + \tfrac{1}{2}(x_k\omega_k)^2 + x_k a_{0k}$$
$$= -\tfrac{1}{2}r^2\omega^2 + \tfrac{1}{2}(\mathbf{r} \cdot \boldsymbol{\omega})^2 + \mathbf{r} \cdot \mathbf{a_0} \qquad (10.6.4)$$

A clue as to what type of term is to be put into the Lagrangian to give the Coriolis force is provided by the analogy to the Lorentz force experienced by a charged particle in an electromagnetic field, Eq. (10.5.6). Treating $\boldsymbol{\omega}$ as the analogue of a magnetic field, a vector potential is introduced by means of

$$\boldsymbol{\omega} = \nabla \times \mathbf{A} \qquad (10.6.5)$$

Remember that $\boldsymbol{\omega}$, considered as a vector field existing throughout the accelerated coordinate system, is a function of only t, and at a given instant has the same value at all points of the coordinate system. It is then easily verified that (10.6.5) is satisfied by

$$\mathbf{A} = \tfrac{1}{2}(\boldsymbol{\omega} \times \mathbf{r}) \qquad (10.6.6)$$

The proposed Lagrangian, therefore, is

$$L = T - m\psi + 2m(\mathbf{A} \cdot \mathbf{u})$$
$$= T + \tfrac{1}{2}mr^2\omega^2 - \tfrac{1}{2}m(\mathbf{r} \cdot \boldsymbol{\omega})^2 - m(\mathbf{r} \cdot \mathbf{a_0}) + m\mathbf{u} \cdot (\boldsymbol{\omega} \times \mathbf{r}) \qquad (10.6.7)$$
$$= \tfrac{1}{2}m\dot{x}_i\dot{x}_i + \tfrac{1}{2}m(x_i x_i)\omega^2 - \tfrac{1}{2}m(x_i\omega_i)^2 - m(x_i a_{0i}) + m\delta_{ijk}\dot{x}_i\omega_j x_k$$

The correctness of this Lagrangian is verified as follows:

$$\frac{\partial L}{\partial x_k} = mx_k\omega^2 - m(x_i\omega_i)\omega_k - ma_{0k} + m\delta_{ijk}\dot{x}_i\omega_j \qquad (10.6.8)$$

$$\frac{\partial L}{\partial \dot{x}_k} = m\dot{x}_k + m\delta_{kij}\omega_i x_j = p_k \qquad (10.6.9)$$

$$\frac{d}{dt}\left(\frac{\partial L}{\partial \dot{x}_k}\right) = m\ddot{x}_k + m\delta_{kij}\omega_i\dot{x}_j + m\delta_{kij}\dot{\omega}_i x_j \qquad (10.6.10)$$

$$\frac{d}{dt}\left(\frac{\partial L}{\partial \dot{x}_k}\right) - \frac{\partial L}{\partial x_k} = m\ddot{x}_k - 2m(\mathbf{u} \times \boldsymbol{\omega})_k - mx_k\omega^2$$
$$+ m\omega_k(\mathbf{r} \cdot \boldsymbol{\omega}) + ma_{0k} - m(\mathbf{r} \times \dot{\boldsymbol{\omega}})_k \qquad (10.6.11)$$

By using (10.6.1) and (10.6.2) and setting $F_k = m\ddot{x}_k$ it is seen that

$$\frac{d}{dt}\left(\frac{\partial L}{\partial \dot{x}_k}\right) - \frac{\partial L}{\partial x_k} = 0 \qquad (10.6.12)$$

Thus (10.6.7) represents a Lagrangian that describes correctly the motion of a particle with respect to an accelerated reference frame.

As indicated, (10.6.9) represents the canonical momenta associated with the Cartesian coordinates x_k. The equations of motions of a particle with respect to an accelerated coordinate system can be expressed in the Hamiltonian form:

$$\dot{p}_k = -\frac{\partial H}{\partial x_k} \qquad \dot{x}_k = \frac{\partial H}{\partial p_k} \qquad (10.6.13)$$

The Hamiltonian is given by

$$\begin{aligned} H &= p_k\dot{x}_k - L \\ &= \tfrac{1}{2}m\dot{x}_k\dot{x}_k - \tfrac{1}{2}mr^2\omega^2 + \tfrac{1}{2}m(\mathbf{r} \cdot \boldsymbol{\omega})^2 + m(\mathbf{r} \cdot \mathbf{a}_0) \qquad (10.6.14) \\ &= T + V \end{aligned}$$

where $V = m\psi$ and ψ is given by (10.6.4). It is possible to think of V as being the potential energy associated with the vector field $\mathbf{a}_0$ and the centrifugal force. If L has no explicit time dependence, specificially if the vector fields $\boldsymbol{\omega}$ and $\mathbf{a}_0$ are independent of time, then H is a constant of the motion. If $\dot{x}_k$ is eliminated by means of (10.6.9), the Hamiltonian expressed in terms of coordinates and generalized momenta is

$$H = \frac{1}{2m}[\mathbf{p} - m(\boldsymbol{\omega} \times \mathbf{r})]^2 - \tfrac{1}{2}mr^2\omega^2 + \tfrac{1}{2}m(\mathbf{r} \cdot \boldsymbol{\omega})^2 + m(\mathbf{r} \cdot \mathbf{a}_0) \quad (10.6.15)$$

If it is noted that

$$(\boldsymbol{\omega} \times \mathbf{r}) \cdot (\boldsymbol{\omega} \times \mathbf{r}) = r^2\omega^2 - (\boldsymbol{\omega} \cdot \mathbf{r})^2 \qquad (10.6.16)$$

the Hamiltonian simplifies to

$$H = \frac{p^2}{2m} - (\mathbf{p} \cdot \boldsymbol{\omega} \times \mathbf{r}) + m(\mathbf{r} \cdot \mathbf{a}_0) \tag{10.6.17}$$

With the Hamiltonian so expressed, the reader can verify that (10.6.13) gives the correct equations of motion.

10.7. VELOCITY OF ESCAPE FROM A ROTATING PLANET

From the point of view of an inertial frame, it is easily shown that the escape velocity of a particle from the surface of a planet of radius r_0 is given by

$$\tfrac{1}{2}mu_0{}^2 = \frac{GmM}{r_0} \tag{10.7.1}$$

where u_0 is the initial speed of the particle *relative to the inertial frame* as it leaves the planet, m is the mass of the particle and M is the mass of the planet. u_0 includes the contribution due to the planet's rotations. Suppose that the escaping particle moves in the equatorial plane. If $u_{\perp}'$ and $u_{\parallel}'$ are the components of the particles initial velocity perpendicular and parallel to the surface of the planet as *measured by observers on the planet*, then

$$u_0{}^2 = u_{\perp}'^2 + (u_{\parallel}' + r_0\omega)^2 \tag{10.7.2}$$

As an exercise in the use of the concepts developed in Section 10.6, these results will be derived from the point of view of an observer stationed on the planet.

The discussion will be confined to the motion of a particle in the equatorial plane. Let a cylindrical coordinate system be fixed in the planet with its origin at the center of the planet and the z-axis pointing along the axis of rotation. The three unit coordinate vectors $\hat{\mathbf{r}}$, $\hat{\boldsymbol{\theta}}$, and $\hat{\mathbf{k}}$ from a right-hand triad. The acceleration $\mathbf{a}_0$ of the origin of our coordinate system is zero. The position and velocity of the particle are given by

$$\mathbf{r} = r\hat{\mathbf{r}} \qquad \mathbf{u} = \dot{r}\hat{\mathbf{r}} + r\dot{\theta}\hat{\boldsymbol{\theta}} \tag{10.7.3}$$

where all quantities are measured with respect to an observer on the planet. Noting that

$$\boldsymbol{\omega} = \omega\hat{\mathbf{k}} \qquad \boldsymbol{\omega} \times \mathbf{r} = r\omega\hat{\boldsymbol{\theta}}$$
$$\boldsymbol{\omega} \times \mathbf{r} \cdot \mathbf{u} = r^2\omega\dot{\theta} \qquad \mathbf{r} \cdot \boldsymbol{\omega} = 0 \tag{10.7.4}$$

the Lagrangian, including the necessary term to account for the effect of the gravitational field of the planet, is found to be

$$L = \tfrac{1}{2}m(\dot{r}^2 + r^2\dot{\theta}^2) + \tfrac{1}{2}mr^2\omega^2 + mr^2\omega\dot{\theta} + \frac{GmM}{r} \tag{10.7.5}$$

The canonical momentum conjugate to the coordinate θ is

$$l = \frac{\partial L}{\partial \dot\theta} = mr^2(\dot\theta + \omega) \tag{10.7.6}$$

and, since θ is a cyclic coordinate, is a constant of the motion.

A second constant of the motion is the Hamiltonian:

$$H = \tfrac{1}{2}m(\dot r^2 + r^2\dot\theta^2) - \tfrac{1}{2}mr^2\omega^2 - \frac{GmM}{r} \tag{10.7.7}$$

Using (10.7.6) to eliminate $\dot\theta$ gives

$$H = \tfrac{1}{2}m\dot r^2 + \frac{l^2}{2mr^2} - l\omega - \frac{GmM}{r} \tag{10.7.8}$$

Since the effective potential

$$V_e(r) = \frac{l^2}{2mr^2} - l\omega - \frac{GmM}{r} \tag{10.7.9}$$

has no relative maxima between $r = 0$ and $r = \infty$, the velocity for escape is found by setting $\dot r = 0$ at $r = \infty$. Equating the Hamiltonians evaluated at $r = r_0$ and $r = \infty$ gives

$$\tfrac{1}{2}m\dot r_0^2 + \frac{l^2}{2mr_0^2} - \frac{GmM}{r_0} = 0 \tag{10.7.10}$$

If l is eliminated by using (10.7.6), it is found that (10.7.10) can be expressed by

$$\tfrac{1}{2}m[\dot r_0^2 + (r_0\dot\theta_0 + r_0\omega)^2] = \frac{GmM}{r_0} \tag{10.7.11}$$

Since $u_\perp' = \dot r_0$ and $u_\parallel' = r_0\dot\theta_0$ are the components of the initial velocity relative to an observer on the planet, (10.7.11) is equivalent to (10.7.1).

10.8. THE FOUCAULT PENDULUM

The Foucault pendulum is essentially a spherical pendulum for which the Coriolis force has been included. The perturbation on the motion due to the Coriolis force is called the *Foucault effect*.

The coordinate system fixed to the surface of the earth and pictured in Fig. 10.5.2 will be used, but the coordinate axes will be labeled x_1, x_2, and x_3. The pendulum is hung from the x_3-axis at a point such that the origin is the equilibrium position. Since the particle never moves very far from the origin, the centrifugal force terms, proportional to $r\omega^2$, will be neglected. An appropriate Lagrangian, therefore, is

$$L = \tfrac{1}{2}mu^2 - mgx_3 + m\mathbf{u}\cdot(\boldsymbol{\omega}\times\mathbf{r}) \tag{10.8.1}$$

where g is the empirically determined acceleration of a freely falling body at the location where the pendulum is set up. The equation of constraint connecting the three coordinates x_1, x_2, and x_3 is

$$x_1^2 + x_2^2 + (x_3 - s)^2 = s^2 \qquad (10.8.2)$$

where s is the length of the pendulum, i.e., the radius of the sphere over which the particle is constrained to move.

For small amplitudes, the term proportional to x_3^2 can be neglected. The equation of constraint (10.8.2) is approximately

$$x_3 = \frac{1}{2s}(x_1^2 + x_2^2) \qquad (10.8.3)$$

If all terms higher than second order are neglected, the approximate Lagrangian is

$$L = \tfrac{1}{2}m(\dot{x}_1^2 + \dot{x}_2^2) - \frac{mg}{2s}(x_1^2 + x_2^2) + m\omega_3(x_1\dot{x}_2 - \dot{x}_1 x_2) \qquad (10.8.4)$$

where $\omega_3 = \omega \cos \gamma$ is the x_3 component of $\boldsymbol{\omega}$. The equations of motion are found to be

$$\ddot{x}_1 + \frac{g}{s}x_1 - 2\omega_3\dot{x}_2 = 0 \qquad (10.8.5)$$

$$\ddot{x}_2 + \frac{g}{s}x_2 + 2\omega_3\dot{x}_1 = 0 \qquad (10.8.6)$$

These equations are recognized to be the equations of motion of a harmonic oscillator moving in the x_1x_2-plane to which have been added terms to account for the Coriolis force. If (10.8.6) is multiplied by $i = \sqrt{-1}$ and added to (10.8.5), the resulting equation can be written in terms of the complex variable $z = x_1 + ix_2$ as

$$\ddot{z} + 2i\omega_3\dot{z} + \frac{g}{s}z = 0 \qquad (10.8.7)$$

This is a linear differential equation with constant coefficients. Its general solution can be written

$$z = e^{-i\omega_3 t}(c_1 e^{it\sqrt{\omega_3^2 + (g/s)}} + c_2 e^{-it\sqrt{\omega_3^2 + (g/s)}}) \qquad (10.8.8)$$

where c_1 and c_2 are constants of integration and are, in general, complex numbers. Since ω^2 is approximately 53×10^{-10} sec^{-2}, it can be neglected in comparison to g/s. The evaluation of c_1 and c_2 in terms of arbitrary initial conditions results in

$$c_1 = \tfrac{1}{2}z_0\left(1 + \frac{\omega_3}{\omega_0}\right) - i\frac{\dot{z}_0}{2\omega_0} \qquad (10.8.9)$$

$$c_2 = \tfrac{1}{2}z_0\left(1 - \frac{\omega_3}{\omega_0}\right) + i\frac{\dot{z}_0}{2\omega_0} \qquad (10.8.10)$$

where $\omega_0 = \sqrt{g/s}$ is the natural frequency of the pendulum when the Coriolis force is absent and z_0 and $\dot{z}_0$ are the complex initial position and velocity.

As a specific example, suppose that the pendulum is drawn aside along the x_1-axis a distance x_0 and released from rest. With $z_0 = x_0$ and $\dot{z}_0 = 0$,

$$c_1 = \tfrac{1}{2}x_0\left(1 + \frac{\omega_3}{\omega_0}\right) \qquad (10.8.11)$$

$$c_2 = \tfrac{1}{2}x_0\left(1 - \frac{\omega_3}{\omega_0}\right) \qquad (10.8.12)$$

The solution (10.8.8) can then be expressed by

$$z = x_0 e^{-i\omega_3 t}\left(\cos\omega_0 t + i\frac{\omega_3}{\omega_0}\sin\omega_0 t\right) \qquad (10.8.13)$$

In the complex z-plane, $x_0 e^{-i\omega_3 t}$ represents a point which moves on a circle of radius x_0 in the clockwise direction and at an angular frequency ω_3. At a latitude of $45°$, this point makes one complete revolution in 1.414 days. The factor $x_0 e^{-i\omega_3 t}$ is multiplied by a rapidly oscillating factor

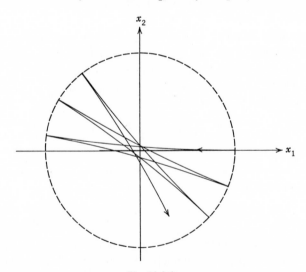

Fig. 10.8.1

so that the motion resembles Fig. 10.8.1 which, of course, is much exagerated. The motion is to be pictured as a plane pendulum which precesses so slowly that it takes 1.414 days to make one complete revolution. At the North pole, the angular velocity of precession is exactly one revolution per day. An observer located in an inertial frame away from the earth would see a simple pendulum always oscillating in the same plane with the earth turning beneath it once every 24 hours. At the equator, $\cos \gamma = 0$ and there is no Foucault effect. In the southern hemisphere, $\cos \gamma < 0$ and a Foucault pendulum precesses in a *counterclockwise direction.*

Remember that a spherical pendulum shows a precessional effect that is independent of the Foucault effect. Reference to Problem 9.5.14 reveals that this precessional velocity is proportional to the area which the particle circumscribes. Therefore, in order not to mask the Foucault effect, the amplitude of the pendulum should be kept small in comparison to its length and the initial conditions chosen such that the motion is as near to a plane pendulum as possible.

10.9. ARTIFICIAL GRAVITY ON SPACE CRAFT

In addition to the inconvenience, there are possibly definite physiological hazards in the prolonged absence of a gravitational field on board a space ship. One possible way of overcoming this problem is to produce artificial gravity by spinning the ship about an axis. The ship could, for example, be built in the form of a large ring, the outer wall of which becomes the "floor" when the ring turns about its axis. The static gravitational field results from the centrifugal acceleration. To get an idea of the angular velocity of spin required to produce a gravitational field equivalent to that on earth, suppose that the outer radius of the ring is $r_0 = 200$ ft as illustrated in Fig. 10.9.1. Setting $r_0\omega^2 = 32$ ft/sec^2 gives $\omega = 0.4$ rad/sec or about one revolution every 16 seconds. The angular velocity requirement is modest, but it is many times greater than the angular velocity of rotation of the earth. The Coriolis effect is proportionally more pronounced.

The floor of the ring-shaped space ship is represented by a circle of radius r_0 in Fig. 10.9.1. The coordinate system is permanently fixed to the ship with the x_3-axis pointing "up," i.e., toward the center of the ring. The x_2-axis is tangent to the ring and the x_1-axis is parallel to the axis of the ring. The angular velocity of rotation can therefore be written as

$$\boldsymbol{\omega} = \hat{\mathbf{i}}\omega \tag{10.9.1}$$

The acceleration of the origin of the coordinate system is

$$\mathbf{a}_0 = a_0\hat{\mathbf{k}} = r_0\omega^2\hat{\mathbf{k}} \tag{10.9.2}$$

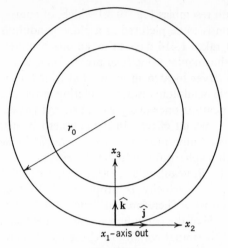

*x*₁-axis out

Fig. 10.9.1

If **r** is the displacement vector of a particle extended from the origin of the accelerated frame, the Lagrangian which gives the equations of motion of a particle with respect to the accelerated coordinate system, i.e., with respect to the astronauts on board the ship, is

$$L = T + \tfrac{1}{2}m[r^2\omega^2 - (\mathbf{r}\cdot\boldsymbol{\omega})^2] - m\mathbf{r}\cdot\mathbf{a}_0 + m\mathbf{u}\cdot\boldsymbol{\omega}\times\mathbf{r} \quad (10.9.3)$$

With $\boldsymbol{\omega}$ and $\mathbf{a}_0$ as given by (10.9.1) and (10.9.2), the Lagrangian works out to

$$L = \tfrac{1}{2}m(\dot{x}_1{}^2 + \dot{x}_2{}^2 + \dot{x}_3{}^2) + \tfrac{1}{2}m\omega^2(x_2{}^2 + x_3{}^2)$$
$$- mr_0x_3\omega^2 + m\omega(\dot{x}_3x_2 - \dot{x}_2x_3) \quad (10.9.4)$$

where it is assumed that there are no constraints so that the particle has three degrees of freedom. The equations of motion as deduced from the Lagrangian are

$$\ddot{x}_1 = 0 \quad (10.9.5)$$

$$\ddot{x}_2 - \omega^2x_2 - 2\omega\dot{x}_3 = 0 \quad (10.9.6)$$

$$\ddot{x}_3 + (r_0 - x_3)\omega^2 + 2\omega\dot{x}_2 = 0 \quad (10.9.7)$$

Equation (10.9.5) is trivial. By introducing the complex variable $z = x_2 + ix_3$, it is found that

$$\ddot{z} + 2i\omega\dot{z} - \omega^2z + i\omega^2r_0 = 0 \quad (10.9.8)$$

is equivalent to (10.9.6) and (10.9.7). The general solution, containing two arbitrary complex constants of integration, is

$$z = ir_0 + (c_1 + c_2t)e^{-i\omega t} \quad (10.9.9)$$

In terms of the complex initial position and velocity, the constants c_1 and c_2 are

$$c_1 = z_0 - i r_0 \tag{10.9.10}$$

$$c_2 = \dot{z}_0 + i \omega z_0 + \omega r_0 \tag{10.9.11}$$

For instance, suppose that a particle is dropped from rest from the ceiling of a room of height h built into the ring-shaped space ship. The initial conditions can be taken as $z_0 = ih$ and $\dot{z}_0 = 0$. It is found that

$$c_1 = i(h - r_0) \qquad c_2 = \omega(r_0 - h)$$
$$z = i r_0 + (r_0 - h)(\omega t - i)e^{-i\omega t} \tag{10.9.12}$$

In those cases where the motion is over a small distance so that $\omega t \ll 1$, It is useful to expand the exponent to obtain an approximate solution:

$$z = -\tfrac{1}{3}(r_0 - h)(\omega t)^3 + \tfrac{1}{30}(r_0 - h)(\omega t)^5 + \cdots$$
$$+ i[h - \tfrac{1}{2}(r_0 - h)(\omega t)^2 + \tfrac{1}{8}(r_0 - h)(\omega t)^4 + \cdots] \tag{10.9.13}$$

Terms of higher order than $(\omega t)^5$ have been neglected and the expansion has been separated into its real and imaginary parts. The quantity

$$g = (r_0 - h)\omega^2 \tag{10.9.14}$$

has the significance that it is the gravitational field strength at the point where the particle is released. Since $z = x_2 + i x_3$, the solution (10.9.13) can be expressed

$$x_2 = -\tfrac{1}{3} g \omega t^3 + \tfrac{1}{30} g \omega^3 t^5 + \cdots \tag{10.9.15}$$

$$x_3 = h - \tfrac{1}{2} g t^2 + \tfrac{1}{8} g \omega^2 t^4 + \cdots \tag{10.9.16}$$

These equations are similar in form to the equations of motion of a particle dropped from a point above the surface of the earth; see Eq. (10.5.14). Notice qualitatively that if a particle falls down the x_3-axis of Fig. 10.9.1, the direction of $\mathbf{u} \times \boldsymbol{\omega}$ is in the direction of the negative x_2-axis, thus confirming the correctness of the negative sign in front of the leading term of Eq. (10.9.15).

Equation (10.9.16) can be solved approximately for the time to yield

$$t = \sqrt{\frac{2(h - x_3)}{g}}\left[1 + \frac{\omega^2}{4g}(h - x_3)\right] \tag{10.9.17}$$

As a numerical example, let $\omega = 0.4 \text{ sec}^{-1}$, $g = 32 \text{ ft/sec}^2$, and $h = 8.0 \text{ ft}$. The difference in the gravitational field between the ceiling and the floor is $h\omega^2 = 1.3 \text{ ft/sec}^2$ so that at the floor the field strength will be 33.3 ft/sec^2. By Eq. (10.9.17), the time required for a particle to fall from ceiling to floor is

$$t = \sqrt{\frac{2 \times 8}{32}}\left(1 + \frac{0.16 \times 8}{4 \times 32}\right) = 0.707(1 + 0.01) = 0.714 \sec \quad (10.9.18)$$

Notice that $\omega t = 0.286$. The first neglected term in the expansion (10.9.13) is of order $(\omega t)^6 = 0.0005$. The horizontal deflection of the particle is found from Eq. (10.9.15):

$$
\begin{aligned}
x_2 &= -\tfrac{1}{3}g\omega t^3(1 - \tfrac{1}{10}\omega^2 t^2) \\
&= -1.56(1 - 0.008) = -1.55 \, \text{ft} \quad (10.9.19)
\end{aligned}
$$

Here the deflection is quite large and easily observed.

10.10. THE ADDITION OF ANGULAR VELOCITIES

As indicated by Eq. (10.2.6), the transformation property of angular velocity is that of a vector. In this section it is shown that angular velocities can be added like ordinary vectors. Figure 10.10.1 shows an inertial frame F and two arbitrary frames F_a and F_b. The velocity of a particle at point P measured relative to F can be expressed

$$\mathbf{u} = \mathbf{u}_a + \mathbf{v}_a + \boldsymbol{\omega}_a \times \mathbf{r}_a \quad (10.10.1)$$

where $\mathbf{u}_a$ is the velocity of the particle as measured by observers in F_a, $\mathbf{v}_a$ is the velocity of the origin of F_a, $\boldsymbol{\omega}_a$ is the instantaneous angular velocity of F_a, and $\mathbf{r}_a$ is the position vector of the particle with respect to the origin of F_a. As a special case of (10.10.1), the velocity of the origin of frame F_b

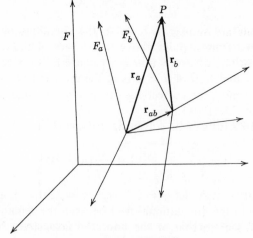

Fig. 10.10.1

can be expressed

$$\mathbf{v}_b = \mathbf{v}_{ba} + \mathbf{v}_a + \boldsymbol{\omega}_a \times \mathbf{r}_{ab} \qquad (10.10.2)$$

where $\mathbf{v}_{ba}$ is the velocity of the origin of F_b as determined by observers in F_a. The velocity of the particle at P can also be expressed by

$$\mathbf{u} = \mathbf{u}_b + \mathbf{v}_b + \boldsymbol{\omega}_b \times \mathbf{r}_b \qquad (10.10.3)$$

By equating (10.10.1) and (10.10.3) and using (10.10.2) to eliminate $\mathbf{v}_b$,

$$\mathbf{u}_a = \mathbf{u}_b + \mathbf{v}_{ba} + \boldsymbol{\omega}_b \times \mathbf{r}_b - \boldsymbol{\omega}_a \times (\mathbf{r}_a - \mathbf{r}_{ab}) \qquad (10.10.4)$$

But, from Fig. 10.10.1,

$$\mathbf{r}_a - \mathbf{r}_{ab} = \mathbf{r}_b \qquad (10.10.5)$$

Therefore,

$$\mathbf{u}_a = \mathbf{u}_b + \mathbf{v}_{ba} + (\boldsymbol{\omega}_b - \boldsymbol{\omega}_a) \times \mathbf{r}_b \qquad (10.10.6)$$

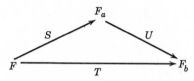

Fig. 10.10.2

This is recognized as the transformation law for particle velocity between the two arbitrary noninertial frames F_a and F_b, and it implies that the angular velocity of F_b with respect to F_a is

$$\boldsymbol{\omega}_{ba} = \boldsymbol{\omega}_b - \boldsymbol{\omega}_a \qquad (10.10.7)$$

that is, angular velocities are to be combined by simple vector addition.

A somewhat more formal approach to the derivation of (10.10.7) can be based on the results of Section 10.3. Let S, T, and U denote the orthogonal transformations connecting the three Cartesian coordinate systems as indicated schematically in Fig. 10.10.2. In matrix form, T is equivalent to the product of U and S: $T = US$. Using $S\tilde{S} = I$, this can be expressed by

$$U = T\tilde{S} \qquad (10.10.8)$$

The compoments of the angular velocity of F_a with respect to F are found from the tensor

$$\Omega_a = \tilde{S}\dot{S} = \begin{pmatrix} 0 & \omega_3 & -\omega_2 \\ -\omega_3 & 0 & \omega_1 \\ \omega_2 & -\omega_1 & 0 \end{pmatrix} \qquad (10.10.9)$$

Similarly, the angular velocity tensor of F_b with respect to F is

$$\Omega_b = \tilde{T}\dot{T} \qquad (10.10.10)$$

The angular velocity tensor of F_b with respect to F_a is

$$\Omega'_{ba} = \tilde{U}\dot{U} \qquad (10.10.11)$$

where, as the prime indicates, the components are given in the coordinate system F_a. To find the components of (10.10.11) in F, we must use the tensor transformation

$$\Omega_{ba} = \tilde{S}\Omega'_{ba}S = \tilde{S}\tilde{U}\dot{U}S \qquad (10.10.12)$$

Substituting for U from (10.10.8) gives

$$\begin{aligned}\Omega_{ba} &= \tilde{S}S\tilde{T}(\dot{T}\tilde{S} + T\dot{\tilde{S}})S \\ &= \tilde{T}\dot{T} + \dot{\tilde{S}}S\end{aligned} \qquad (10.10.13)$$

where $\tilde{S}S = I$ and $\tilde{T}T = I$ are used. Noting that

$$\dot{\tilde{S}}S = -\tilde{S}\dot{S} = -\Omega_a \qquad (10.10.14)$$

it is seen that

$$\Omega_{ba} = \Omega_b - \Omega_a \qquad (10.10.15)$$

which is equivalent to (10.10.7)

10.11. KINEMATICS OF RIGID BODY MOTION

Sections 8.11, 8.12, and 8.13 treat the motion of a rigid body in two dimensions. In this section, we begin the extension of the theory to three-dimensional motion. It is convenient to fix within the rigid body a rectangular Cartesian coordinate system which is called the *body set of axes*. An inertial frame relative to which the rigid body moves is designated as the space set of axes. The velocity of a point of the rigid body relative to the body set of axes is zero. By setting $\mathbf{u}' = 0$ in (10.2.11) there results

$$\mathbf{u} = (\boldsymbol{\omega} \times \mathbf{r}') + \mathbf{v} \qquad (10.11.1)$$

which gives the velocity of a point of the rigid body relative to the space set of axes. In (10.11.1), $\mathbf{r}'$ is the displacement of the point in question from the origin of the body set of axes and $\mathbf{v}$ is the velocity of the origin of the body set of axes. Equation (10.11.1) is identical to Eq. (8.13.6) which was derived for the special case of two dimensional motion.

Let us see if there is instantaneously a point of the rigid body with zero velocity. Setting $\mathbf{u} = 0$ and writing out the components of (10.11.1) gives

$$\begin{aligned}v_1 &= (x_2 - X_2)\omega_3 - (x_3 - X_3)\omega_2 \\ v_2 &= -(x_1 - X_1)\omega_3 + (x_3 - X_3)\omega_1 \\ v_3 &= (x_1 - X_1)\omega_2 - (x_2 - X_2)\omega_1\end{aligned} \qquad (10.11.2)$$

where X_i represents the coordinates of the origin of the body set of axes with respect to the space set, $x_i - X_i$ are the components of $\mathbf{r}'$, and ω_i are the components of $\boldsymbol{\omega}$, all expressed in the space set of axes. Equations (10.11.2) are to be regarded as a set of three linear equations in the three unknowns $x_i - X_i$. Usually, however, there is no solution because the determinant of the coefficients vanishes:

$$\begin{vmatrix} 0 & \omega_3 & -\omega_2 \\ -\omega_3 & 0 & \omega_1 \\ \omega_2 & -\omega_1 & 0 \end{vmatrix} = 0 \qquad (10.11.3)$$

The conclusion is that, in general, we cannot expect to find a point of the body coordinates which is momentarily stationary. There are exceptions. One occurs when the motion is two dimensional. This is discussed in Section 8.13 and leads to the concept of body and space centrodes.

As another exceptional case, suppose that the rigid body is constrained to move in such a way that one point is always fixed. For convenience, this point will be chosen as the origin of both the space and the body set of axes. Then $X_i = 0$ and $v_i = 0$ and solutions of (10.11.2) exist which obey

$$\frac{x_1}{\omega_1} = \frac{x_2}{\omega_2} = \frac{x_3}{\omega_3} \qquad (10.11.4)$$

Thus there are an infinite number of solutions of (10.11.2) which fall on a straight line passing through the common origins of the body and space axes. This line is in the same direction as $\boldsymbol{\omega}$ and all points of the rigid body

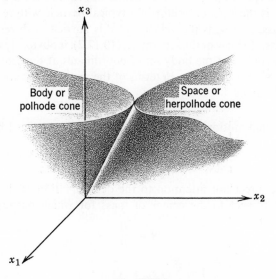

Fig. 10.11.1

which fall on it are instantaneously at rest. It is an *instantaneous axis of rotation*. Since $\boldsymbol{\omega}$ is a function of time, the instantaneous axis of rotation traces out conical surfaces both in the space and body set of coordinates. The space cone is called the *herpolhode* and the body cone is called the *polhode*. The motion of the rigid body with one point fixed is described by saying that the polhode cone rolls without slipping on the herpolhode cone. If the original fixed point is allowed to recede to infinity, the cones degenerate into cylinders, cross sections of which become the body and space centrodes to two dimensional motion.

10.12. EULER'S EQUATIONS

Let $\mathbf{N}_p$ be the torque exerted on a rigid body and $\mathbf{l}_p$ be its angular momentum, both measured with respect to some fixed point P by observers in the space set of axes relative to which the motion of the rigid body is to be determined. The basic equation of motion of the rigid body is

$$\mathbf{N}_p = \frac{d\mathbf{l}_p}{dt} \tag{10.12.1}$$

According to Eq. (8.3.9), the angular momentum of the rigid body, regarded as a system of particles, separates into two parts

$$\mathbf{l}_p = \mathbf{R}_c \times M\mathbf{v}_c + \sum_{\alpha=1}^{N} (\mathbf{r}_\alpha' \times m_\alpha \mathbf{w}_\alpha) \tag{10.12.2}$$

where $\mathbf{R}_c$ is the position vector of the center of mass, M is the total mass of the body, $\mathbf{r}_\alpha'$ is the position vector of a typical particle with respect to the center of mass, and $\mathbf{w}_\alpha$ is the velocity of this particle with respect to the center of mass. In view of the separation (10.12.2), it obviously is expedient to choose the origin of the body set of coordinates at the center of mass. The velocity $\mathbf{w}_\alpha$ can be written in terms of the angular velocity of rotation of the rigid body as

$$\mathbf{w}_\alpha = \boldsymbol{\omega} \times \mathbf{r}_\alpha' \tag{10.12.3}$$

The translational motion of the center of mass is determined by

$$\mathbf{F} = M\frac{d\mathbf{v}_c}{dt} \tag{10.12.4}$$

We therefore direct our attention to the determination of the rotational motion with respect to the center of mass for which purpose we need

$$\mathbf{N} = \frac{d\mathbf{l}}{dt} \tag{10.12.5}$$

$$\mathbf{l} = \sum_{\alpha=1}^{N} m_\alpha \mathbf{r}_\alpha' \times (\boldsymbol{\omega} \times \mathbf{r}_\alpha') \tag{10.12.6}$$

The torque and angular momentum as given by (10.12.5) and (10.12.6) are measured by observers in the space coordinate system and are taken about the center of mass of the rigid body. If the rigid body is constrained in such a way that one point is fixed, the body coordinate system can be chosen to have its origin at the fixed point. Equations (10.12.5) and (10.12.6) apply, except that **N** and **l** must be taken about the fixed point.

In either case, the angular momentum vector (10.12.6) can be resolved into its components, either along the space set or along the body set of axes. For reasons which will quickly become apparent, we choose to deal with **l** expressed in the body coordinate system. Thus,

$$\mathbf{l} = \sum_{\alpha=1}^{N} m_\alpha [\boldsymbol{\omega} r_\alpha'^2 - \mathbf{r}_\alpha (\mathbf{r}_\alpha' \cdot \boldsymbol{\omega})]$$

$$= \sum_{\alpha=1}^{N} m_\alpha [\omega_i' \hat{\mathbf{e}}_i' r_\alpha'^2 - (x_{\alpha i} \hat{\mathbf{e}}_i')(x_{\alpha j}' \omega_j')]$$

$$= \sum_{\alpha=1}^{N} m_\alpha [\delta_{ij} r_\alpha'^2 - x_{\alpha i}' x_{\alpha j}'] \omega_j' \hat{\mathbf{e}}_i' \qquad (10.12.7)$$

$$= I_{ij}' \omega_j' \hat{\mathbf{e}}_i'$$

The set of nine quantities

$$I_{ij}' = \sum_{\alpha=1}^{N} m_\alpha (\delta_{ij} r_\alpha'^2 - x_{\alpha i}' x_{\alpha j}') \qquad (10.12.8)$$

are the components of a symmetric tensor known as the *moment of inertia tensor*. The diagonal elements are known as *moments of inertia* and the off-diagonal elements are called *products of inertia*. If the rigid body is continuous rather than composed of discrete particles, then (10.12.8) must be replaced by

$$I_{ij}' = \int (\delta_{ij} r_\alpha'^2 - x_{\alpha i}' x_{\alpha j}') \rho \, d\Sigma' \qquad (10.12.9)$$

where ρ is the mass density of the body. Since the body set of coordinates is fixed for all time within the rigid body, the components (10.12.9) are constants. The components of the moment of inertia tensor in the *space set* of axes are given by

$$I_{ij} = S_{ki} S_{lj} I_{kl}' \qquad (10.12.10)$$

and are *time dependent* due to the time dependence of S_{ki}. It was to avoid this complication that **l** was originally expressed in the body set of coordinates. Considered as a vector field, **l** is like $\boldsymbol{\omega}$ in that it depends on time only and not on position in space. The components of **l** in the space and body sets of axes are connected by

$$l_i' = S_{ij} l_j \qquad (10.12.11)$$

To find the torque, the angular momentum as given by (10.12.7) is differentiated with respect to the time:

$$\mathbf{N} = \frac{d\mathbf{l}}{dt} = I'_{ij}\dot{\omega}_j'\hat{\mathbf{e}}_i' + I'_{ij}\omega_j'\delta_{ikl}\hat{\mathbf{e}}_k'\omega_l'$$

$$= (I'_{ij}\dot{\omega}_j' + \delta_{ijk}\omega_j'I'_{kl}\omega_l')\hat{\mathbf{e}}_i' \tag{10.12.12}$$

where (10.2.8) is used for the derivative of the unit vectors. The components of $\mathbf{N}$ in the body coordinates, therefore, are

$$N_i' = I'_{ij}\dot{\omega}_j' + \delta_{ijk}\omega_j'I'_{kl}\omega_l' \tag{10.12.13}$$

a result which is commonly known as *Euler's equations of motion*.

So far, the only restriction placed on the body coordinate system is that its origin be either at the center of mass or at a fixed point. The orientation of the axes within the rigid body is still arbitrary. The moment of inertia tensor is symmetric and, as is shown in Section 3.4, there exists an orientation of the body coordinate system such that the moment of inertia tensor is diagonal. The use of a diagonal representation allows a considerable simplification of Euler's equations to be made:

$$N_1' = I_1'\dot{\omega}_1' + \omega_2'\omega_3'(I_3' - I_2') \tag{10.12.14}$$

$$N_2' = I_2'\dot{\omega}_2' + \omega_3'\omega_1'(I_1' - I_3') \tag{10.12.15}$$

$$N_3' = I_3'\dot{\omega}_3' + \omega_1'\omega_2'(I_2' - I_1') \tag{10.12.16}$$

where I_1', I_2', and I_3' designate the three diagonal elements of the moment of inertia tensor; since we are now in a diagonal representation they are actually the eigenvalues.

The kinetic energy of rotation of a rigid body about the center of mass or about a fixed point is

$$T = \tfrac{1}{2}\sum_{\alpha=1}^{N} m_\alpha w_\alpha^2 = \tfrac{1}{2}\sum_{\alpha=1}^{N} m_\alpha(\boldsymbol{\omega} \times \mathbf{r}_\alpha') \cdot (\boldsymbol{\omega} \times \mathbf{r}_\alpha') \tag{10.12.17}$$

This *does not* include the kinetic energy of translation. It is readily proved that (10.12.17) can be written

$$T = \tfrac{1}{2}\sum_{\alpha=1}^{N} m_\alpha[\delta_{ij}r_\alpha'^2 - x_{i\alpha}'x_{j\alpha}']\omega_i'\omega_j'$$

$$= \tfrac{1}{2}I'_{ij}\omega_i'\omega_j' \tag{10.12.18}$$

From Euler's equations (10.12.13), since $\omega_i'\delta_{ijk}\omega_j' = 0$, we have

$$N_i'\omega_i' = I'_{ij}\dot{\omega}_j'\omega_i' = \frac{dT}{dt} \tag{10.12.19}$$

which says that the rate at which the net torque does work on the rigid body is equal to its time rate of change of kinetic energy.

It is fairly easy to show that Euler's equations (10.12.13) hold also in the space coordinates. The components of the angular momentum are given by

$$l_i = I_{ij}\omega_j \tag{10.12.20}$$

but now the components of the moment of inertia tensor are time dependent:

$$\dot{l}_i = \dot{I}_{ij}\omega_j + I_{ij}\dot{\omega}_j \tag{10.12.21}$$

By differentiation of (10.12.10),

$$\dot{I}_{ij} = S_{ki}\dot{S}_{lj}I'_{kl} + \dot{S}_{ki}S_{lj}I'_{kl} \tag{10.12.22}$$

By means of the inverse of (10.12.10),

$$\begin{aligned} \dot{I}_{ij} &= S_{ki}\dot{S}_{lj}S_{kr}S_{ls}I_{rs} + \dot{S}_{ki}S_{lj}S_{kr}S_{ls}I_{rs} \\ &= \dot{S}_{lj}S_{ls}I_{is} + \dot{S}_{ki}S_{kr}I_{rj} \end{aligned} \tag{10.12.23}$$

By means of (10.3.10),

$$\begin{aligned} \dot{I}_{ij} &= \Omega_{sj}I_{is} + \Omega_{ri}I_{rj} \\ &= \delta_{sjn}\omega_n I_{is} + \delta_{rin}\omega_n I_{rj} \end{aligned} \tag{10.12.24}$$

Thus (10.12.21) becomes

$$\begin{aligned} \dot{l}_i &= I_{ij}\dot{\omega}_j + (\delta_{sjn}\omega_n I_{is} + \delta_{rin}\omega_n I_{rj})\omega_j \\ &= I_{ij}\dot{\omega}_j + \delta_{rin}I_{rj}\omega_j\omega_n \end{aligned} \tag{10.12.25}$$

$$N_i = \dot{l}_i = I_{ij}\dot{\omega}_j + \delta_{inr}\omega_n I_{rj}\omega_j$$

Hence, Euler's equations in precisely the same form hold with all quantities expressed in either the body or the space set of coordinates. This simple covariance of the theory breaks down when an attempt is made to set up Euler's equations in a third coordinate system which is intermediate to the space and body coordinates and is rotating at, say, an angular velocity χ with respect to the space coordinates. The difficulty is connected with the breakdown of the simple transformation law (10.2.25) for the components of the angular acceleration which holds *only* for transformations between body and space coordinates.

10.13. APPLICATIONS OF EULER'S EQUATIONS

A uniform thin disc of radius a and mass M is mounted on a shaft which turns at a constant angular velocity. The shaft passes through the center of the disc, but the disc is tilted at an angle θ with respect to the shaft as illustrated in Fig. 10.13.1. We will use Euler's equations to calculate the torques which the bearings that support the shaft must sustain. A body

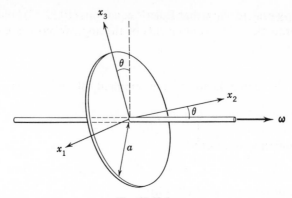

Fig. 10.13.1

set of coordinates is introduced with its origin at the center of the disc. The x_2-axis is perpendicular to the disc and is at an angle θ with respect to the shaft. The x_1-axis is perpendicular to the shaft. As long as no confusion will result, primes will be omitted from the quantities which refer to the body coordinates. The components of the angular velocity of rotation with respect to the body coordinates are

$$\omega_1 = 0 \qquad \omega_2 = \omega \cos \theta \qquad \omega_3 = -\omega \sin \theta \qquad (10.13.1)$$

Writing $d\sigma$ as an element of area on the disc and

$$\lambda = \frac{M}{\pi a^2}$$

as the mass per unit area, the first moment of inertia, from (10.12.9), is

$$I_1 = \int (r^2 - x_1{}^2)\lambda \, d\sigma \qquad (10.13.2)$$

Introducing polar coordinates on the surface of the disc as in Fig. 10.13.2,

$$\begin{aligned}
I_1 &= \lambda \int_{r=0}^{a} \int_{\phi=0}^{2\pi} (r^2 - r^2 \cos^2 \phi) r \, dr \, d\phi \\
&= \lambda \int_0^a r^3 \, dr \int_0^{2\pi} \sin^2 \phi \, d\phi \qquad (10.13.3) \\
&= \tfrac{1}{4}\lambda \pi a^4 = \tfrac{1}{4}Ma^2
\end{aligned}$$

Similarly,

$$I_3 = \tfrac{1}{4}Ma^2 \qquad (10.13.4)$$

Since the disc is thin, we get for the second moment of inertia

$$I_2 = \int r^2 \lambda \, d\sigma = \tfrac{1}{2}Ma^2 \qquad (10.13.5)$$

The products of inertia are all found to be zero, e.g.,

$$\begin{aligned} I_{13} &= -\lambda \int x_1 x_3 \, d\sigma \\ &= -\lambda \int_{r=0}^{a} \int_{\phi=0}^{2\pi} r^2 \sin\phi \cos\phi r \, dr \, d\phi = 0 \end{aligned} \tag{10.13.6}$$

Thus the body coordinate system of Fig. 10.13.1 turns out to be a principal axis system. It is frequently possible to choose a principal axis system for a given object by inspection or by use of symmetry considerations.

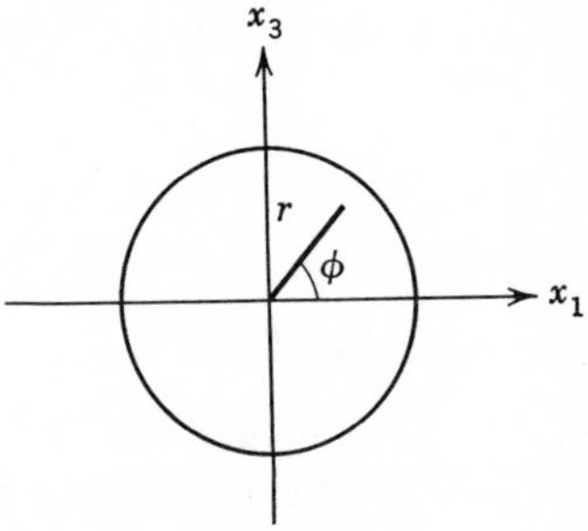

Fig. 10.13.2

Since $\dot{\boldsymbol{\omega}} = 0$, Euler's equations (10.12.14), (10.12.15), and (10.12.16) give

$$\begin{aligned} N_1 &= \tfrac{1}{4} M a^2 \omega^2 \sin\theta \cos\theta, \\ N_2 &= N_3 = 0 \end{aligned} \tag{10.13.7}$$

Thus in the body coordinate system there exists one nonzero component of torque. A space set of axes (x_{01}, x_{02}, x_{03}) can be introduced, as in Fig. 10.13.3, with the x_{02}-axis in the direction of $\boldsymbol{\omega}$. We find that the components of torque in the space system are

$$N_{01} = N_1 \cos\omega t \qquad N_{02} = 0 \qquad N_{03} = -N_1 \sin\omega t \tag{10.13.8}$$

The components of the angular momentum vector in the body set of axes are

$$\begin{aligned} l_1 &= I_1 \omega_1 = 0 \\ l_2 &= I_2 \omega_2 = \tfrac{1}{2} M a^2 \omega \cos\theta \\ l_3 &= I_3 \omega_3 = -\tfrac{1}{4} M a^2 \omega \sin\theta \end{aligned} \tag{10.13.9}$$

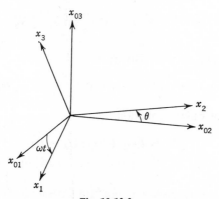

Fig. 10.13.3

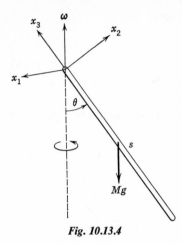

Fig. 10.13.4

The angular momentum vector lies in the plane formed by $\boldsymbol{\omega}$ and the x_2-axis, and relative to the space set of axes it traces out a cone about the axis of rotation.

As a second example, suppose that a thin rod of length s and mass M is pinned at one end to a rotating shaft as in Fig. 10.13.4. The problem is to find the equilibrium value of θ for a given angular velocity of rotation, assumed constant. Choose a body system of coordinates with origin at the pin, which is a fixed point. The x_2-axis is perpendicular to the plane formed by the axis of rotation, and the rod and the x_3-axis coincides with the rod.

Let $\mu = M/s$ be the mass per unit length of the rod. The moments of inertia are

$$I_1 = \int_{x_3=-s}^{0} (r^2 - x_1^2)\mu \, dx_3 = \int_{-s}^{0} x_3^2 \mu \, dx_3 = \tfrac{1}{3}\mu s^3 = \tfrac{1}{3}Ms^2 \quad (10.13.10)$$

$$I_2 = \tfrac{1}{3}Ms^2 \qquad I_3 = 0$$

The products of inertia are all zero.

As shown in Section 8.3, all the dynamical effects of a uniform gravitational field acting on a rigid body are equivalent to a single force equal to the weight of the body acting at the center of mass. From Fig. 10.13.4, the components of torque about the pin produced by the gravitational field are

$$N_2 = 0 \qquad N_1 = -Mg\frac{s}{2}\sin\theta \qquad N_3 = 0 \qquad (10.13.11)$$

The components of angular velocity are

$$\omega_2 = \omega \sin\theta \qquad \omega_1 = 0 \qquad \omega_3 = \omega \cos\theta \qquad (10.13.12)$$

By application of Euler's equation,

$$N_1 = I_1\dot\omega_1 + \omega_2\omega_3(I_3 - I_2) \qquad (10.13.13)$$

it is found that

$$-Mg\frac{s}{2}\sin\theta = -\omega^2 \sin\theta \cos\theta \cdot \frac{1}{3}Ms^2$$

Solving for $\cos \theta$ yields

$$\cos \theta = \frac{3g}{2s\omega^2} \qquad (10.13.14)$$

which gives the equilibrium value of θ in terms of the angular velocity of rotation of the shaft.

As a third example, suppose that a propeller-driven airplane is flying in a circle of radius s. We want to know the forces which the bearings of the propeller shaft must sustain. As illustrated in Fig. 10.13.5, a body coordinate system is fixed in the propeller which is assumed to have four

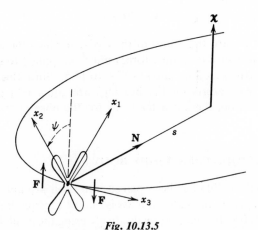

Fig. 10.13.5

blades. The propeller turns at a constant angular velocity $\dot{\psi}\hat{\mathbf{k}}$ *with respect to the airplane*, and the airplane itself is moving in a circle with a constant angular velocity $\boldsymbol{\chi}$ with respect to an inertial frame which can be pictured as the earth below. The angular velocity of the plane expressed in the body coordinates of the propeller is

$$\boldsymbol{\chi} = \chi \sin \psi \hat{\mathbf{i}} + \chi \cos \psi \hat{\mathbf{j}} \qquad (10.13.15)$$

It was proved in Section 10.10 that angular velocities are to be added like vectors. Therefore, the angular velocity of the propeller *with respect to the earth but resolved into its components along the body axes* is

$$\boldsymbol{\omega} = \boldsymbol{\chi} + \dot{\psi}\hat{\mathbf{k}} = \chi \sin \psi \hat{\mathbf{i}} + \chi \cos \psi \hat{\mathbf{j}} + \dot{\psi}\hat{\mathbf{k}} \qquad (10.13.16)$$

A little consideration shows that the moments of inertia obey

$$I_1 = I_2, \qquad I_3 = 2I_1 \qquad (10.13.17)$$

and that the products of inertia are zero. Since $\dot\psi$ is a constant, Euler's equations give the following results:

$$N_1 = I_1\dot\omega_1 + \omega_2\omega_3(I_3 - I_2) = 2I_1\chi\dot\psi \cos \psi \qquad (10.13.18)$$

$$N_2 = I_2\dot\omega_2 + \omega_3\omega_1(I_1 - I_3) = -2I_1\chi\dot\psi \sin \psi \qquad (10.13.19)$$

$$N_3 = I_3\dot\omega_3 + \omega_1\omega_2(I_2 - I_1) = 0 \qquad (10.13.20)$$

It is seen from Fig. 10.13.5 that N_1 and N_2 are the x_1 and x_2 components of a torque of constant magnitude

$$N = 2I_1\chi\dot\psi \qquad (10.13.21)$$

that is directed toward the center of the circle in which the airplane flies. In order to maintain this torque, forces must be applied to the propeller shaft by the bearings as illustrated in Fig. 10.13.5. Since the forces which the propeller shaft exerts on the bearings are equal in magnitude but opposite in direction, there is a tendency for the nose of the airplane to lilt upward.

10.14. THE TORQUE-FREE RIGID BODY

If an artificial satellite has been placed into orbit around the earth, its motion with respect to its center of mass is that of a torque-free rigid body. We will first analyze this motion qualitatively using the available constants of the motion and then present an exact solution of the Euler equations in terms of Jacobian elliptic functions.

The origins of both space and body axes can be taken as the center of mass. If an artificial satellite is under discussion, the space set of axes is not a true inertial frame but is only locally inertial. This makes no difference unless the dimensions of the satellite are very large. Since there are no torques, the angular momentum vector $\mathbf{l}$ is a constant. The space set of axes is oriented in such a way that $\mathbf{l}$ coincides with the x_{03}-axis, Fig. 10.14.1. Another constant of the motion is the total energy which in this case is all kinetic:

$$T = \tfrac{1}{2}I_{ij}\omega_i\omega_j = \tfrac{1}{2}(\mathbf{l} \cdot \boldsymbol\omega) = \tfrac{1}{2}l\omega \cos \alpha \qquad (10.14.1)$$

where α is the angle between $\mathbf{l}$ and $\boldsymbol\omega$. Since $\mathbf{l}$ is a constant, the projection of $\boldsymbol\omega$ onto $\mathbf{l}$, $\omega \cos \alpha$, is also a constant. Both ω and α can vary, however.

Since the moment of inertia tensor has three positive eigenvalues, the quadratic form

$$\phi = I_{ij}x_ix_j \qquad (10.14.2)$$

is positive definite. This means that when the equation

$$\phi = \text{const.} \tag{10.14.3}$$

is plotted out in the body coordinate system (x_1, x_2, x_3) of Fig. 10.14. an ellipsoid is obtained. This ellipsoid is known as the *moment of inertia ellipsoid*. If the reader will recall the geometry of the eigenvalue problem as discussed in Section 3.4, he will realize that if the body coordinate system

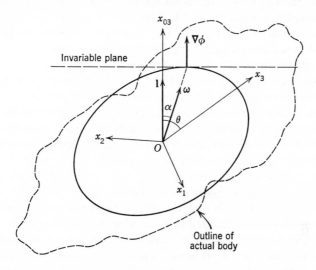

Fig. 10.14.1

is made to coincide with the principle axes of the ellipsoid, then the moment of inertia tensor will be diagonal. Notice that the center of mass of the rigid body, the origins of both body and space coordinates, and the center of the moment of inertia ellipsoid all coincide.

Suppose that some convenient value of the constant in (10.14.3) has been chosen (the actual numerical value is not important) and that the resulting ellipsoid has been plotted within the rigid body. The gradient of the scalar function (10.14.2) is

$$\nabla\phi = 2I_{ij}x_j\hat{\mathbf{e}}_i \tag{10.14.4}$$

and, as we know, the gradient vector constructed at any point of the surface $\phi = \text{const.}$ is perpendicular to that surface. The angular velocity vector varies with time but suppose that at a given instant we construct a straight line from the origin that is in the same direction as $\boldsymbol{\omega}$. The equation of such a line is

$$x_j = c\omega_j \tag{10.14.5}$$

where c is a constant. The gradient (10.14.4) evaluated at the point where (10.14.5) pierces the surface of the ellipsoid is

$$\nabla \phi = 2cI_{ij}\omega_j \hat{\mathbf{e}}_i = 2c\mathbf{l} \tag{10.14.6}$$

and is parallel to $\mathbf{l}$. It is possible, therefore, to construct a plane which is tangent to the ellipsoid at the point where (10.14.5) pierces it and is also perpendicular to the direction of $\mathbf{l}$. This plane is known as the invariable plane and is illustrated in Fig. 10.14.1. As noted previously the projection of $\boldsymbol{\omega}$, and hence the projection of the displacement vector from O to the point where the invariable plane is tangent to the ellipsoid, is a constant. As the rigid body moves, the invariable plane remains fixed in space and always tangent to the moment of inertia ellipsoid.

The center of mass is a fixed point with respect to the space set of axes. It was shown in Section 10.11 that a line drawn through a fixed point of a rigid body and in the same direction as $\boldsymbol{\omega}$, i.e., the line given by (10.14.5) in this case, is an instantaneous axis of rotation. Every point of this line, including the point where the invariable plane is tangent to the moment of inertia ellipsoid, is momentarily stationary. This means that the motion of the rigid body is such that the moment of inertia ellipsoid rolls without slipping on the invariable plane while its center is fixed.

We turn now to an analytical procedure for obtaining $\boldsymbol{\omega}$ as a function of time for the torque-free rigid body. Specifically, we seek solutions of the system of differential equations

$$I_1\dot{\omega}_1 + \omega_2\omega_3(I_3 - I_2) = 0$$
$$I_2\dot{\omega}_2 + \omega_3\omega_1(I_1 - I_3) = 0 \tag{10.14.7}$$
$$I_3\dot{\omega}_3 + \omega_1\omega_2(I_2 - I_1) = 0$$

where it is assumed that the body coordinates coincide with the principle axes. There are three expecially simple sets of solutions:

$$\omega_1 = \text{const.} \qquad \omega_2 = 0 \qquad \omega_3 = 0 \tag{10.14.8}$$
$$\omega_1 = 0 \qquad \omega_2 = \text{const.} \qquad \omega_3 = 0 \tag{10.14.9}$$
$$\omega_1 = 0 \qquad \omega_2 = 0 \qquad \omega_3 = \text{const.} \tag{10.14.10}$$

Physically, these solutions mean that it is possible for an isolated rigid to spin about any one of its principle directions at constant angular velocity.

The cyclic repitition of the three functions ω_1, ω_2, and ω_3 in (10.14.7) indicates that the solutions must have the property that the derivative of any one of them is proportional to the product of the other two. Reference to the first three entries in Appendix C reveals that the Jacobian elliptic functions *snu, cnu,* and *dnu* have this property. One possible set of solutions is

$$\omega_1 = Acn[a(t - t_0)] \qquad \omega_2 = Bsn[a(t - t_0)] \qquad \omega_3 = Cdn[a(t - t_0)]$$
$$(10.14.11)$$

where A, B, C, a, and t_0 are constants. Substitution of the trial solutions (10.14.11) into the differential equations (10.14.7) yields

$$-I_1 Aa + BC(I_3 - I_2) = 0 \qquad (10.14.12)$$

$$I_2 Ba + AC(I_1 - I_3) = 0 \qquad (10.14.13)$$

$$-I_3 Cak^2 + AB(I_2 - I_1) = 0 \qquad (10.14.14)$$

where k is the modulus.

The equations of motion (10.14.7) are a set of three first order differential equations and, it is therefore to be expected that the solutions will contain three undetermined constants. One of these is t_0. Of the five remaining parameters, A, B, C, a, and k, any two can be regarded as undetermined constants of integration and the remaining three expressed in terms of them. We will solve (10.14.12), (10.14.13), and (10.14.14) for a, B, and k in terms of A and C which, in turn, must be regarded determined by the initial conditions. First write (10.14.12) and (10.14.13) as

$$I_1 Aa = -BC(I_2 - I_3)$$
$$AC(I_1 - I_3) = -I_2 Ba$$

then divide one equation by the other and solve for a to get

$$a = |C| \sqrt{\frac{(I_1 - I_3)(I_2 - I_3)}{I_1 I_2}} \qquad (10.14.15)$$

Write (10.14.13) and (10.14.14) as

$$I_2 Ba = -AC(I_1 - I_3)$$
$$I_3 Cak^2 = -AB(I_1 - I_2)$$

Multiply the equations together and solve for ak:

$$ak = |A| \sqrt{\frac{(I_1 - I_3)(I_1 - I_2)}{I_2 I_3}} \qquad (10.14.16)$$

Now combine (10.14.15) and (10.14.16) to get

$$k = \left| \frac{A}{C} \right| \sqrt{\frac{I_1(I_1 - I_2)}{I_3(I_2 - I_3)}} \qquad (10.14.17)$$

Finally, use (10.14.15) and (10.14.17) to eliminate a and k from (10.14.14):

$$B = -A \frac{C}{|C|} \sqrt{\frac{I_1(I_1 - I_3)}{I_2(I_2 - I_3)}} \qquad (10.14.18)$$

Inspection of (10.14.15), (10.14.17), and (10.14.18) reveals that the set of solutions given by (10.14.11) is valid, provided that the moments of inertia are ordered as

$$I_1 > I_2 > I_3 \qquad (10.14.19)$$

Assuming that the moments of inertia are all different, it is always possible to label the body coordinates so that this condition is satisfied.

It is to be remembered that the modulus (10.14.17) must obey $0 \leqslant k < 1$ which placed a restriction on the constants of integration A and C. There is a second set of solutions which satisfies the condition (10.14.19), namely,

$$\omega_1 = A \; dn[a(t - t_0)] \qquad \omega_2 = Bsn[a(t - t_0)] \qquad \omega_3 = Ccn[a(t - t_0)]$$
$$(10.14.20)$$

The reader can substitute this set of solutions into the differential equations (10.14.7), derive expressions for the constants, and verify that

$$k = \frac{1}{\dfrac{|A|}{|C|} \sqrt{\dfrac{I_1(I_1 - I_2)}{I_3(I_2 - I_3)}}} \qquad (10.14.21)$$

For given initial conditions, i.e., given values of A and C, one or the other of (10.14.17) and (10.14.21) will satisfy $0 \leqslant k < 1$.

If any two of the moments of inertia are equal, solutions can be found in terms of trigonometric functions. Obtaining these solutions is left as a problem.

The solutions obtained here express the components of the angular velocity in the body coordinates as functions of time. Lacking are expressions which give the actual position of the rigid body as a function of time with respect to the space set of axes. In the next section, a set of parameters known as the Euler angles, which are suitable for this purpose, will be discussed.

10.15. THE EULER ANGLES

The object of this section is to derive a representation for an orthogonal transformation in terms of three parameters which can be conveniently used to specify the rotation of a rigid body.

In Fig. 10.15.1, x_i denotes space coordinates and x_i' body coordinates; $\bar{x}_i$ and $\bar{\bar{x}}_i$ are intermediate coordinate systems. The coordinate system

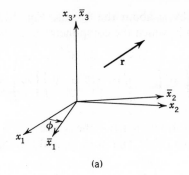

(a)

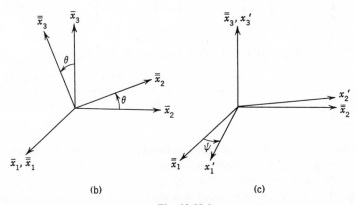

(b)   (c)

Fig. 10.15.1

labelled $\bar{x}_i$ is obtained by rotation through an angle ϕ about the space x_3-axis. Let **r** be a displacement vector between any two points in space. The components of **r** in the two systems x_i and $\bar{x}_i$ are connected by

$$\begin{pmatrix} \bar{r}_1 \\ \bar{r}_2 \\ \bar{r}_3 \end{pmatrix} = \begin{pmatrix} \cos\phi & \sin\phi & 0 \\ -\sin\phi & \cos\phi & 0 \\ 0 & 0 & 1 \end{pmatrix} \begin{pmatrix} r_1 \\ r_2 \\ r_3 \end{pmatrix} = S_1 \begin{pmatrix} r_1 \\ r_2 \\ r_3 \end{pmatrix} \qquad (10.15.1)$$

A third coordinate system is now obtained by a rotation through an angle θ about the $\bar{x}_1$-axis as in Fig. 10.15.1b. The components of **r** in the $\bar{\bar{x}}_i$ system are

$$\begin{pmatrix} \bar{\bar{r}}_1 \\ \bar{\bar{r}}_2 \\ \bar{\bar{r}}_3 \end{pmatrix} = \begin{pmatrix} 1 & 0 & 0 \\ 0 & \cos\theta & \sin\theta \\ 0 & -\sin\theta & \cos\theta \end{pmatrix} \begin{pmatrix} \bar{r}_1 \\ \bar{r}_2 \\ \bar{r}_3 \end{pmatrix} = S_2 \begin{pmatrix} \bar{r}_1 \\ \bar{r}_2 \\ \bar{r}_3 \end{pmatrix} \qquad (10.15.2)$$

Finally, rotation by ψ about the $\bar{x}_3$-axis, Fig. 10.15.1c, gives the body coordinate system in which the components of $\mathbf{r}$ are

$$\begin{pmatrix} r_1' \\ r_2' \\ r_3' \end{pmatrix} = \begin{pmatrix} \cos \psi & \sin \psi & 0 \\ -\sin \psi & \cos \psi & 0 \\ 0 & 0 & 1 \end{pmatrix} \begin{pmatrix} \bar{r}_1 \\ \bar{r}_2 \\ \bar{r}_3 \end{pmatrix} = S_3 \begin{pmatrix} \bar{r}_1 \\ \bar{r}_2 \\ \bar{r}_3 \end{pmatrix} \qquad (10.15.3)$$

The overall transformation from the space to the body coordinates is found by combining (10.15.1), (10.15.2), and (10.15.3)

$$\begin{pmatrix} r_1' \\ r_2' \\ r_3' \end{pmatrix} = S_3 S_2 S_1 \begin{pmatrix} r_1 \\ r_2 \\ r_3 \end{pmatrix} = S \begin{pmatrix} r_1 \\ r_2 \\ r_3 \end{pmatrix} \qquad (10.15.4)$$

By performing the matrix multiplication $S = S_3 S_2 S_1$ we find

$$\begin{pmatrix} \cos \psi \cos \phi - \cos \theta \sin \phi \sin \psi & \cos \psi \sin \phi + \cos \theta \cos \phi \sin \psi & \sin \psi \sin \theta \\ -\sin \psi \cos \phi - \cos \theta \sin \phi \cos \psi & -\sin \psi \sin \phi + \cos \theta \cos \phi \cos \psi & \cos \psi \sin \theta \\ \sin \theta \sin \phi & -\sin \theta \cos \phi & \cos \theta \end{pmatrix}$$

$$(10.15.5)$$

The orthogonal transformation S is a function of the three independent parameters ϕ, θ, and ψ which are known as the *Euler angles*. Since S is obtained by a succession of rotations, it is necessarily proper. An arbitrary proper orthogonal transformation can be expressed in terms of the Euler angles by means of (10.15.5).

If x_i and x_i' are space and body coordinates, then the Euler angles are a suitable set of generalized coordinates for the description of the rotational motion of a rigid body, either about a fixed point of the rigid body or about the center of mass. If a rigid body is constrained to move about a fixed point, there are three degrees of freedom necessitating three independent coordinates, such as the Euler angles, to describe the motion. In the general motion involving both translation and rotation, there are a total of six degrees of freedom. A total of six coordinates are needed and, for example, they could be the three rectangular coordinates of the center of mass and the three Euler angles to give the rotation of the body about the center of mass.

The components of the angular velocity of a rigid body in terms of the Euler angles in the body and space coordinates can be obtained from equations (10.3.12) and (10.3.13), but the calculation is very tedious. Instead, our calculation of the components of angular velocity in the body coordinate system will rest on the result of Section 10.10, namely, that angular velocities add like vectors. Referring to Fig. 10.15.1a, the rotation

ϕ produces an angular velocity $\dot{\phi}$ along the x_3-axis. In the $\bar{x}_i$ coordinate system, this is represented by

$$(0, 0, \dot{\phi}) \qquad (10.15.6)$$

The rotation θ in Fig. 10.15.1b produces a component of angular velocity $\dot{\theta}$ in the $\bar{x}_1$-direction. Resolution of $\dot{\phi}$ into its $\bar{x}_2$ and $\bar{x}_3$ components yields the combined effects of the θ and ϕ rotations expressed in the $\bar{x}_i$ coordinate systems:

$$(\dot{\theta}, \dot{\phi} \sin \theta, \dot{\phi} \cos \theta) \qquad (10.15.7)$$

The rotation ψ in Fig. 10.15.1c produces a component $\dot{\psi}$ in the x_3'-direction. The combined effects of all three rotations expressed in the body coordinates is then

$$\omega_1' = \dot{\theta} \cos \psi + \dot{\phi} \sin \theta \sin \psi$$
$$\omega_2' = \dot{\phi} \sin \theta \cos \psi - \dot{\theta} \sin \psi \qquad (10.15.8)$$
$$\omega_3' = \dot{\psi} + \dot{\phi} \cos \theta$$

The components of $\boldsymbol{\omega}$ in the space coordinates can be obtained by a similar geometrical procedure or by means of the vector transformation

$$\omega = \tilde{S}\omega' \qquad (10.15.9)$$

Where ω and ω' represent column matrices. The result is

$$\omega_1 = \dot{\theta} \cos \phi + \dot{\psi} \sin \theta \sin \phi$$
$$\omega_2 = \dot{\theta} \sin \phi - \dot{\psi} \sin \theta \cos \phi \qquad (10.15.10)$$
$$\omega_3 = \dot{\phi} + \dot{\psi} \cos \theta$$

A quantity of importance is the kinetic energy of rotation of a rigid body. For the special case where $I_1' = I_2'$, the kinetic energy in terms of the Euler angles is

$$T = \tfrac{1}{2}I_1'\omega_1'^2 + \tfrac{1}{2}I_2'\omega_2'^2 + \tfrac{1}{2}I_3'\omega_3'^2$$
$$= \tfrac{1}{2}I_1'(\dot{\theta}^2 + \dot{\phi}^2 \sin^2 \theta) + \tfrac{1}{2}I_3'(\dot{\psi} + \dot{\phi} \cos \theta)^2 \qquad (10.15.11)$$

10.16. THE TORQUE-FREE RIGID BODY *(Continued)*

In this section, the Euler angles as functions of time for the torque-free rigid body for the special case $I_1 = I_2$ will be computed. This means that a cross section of the moment of inertia ellipsoid in Fig. 10.14.1 taken perpendicular to the body x_3-axis is a circle. A little study of Figs. 10.14.1 and 10.15.1 will convince the reader that θ is the angle between the x_{03}- and x_3-axes and, therefore, is a constant. (The notation in Figs. 10.14.1

and 10.15.1 is slightly different. In Fig. 10.14.1 space and body axes are denoted x_{0i} and x_i, respectively, whereas in Fig. 10.15.1 they are denoted x_i and x_i'.) Moreover, ϕ is the angular rate at which the body x_3-axis precesses about the space x_{03}-axis and ψ is the angular velocity of spin about the x_3-axis. Both ϕ and ψ are constants. Solutions of the Euler equations (10.14.7) for the case $I_1 = I_2$ can be verified to be

$$\omega_1 = A \sin at \qquad \omega_2 = A \cos at$$

$$\omega_3 = \text{const.} \qquad a = \omega_3 \frac{I_1 - I_3}{I_1} \qquad (10.16.1)$$

where it is assumed that $\omega_3 > 0$. Now, using (10.15.8) (primes omitted) and the fact that $\dot{\theta} = 0$,

$$\omega_1 = \dot{\phi} \sin \theta \sin \psi = A \sin at$$

$$\omega_2 = \dot{\phi} \sin \theta \cos \psi = A \cos at \qquad (10.16.2)$$

$$\omega_3 = \dot{\psi} + \dot{\phi} \cos \theta$$

It is seen that

$$A = \dot{\phi} \sin \theta \qquad \psi = at$$

$$\omega_3 = A \frac{I_1}{I_3} \cot \theta \qquad a = A \frac{I_1 - I_3}{I_3} \cot \theta \qquad (10.16.3)$$

10.17. THE SYMMETRIC TOP WITH ONE POINT FIXED

A rigid body consists of a large number of particles connected by holonomic constraints. The Lagrange formulation of mechanics, therefore, applies and offers an alternative method of obtaining the equations of motion of a rigid body. In this section the equations of motion of a symmetrical toy top will be obtained by the Lagrange method. The Euler angles provide a convenient set of generalized coordinates. As illustrated in Fig. 10.17.1, the body x_3-axis coincides with the symmetry axis of the top so that $I_1 = I_2$. The symmetry axis is sometimes called the *figure axis*. The point of contact of the top with the floor is a fixed pivot point and is chosen to be the origin of both body and space coordinates. The Euler angle θ is the angle between the figure axis of the top and the space x_{03}-axis. Let a line be drawn in the $x_{01}x_{02}$-plane perpendicular to the plane formed by the x_3- and x_{03}-axes. This line is called the *line of nodes*. The Euler angle ϕ is the angle between the x_{01} axis and the line of nodes. Finally, the Euler angle ψ gives the rotation of the top about its figure axis.

If s is the distance from the pivot to the center of mass, then

$$V = Mgs \cos \theta \qquad (10.17.1)$$

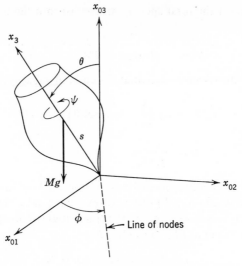

Fig. 10.17.1

is the potential energy of the system. Employing the expression (10.15.11) for the kinetic energy yields the Lagrangian

$$L = \frac{I_1}{2}(\dot{\theta}^2 + \dot{\phi}^2 \sin^2 \theta) + \frac{I_3}{2}(\dot{\psi} + \dot{\phi} \cos \theta)^2 - Mgs \cos \theta \quad (10.17.2)$$

The coordinates ϕ and ψ are cyclic resulting in the two constants of the motion

$$h = \frac{\partial L}{\partial \dot{\psi}} = I_3(\dot{\psi} + \dot{\phi} \cos \theta) = I_3 \omega_3 \quad (10.17.3)$$

$$l = \frac{\partial L}{\partial \dot{\phi}} = (I_1 \sin^2 \theta + I_3 \cos^2 \theta)\dot{\phi} + I_3 \dot{\psi} \cos \theta \quad (10.17.4)$$

The elimination of $\dot{\psi}$ between (10.17.3) and (10.17.4) results in

$$\dot{\phi} = \frac{l - h \cos \theta}{I_1 \sin^2 \theta} \quad (10.17.5)$$

The total energy provides a third constant of the motion:

$$\mathscr{E} = \frac{I_1}{2}(\dot{\theta}^2 + \dot{\phi}^2 \sin^2 \theta) + \frac{I_3}{2}(\dot{\psi} + \dot{\phi} \cos \theta)^2 + Mgs \cos \theta \quad (10.17.6)$$

The coordinates ϕ and ψ can be eliminated from the energy by means of (10.17.3) and (10.17.5):

$$\mathscr{E} = \tfrac{1}{2}I_1 \dot{\theta}^2 + \frac{h^2}{2I_3} + \frac{(l - h \cos \theta)^2}{2I_1 \sin^2 \theta} + Mgs \cos \theta \quad (10.17.7)$$

The expression for the total energy can be cast into the form

$$\epsilon \equiv \frac{\mathscr{E}}{Mgs} - \frac{h^2}{2I_3 Mgs} = \frac{I_1}{2Mgs} \dot\theta^2 + V_e(\theta) \tag{10.17.8}$$

where $V_e(\theta)$ is the effective potential (in units of Mgs) given by

$$V_e(\theta) = \mu\left(\frac{1}{\sin\theta} - \frac{h}{l}\cot\theta\right)^2 + \cos\theta \tag{10.17.9}$$

$$\mu = \frac{l^2}{2I_1 Mgs} \tag{10.17.10}$$

The study of the effective potential is complicated by its dependence on two parameters μ and h/l. In order to see what kind of numerical values these parameters can have, suppose that the initial conditions are such that $\dot\phi_0 = 0$. Then from (10.17.3) and (10.17.4)

$$l = I_3 \dot\psi_0 \cos\theta_0 = h\cos\theta_0 \tag{10.17.11}$$

Thus h/l depends on the initial value of θ. If $\theta_0 \neq \pi/2$, l, and hence μ, can be large or small, depending on how much initial spin the top is given. A trivial case is $\mu = 0$ in which case $V_e(\theta) = \cos\theta$. This is just the potential function for a plane pendulum.

In order to investigate the behavior of the top more generally, several plots of the effective potential have been made for different values of μ and h/l and appear in figures (10.17.2), (10.17.3), and (10.17.4). In Fig. 10.17.2, μ is given the value 1.00 corresponding to a certain choice for the orbital angular momentum l. The values $h/l = 2$ and $h/l = \frac{1}{2}$ then correspond to two different values of the "spin" angular momentum h of the top about its figure axis. The effective potential is a simple function going to ∞ at both $\theta = 0$ and $\theta = 180°$ and having a single minimum. If the constant ϵ in equation (10.17.8) is exactly equal to the minimum value of $V_e(\theta)$, the top precesses uniformly about the space x_{03}-axis at a constant value of θ. The constant ϵ is essentially the total energy reduced by the "spin energy." Since (10.17.8) has been divided by Mgs, we are essentially working with energy in units of Mgs. More generally, the figure axis of the top oscillates between two definite turning points, an effect which is called *nutation*. The value $h/l = 2$ corresponds to a fairly high value of spin angular momentum. The equilibrium value of θ falls at about 65°. The value $h/l = \frac{1}{2}$ gives a smaller value of spin angular momentum relative to orbital angular momentum, and the effective potential for this case resembles that of a spherical pendulum; the effective potential for a spherical pendulum is sketched in Fig. 6.6.2. (Note difference in definition of θ!) If $h/l = 0$, corresponding to no spin

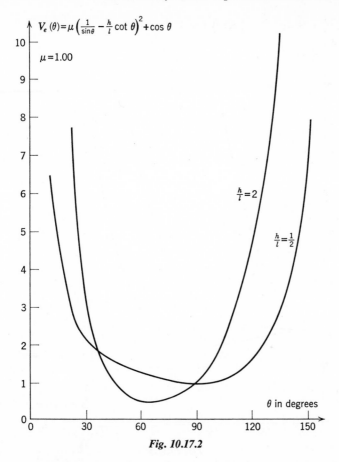

Fig. 10.17.2

angular momentum, the motion of the top becomes identical to that of a spherical pendulum.

A special case of $V_e(\theta)$ occurs when $h/l = 1$ exactly. This is the one value of the parameter h/l where $V_e(\theta)$ is finite at $\theta = 0$:

$$\lim_{\theta \to 0} V_e(\theta) = \lim_{\theta \to 0} \mu \left(\frac{1 - \cos \theta}{\sin \theta} \right)^2 + \cos \theta = 1 \qquad (10.17.12)$$

Figure 10.17.3 shows $V_e(\theta)$ for $h/l = 1$ and three different values of μ. For $\mu = 1$, $V_e(\theta)$ has a relative maximum at $\theta = 0$ and a minimum at about $67°$, indicating that $\theta = 0$ is a position of unstable equilibrium. The angular momentum of the top (both spin and orbital since $h = l$) is too small for it to stand straight up. If started at $\theta = 0$, it will topple over and not recover itself until θ reaches about $90°$. For $\mu = 10$, corresponding to much higher values of h and l, there is a stable equilibrium at

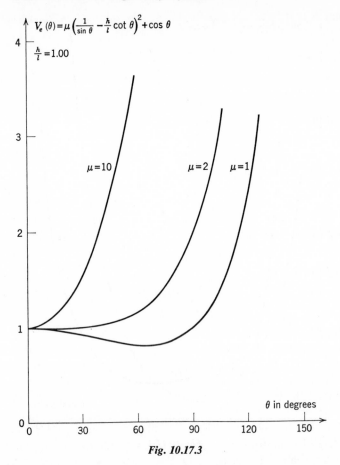

Fig. 10.17.3

$\theta = 0$, meaning that the top can stand straight up in a stable manner. This is generally called a *sleeping top*.

There must be some critical value of μ where the transition between two cases as typified by $\mu = 1$ and $\mu = 10$ occurs. Specifically, for what value of μ does the relative minimum in $V_e(\theta)$ disappear? Setting $h/l = 1$ in (10.17.9) and differentiating yields

$$\frac{dV_e}{d\theta} = 2\mu \frac{(1 - \cos \theta)^2}{\sin^3 \theta} - \sin \theta \tag{10.17.13}$$

Setting $dV_e/d\theta = 0$ results in

$$2\mu(1 - \cos \theta)^2 = \sin^4 \theta \tag{10.17.14}$$

Note that $\theta = 0$ is a solution for all μ. By taking the square root of (10.17.14),

$$\pm\sqrt{2\mu}(1 - \cos\theta) = \sin^2\theta = (1 - \cos^2\theta) = (1 + \cos\theta)(1 - \cos\theta)$$

$$\pm\sqrt{2\mu} = 1 + \cos\theta \qquad (10.17.15)$$

The positive root applies since $1 + \cos\theta \geq 0$. The position of the minimum in $V_e(\theta)$, therefore, is given by

$$\cos\theta = \sqrt{2\mu} - 1 \qquad (10.17.16)$$

Since $\cos\theta \leq 1$, the minimum exists if $\sqrt{2\mu} - 1 < 1$ or

$$\mu < 2 \qquad (10.17.17)$$

A plot of $V_e(\theta)$ for the critical value $\mu = 2$ is given in Fig. 10.17.2. Referring to the definition of μ as given by (10.17.10), $\mu = 2$ gives

$$l^2 = h^2 = 4I_1Mgs \qquad (10.17.18)$$

By means of (10.17.3)

$$\omega_3{}^2 = \frac{4I_1Mgs}{I_3{}^2} \qquad (10.17.19)$$

We should hasten to say that it is dangerous to draw general conclusions about the mechanical behavior of a system for special values of the parameters where the behavior of $V_e(\theta)$ is qualitatively different from all other values. In a practical sense, we could not determine the initial conditions of the top to be such that $h = l$ *precisely*. The fact that $V_e(\theta)$ jumps from unity at $\theta = 0$ for $h = l$ to ∞ for $h \cong l$ is in itself a kind of instability. For this reason, $V_e(\theta)$ has been redrawn in Fig. 10.17.4 for the same values of μ but for $h = 1.01\, l$. In all cases, there is a single minimum in $V_e(\theta)$ for some θ between 0 and 180°. There is, however, a definite transition at $\mu = 2$ in that the equilibrium position moves suddenly very close to the vertical axis. Equation (10.17.19), therefore, remains as a useful criterion for determining the minimum angular velocity of spin required for a top to remain in a (nearly!) upright position.

We now turn to the problem of determining the uniform angular velocity of precession of the top when $\theta = $ const. corresponding to the minimum of the effective potential. The condition $dV_e/d\theta = 0$ gives

$$2\mu\left(1 - \frac{h}{l}\cos\theta\right)\left(\frac{h}{l} - \cos\theta\right) = \sin^4\theta = (1 - \cos^2\theta)^2 \quad (10.17.20)$$

This represents a quartic equation for $\cos\theta$, one of the roots of which gives the minimum of $V_e(\theta)$ lying between $\theta = 0$ and $\theta = 180°$. Our plots of $V_e(\theta)$ in Figs. 10.17.2 and 10.17.4 indicate that there is only one

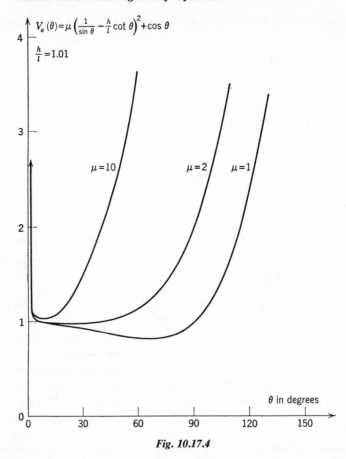

Fig. 10.17.4

root in this range. The problem of finding this root explicitly from (10.17.20) is formidable except in the special case $h = l$ already examined.

If in (10.17.20) we substitute (10.17.10) for μ and then eliminate l, using (10.17.5), the result is fairly simple quadratic equation for $\dot{\phi}$:

$$\dot{\phi}^2 I_1 \cos \theta - \dot{\phi} h + Mgs = 0 \qquad (10.17.21)$$

Consider first the simple case where the equilibrium value of θ is 90°. Then

$$\dot{\phi} = \frac{Mgs}{h} \qquad (10.17.22)$$

The meaning of this result is as follows: Let the figure axis of the top be held in a horizontal position. Give the top a spin angular momentum *h* and then give it a push in the ϕ-direction, such that its initial value of $\dot{\phi}$ is given by (10.17.22). The top will then continue to move with $\theta = 90°$ and with $\dot{\phi}$ given by the constant value (10.17.22). More generally, the root of (10.17.21) which reduces to (10.17.22) when $\theta = 90°$ is

$$\dot{\phi} = \frac{h}{2I_1 \cos \theta}\left(1 - \sqrt{1 - \frac{4I_1 Mgs \cos \theta}{h^2}}\right) \qquad (10.17.23)$$

If the second term under the radical is small either because $\theta \cong 90°$ or because the spin angular momentum is large, then

$$\dot{\phi} \cong \frac{h}{2I_1 \cos \theta}\left(1 - \left[1 - \frac{2I_1 Mgs \cos \theta}{h^2}\right]\right) = \frac{Mgs}{h} \qquad (10.17.24)$$

If *h* is large, it is possible for the top to precess uniformly at the angular velocity (10.17.24) for any value of θ.

Obtaining an exact analytical solution for the θ motion is no more difficult than it was for a spherical pendulum. Omitting details, but following exactly the same procedure that led to Eq. (6.6.29), we obtain the equation of motion

$$\ddot{x} + \alpha x - \beta x^2 = 0 \qquad (10.17.25)$$

where

$$x = \cos \theta - c \qquad (10.17.26)$$

$$c = \frac{\mu h^2}{3l^2} + \frac{\epsilon}{3} - \sqrt{\left(\frac{\mu h^2}{3l^2} + \frac{\epsilon}{3}\right)^2 + \frac{1}{3} - \frac{2\mu h}{3l}} \qquad (10.17.27)$$

$$\alpha = \frac{6Mgs}{I_1}\sqrt{\left(\frac{\mu h^2}{3l^2} + \frac{\epsilon}{3}\right)^2 + \frac{1}{3} - \frac{2\mu h}{3l}} \qquad (10.17.28)$$

$$\beta = \frac{3Mgs}{I_1} \qquad (10.17.29)$$

The periodic solutions of (10.17.25) are desired and these can be found in Section 6.5. After the solution is obtained and $\cos \theta$ is expressed in terms of Jacobian elliptic functions, $\phi(t)$ is obtained by integrating (10.17.5), but this involves a very bad integral.

We have recourse to the Fourier series solution of (10.17.25), Eqs. (4.13.35) and (4.13.36). The solution for ϕ is then obtained just as Eq. (6.6.55) was obtained for the spherical pendulum. A calculation based on the numerical values $\mu = 1.00$, $h/l = 2.00$, $\epsilon = 1.00$ results in

$$\cos \theta = 0.356 + 0.349 \cos \omega t - 0.008 \cos 2\omega t \qquad (10.17.30)$$

$$\phi = 0.127 \,\omega t - 0.355 \sin \omega t - 0.0156 \sin 2\omega t \qquad (10.17.31)$$

For details, refer to Problem 10.17.34. In order to get an idea of the motion of the figure axis of the top, ϕ versus θ has been graphed on a

$$\cos \theta = 0.356 + 0.349 \cos \omega t - 0.008 \cos 2\omega t$$
$$\phi = 0.127 \omega t - 0.355 \sin \omega t - 0.0156 \sin 2\omega t$$

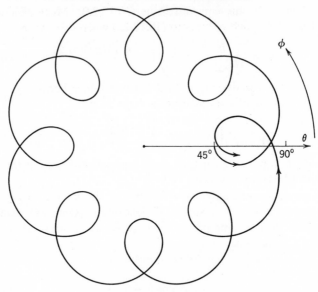

Fig. 10.17.5

polar diagram in Fig. 10.17.5. The turning points are roughly at $\theta = 45°$ and $\theta = 90°$. It would of course be more realistic to plot ϕ versus θ on a spherical rather than a flat surface. Other qualitatively different types of motion can occur. For instance, if the initial values of $\dot\theta$ and $\dot\phi$ are both zero, there will be cusps in the motion at the inner turning point. The possible types of motion are qualitatively the same as are given by a charged particle in a magnetron; see Fig. 7.5.1.

10.18. TORQUES AND GENERALIZED FORCES

Suppose that a rigid body has three degrees of freedom about its center of mass or about a fixed point. Let some convenient set of parameters such as the Euler angles be chosen to serve as generalized coordinates. (Another possible set would be the parameters a, b, and c of Problem 10.15.32.) The rotational kinetic energy of the rigid body is

$$T = \tfrac{1}{2} I_{ij} \omega_i \omega_j \tag{10.18.1}$$

and the generalized force components follow from

$$\frac{d}{dt}\left(\frac{\partial T}{\partial \dot q_i}\right) - \frac{\partial T}{\partial \dot q_i} = Q_i \tag{10.18.2}$$

If the external forces are derivable from a potential, then $Q_i = -\partial V/\partial q_i$ and Lagrange's equations follow. An example which is extensively discussed in Section 10.17 is the toy top with one point fixed. In the study of the mechanics of a single particle, it is relatively easy to establish a connection between the generalized force components and the physical components of force acting on the particle. For a more complicated system such as the rigid body under consideration, the physical meaning of the Q_i is no longer so simple. It is the purpose of this section to establish the general relation between the Q_i as given by (10.18.2) and the torques as given by the Euler equations

$$N_i = I_{ij}\dot{\omega}_j + \delta_{ijk}\omega_j I_{kl}\omega_l \tag{10.18.3}$$

The theory obeys the principle of covariance. The three basic equations (10.18.1), (10.18.2), and (10.8.3) are valid in either the body coordinates or the space coordinates. If the choice is made to work in the space coordinates, the components of the moment of inertia tensor are functions of the generalized coordinates. It is established in Section 10.12 that the rate at which the external torques do work on the rigid body is

$$\frac{dT}{dt} = N_i\omega_i \tag{10.18.4}$$

It is also true that

$$\frac{dT}{dt} = Q_i\dot{q}_i \tag{10.18.5}$$

a result which follows most easily from a consideration of the *rectangular* components of force which act on *each particle* of the rigid body. Equation (10.18.5) then follows directly from Eq. (6.7.12). Thus,

$$\frac{dT}{dt} = N_i\omega_i = Q_i\dot{q}_i \tag{10.18.6}$$

If the choice is made to work with the components of **ω** in the body coordinates, then by (10.3.12),

$$\omega_i = \tfrac{1}{2}\delta_{ijk}\dot{S}_{jl}S_{kl} \tag{10.18.7}$$

If we write

$$\dot{S}_{jl} = \frac{\partial S_{jl}}{\partial q_m}\dot{q}_m \tag{10.18.8}$$

then

$$\omega_i = \tfrac{1}{2}\delta_{ijk}S_{kl}\frac{\partial S_{jl}}{\partial q_m}\dot{q}_m \tag{10.18.9}$$

showing that the components of angular velocity are always linear combinations of the generalized velocity components. By differentiation of (10.18.9),

$$\frac{\partial \omega_i}{\partial \dot{q}_m} = \tfrac{1}{2} \delta_{ijk} S_{kl} \frac{\partial S_{jl}}{\partial q_m} \tag{10.18.10}$$

so that (10.18.9) can be expressed as

$$\omega_i = \frac{\partial \omega_i}{\partial \dot{q}_m} \dot{q}_m \tag{10.18.11}$$

We now substitute this result into (10.18.6) to get

$$N_i \frac{\partial \omega_i}{\partial \dot{q}_m} \dot{q}_m = Q_m \dot{q}_m \tag{10.18.12}$$

Since the q_i are a set of three independent coordinates, (10.18.12) implies

$$N_i \frac{\partial \omega_i}{\partial \dot{q}_m} = Q_m \tag{10.18.13}$$

This is the sought-after connection between torques and generalized forces. Exactly the same relation holds if N_i and ω_i are components of **N** and **ω** with respect to the space set of axes. The Q_i are not components of anything in either the space or the body coordinates, but belong to the abstract "generalized" coordinate system as expressed by the parameters q_i.

The transformation equations (10.18.11) and (10.18.13) can be inverted. If (10.18.11) is solved for $\dot{q}_m$ in terms of ω_i, there results the set of linear equations

$$\dot{q}_m = A_{mi} \omega_i \tag{10.18.14}$$

Since by (10.18.10) $\partial \omega_i / \partial \dot{q}_m$ depends on only the generalized coordinates q_i, the same will be true of the coefficients A_{mi}. By differentiation of (10.18.14),

$$\frac{\partial \dot{q}_m}{\partial \omega_i} = A_{mi} \tag{10.18.15}$$

So that

$$\dot{q}_m = \frac{\partial \dot{q}_m}{\partial \omega_i} \omega_i \tag{10.18.16}$$

Now, by combining (10.18.11) and (10.18.16),

$$\dot{q}_m = \frac{\partial \dot{q}_m}{\partial \omega_i} \frac{\partial \omega_i}{\partial \dot{q}_n} \dot{q}_n \tag{10.18.17}$$

which implies

$$\frac{\partial \dot{q}_m}{\partial \omega_i} \frac{\partial \omega_i}{\partial \dot{q}_n} = \delta_{mn} \qquad (10.18.18)$$

The solution of (10.18.13) for the N_i in terms of Q_m yields

$$N_i = a_{im} Q_m \qquad (10.18.19)$$

where, by the same kind of procedure that led to (10.18.18),

$$a_{im} \frac{\partial \omega_j}{\partial \dot{q}_m} = \delta_{ij} \qquad (10.18.20)$$

If (10.18.20) is multiplied by $(\partial \dot{q}_n / \partial \omega_j)$ and (10.18.18) is used,

$$a_{im} \frac{\partial \omega_j}{\partial \dot{q}_m} \frac{\partial \dot{q}_n}{\partial \omega_j} = \delta_{ij} \frac{\partial \dot{q}_n}{\partial \omega_j}$$

$$a_{im} \delta_{nm} = a_{in} = \frac{\partial \dot{q}_n}{\partial \omega_i} \qquad (10.18.21)$$

Thus,

$$N_i = \frac{\partial \dot{q}_m}{\partial \omega_i} Q_m \qquad (10.18.22)$$

The components of torque N_i have been defined in such a way that they obey a right-hand rule. In labeling a set of parameters to serve as generalized coordinates, a problem exists in that the ordering must be such as to follow a right-hand rule. In the case of the Euler angles, it is possible to follow the rotations in Fig. 10.15.1 and construct a triad of unit vectors which point in the directions of the three rotations as would be determined by a right-hand screw rule. This results in Fig. 10.18.1. If we choose $\psi = q_3$, it is necessary that $\theta = q_1$ and $\phi = q_2$. If geometrical intuition fails, we can compute the Jacobian of the transformation (10.18.13) [or its inverse (10.18.22)] and see if it is positive or negative. If a negative Jacobian results, we must relabel the coordinate system.

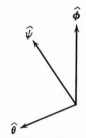

Fig. 10.18.1

As a specific example, the transformation (10.18.13) when the q_i are the Euler angles is

$$Q_1 = N_1 \cos \psi - N_2 \sin \psi$$
$$Q_2 = N_1 \sin \theta \sin \psi + N_2 \sin \theta \cos \psi + N_3 \cos \theta \qquad (10.18.23)$$
$$Q_3 = N_3$$

where N_i are the body components of **N** and the partial derivatives have been computed from (10.15.8). The Jacobian of the transformation

(10.18.23) turns out to be

$$J = \det \left(\frac{\partial \omega_i}{\partial \dot{q}_m} \right) = \sin \theta \qquad (10.18.24)$$

Thus $J > 0$ since $0 < \theta < \pi$. The direct inversion of (10.18.23) results in

$$N_1 = Q_1 \cos \psi + Q_2 \frac{\sin \psi}{\sin \theta} - Q_3 \sin \psi \cot \theta$$

$$N_2 = -Q_1 \sin \psi + Q_2 \frac{\cos \psi}{\sin \theta} - Q_3 \cos \psi \cot \theta \qquad (10.18.25)$$

$$N_3 = Q_3$$

For the symmetric top treated in Section 10.17, all forces other than the forces which are exerted at the pivot point are obtained from the potential $V = Mgs \cos \theta$. The generalized force components, therefore, are

$$Q_1 = -\frac{\partial V}{\partial \theta} = Mgs \sin \theta \qquad Q_2 = Q_3 = 0 \qquad (10.18.26)$$

From (10.18.25), the body components of torque are

$$N_1 = Mgs \sin \theta \cos \psi$$
$$N_2 = -Mgs \sin \theta \sin \psi \qquad (10.18.27)$$

As another example, consider the thin rod pinned to the rotating shaft pictured in Fig. 10.13.4. This is an example of a *constrained system*, the constraints in terms of the Euler angles being

$$\psi = \dot{\psi} = 0 \qquad \dot{\phi} = \omega = \text{const.} \qquad (10.18.28)$$

The equilibrium value of the angle θ was found in Section 10.13. Assume now a more general situation where θ is varying as a function of time and pose the following question: What torque does the pin exert on the rod about the x_3-axis? Assume a small but nonzero value for the moment of inertia I_3. From Eq. (10.18.23),

$$\frac{d}{dt} \left(\frac{\partial T}{\partial \dot{\psi}} \right) - \frac{\partial T}{\partial \psi} = Q_3 = N_3 \qquad (10.18.29)$$

The appropriate form of the kinetic energy is given by (10.15.11). The differentiations indicated in equation (10.18.29) must be carried out *before* the introduction of the equations of constraint (10.18.28)! Otherwise, the result $N_3 = 0$ is obtained. This is because N_3 is a *constraining* torque and, as we know, the explicit introduction of the equations of constraint into the kinetic energy removes the forces of constraint from the formalism. Equations (10.15.11) and (10.18.29) yield

$$Q_3 = I_3(\ddot{\psi} + \ddot{\phi} \cos \theta - \dot{\phi}\dot{\theta} \sin \theta) \qquad (10.18.30)$$

If the equations of constraint (10.18.28) are now introduced, the result is

$$N_3 = -I_3\omega\dot\theta \sin\theta \qquad (10.18.31)$$

We could, of course, get this result directly from Euler's equations.

The rod pinned to a rotating shaft has one other feature which makes it an instructive example. The equations of constraint (10.18.28) are nonholonomic. If it is desired to find the θ motion, the constraints can first be introduced explicitly into the kinetic energy (10.15.11) yielding

$$T = \tfrac{1}{2}I_1\dot\theta^2 + \tfrac{1}{2}I_1 \sin^2\theta\omega^2 \qquad (10.18.32)$$

where $I_3 \cong 0$ is assumed. The potential energy is

$$V = mg\frac{s}{2}(1 - \cos\theta) \qquad (10.18.33)$$

so that the Lagrangian is

$$L = \tfrac{1}{2}I_1\dot\theta^2 + \tfrac{1}{2}I_1 \sin^2\theta\omega^2 - mg\frac{s}{2}(1 - \cos\theta) \qquad (10.18.34)$$

The generalized momentum conjugate to θ is

$$p_\theta = \frac{\partial L}{\partial\dot\theta} = I_1\dot\theta \qquad (10.18.35)$$

Since L has no explicit time-dependence, the Hamiltonian is a constant of the motion:

$$\begin{aligned}H &= p_\theta\dot\theta - L \\ &= \tfrac{1}{2}I_1\dot\theta^2 - \tfrac{1}{2}I_1 \sin^2\theta\omega^2 + mg\frac{s}{2}(1 - \cos\theta)\end{aligned} \qquad (10.18.36)$$

On the other hand, the *total energy* is given by

$$\mathscr{E} = T + V = \tfrac{1}{2}I_1\dot\theta^2 + \tfrac{1}{2}I_1 \sin^2\theta\omega^2 + mg\frac{s}{2}(1 - \cos\theta) \qquad (10.18.37)$$

and is *not* identical to the Hamiltonian and is *not* conserved! This is because the time-dependent constraints do work on the system! Since H *is* a constant, it is convenient to write (10.18.36) as

$$H = \tfrac{1}{2}I_1\dot\theta^2 + V_e(\theta) \qquad (10.18.38)$$

and study the system qualitatively using the effective potential

$$V_e(\theta) = -\frac{1}{2}I_1 \sin^2\theta\omega^2 + mg\frac{s}{2}(1 - \cos\theta) \qquad (10.18.39)$$

The continuation of this example is left as a problem.

10.19. THE MOTION OF A RIGID BODY WITH RESPECT TO AN ACCELERATED REFERENCE FRAME

Suppose that we have a toy top or a gyroscope and we wish to know the perturbation on the motion due to the rotation of the earth. It is a fairly simple matter to write down a Lagrangian for this purpose. Since $\boldsymbol{\omega}$ has been used to represent the angular velocity of rotation of a rigid body, $\boldsymbol{\chi}$ will be used to represent the angular velocity of the reference frame relative to which the motion of the rigid body is desired. From Eq. (10.6.7), replacing $\boldsymbol{\omega}$ with $\boldsymbol{\chi}$, the Lagrangian for a single particle of the body is

$$L_\alpha = T_\alpha + \tfrac{1}{2} m_\alpha (r_\alpha^2 \delta_{ij} - x_{\alpha i} x_{\alpha j}) \chi_i \chi_j + \boldsymbol{\chi} \cdot \mathbf{l}_\alpha - V_\alpha \qquad (10.19.1)$$

where

$$m_\alpha \mathbf{u}_\alpha \cdot (\boldsymbol{\chi} \times \mathbf{r}_\alpha) = \boldsymbol{\chi} \cdot \mathbf{r}_\alpha \times m_\alpha \mathbf{u}_\alpha = \boldsymbol{\chi} \cdot \mathbf{l}_\alpha \qquad (10.19.2)$$

and the term $m_\alpha(\mathbf{r}_\alpha \cdot \mathbf{a}_0)$ has been included in V_α. In any rigid body problem, $-\mathbf{a}_0$ can be treated as a uniform gravitational field acting on the body. If the rigid body has a fixed point, then $\mathbf{r}_\alpha$ is the displacement vector from the fixed point to the particle under consideration. "Fixed point" now means fixed with respect to the observer in the accelerated frame. If there is no fixed point, we must write

$$\mathbf{r}_\alpha = \mathbf{R} + \mathbf{r}_\alpha' \qquad (10.19.3)$$

where $\mathbf{R}$ is the position vector of the center of mass and $\mathbf{r}_\alpha'$ is the position vector of the typical particle with respect to the center of mass. The Lagrangian for the rigid body as a whole can be found from

$$L = \sum_{\alpha=1}^{N} L_\alpha \qquad (10.19.4)$$

where the sum extends over all particles of the body. If we recall that the kinetic energy and the angular momentum of a rigid body separate into two parts as

$$T = T_c + T_{\text{rot}} \qquad \mathbf{l} = \mathbf{l}_c + \mathbf{l}_{\text{rot}} \qquad (10.19.5)$$

and that

$$\sum_{\alpha=1}^{N} m_\alpha x_{\alpha i}' = 0 \qquad (10.19.6)$$

we see that the Lagrangian can be expressed by

$$L = L_c + L_{\text{rot}} - V \qquad (10.19.7)$$

where

$$L_c = T_c + \tfrac{1}{2} M (R^2 \delta_{ij} - X_i X_j) \chi_i \chi_j + \boldsymbol{\chi} \cdot \mathbf{l}_c \qquad (10.19.8)$$

$$L_{\text{rot}} = T_{\text{rot}} + \tfrac{1}{2} I_{ij} \chi_i \chi_j + \boldsymbol{\chi} \cdot \mathbf{l}_{\text{rot}} \qquad (10.19.9)$$

As usual,

$$I_{ij} = \sum_{\alpha=1}^{N} m_\alpha (r_\alpha'^2 \delta_{ij} - x_{\alpha i}' x_{\alpha j}') \tag{10.19.10}$$

are the components of the moment of inertia tensor with respect to the center of mass. We have recovered the theorem which states that the motion of the center of mass of a body can be found by treating it as a single particle of mass M and that the rotational motion with respect to the center of mass can be treated separately! Of course, if the rigid body has a fixed point, the Lagrangian can be more simply expressed as

$$L = T + \tfrac{1}{2} I_{ij} \chi_i \chi_j + \boldsymbol{\chi} \cdot \mathbf{l} - V \tag{10.19.11}$$

where I_{ij} are components of inertia about the fixed point. $T = \tfrac{1}{2} I_{ij} \omega_i \omega_j$ is the rotational kinetic energy and $\mathbf{l}$ is the angular momentum, both measured with respect to the fixed point. The quantity $\tfrac{1}{2} I_{ij} \chi_i \chi_j$ will be called the *centrifugal term* and $\boldsymbol{\chi} \cdot \mathbf{l}$ the *Coriolis term*. Since the terms in the Lagrangian are invariants, they can be computed either in the space or the body coordinates—or even in a intermediate coordinate system should that be convenient. If the choice is made to work in space coordinates, the components I_{ij} come out to be functions of the Euler angles or other parameters being used as generalized coordinates. By "space coordinates" is now meant a coordinate system fixed with respect to an observer in the accelerated frame.

An example which will serve to fix these concepts in mind is the *gyrocompass*. Let a coordinate system denoted (x_{01}, x_{02}, x_{03}) be fixed to the earth at colatitude γ (Fig. 10.5.2). The x_{01}-axis points north and the x_{02}-axis points west, Fig. 10.19.1. A symmetric spinning body is mounted in gimbals such that its symmetry axis is constrained to move in the horizon plane (the $x_{01} x_{02}$-plane of Fig. 10.19.1) and such that its center of mass is a fixed point. Due to the constraint, there are two degrees of freedom which are chosen to be the Euler angles ϕ and θ. The angle ϕ measures the position of the symmetry axis (the x_1-axis of Fig. 10.19.1) with respect to a northerly direction and θ measures the spin of the body about the x_1-axis. The equation of constraint is then

$$\psi = 0 \tag{10.19.12}$$

The coordinate system labeled x_i' in Fig. 10.19.1 is the actual body coordinate system, but due to the symmetry, the moment of inertia tensor is already constant in the intermediate coordinate system x_i. We will therefore choose to work in the intermediate system.

In the x_i coordinates, the components of the angular velocity of rotation of the rigid body are simply

$$\omega_1 = \dot{\theta} \qquad \omega_2 = 0 \qquad \omega_3 = \dot{\phi} \tag{10.19.13}$$

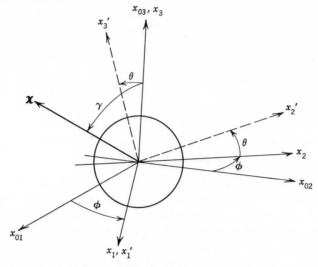

Fig. 10.19.1

The inertia tensor is diagonal and $I_2 = I_3$. The kinetic energy is

$$T = \tfrac{1}{2}I_1\dot\theta^2 + \tfrac{1}{2}I_2\dot\phi^2 \tag{10.19.14}$$

The components of angular momentum are

$$l_1 = I_1\dot\theta \qquad l_2 = 0 \qquad l_3 = I_2\dot\phi \tag{10.19.15}$$

From the geometry of Fig. 10.19.1, the components of the angular velocity of rotation $\boldsymbol{\chi}$ of the earth expressed in the x_i coordinates are

$$\chi_1 = \chi \sin\gamma \cos\phi \qquad \chi_2 = -\chi \sin\gamma \sin\phi \qquad \chi_3 = \chi \cos\gamma \tag{10.19.16}$$

Thus

$$I_{ij}\chi_i\chi_j = \chi^2 \sin^2\gamma(I_1 \cos^2\phi + I_2 \sin^2\phi) + I_2\chi^2 \cos^2\gamma \tag{10.19.17}$$

$$\mathbf{l}\cdot\boldsymbol{\chi} = I_1\dot\theta\chi \sin\gamma \cos\phi + I_2\dot\phi\chi \cos\gamma \tag{10.19.18}$$

The Lagrangian for the gyrocompass, therefore, is

$$L = \tfrac{1}{2}I_1\dot\theta^2 + \tfrac{1}{2}I_2\dot\phi^2 + \tfrac{1}{2}\chi^2 \sin^2\gamma(I_1 \cos^2\phi + I_2 \sin^2\phi)$$
$$+ I_1\dot\theta\chi \sin\gamma \cos\phi + I_2\dot\phi\chi \cos\gamma \tag{10.19.19}$$

where the constant term $I_2\chi^2 \cos^2\gamma$ has been discarded. No potential term involving the gravitational field of the earth appears since the gyrocompass is balanced at its center of mass. The coordinate θ is cyclic leading to

$$\frac{\partial L}{\partial\dot\theta} = I_1\dot\theta + I_1\chi \sin\gamma \cos\phi = h = \text{const.} \tag{10.19.20}$$

By means of

$$\frac{d}{dt}\left(\frac{\partial L}{\partial \dot{\phi}}\right) - \frac{\partial L}{\partial \phi} = 0 \qquad (10.19.21)$$

the equation of motion

$$I_2\ddot{\phi} + I_1\dot{\theta}\chi \sin \gamma \sin \phi + (I_1 - I_2)\chi^2 \sin^2 \gamma \sin \phi \cos \phi = 0 \quad (10.19.22)$$

results. The elimination of $\dot{\theta}$ by means of (10.19.20) results in

$$I_2\ddot{\phi} + h\chi \sin \gamma \sin \phi - I_2\chi^2 \sin^2 \gamma \sin \phi \cos \phi = 0 \quad (10.19.23)$$

Since L has no explicit time dependence, another constant of the motion is provided by the Hamiltonian:

$$\begin{aligned} H &= p_\theta\dot{\theta} + p_\phi\dot{\phi} - L \\ &= \tfrac{1}{2}I_1\dot{\theta}^2 + \tfrac{1}{2}I_2\dot{\phi}^2 - \tfrac{1}{2}\chi^2 \sin^2 \gamma(I_1 \cos^2 \phi + I_2 \sin^2 \phi) \end{aligned} \qquad (10.19.24)$$

For practical applications where motion with respect to the earth is concerned, terms proportional to χ^2 can be neglected. This could have been done at the start by neglecting the centrifugal term in the Lagrangian (10.19.11). However, all equations have so far been kept exact in order to illustrate the method in full generality. Neglecting the χ^2 term in (10.19.23) gives

$$I_2\ddot{\phi} + h\chi \sin \gamma \sin \phi = 0 \qquad (10.19.25)$$

This is identical to the equation of motion of a simple plane pendulum and shows that provided $h > 0$ the northerly direction is a position of stable equilibrium. If the initial spin of the gyroscope is in the opposite direction, $h < 0$ and the gyroscope tends to point south instead of north. A gyroscope once pointed north tends to remain in this direction. This is essentially the principle of the gyrocompass.

As another example, suppose that observers in the ring-shaped space ship discussed in Section 10.9 have a gyroscope which is mounted in gimbals such that its center of mass is stationary with respect to the ship but is otherwise unconstrained. Let a coordinate system denoted x_{0i} be fixed to the floor of one of the rooms of the ship and let the x_{03} axis point in the same direction as the angular velocity of rotation χ of the ship. *Relative to an inertial frame*, the angular velocity of the gyroscope is $\chi + \omega$ where ω is its angular velocity with respect to the ship. Relative to the inertial frame, the torque about the center of mass is zero. Therefore, from Euler's equations,

$$\begin{aligned} 0 &= I_1(\dot{\omega}_1 + \dot{\chi}_1) + (\omega_2 + \chi_2)(\omega_3 + \chi_3)(I_3 - I_2) \\ 0 &= I_2(\dot{\omega}_2 + \dot{\chi}_2) + (\omega_3 + \chi_3)(\omega_1 + \chi_1)(I_1 - I_3) \\ 0 &= I_3(\dot{\omega}_3 + \dot{\chi}_3) + (\omega_1 + \chi_1)(\omega_2 + \chi_2)(I_2 - I_1) \end{aligned} \qquad (10.19.26)$$

where all quantities are expressed in the body frame of the gyroscope. Thus $\omega_i + \chi_i$ obey the equations of a torque free body. In particular, if $I_1 = I_2$, we have from (10.16.1),

$$\omega_1 + \chi_1 = A \sin at \qquad \omega_2 + \chi_2 = A \cos at$$

$$\omega_3 + \chi_3 = \text{const.} \qquad a = (\omega_3 + \chi_3)\frac{I_1 - I_3}{I_1} \qquad (10.19.27)$$

In the frame fixed to the ship, $\chi_{0i} = (0, 0, \chi)$. By means of (10.15.5) the components of χ with respect to the body frame are found to be

$$\chi_1 = \chi \sin \psi \sin \theta \qquad \chi_2 = \chi \cos \psi \sin \theta \qquad \chi_3 = \chi \cos \theta \quad (10.19.28)$$

where ψ and θ are Euler angles of the body frame with respect to the frame x_{01} fixed to the ship. If ω_i are also expressed in terms of Euler angles there results

$$\dot{\theta} \cos \psi + [\dot{\phi} + \chi] \sin \theta \sin \psi = A \sin at$$
$$-\dot{\theta} \sin \psi + [\dot{\phi} + \chi] \sin \theta \cos \psi = A \cos at \qquad (10.19.29)$$

$$\dot{\psi} + [\dot{\phi} + \chi] \cos \theta = \text{const} = \frac{h}{I_3}$$

Thus, in essence, the motion with respect to the space ship is that of a free body on which is superimposed a uniform precession at an angular rate χ about the x_{03}-axis. It is not possible for us to conclude here that $\dot{\theta} = 0$ as we did in Eq. (10.16.2) since now the x_{03}-axis goes in the direction of χ and this will not, in general, be the same as the direction of the angular momentum l of the gyroscope.

10.20. A GENERALIZATION OF EULER'S EQUATIONS

Euler's equations in the form

$$N_i = I_{ij}\dot{\omega}_j + \delta_{ijk}\omega_j I_{kl}\omega_l \qquad (10.20.1)$$

are valid in either the body or the space coordinates. Between body and space coordinates, all quantities which appear in (10.20.1) have a normal transformation property, e.g.,

$$N_i' = S_{ij}N_j \qquad \dot{\omega}_j' = S_{jk}\dot{\omega}_k, \text{ etc.} \qquad (10.20.2)$$

A complication exists, however, if it is desired to compute torques in an intermediate frame, such as the frame labeled (x_1, x_2, x_3) in Fig. 10.19.1.

Let us write the angular momentum of the rigid body as

$$l = I_{ij}\omega_j\hat{e}_i \qquad (10.20.3)$$

where the $\hat{e}_i$ are the basis vectors of an intermediate coordinate system the angular velocity of which is λ with respect to the space axes. Assume that the I_{ij} are constants. Again, the gyrocompass problem illustrated in Fig. 10.19.1 gives an example where this can happen. The differentiation of (10.20.3) leads to

$$\mathbf{N} = \frac{d\mathbf{l}}{dt} = I_{ij}\dot{\omega}_j\hat{\mathbf{e}}_i + I_{ij}\omega_j\delta_{ikl}\hat{\mathbf{e}}_k\lambda_l$$
$$= I_{ij}\dot{\omega}_j\hat{\mathbf{e}}_i + I_{kj}\omega_j\delta_{kil}\lambda_l\hat{\mathbf{e}}_i \qquad (10.20.4)$$

where (10.2.8) is used. The components of torque, therefore, are

$$N_i = I_{ij}\dot{\omega}_j + \delta_{ilk}\lambda_l I_{kj}\omega_j \qquad (10.20.5)$$

If, as is usually the case, the moment of inertia tensor is diagonal,

$$
\begin{aligned}
N_1 &= I_1\dot{\omega}_1 + \lambda_2\omega_3 I_3 - \lambda_3\omega_2 I_2 \\
N_2 &= I_2\dot{\omega}_2 + \lambda_3\omega_1 I_1 - \lambda_1\omega_3 I_3 \\
N_3 &= I_3\dot{\omega}_3 + \lambda_1\omega_2 I_2 - \lambda_2\omega_1 I_1
\end{aligned}
\qquad (10.20.6)
$$

No complication of this type exists with the Lagrange formulation since the Lagrangian involves no angular accelerations and is composed of true invariant scalars constructed out of angular velocities.

Equations (10.20.6) can be applied to the gyrocompass problem. Let us choose as our space coordinates an inertial frame away from the earth. In this way, the complication of having to introduce the torques produced by Coriolis and centrifugal forces can be avoided. The calculation will be done in the intermediate coordinates x_i of Fig. 10.19.1. By $\boldsymbol{\omega}$ is now meant the angular velocity of the gyrocompass with respect to the inertial frame outside the earth. The components of $\boldsymbol{\omega}$ in the intermediate coordinates are

$$
\begin{aligned}
\omega_1 &= \dot{\theta} + \chi \sin \gamma \cos \phi \\
\omega_2 &= -\chi \sin \gamma \sin \phi \\
\omega_3 &= \dot{\phi} + \chi \cos \gamma
\end{aligned}
\qquad (10.20.7)
$$

The components of the angular velocity of the intermediate coordinate system with respect to the inertial frame are

$$
\begin{aligned}
\lambda_1 &= \chi \sin \gamma \cos \phi \\
\lambda_2 &= -\chi \sin \gamma \sin \phi \\
\lambda_3 &= \dot{\phi} + \chi \cos \gamma
\end{aligned}
\qquad (10.20.8)
$$

The torques N_1 and N_3 are zero whereas N_2 is an unknown torque of constraint. Equations (10.20.6) yield

$$I_1\omega_1 = I_1\dot{\theta} + \chi I_1 \sin \gamma \cos \phi = h = \text{const.} \qquad (10.20.9)$$

$$
\begin{aligned}
N_2 &= -I_2\chi\dot{\phi} \sin \gamma \cos \phi + (\dot{\phi} + \chi \cos \gamma)\dot{\theta} I_1 \\
&\quad + (I_1 - I_2)(\chi \sin \gamma \cos \phi)(\dot{\phi} + \chi \cos \gamma)
\end{aligned}
\qquad (10.20.10)
$$

$$I_2\ddot{\phi} + (I_1 - I_2)\chi^2 \sin^2 \gamma \sin \phi \cos \phi + I_1\dot{\theta}\chi \sin \gamma \sin \phi = 0 \qquad (10.20.11)$$

Equations (10.20.9) and (10.20.11) are identical to Eqs. (10.19.20) and
(10.19.22) obtained by the Lagrange method. Equation (10.20.10) is a new
equation and gives a method for computing the torque which the bearings
of the gyrocompass must sustain.

PROBLEMS

10.1.1   A glass of water slides without friction down an inclined plane. What
is the orientation of the liquid surface?

10.1.2   A bucket is filled with water and is hung by a rope to form a plane
pendulum. Described the behavior of the water in the bucket after it has been
set swinging.

10.1.3   A particle released near the surface of the earth has a Cartesian
coordinate system rigidly attached to it at the origin. The y-axis points directly
away from the center of the earth. Two observers, one on the y-axis and one on
x-axis, are each a distance of one earth radius away from the origin release test
particles. What initial acceleration do they observe?

10.1.4   Protons are traveling in a circular orbit of radius 10 meters in a
synchrotron at a velocity of 3×10^9 cm/sec. How far do the protons fall in a
single revolution because of the gravitational field of the earth? Find the ratio
of the magnetic force necessary to keep the protons in a circular orbit to the
gravitational force.

10.3.5   What are the matrix representations of Eqs. (10.3.5), (10.3.9), and
(10.3.10)? If the x_3- and x_3'-axes of F and F' remain always parallel, the matrix
representation of the orthogonal transformation connecting F and F' is

$$S = \begin{pmatrix} \cos\theta & \sin\theta & 0 \\ -\sin\theta & \cos\theta & 0 \\ 0 & 0 & 1 \end{pmatrix}$$

Calculate the matrices Ω' and Ω, using (10.3.5) and (10.3.10), and then verify the
transformation (10.3.9).

10.3.6   If ω represents a column matrix the components of which are ω_1, ω_2
and ω_3, show that $\Omega\omega = 0$ and $\dot{S}\omega = 0$.

10.3.7   Is Ω a singular or a nonsingular matrix?

10.5.8   Find the variation in the gravitational field which a particle experiences
as it is moved from the side of the earth toward the sun to the side directly away
from it. Do the same calculation for the effect of the moon. See Appendix D
for astronomical data. This variation in the gravitational field at the surface of
the earth, while small, is responsible for the existence of tides.

10.5.9   Neglect the effect due to the sun and consider the earth and the moon
to be an isolated system. The movement of water due to the tides (see Problem
10.5.8) dissipates energy at the expense of the rotational kinetic energy of the
earth. What effect does this have on the earth-moon separation? After the

spin energy of the earth is completely exhausted, what will be the mean separation of the earth and the moon?

10.5.10 At the earth's equator, what fraction of the measured value of the acceleration due to gravity is due to the centrifugal acceleration?

10.5.11 A projectile is fired due south from a gun located at 45° latitude. The velocity of projection is 100 ft/sec and the angle of elevation of the gun above the horizon plane is 45°. Find the correction to the point of impact due to the Coriolis force.

10.5.12 A 10-ton freight car is headed due north at 45° latitude on the earth's surface at a velocity of 100 ft/sec. What is the sidewise force on the railroad tracks due to the Coriolis force?

10.5.13 If a low-pressure area develops in the earth's atmosphere, the surrounding air tends to move toward it. Qualitatively, how does the Coriolis effect tend to modify this air flow?

10.6.14 Verify that the Hamiltonian (10.6.17) gives the correct equations of motion.

10.7.15 The acceleration due to gravity at the equator of the plant Uranus, exclusive of the centrifugal acceleration, is about 950 cm/sec^2, a value which is close to that on earth. The period of rotation of the planet is quite short, being about 10.7 hr. The mean radius of Uranus is 23.8 × 10^3 km or about 3.73 earth radii. What would be the actual measured value of the acceleration of a falling object at the equator of Uranus? A rocket is to be fired vertically from the surface of Uranus at the equator. What is the minimum velocity for escape measured relative to the planet? What would be the initial direction of the rocket as seen by an observer in an inertial frame?

10.7.16 Find the Lagrangian for a satellite launched into an arbitrary orbit around the earth from the point of view of observers on the earth. Use spherical coordinates with origin at the earth's center and fixed with respect to the earth. Find two constants of the motion. What does the orbit look like for the special initial condition $\dot{\phi} = -\omega$?

10.8.17 What area does a spherical pendulum 1 meter long located at a latitude of 45° have to circumscribe in order that the normal precessional velocity of the pendulum shall equal the precessional velocity due to the Coriolis force? See Problem (9.5.14).

10.8.18 Write down the Lagrangian for a spherical pendulum in spherical polar coordinates (see Fig. 6.6.1) which includes a term to account for the Coriolis force.

10.9.19 Separate the exact solution (10.9.12) into its real and imaginary parts.

10.9.20 Solve the same problem as was treated in Section 10.9 as follows: Set the Lagrangian up in cylindrical coordinates r, θ, z with the origin at the center of the ring and the z-axis in the same direction as ω. Find the differential equations of motion and solve them for r and θ as functions of time subject to the initial conditions that the particle is released from rest relative to the rotating frame at $r = r_0$ and $\theta = 0$. *Hint:* The solution with respect to the *inertial frame* x, y of the figure is trivial. The transformation connecting the polar coordinates

r, θ of the rotating frame and the rectangular coordinates x, y of the inertial frame is

$$x = r \cos (\theta + \omega t) \qquad y = r \sin (\theta + \omega t)$$

Plot the ratio r/r_0 as a function of θ on a polar diagram.

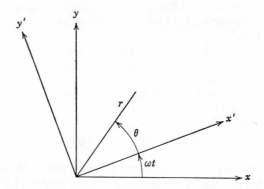

10.9.21   Observers on board the space ship discussed in Section 10.9 are doing the elementary experiment of allowing a particle to slide without friction down an inclined plane. Let the plane be sloped toward the direction of ω so that the angle between ω and the plane is the same as the angle of the plane (see figure). Find the equations of motion and an approximate solution for the case where the particle starts from rest at the origin of the coordinate system pictured in the figure. Treat $-m\mathbf{a}_0$ as the measured value of the static weight of a particle at the origin. The complex variable method does not work in this case. It will be necessary to use an iteration procedure. Write the solutions as

$$x_1 = f_1 + f_2\omega + f_3\omega^2 + \cdots$$
$$x_2 = g_1 + g_2\omega + g_3\omega^2 + \cdots$$

Substitute into the differential equations and equate the coefficients of like powers of ω to zero. Retain terms to order ω^2 in the calculation.

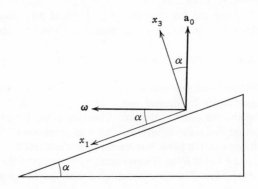

10.9.22 On board the space ship of Section 10.9, there are pipes through which fluid is flowing at a velocity of 100 ft/sec. Suppose that at a particular point, the static gravitational field has the value 32 ft/sec². What is the greatest amount that the weight of the fluid in the pipes can increase due to the Coriolis force? How much can the weight of the fluid decrease? Use the value $\omega = 0.4$ sec^{-1}.

10.11.23 A wheel is rotating at a constant angular velocity ω_a with respect to an inertial frame. Mounted on this wheel is a second wheel, the axes of which make an angle θ with respect to the axis of the first wheel. The axes of both wheels intersect at a common point. Relative to an observer on the first wheel, the second wheel turns at an angular velocity ω_{ba}. What are the components of the angular velocity of the second wheel with respect to the inertial frame? Locate the herpolhode of the second wheel.

10.12.24 A rigid body consists of two uniform thin rectangular plates of width s and length $2s$ located in the $x_1 x_2$ plane, as illustrated. Regarding the coordinate system in the figure as body coordinates, compute the moments and products of inertia. Let M be the total mass of the system. Find the eigenvalues and the eigenvectors of the moment of inertia tensor. Find a set of body coordinates in which the moment of inertia tensor is diagonal.

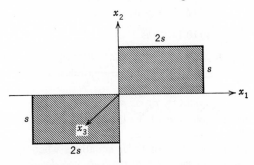

10.13.25 Transform the components of angular momentum (10.13.9) to the space set of axes and verify that

$$N_{01} = \frac{dl_{01}}{dt} \qquad N_{02} = \frac{dl_{02}}{dt} \qquad N_{03} = \frac{dl_{03}}{dt}$$

where N_{01}, N_{02}, and N_{03} are the components of torque in the space set of axes as

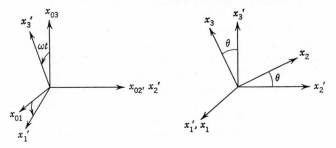

given by (10.13.8). *Hint:* Find the transformation between space and body coordinates by combining the two transformations indicated in the figure. Compute the components of the tensor Ω in the body coordinates. *Hint:* Write Eq. (10.3.5) in matrix form.

10.13.26   Find the equilibrium value of the angle θ in Fig. 10.13.4 by working with the origin of the body coordinate system at the center of mass instead of at the pin. *Hint:* In order to compute the torques about the center of mass, it is necessary to know the forces exerted on the rod at the pin. For this purpose, use (10.12.4).

10.13.27   Find the equilibrium value of the angle θ in Fig. 10.13.4. Treat as a problem in statics from the point of view of an observer in the rotating frame of reference of the rod.

10.14.28   Show that the constants A and C of the solution (10.14.11) when expressed in terms of the kinetic energy and angular momentum are

$$A = \pm \sqrt{\frac{2TI_1 - l^2}{I_3(I_1 - I_3)}} \qquad C = \pm \sqrt{\frac{l^2 - 2TI_3}{I_1(I_1 - I_3)}}$$

Note that this result implies

$$2TI_1 - l^2 > 0 \qquad 2TI_3 - l^2 < 0$$

Show that the modulus (10.14.17) is

$$k = \frac{\sqrt{l^2 - 2TI_3}\,\sqrt{I_1 - I_2}}{\sqrt{2TI_1 - l^2}\,\sqrt{I_2 - I_3}}$$

and that $0 \le k < 1$ requires $2\,TI_2 - l^2 > 0$.

10.14.29   Find the solutions of the differential equations for the torque-free rigid body when $I_1 = I_2$.

10.14.30   Find the solutions of Euler equations when $I_1 = I_2$ and the first component of torque, *as measured in the body coordinates*, is not zero but is constant. Assume $\omega_3 > 0$ for simplicity.

10.14.31   An artificial satellite which is approximately cylindrical in shape has moments of inertia $I_1 = I_2 = 10^4$ slug ft^2 and $I_3 = 0.4 \times 10^4$ slug ft^2 in a principle axis coordinate system. Initially, the satellite is not rotating. There are two rockets mounted directly opposite one another parallel and antiparallel to the x_1 body axis such that they cause a torque in the positive x_3-direction. These rockets have a thrust of 10 lb and are 4 ft from the center of mass. They are fired for 100 sec and then shut off. Now rockets, which are at opposite ends of the satellite mounted parallel and antiparallel to the body x_2-axis such that they produce a torque in the positive x_1-direction, are fired for 2.62 sec. These rockets each produce 300 lb of thrust and are 10 ft from the center of mass. Assume that the amount of material ejected by the rockets is small enough so that the moments of inertia are not affected. Find the components of angular velocity in the body set of axes for the final motion. What is the angle between ω and l? What is the angle between ω and the body x_3-axis? At what angular rate does $\boldsymbol{\omega}$ precess about the body x_3-axis? At what angular rate does $\boldsymbol{\omega}$ precess about $\mathbf{l}$?

10.15.32 If A is an antisymmetric 3×3 matrix, show that $S = (I - A) \times (I + A)^{-1}$ is orthogonal. Is S proper or improper? By writing

$$A = \begin{pmatrix} 0 & -c & -b \\ c & 0 & -a \\ b & a & 0 \end{pmatrix}$$

find a representation for S in terms of the three parameters a, b, and c.

10.17.33 For the special values of the parameters

$$\mu = 1 \qquad \frac{h}{l} = 1 \qquad \epsilon = 1$$

Eq. (17.25) gives the exact simple solution

$$x = -\tfrac{1}{3} + \tanh^2 at \qquad a = \sqrt{\frac{Mgs}{2I_1}}$$

Verify this statement and explain the significance of the solution. Check that $\theta \to 0$ as $t \to \infty$. Refer to Fig. 10.17.2 for a plot of the effective potential. Show that there is a turning point in the motion exactly at $\theta = 90°$. Show that the ϕ-motion is given by

$$\tan (\phi - at) = \tanh at$$

assuming $\phi = 0$ at $t = 0$. Find an expression that gives $\dot\psi$ explicitly as a function of time.

10.17.34 The initial conditions for the motion of a top are such that

$$\epsilon = 1 \qquad \frac{h}{l} = 2 \qquad \mu = 1.00$$

Locate the turning points for the motion of the figure axis. (The effective potential for these values of the parameters is plotted in Fig. 10.17.1.) Obtain Eqs. (10.17.30) and (10.17.31).

10.18.35 Compute the transformations analogous to (10.18.23) and (10.18.25) which have the components of $\mathbf{N}$ expressed in the *space coordinates*. Find the space components of $\mathbf{N}$ about the pivot point of a top which are produced by the gravitational field. Check to see if this is what you would guess them to be by simple geometrical arguments.

10.18.36 Derive Eq. (10.18.31) directly from Euler's equations.

10.18.37 With $I = \tfrac{1}{3}Ms^2$, sketch the general form of the effective potential (10.18.39) and interpret. Let $\alpha = 3g/(25\omega^2)$ and consider the two cases $\alpha > 1$ and $\alpha < 1$.

10.19.38 Show that the Lagrangian (10.19.11) can be expressed

$$L = \tfrac{1}{2}I_{ij}(\chi_i + \omega_i)(\chi_j + \omega_j) - V$$

What does this mean?

10.19.39   Observers in the space ship discussed in Section 10.9 have a gyroscope mounted on gymbals such that its center of mass is a fixed point with respect to the space ship. Let a coordinate system be fixed to the floor of one of the rooms of the ship with its x_{03}-axis pointing in the same direction as the angular velocity of rotation χ of the ship and its x_{02}-axis pointing toward the center of the ring. The symmetry axis of the gyroscope is constrained to move in the $x_{02}x_{03}$-plane. Find the second order differential equation which describes the oscillations of the axis of the gyroscope.

10.20.40   The angular velocity of a frame x_i' relative to a frame x_i is λ. Show that the transformation of angular velocity between the two frames is

$$\dot{\omega}_i' = S_{ij}(\dot{\omega}_j + \delta_{jkl}\lambda_l\omega_k)$$
$$\dot{\omega}_j = S_{ij}(\dot{\omega}_i' + \delta_{ilk}\lambda_l'\omega_k')$$

Hint: Write $\omega_i' = S_{ij}\omega_j$, differentiate, and apply (10.3.5) and (10.3.8) with λ_k' in place of ω_k'.

10.20.41   Referring to Eqs. (10.19.26), show that

$$\dot{\chi}_1 = \chi_2\omega_3 - \chi_3\omega_2 \qquad \dot{\chi}_2 = \chi_3\omega_1 - \chi_1\omega_3 \qquad \dot{\chi}_3 = \chi_1\omega_2 - \chi_2\omega_1$$

(*Hint:* Write $\chi_i = S_{ij}\chi_{0j}$, differentiate and use the fact that χ_{0j} are constants.) Show that Eqs. (10.19.26) can be expressed in the form

$$I_1\dot{\omega}_1 + \omega_2\omega_3(I_3 - I_2) = \chi_2\chi_3(I_2 - I_3) + \chi_2\omega_3(I_2 - I_3 - I_1) - \chi_3\omega_2(I_3 - I_2 - I_1)$$

and explain the meaning of the various terms.

10.20.42   What is the appropriate modification of the result of Problem 10.20.41 if constraining torques are present? A spinning body, such as the armature of an electric motor, is constrained so that its axis of rotation is stationary with respect to a space ship that is turning at an angular velocity χ. Show that the torque of constraint is

$$\mathbf{N} \cong \boldsymbol{\chi} \times \mathbf{l}$$

where $\mathbf{l}$ is the angular momentum of the body with respect to the space ship and centrifugal terms (i.e., terms proportional to χ^2) are neglected. A motor turns at 3600 rpm and has a moment of inertia of 1/36 slug-ft². Find the greatest torque which the bearings of the motor must sustain if $\chi = 0.4$ rad/sec.

11

The Calculus of Variations

This chapter contains an introduction to the calculus of variations and applications to problems in mechanics.

11.1. INTRODUCTORY PROBLEMS

Historically, the *brachistochrone problem*, which we now briefly describe, was the first to be treated by the methods of the calculus of variations. In Fig. 11.1.1, there is a uniform gravitational field in the y-direction. A particle is constrained to move from point 1 to 2 along a path $y(x)$. How should the function $y(x)$ be chosen to minimize the time required? The word *brachistochrone* derives from the Greek *brachistos* (shortest) and *chronos* (time). It is assumed that there is no friction so that the forces are those due to the gravitational field and the workless force of constraint introduced by $y(x)$. Imagine, for example, that the particle is a bead that slides without friction on a wire bent into the form of $y(x)$ between the fixed end points 1 and 2. The time required for the particle to move from point 1 to point 2 can be expressed by

$$t_{12} = \int_1^2 \frac{ds}{u} \qquad (11.1.1)$$

where u is the speed of the particle. From conservation of energy

$$u = \sqrt{u_1^2 + 2g(y - y_1)} \qquad (11.1.2)$$

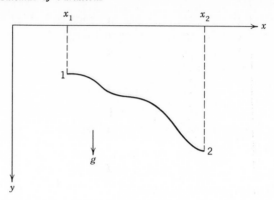

Fig. 11.1.1

where u_1 is the initial speed at 1. The line element ds can be expressed

$$ds = \sqrt{dx^2 + dy^2} = \sqrt{1 + (y')^2}\, dx \qquad (11.1.3)$$

where $y' = dy/dx$. Thus, the problem is to choose $y(x)$ such that

$$t_{12} = \int_1^2 \frac{\sqrt{1 + (y')^2}}{\sqrt{u_1{}^2 + 2g(y - y_1)}}\, dx \qquad (11.1.4)$$

has the smallest possible value. We will solve this problem in detail after the necessary mathematical apparatus has been set up.

The calculus of variations is applicable to the solution of the problem of finding the geodesics in a curved space. Suppose that our space is metricized by functions g_{ij} so that an element of arc length is given by

$$ds^2 = g_{ij}\, dq_i\, dq_j \qquad (11.1.5)$$

Let a curve connecting two given points in the space be given parametrically by

$$q_i = q_i(\lambda) \qquad (11.1.6)$$

where λ is any convenient parameter. The total distance between the two points measured along the curve is given by

$$s_{12} = \int_1^2 \sqrt{g_{ij}q_i' q_j'}\, d\lambda \qquad (11.1.7)$$

where $q_i' = dq_i/d\lambda$. The *geodesic* between the two points is by definition the curve which gives the shortest distance. In order to find the geodesic, we seek the functions $q_i(\lambda)$ which minimize (11.1.7). In the ordinary Euclidean space of three dimensions the answer is a straight line.

Consider a surface over which the coordinates are q_1, q_2. This surface can be visualized as a two-dimensional subspace embedded in the three space. For instance, if the surface is a sphere the coordinates which specify position may be chosen to be the spherical coordinate angles θ and ϕ. The element of distance on the surface is

$$ds^2 = r^2(d\theta^2 + \sin^2 \theta \, d\phi^2) \tag{11.1.8}$$

where r is a constant. This subspace is intrinsically curved or *non-Euclidean* in the sense that no change of variable exists which will put the metric (11.1.8) into the form

$$ds^2 = dx_1{}^2 + dx_2{}^2 \tag{11.1.9}$$

The geodesics turn out to be great circles.

11.2. THE EULER-LAGRANGE EQUATION

In this section we consider the general problem of choosing the function $y(x)$ such that

$$I = \int_1^2 f(x, y, y') \, dx \tag{11.2.1}$$

has an extremum, i.e,, is either maximum or minimum. The function $f(x, y, y')$ is given; $y(x)$ is required to pass through the points 1 and 2. The brachistochrone discussed in Section 11.1 is an example.

A basic lemma which will be required is: If $G(x)$ is piecewise continuous in $x_1 \leq x \leq x_2$ and

$$\int_{x_1}^{x_2} G(x)\eta(x) \, dx = 0 \tag{11.2.2}$$

for *all* choices of $\eta(x)$, then $G(x) = 0$ identically in $x_1 \leq x \leq x_2$. Figure 11.2.1 shows two choices for $G(x)$. In each case, it is possible to choose $\eta(x)$ so that the integrand of (11.2.2) is positive definite, meaning that the integral is not zero. Thus the functions pictured in Fig. 11.2.1 are eliminated as possibilities for $G(x)$. It is evident that all choices for $G(x)$ can be eliminated by this argument except $G(x) = 0$.

Suppose that $y(x)$ is the actual extremizing arc for the integral (11.2.1). All arcs in the neighborhood of $y(x)$, so-called *comparison functions*, can be expressed as

$$Y(x) = y(x) + \epsilon\eta(x) \tag{11.2.3}$$

where ϵ is a parameter and $\eta(x)$ is arbitrary except for the requirement

$$\eta(x_1) = \eta(x_2) = 0 \tag{11.2.4}$$

This condition insures that all of the possible comparison functions pass through the same end-points 1 and 2. In other words, out of all the arcs

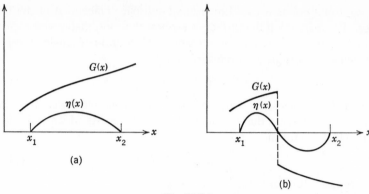

Fig. 11.2.1

(11.2.3) which pass through definite end-points denoted 1 and 2, we seek that arc which extremizes (11.2.1). For an arbitrary arc (11.2.3), the integral (11.2.1) is

$$I(\epsilon) = \int_1^2 f(x, Y, Y') \, dx \qquad (11.2.5)$$

A necessary condition that $y(x)$ be an extremizing function is that

$$\frac{dI(\epsilon)}{d\epsilon} \bigg|_{\epsilon=0} = 0 \qquad (11.2.6)$$

By differentiating (11.2.5) with respect to ϵ,

$$\frac{dI}{d\epsilon} = \int_1^2 \left(\frac{\partial f}{\partial Y} \eta(x) + \frac{\partial f}{\partial Y'} \eta'(x) \right) dx \qquad (11.2.7)$$

Using (11.2.6), and noting that $Y(x) = y(x)$ and $Y'(x) = y'(x)$ when $\epsilon = 0$, results in

$$0 = \int_1^2 \left(\frac{\partial f}{\partial y} \eta(x) + \frac{\partial f}{\partial y'} \eta'(x) \right) dx \qquad (11.2.8)$$

Integrating by parts and using (11.2.4) yields

$$0 = \int_1^2 \left(\frac{\partial f}{\partial y} - \frac{d}{dx} \frac{\partial f}{\partial y'} \right) \eta(x) \, dx \qquad (11.2.9)$$

Since $\eta(x)$ is arbitrary except for (11.2.4) the basic lemma (11.2.2) requires

$$\frac{\partial f}{\partial y} - \frac{d}{dx} \frac{\partial f}{\partial y'} = 0 \qquad (11.2.10)$$

which is known as the *Euler-Lagrange* equation. The resemblance of

(11.2.10) to Lagrange's equations already encountered in mechanics is obvious. The given function $f(x, y, y')$ is the "Lagrangian" and (11.2.10) provides a differential equation which must be obeyed by the extremizing function $y(x)$. If f has no explicit dependence on x, i.e., if $\partial f/\partial x = 0$, then the "Hamiltonian"

$$h = \frac{\partial f}{\partial y'} y' - f \tag{11.2.11}$$

is a constant and provides a first integral of (11.2.10).

11.3. THE BRACHISTOCHRONE PROBLEM

We will apply the results of Section 11.2 to the completion of the brachistochrone problem begun in Section 11.1. From (11.1.4),

$$f(x, y, y') = \sqrt{\frac{1 + (y')^2}{u_1{}^2 + 2gy}} \tag{11.3.1}$$

where, for simplicity, y_1 is taken as zero. Since (11.3.1) has no explicit dependence on x, (11.2.11) with $h = $ const. applies:

$$h = -\frac{1}{\sqrt{1 + (y')^2}\sqrt{u_1{}^2 + 2gy}} \tag{11.3.2}$$

It is convenient to introduce a parameter ϕ defined by

$$y' = -\tan \frac{\phi}{2} \tag{11.3.3}$$

then

$$\sqrt{1 + (y')^2} = \frac{1}{\cos (\phi/2)} \tag{11.3.4}$$

and, by means of (11.3.2),

$$y = \frac{1}{2gh^2} \cos^2 \frac{\phi}{2} - \frac{u_1{}^2}{2g} \tag{11.3.5}$$

$$dy = -\frac{1}{2gh^2} \cos \frac{\phi}{2} \sin \frac{\phi}{2} d\phi \tag{11.3.6}$$

Combining (11.3.6) and (11.3.3) results in

$$dx = \frac{1}{2gh^2} \cos^2 \frac{\phi}{2} d\phi = \frac{1}{4gh^2} (1 + \cos \phi) d\phi$$

$$x = \frac{1}{4gh^2} (\phi + \sin \phi) + a \tag{11.3.7}$$

where a is a constant of integration. If we set $b = 1/(4gh^2)$, (11.3.5) and (11.3.7) can be written

$$y = b(1 + \cos \phi) - \frac{u_1^2}{2g} \qquad (11.3.8)$$

$$x = b(\phi + \sin \phi) + a \qquad (11.3.9)$$

which gives the solution of the brachistochrone in parametric form.

In order to interpret (11.3.8) and (11.3.9), consider the initial conditions $u_1 = 0$ and $a = 0$ which means that the particle starts from rest at $y = 0$, $x = -b\pi$, $\phi = -\pi$.

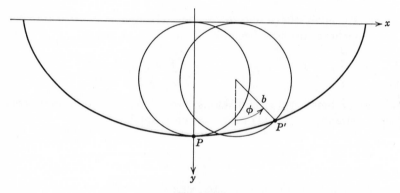

Fig. 11.3.1

Consider, as in Fig. 11.3.1, a circle of radius b which rolls without slipping on the x-axis. A point P on the rim of the circle moves from the y-axis, coordinates $(0, 2b)$, to P' when the circle has rolled through an angle ϕ. From the geometry, the coordinates of P' are

$$x = b\phi + b \sin \phi \qquad y = b + b \cos \phi \qquad (11.3.10)$$

These are the parametric equations of a cycloid and are identical to (11.3.8) and (11.3.9) with $u_1 = 0$ and $a = 0$.

It is true that the cycloidal arc is a necessary condition for the time of travel of a particle between two points to be an extremum. But is this extremum a maximum or a minimum? Could it even be that we have found an inflection point? It is fairly clear in the case of the brachistochrone that the cycloidal arc is a path of least time since a path connecting the two points can always be constructed over which the time of travel is arbitrarily long. In other problems, the answer is not quite so clear and a more detailed analysis is required. A rigorous account of several problems is found in Bliss.

11.4. VARIATION OF END POINTS

Suppose that the brachistochrone problem is modified by requiring that the path of least time be found between a fixed point and a given curve $F(x, y) = 0$ as illustrated in Fig. 11.4.1. In other words, of all the curves which can be drawn from the fixed point 1 to the given curve $F(x, y,) = 0$, which one is the path of least time for a particle? As usual,

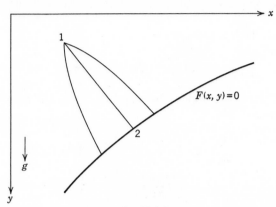

Fig. 11.4.1

the minimizing arc is denoted $y(x)$ and all other possible comparison arcs are represented by

$$Y(x) = y(x) + \epsilon\eta(x) \tag{11.4.1}$$

where $\eta(x)$ is arbitrary and is now not required to be zero at the second end point. Let X_2, Y_2 refer to the coordinates of the intersection of a comparison arc and the given curve $F(x, y) = 0$. For a *given* $\eta(x)$, X_2 and Y_2 depend *only* on the parameter ϵ, i.e., by allowing ϵ to vary, the point at which $Y(x)$ intersects $F(x, y) = 0$ changes. By differentiation of $F(X_2, Y_2) = 0$ with respect to ϵ,

$$0 = \frac{\partial F}{\partial X_2}\frac{dX_2}{d\epsilon} + \frac{\partial F}{\partial Y_2}\frac{dY_2}{d\epsilon} \tag{11.4.2}$$

At the point of intersection,

$$Y_2(\epsilon) = y(X_2) + \epsilon\eta(X_2) \tag{11.4.3}$$

Hence,

$$\frac{dY_2}{d\epsilon} = y'(X_2)\frac{dX_2}{d\epsilon} + \eta(X_2) + \epsilon\eta'(X_2)\frac{dX_2}{d\epsilon} \tag{11.4.4}$$

By combining (11.4.2) and (11.4.4) and setting $\epsilon = 0$ we get, at the point

492   *The Calculus of Variations*

where the minimizing arc $y(x)$ intersects $F(x, y) = 0$,

$$\frac{dX_2}{d\epsilon}\bigg|_{\epsilon=0} = -\frac{\eta(x_2)\dfrac{\partial F}{\partial y_2}}{\dfrac{\partial F}{\partial x_2} + \dfrac{\partial F}{\partial y_2}y'(x_2)} \tag{11.4.5}$$

where $\partial F/\partial x_2$ and $\partial F/\partial y_2$ are the partial derivatives of F and $y'(x_2)$ is the slope of the minimizing arc all evaluated at the intersection of $y(x)$ with $F(x, y) = 0$.

The integral which is to be minimized is

$$I(\epsilon) = \int_1^{X_2(\epsilon)} f(x, Y, Y')\, dx \tag{11.4.6}$$

where, as indicated, the upper limit depends on ϵ. Differentiation with respect to ϵ gives

$$\begin{aligned}
\frac{dI}{d\epsilon} &= \frac{dX_2}{d\epsilon}f(x_2, Y_2, Y_2') + \int_1^2 \frac{d}{d\epsilon}f(x, Y, Y')\, dx \\
&= \frac{dX_2}{d\epsilon}f_2 + \int_1^2 \left(\frac{\partial f}{\partial Y}\eta + \frac{\partial f}{\partial Y'}\eta'\right) dx
\end{aligned} \tag{11.4.7}$$

Integrating the second term in the integral by parts and setting $\epsilon = 0$ yields

$$0 = \frac{dX_2}{d\epsilon}\bigg|_{\epsilon=0} f_2 + \frac{\partial f}{\partial y'}\bigg|_2 \eta_2 + \int_1^2 \left(\frac{\partial f}{\partial y} - \frac{d}{dx}\frac{\partial f}{\partial y'}\right)\eta\, dx \tag{11.4.8}$$

where use is made of the fact that $\eta_1 = 0$, i.e., that all comparison arcs must pass through the initial end point. Since $\eta(x)$ is an arbitrary function, (11.4.8) implies

$$\frac{\partial f}{\partial y} - \frac{d}{dx}\frac{\partial f}{\partial y'} = 0 \tag{11.4.9}$$

in other words, that the Euler-Lagrange equation still holds and

$$\frac{dX_2}{d\epsilon}\bigg|_{\epsilon=0} f_2 + \frac{\partial f}{\partial y'}\bigg|_2 \eta_2 = 0 \tag{11.4.10}$$

which is a subsiduary condition that solutions of the Euler-Lagrange equation must now obey. Combining (11.4.5) and (11.4.10) results in

$$\frac{\partial f}{\partial y'}\bigg|_2 \left(\frac{\partial F}{\partial x_2} + \frac{\partial F}{\partial y_2}y_2'\right) = f_2 \frac{\partial F}{\partial y_2} \tag{11.4.11}$$

which is known as the *end-point condition*.

For the brachistochrone problem, the minimizing arc is still a cycloid since the Euler-Lagrange equation is still obeyed. For the function f as given by (11.3.1), the end-point condition (11.4.11) reduces to

$$dy \frac{\partial F}{\partial x_2} - dx \frac{\partial F}{\partial y_2} = 0 \qquad (11.4.12)$$

which means simply that the minimizing cycloid makes a perpendicular intersection with $F(x, y) = 0$.

For further development of the problem of variation of end points, the reader is referred to Weinstock, Reference 1 at the end of this chapter.

11.5. ISOPERIMETRIC PROBLEMS

Suppose that a flexible chain is hung between two fixed supports as in Fig. 11.5.1 and it is desired to find its shape. Since the chain is hanging stationary in stable equilibrium, we can assume that its potential energy

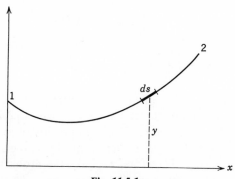

Fig. 11.5.1

is a minimum. If the mass density of the chain is μ per unit length, the potential energy of an element of length ds is

$$dV = \mu gy \, ds = \mu gy \sqrt{1 + (y')^2} \, dx \qquad (11.5.1)$$

The total potential energy is

$$V = \int_1^2 \mu gy \sqrt{1 + (y')^2} \, dx \qquad (11.5.2)$$

There is, however, a constraint. The chain has a fixed length

$$s = \int_1^2 ds = \int_1^2 \sqrt{1 + (y')^2} \, dx \qquad (11.5.3)$$

meaning that out of all the possible curves *of the same length which* connect the end points 1 and 2 we must pick the one which minimizes (11.5.2). It is now not possible to express the comparison arcs as

$$Y(x) = y(x) + \epsilon\eta(x)$$

because, for a given $\eta(x)$, the variation of the parameter ϵ would, in general, produce a change in the length of the chain. The comparison arcs, therefore, are made to depend on *two* parameters ϵ_1 and ϵ_2:

$$Y(x) = y(x) + \epsilon_1\eta_1(x) + \epsilon_2\eta_2(x) \tag{11.5.4}$$

the idea being that if ϵ_1 is varied, ϵ_2 can always be varied in such a way that the length of the chain is maintained. ϵ_1 and ϵ_1 are thus *not* independent variables. For given functions η_1 and η_2, a variation in ϵ_1 requires a unique and compensating variation in ϵ_2.

In general terms, the type of problem which we are considering can be stated as follows: Find the arc $y(x)$ which passes through fixed end-points and minimizes

$$I = \int_1^2 f(x, Y, Y') \, dx \tag{11.5.5}$$

subject to the constraint

$$J = \int_1^2 g(x, Y, Y') \, dx = \text{const.} \tag{11.5.6}$$

Problems of this type are called *isoperimetric problems* after the well-known problem of finding the closed curve of given perimeter for which the area enclosed is a maximum. The use of comparison arcs of the form (11.5.4) means that the integrals I and J are to be regarded as ordinary functions of ϵ_1 and ϵ_2. Therefore, we have the problem of extremizing $I(\epsilon_1, \epsilon_2)$, subject to the constraint $J(\epsilon_1, \epsilon_2) = \text{const.}$ Thus,

$$dI = \frac{\partial I}{\partial \epsilon_1} d\epsilon_1 + \frac{\partial I}{\partial \epsilon_2} d\epsilon_2 \tag{11.5.7}$$

$$dJ = \frac{\partial J}{\partial \epsilon_1} d\epsilon_1 + \frac{\partial J}{\partial \epsilon_2} d\epsilon_2 \tag{11.5.8}$$

If ϵ_1 and ϵ_2 were independent variables, (11.5.7) would imply $\partial I/\partial \epsilon_1 = 0$ and $\partial I/\partial \epsilon_2 = 0$. Since this is not the case, we multiply the differential equation of constraint (11.5.8) by a parameter λ, known as a Lagrange multiplier, and add the result to (11.5.7):

$$0 = \left(\frac{\partial I}{\partial \epsilon_1} + \lambda \frac{\partial J}{\partial \epsilon_1}\right) d\epsilon_1 + \left(\frac{\partial I}{\partial \epsilon_2} + \lambda \frac{\partial J}{\partial \epsilon_2}\right) d\epsilon_2 \tag{11.5.9}$$

We require that λ be such that

$$\frac{\partial I}{\partial \epsilon_1} + \lambda \frac{\partial J}{\partial \epsilon_1} = 0 \qquad (11.5.10)$$

It then follows from (11.5.9) that

$$\frac{\partial I}{\partial \epsilon_2} + \lambda \frac{\partial J}{\partial \epsilon_2} = 0 \qquad (11.5.11)$$

Writing $I^* = I + \lambda J, f^* = f + \lambda g$, our problem can be restated as

$$I^* = \int_1^2 f^*(x, Y, Y') \, dx \qquad (11.5.12)$$

$$\frac{\partial I^*}{\partial \epsilon_1} = 0 \qquad \frac{\partial I^*}{\partial \epsilon_2} = 0 \qquad (11.5.13)$$

It is found from (11.5.13) and the fact that η_1 and η_2 vanish at the end points that

$$0 = \int_1^2 \left(\frac{\partial f^*}{\partial y} - \frac{d}{dx} \frac{\partial f^*}{\partial y'} \right) \eta_1 \, dx, \qquad 0 = \int_1^2 \left(\frac{\partial f^*}{\partial y} - \frac{d}{dx} \frac{\partial f^*}{\partial y'} \right) \eta_2 \, dx \qquad (11.5.14)$$

Since η_1 and η_2 are both arbitrary functions, f^* must obey the Euler-Lagrange differential equation:

$$\frac{\partial f^*}{\partial y} - \frac{d}{dx} \frac{\partial f^*}{\partial y'} = 0 \qquad (11.5.15)$$

By redefining the Lagrange multiplier as $\lambda \mu g$, the function f^* for the hanging chain problem can be expressed by

$$f^* = \text{const.} \, (y + \lambda)\sqrt{1 + (y')^2} \qquad (11.5.16)$$

provided, of course, that the linear density μ is a constant. Since f^* has no explicit dependence on x, we can use

$$y' \frac{\partial f^*}{\partial y'} - f^* = \text{const.} \qquad (11.5.17)$$

in place of the Euler-Lagrange equation. This gives the differential equation

$$-a\sqrt{1 + (y')^2} + y + \lambda = 0 \qquad (11.5.18)$$

where a is a constant. It is readily verified that the solution

$$y = -\lambda + a \cosh \left(\frac{x - x_0}{a} \right) \qquad (11.5.19)$$

where x_0 is a constant of integration. Equation (11.5.19) is the equation of a *catenary*.

The three undetermined constants λ, a, and x_0 are determined as follows: First, the length of the chain is

$$s = \int_1^2 \sqrt{1 + (y')^2}\, dx = \int_1^2 \cosh\left(\frac{x - x_0}{a}\right) dx$$

$$s = a \sinh\left(\frac{x_2 - x_0}{a}\right) - a \sinh\left(\frac{x_1 - x_0}{a}\right)$$

(11.5.20)

Since s is presumed known, (11.5.20) gives one condition on the constants. The fact that the chain passes through known end points gives two additional equations that are then sufficient to determine all constants.

11.6. THE SYMBOL OF VARIATION

It will be the purpose of this section to introduce a shorthand notation which is in widespread use. Consider the problem of extremizing the integral (11.2.5). Take Eq. (11.2.7), multiply by $d\epsilon$, and write $y(x)$ in place of $Y(x)$:

$$\frac{dI}{d\epsilon}\, d\epsilon = \int_1^2 \left(\frac{\partial f}{\partial y}\, \eta\, d\epsilon + \frac{\partial f}{\partial y'}\, \eta'\, d\epsilon\right) dx$$

(11.6.1)

One defines the *variation* of the integral by

$$\delta I = \frac{dI}{d\epsilon}\, d\epsilon$$

(11.6.2)

and the *variation* of the arc $y(x)$ and its first derivative by

$$\delta y = \eta d\epsilon \qquad \delta y' = \eta'\, d\epsilon$$

(11.6.3)

Note that as a consequence of (11.6.3),

$$\delta y' = \frac{d}{dx}\, \delta y$$

(11.6.4)

We then write in place of (11.6.1)

$$\delta I = \int_1^2 \left(\frac{\partial f}{\partial y}\, \delta y + \frac{\partial f}{\partial y'}\, \delta y'\right) dx$$

(11.6.5)

and speaks of δI as being the variation in the integral produced by the variation δy of an arc connecting the fixed end points 1 and 2. Figure 11.6.1 shows two possible arcs and the variation (assumed infinitesimal) at a point with coordinate x. Since the variation vanishes at the end points,

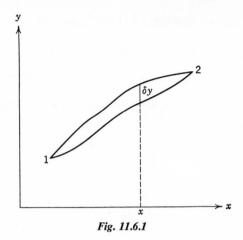

Fig. 11.6.1

the integration of the second term of (11.6.5) by parts results in

$$\delta I = \int_1^2 \left(\frac{\partial f}{\partial y} - \frac{d}{dx} \frac{\partial f}{\partial y'} \right) \delta y \, dx \tag{11.6.6}$$

Since the variation δy is arbitrary, the vanishing of δI requires that the Euler-Lagrange equation be satisfied.

The variational notation has conveniences, but if some question should arise as to the validity of a procedure, it may be best to fall back on the more cumbersome, but precise, representation of a comparison function by means of (11.2.3).

11.7. HAMILTON'S PRINCIPLE

Suppose that a mechanical system of f degrees of freedom is given and that a Lagrangian for the system is known. Then Hamilton's principle is:

The motion of the system from time t_1 to time t_2 is such that

$$I = \int_1^2 L(q_k, \dot{q}_k, t) \, dt \tag{11.7.1}$$

is an extremum, i.e., out of all the possible paths by which a point in the configuration space of the system could travel from its position at time t_1 to its position at time t_2, it will actually travel along that path for which (11.7.1) is an extremum.

The integrand now depends on many functions q_k of the parameter t instead of just one function of the parameter x as in previous examples.

The variation of the integral (11.7.1) results in

$$\delta I = \int_1^2 \left(\frac{\partial L}{\partial q_k} \delta q_k + \frac{\partial L}{\partial \dot{q}_k} \delta \dot{q}_k \right) dt \tag{11.7.2}$$

where, of course, the repeated subscript implies a sum from 1 to f. Integration by parts and the use of the fact that the variations vanish at the fixed end points results in

$$\delta I = \int_1^2 \left(\frac{\partial L}{\partial q_k} - \frac{d}{dt} \frac{\partial L}{\partial \dot{q}_k} \right) \delta q_k \, dt \tag{11.7.3}$$

The coordinates q_k are *independent* meaning that the variation δq_1, $\delta q_2, \ldots, \delta q_f$ can be made *completely independently* of one-another. By an obvious extension of the basic lemma of section 2, $\delta I = 0$ requires

$$\frac{\partial L}{\partial q_k} - \frac{d}{dt} \frac{\partial L}{\partial \dot{q}_k} = 0 \tag{11.7.4}$$

that is, Lagrange's equations in their usual form.

Nothing new has been learned here but, as will be seen in Section 11.9, the calculus of variations provides a powerful technique for handling the problem of constraints.

11.8. GEODESICS

Before proceeding with our application of variational techniques to problems in mechanics, we briefly take up the problem of finding the geodesics in an arbitrary space already mentioned in Section 11.1. In particular, we seek the functions $q_k(\lambda)$ which extremize (11.1.7). This is really the same problem as discussed in Section 11.7, except λ is not the time. Thus the geodesics are to be found from

$$\frac{d}{d\lambda} \left[\frac{\partial}{\partial q_k'} (\sqrt{g_{ij} q_i' q_j'}) \right] - \frac{\partial}{\partial q_k} (\sqrt{g_{ij} q_i' q_j'}) = 0 \tag{11.8.1}$$

Performing the indicated differentiations with respect to q_k' and q_k results in

$$\frac{d}{d\lambda} \left[\frac{1}{\left(\frac{ds}{d\lambda}\right)} g_{ki} \left(\frac{dq_i}{d\lambda}\right) \right] - \frac{1}{2} \frac{1}{\left(\frac{ds}{d\lambda}\right)} \frac{\partial g_{ij}}{\partial q_k} \frac{dq_i}{d\lambda} \frac{dq_j}{d\lambda} = 0 \tag{11.8.2}$$

where, after the differentiations are performed, we have put

$$\frac{ds}{d\lambda} = \sqrt{g_{ij} q_i' q_j'} \tag{11.8.3}$$

It is very often conveneint to use the arc-length s itself as the parameter rather than the arbitrary λ. Thus, (11.8.2) can be written

$$\frac{d}{ds}\left[g_{ki}\frac{dq_i}{ds}\right]\frac{ds}{d\lambda} - \frac{1}{2}\frac{\partial g_{ij}}{\partial q_k}\frac{dq_i}{ds}\frac{dq_j}{d\lambda} = 0$$

or, finally,

$$\frac{d}{ds}\left[g_{ki}\frac{dq_i}{ds}\right] - \frac{1}{2}\frac{\partial g_{ij}}{\partial q_k}\frac{dq_i}{ds}\frac{dq_j}{ds} = 0 \qquad (11.8.4)$$

The reader should compare (11.8.4) with (2.20.23). If the acceleration is set equal to zero in (2.20.23), it becomes identical in form to (11.8.4). Thus a particle on which no forces act follows a geodesic. As we know, this is just a simple straight line in ordinary three-dimensional space.

In the theory of general relativity, we deal with a four-dimensional curved space-time continuum in which the properties of the gravitational field are incorporated into the metric tensor. Point masses, on which no external forces act of a nature other than gravitational, are then assumed to follow geodesics. The so-called proper time of a particle, which is essentially the time which a clock carried by the particle records, becomes the arc length in this space. The equation of motion of the particle is then precisely identical to (11.8.4)! Of course, the subscripts take on the values 1 through 4.

It is worth noting that if a scalar function is defined by

$$\psi = \tfrac{1}{2}g_{ij}q_i'q_j' \qquad q_i' = \frac{dq_i}{ds} \qquad (11.8.5)$$

then (11.8.4) can be expressed

$$\frac{d}{ds}\frac{\partial\psi}{\partial q_k'} - \frac{\partial\psi}{\partial q_k} = 0 \qquad (11.8.6)$$

The scalar ψ is, of course, analogous to the kinetic energy of particle mechanics.

11.9. CONSTRAINTS IN MECHANICAL SYSTEMS

Suppose that a mechanical system is described by n generalized coordinates and that there are k equations of constraint of the form

$$f_\alpha(q_1,\ldots,q_n,t) = 0 \qquad (11.9.1)$$

so that the actual number of degrees of freedom of the system is

$$f = n - k \qquad (11.9.2)$$

So far in our treatment of mechanics problems, we have used the equations of constraint to eliminate some of the coordinates from the Lagrangian so that it is expressed in terms of precisely f-coordinates before the equations of motion are derived. This works quite smoothly in most simple problems, especially if a little care has been used in the choice of a coordinate system. If the system has a large number of coordinates, the problem of solving the system (11.9.1) for k coordinates in terms of the remaining ones may in itself be quite formidable. We develop here a scheme whereby this becomes unnecessary.

The systems is described by a Lagrangian so that the motion is to be obtained by extremizing

$$I = \int_1^2 L(q_1, \ldots, q_n, \dot{q}_1, \ldots, \dot{q}_n, t)\, dt \qquad (11.9.3)$$

subject to the constraints (11.9.1). The variation of (11.9.3) leads to

$$\delta I = 0 = \int_1^2 \left(\frac{\partial L}{\partial q_i} - \frac{d}{dt} \frac{\partial L}{\partial \dot{q}_i} \right) \delta q_i\, dt \qquad (11.9.4)$$

but it is *not* possible to conclude that the coefficient of each variation δq_i in the integrand vanishes, as happened in passing from Eq. (11.7.3) to (11.7.4). This is because the variations $\delta q_i, \ldots, \delta q_n$ are *not* independent of one another but rather are connected by differential equations of constraint which are obtained by varying (11.9.1):

$$\frac{\partial f_\alpha}{\partial q_i} \delta q_i = 0 \qquad (11.9.5)$$

Let $\lambda_\alpha(t)$ denote k-undetermined functions of t. Then

$$\lambda_\alpha \frac{\partial f_\alpha}{\partial q_i} \delta q_i = 0 \qquad (11.9.6)$$

where the sum over α runs from 1 to k and the sum over i runs from 1 to n. By integrating (11.9.6) over the path of the system,

$$0 = \int_1^2 \lambda_\alpha \frac{\partial f_\alpha}{\partial q_i} \delta q_i\, dt \qquad (11.9.7)$$

By adding (11.9.4) and (11.9.7),

$$0 = \int_1^2 \left[\frac{\partial L}{\partial q_i} - \frac{d}{dt} \frac{\partial L}{\partial \dot{q}_i} + \lambda_\alpha \frac{\partial f_\alpha}{\partial \dot{q}_i} \right] \delta q_i\, dt \qquad (11.9.8)$$

Let the k-functions $\lambda_\alpha(t)$ be chosen such that the coefficients of the variations $\delta q_1, \ldots, \delta q_k$ vanish identically in the integrand of (11.9.8):

$$\frac{\partial L}{\partial q_i} - \frac{d}{dt} \frac{\partial L}{\partial \dot{q}_i} + \lambda_\alpha \frac{\partial f_\alpha}{\partial q_i} = 0 \qquad (11.9.9)$$

Thus (11.9.8) becomes

$$0 = \int_1^2 \sum_{i=k+1}^{n} \left[\frac{\partial L}{\partial q_i} - \frac{d}{dt} \frac{\partial L}{\partial \dot{q}_i} + \lambda_\alpha \frac{\partial f_\alpha}{\partial q_i} \right] \delta q_i \, dt \qquad (11.9.10)$$

The remaining f-variations $\delta q_{k+1}, \ldots, \delta q_n$ can be performed independently of one another! Thus the coefficients of these variations in (11.9.10) must also vanish identically! Hence, we have for *all* the coordinates $q_1, \ldots, q_n$:

$$\frac{d}{dt} \frac{\partial L}{\partial \dot{q}_i} - \frac{\partial L}{\partial q_i} = \lambda_\alpha a_{\alpha i} \qquad a_{\alpha i} = \frac{\partial f_\alpha}{\partial q_i} \qquad (11.9.11)$$

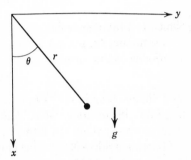

As we know from our previous work, the Lagrange multipliers $\lambda_\alpha(t)$ are related to the forces of constraint acting on the system. The equations of motion (11.9.11) plus the original equations of constraint supply altogether $n + k$ equations for the determination of the n-functions $q_i(t)$ and the k-functions $\lambda_\alpha(t)$.

As a simple example, suppose that a particle moves in a uniform gravitational field in a vertical plane. In terms of plane polar coordinates, Fig. 11.9.1, the potential energy is

Fig. 11.9.1

$$V = -mgx = -mgr \cos \theta.$$

The Lagrangian, therefore, is

$$L = \tfrac{1}{2}m(\dot{r}^2 + r^2\dot{\theta}^2) + mgr \cos \theta \qquad (11.9.12)$$

Let $q_1 = r$, $q_2 = \theta$. The motion of a simple plane pendulum can be obtained by introducing the constraint

$$f_1 = r - s = 0 \qquad (11.9.13)$$

where s is a constant. The equations of motion are

$$\frac{d}{dt} \frac{\partial L}{\partial \dot{r}} - \frac{\partial L}{\partial r} = \lambda_1 \frac{\partial f_1}{\partial r} \qquad (11.9.14)$$

$$\frac{d}{dt} \frac{\partial L}{\partial \dot{\theta}} - \frac{\partial L}{\partial \theta} = \lambda_1 \frac{\partial f_1}{\partial \theta} \qquad (11.9.15)$$

Noting that

$$\frac{\partial f_1}{\partial r} = 1 \qquad \frac{\partial f_1}{\partial \theta} = 0 \qquad (11.9.16)$$

equations (11.9.14) and (11.9.15) yield

$$m\ddot{r} - mr\dot{\theta}^2 - mg\cos\theta = \lambda_1 \qquad (11.9.17)$$

$$mr^2\ddot{\theta} + 2mr\dot{r}\dot{\theta} + mgr\sin\theta = 0 \qquad (11.9.18)$$

which, along with (11.9.13), provide three equations for the determination of the three functions $r(t)$, $\theta(t)$, and $\lambda_1(t)$. Thus substitution of the constraint (11.9.13) into (11.9.17) and (11.9.18) results in

$$-ms\dot{\theta}^2 - mg\cos\theta = \lambda_1$$

$$\ddot{\theta} + \frac{g}{s}\sin\theta = 0 \qquad (11.9.19)$$

which is a familiar result.

It sometimes happens that constraints are specified in terms of differential equations of the form

$$a_{\alpha i}\delta q_i = 0 \qquad (11.9.20)$$

and that no functions f_α exist such that $a_{\alpha i} = \partial f_\alpha/\partial q_i$. Such constraints are said to be *nonintegrable*. Thus it is actually not possible to express the Lagrangian explicitly in terms of f-independent coordinates. The method of Lagrange multipliers as developed in this section is, however, applicable to this type of nonholonomic constraint.

PROBLEMS

11.1.1   Suppose that a ball bearing *rolls* down a track so that the center of mass passes through two given points 1 and 2. How is the integral (11.1.4) for the time to be modified? Assume for simplicity that the initial position and velocity of the center of mass are zero.

11.2.2   Consider the class of curves $Y(x)$ which pass through the points 1 and 2 in the xy-plane, each having the same slope at these points. Out of this class of admissible arcs it is desired to pick the one which extremizes

$$I = \int_1^2 f(x, y, y', y'')\,dx$$

Find the differential equation obeyed by f.

11.3.3   Suppose that a particle is constrained to move without friction on the cycloidal arc (11.3.10) and that there is a uniform gravitational field in the y-direction. Find the equation of motion of the particle and its solution. Show that the motion is strictly isochronous, i.e., that the period is independent of the amplitude of the motion. This is the *tautochrone problem* (Greek: *tauto*, the same; *chronous*, time). *Hint:* Let ϕ be a generalized coordinate. In the differential equation of motion make the change of dependent variable $\theta = 2\sin\phi/2$.

11.3.4   Evaluate the parameters a and b to slide-rule accuracy in Eqs. (11.3.8) and (11.3.9) for a particle which starts from rest at the origin and passes through the points $x = 20\pi$ cm, $y = 20$ cm.

11.4.5   What is the general rule for differentiating

$$I(\epsilon) = \int_{x_1(\epsilon)}^{x_2(\epsilon)} f(x, \epsilon)\, dx$$

with respect to ϵ? See any test on advanced calculus.

11.4.6   A particle is to start from rest at the origin. Find the constants a and b in (11.3.8) and (11.3.9) which will give the path of least time to a vertical line at $x = 50$ cm.

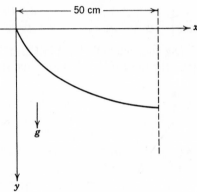

11.7.7   Suppose that

$$I = \int_1^2 f(x, y, \dot{x}, \dot{y})\, dt$$

is to be extremized, subject to

$$J = \int_1^2 g(x, y, \dot{x}, \dot{y})\, dt = \text{const.}$$

Show that $f^* = f + \lambda g$ obeys the Euler-Lagrange equations

$$\frac{d}{dt}\frac{\partial f^*}{\partial \dot{x}} - \frac{\partial f}{\partial x} = 0 \qquad \frac{d}{dt}\frac{\partial f^*}{\partial \dot{y}} - \frac{\partial f^*}{\partial y} = 0$$

Assume fixed end points.

11.7.8   Show that the area of a closed curve in the xy-plane can be expressed

$$\sigma = \tfrac{1}{2} \oint (x\, dy - y\, dx) = \tfrac{1}{2} \oint (x\dot{y} - y\dot{x})\, dt$$

where the integral is taken counterclockwise around the curve and $x(t)$, $y(t)$ are parametric representations of the curve.

11.7.9   What shape should a closed curve of given length be in order to enclose the greatest possible area?

11.8.10   Referring to equations (11.8.5) and (11.8.6), find the function ψ expressed in terms of the polar angles θ and ϕ which will give the geodesics on a sphere. What function is analogous to the Hamiltonian in this problem? Is it a constant? Show that the differential equation for the geodesics is

$$\frac{d\theta}{\sin\theta\sqrt{a\sin^2\theta - 1}} = d\phi$$

where a is a constant. Complete the solution, showing that the geodesics are great circles. *Hint:* the substitution $u = \cot\theta$ is useful.

11.8.11   Show that if a new parameter $\lambda = \lambda(s)$ is introduced into (11.8.4) then

$$\frac{d}{d\lambda}\left[g_{ij}\frac{dq_j}{d\lambda}\right] - \frac{1}{2}\frac{\partial g_{kl}}{\partial q_i}\frac{dq_k}{d\lambda}\frac{dq_l}{d\lambda} = -g_{ij}\frac{dq_j}{d\lambda}\frac{d^2\lambda}{ds^2}\left(\frac{ds}{d\lambda}\right)^2$$

11.9.12   How is the Lagrange multiplier $\lambda_1(t)$ in Eq. (11.9.19) related to the tension in the string (i.e., the force of constraint) of the simple pendulum?

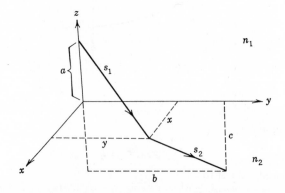

11.9.13   *Fermat's Principle:*   Let the xy-plane of the figure separate two optical media of indices of refraction n_1 and n_2. A light ray starts from $(0, 0, a)$, strikes the interface at $(0, x, y)$ and continues on to $(0, b, -c)$ as shown. Derive the laws of refraction by assuming that the actual path followed by the light ray between the fixed end points extremizes the time. (Recall that the index of refraction of a medium is defined by $n = c/u$ where c is the velocity of light in vacuum and u is its velocity in the medium.)

REFERENCES

1. Weinstock, Robert, *Calculus of Variations*, McGraw-Hill Book Co., New York 1952.
2. Bliss, G. A., *Calculus of Variations*, Open Court Publishing Company, LaSalle Illinois, 1944.

12

Vibrating Systems

In Chapter 4, both linear and nonlinear vibrations of a single particle moving in one dimension were studied. In Section 8.10, the theoretical foundation was laid for the treatment of linear vibrations in a system of many degrees of freedom. In this chapter, several problems of gradually increasing complexity are treated, culminating with a discussion of one-dimensional continuous systems.

12.1. THE DIATOMIC MOLECULE

Frequently used as a semiempirical representation of the overall interaction between atoms in a molecule is the Morse potential

$$V(r) = V_0[1 - e^{-a(r-r_0)}]^2 \tag{12.1.1}$$

where V_0 and a are constants, r is the separation between atomic nuclei, and r_0 is the equilibrium separation. A typical Morse potential is sketched in Fig. 12.1.1.

The Lagrangian which gives the rotational and vibrational motion of the two atoms in a diatomic molecule is

$$L = \tfrac{1}{2}m(\dot{r}^2 + r^2\dot{\theta}^2) - V(r) \tag{12.1.2}$$

Where

$$m = \frac{m_1 m_2}{m_1 + m_2} \tag{12.1.3}$$

is the reduced mass. The atoms are being considered as units, no attempt being made to account for the motions of the individual electrons. To find the *linear* approximation to the motion of a system of this kind, the potential function is expanded about the equilibrium values of the coordinates as

$$V(r) \cong \tfrac{1}{2}k(r - r_0)^2 + \cdots \qquad (12.1.4)$$

and the components of the metric tensor are given their values at the equilibrium configuration of the system. The approximate Lagrangian is then

$$L = \tfrac{1}{2}m(\dot{r}^2 + r_0{}^2\dot{\theta}^2) - \tfrac{1}{2}k(r - r_0)^2 \qquad (12.1.5)$$

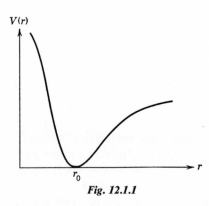

Fig. 12.1.1

The equations of motion are

$$m\ddot{r} + k(r - r_0) = 0 \qquad \ddot{\theta} = 0 \qquad (12.1.6)$$

which are of the general form of equations (8.10.21). The general motion is simply a superposition of the rotation of the molecule, considered as a rigid rotator, and the vibration of the two nuclei considered as a simple harmonic oscillator. This superposition is typical of linear systems. In a higher approximation, a coupling between the vibrational and rotational motion, due to the stretching of the molecule from centrifugal force, would appear.

In Chapter 9, we learned how to do a much better job on a simple system such as this one, but the treatment given here parallels closely the wave-mechanical treatment. It is left as a problem to show that the Bohr-Sommerfeld quantization rule gives for the energy levels

$$\mathscr{E}_{n_r n_\theta} = n_r \frac{h}{2\pi}\sqrt{\frac{k}{m}} + \frac{n_\theta{}^2 h^2}{8\pi^2 m r_0{}^2} \qquad (12.1.7)$$

The Schrödinger wave equation applied to the same problem gives a slightly different (and correct) answer:

$$\mathcal{E}_{n_r n_\theta} = (n_r + \tfrac{1}{2}) \frac{h}{2\pi} \sqrt{\frac{k}{m}} + \frac{n_\theta(n_\theta + 1)h^2}{8\pi^2 m r_0^2} \tag{12.1.8}$$

If the wave-mechanical treatment is done more accurately, coupling terms between rotation and vibration appear just as in the classical theory. The interested reader should consult Pauling and Wilson, Reference 3 at the end of this chapter.

12.2. THE CARBON DIOXIDE MOLECULE

The carbon dioxide molecule consists of two oxygen molecules of mass m_1 joined to a carbon atom of mass m_2. The atoms are arranged along a straight line as in Fig. 12.2.1. The interaction of the oxygen molecules

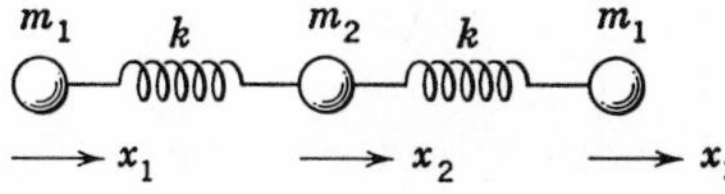

Fig. 12.2.1

with one another will be neglected. Thus we consider only *nearest neighbor* interaction. In order to find the vibrational motion along the line of the molecule, we introduce three coordinates x_1, x_2, and x_3, each measured from the equilibrium position of a particle—in other words $x_1 = x_2 = 0$ means that m_1 and m_2 are a distance r_0 apart. Following the procedure of Section 12.1, the interaction is approximated by a harmonic oscillator potential which gives the Lagrangian

$$L = \tfrac{1}{2}m_1(\dot{x}_1^2 + \dot{x}_3^2) + \tfrac{1}{2}m_2\dot{x}_2^2 - \tfrac{1}{2}k(x_2 - x_1)^2 - \tfrac{1}{2}k(x_3 - x_2)^2 \tag{12.2.1}$$

The equations of motion are found to be

$$m_1\ddot{x}_1 - k(x_2 - x_1) = 0$$
$$m_2\ddot{x}_2 + k(x_2 - x_1) - k(x_3 - x_2) = 0 \tag{12.2.2}$$
$$m_1\ddot{x}_3 + k(x_3 - x_2) = 0$$

As we know from Section 8.10, the solutions can be expected to be of the form

$$x_i = A_i \cos(\omega t - \alpha) \tag{12.2.3}$$

which is substituted into (12.2.2) to yield

$$(-m_1\omega^2 + k)A_1 - kA_2 = 0$$
$$-kA_1 + (-m_2\omega^2 + 2k)A_2 - kA_3 = 0 \tag{12.2.4}$$
$$-kA_2 + (-m_1\omega^2 + k)A_3 = 0$$

The secular equation is

$$\begin{vmatrix} -m_1\omega^2 + k & -k & 0 \\ -k & -m_2\omega^2 + 2k & -k \\ 0 & -k & -m_1\omega^2 + k \end{vmatrix} = 0 \qquad (12.2.5)$$

The eigenvalues are readily found to be

$$\omega_1 = 0 \qquad \omega_2 = \sqrt{k/m_1} \qquad \omega_3 = \sqrt{(k/m_1) + (2k/m_2)} \qquad (12.2.6)$$

The eigenvectors corresponding to the three eigenvalues are

$$X_1 = a_1 \begin{pmatrix} 1 \\ 1 \\ 1 \end{pmatrix} \qquad X_2 = a_2 \begin{pmatrix} 1 \\ 0 \\ -1 \end{pmatrix} \qquad X_3 = a_3 \begin{pmatrix} 1 \\ -\dfrac{2m_1}{m_2} \\ 1 \end{pmatrix} \qquad (12.2.7)$$

where a_i are three undetermined constants. The appearance of the eigenvalue $\omega_1 = 0$ means that one of the possible solutions is nonperiodic. The vibrational modes contain two undetermined phase angles, giving a total of five undetermined constants where six are expected. Physically, we know that the molecule can undergo uniform translation, and this gives a clue as to the nature of the missing portion of the solution. The complete solution corresponding to the eigenvalue $\omega_1 = 0$ is

$$\begin{pmatrix} x_1 \\ x_2 \\ x_3 \end{pmatrix} = (a_1 + b_1 t) \begin{pmatrix} 1 \\ 1 \\ 1 \end{pmatrix} \qquad (12.2.8)$$

Hence, the general solution with six undetermined constants is

$$\begin{pmatrix} x_1 \\ x_2 \\ x_3 \end{pmatrix} = (a_1 + b_1 t) \begin{pmatrix} 1 \\ 1 \\ 1 \end{pmatrix} + a_2 \begin{pmatrix} 1 \\ 0 \\ -1 \end{pmatrix} \cos(\omega_2 t - \alpha_2)$$

$$(12.2.9)$$

$$+ a_3 \begin{pmatrix} 1 \\ -\dfrac{2m_1}{m_2} \\ 1 \end{pmatrix} \cos(\omega_3 t - \alpha_3)$$

and is seen to be a superposition of the uniform translation and the two vibrational modes.

12.3. A SIMPLE SYSTEM WITH DISSIPATION

Consider the mechanical system pictured in Fig. 8.9.2 and its electrical and acoustical analogues shown in Figs. 8.9.4 and 8.9.5. We consider here the modification in the motion which is brought about by the addition of viscous frictional forces in the case of the mechanical and acoustical systems, and resistors in series with the inductors in the case of the electrical system. If viscous frictional forces proportional to the velocities of the masses are included in the Lagrangian equations (8.9.13) they become

$$\frac{d}{dt}\left(\frac{\partial L}{\partial \dot{x}_1}\right) - \frac{\partial L}{\partial x_1} = -b_1\dot{x}_1 \qquad \frac{d}{dt}\left(\frac{\partial L}{\partial \dot{x}_2}\right) - \frac{\partial L}{\partial x_2} = -b_2\dot{x}_2 \qquad (12.3.1)$$

The resulting equations of motion are

$$\begin{aligned}
m_1\ddot{x}_1 + k_1x_1 - k_3(x_2 - x_1) + b_1\dot{x}_1 &= 0 \\
m_2\ddot{x}_2 + k_2x_2 + k_3(x_2 - x_1) + b_2\ddot{x}_2 &= 0
\end{aligned} \qquad (12.3.2)$$

For simplicity, only the special case $m_1 = m_2 = m$, $b_1 = b_2 = b$, $k_1 = k_2 = k$ will be solved in detail. Since the system is linear, the anticipated form of the solution is

$$x_1 = A_1e^{\alpha t} \qquad x_2 = A_2e^{\alpha t} \qquad (12.3.3)$$

The substitution of (12.3.3) into (12.3.2) yields

$$\begin{aligned}
(m\alpha^2 + b\alpha + k + k_3)A_1 - k_3A_2 &= 0 \\
-k_3A_1 + (m\alpha^2 + b\alpha + k + k_3)A_2 &= 0
\end{aligned} \qquad (12.3.4)$$

The secular equation, therefore, is

$$(m\alpha^2 + b\alpha + k + k_3)^2 = k_3^2 \qquad (12.3.5)$$

Its roots are found to be

$$\alpha_{1,2} = -\frac{b}{2m} \pm \sqrt{\left(\frac{b}{2m}\right)^2 - \frac{k}{m}} \qquad (12.3.6)$$

$$\alpha_{3,4} = -\frac{b}{2m} \pm \sqrt{\left(\frac{b}{2m}\right)^2 - \frac{k + 2k_3}{m}} \qquad (12.3.7)$$

The eigenvectors corresponding to the roots (12.3.6) and (12.3.7) follow from (12.3.4):

$$X_{1,2} = \text{const.} \begin{pmatrix} 1 \\ 1 \end{pmatrix} \qquad X_{3,4} = \text{const.} \begin{pmatrix} 1 \\ -1 \end{pmatrix} \qquad (12.3.8)$$

If, for instance, the damping is small so that

$$\alpha_{1,2} = -\frac{b}{2m} \pm i\omega_1 \qquad \omega_1 = \sqrt{\frac{k}{m} - \left(\frac{b}{2m}\right)^2}$$

$$\alpha_{3,4} = -\frac{b}{2m} \pm i\omega_2 \qquad \omega_2 = \sqrt{\frac{k + 2k_3}{m} - \left(\frac{b}{2m}\right)^2} \qquad (12.3.9)$$

Then the complete solution containing four arbitrary constants can be expressed

$$\begin{pmatrix} x_1 \\ x_2 \end{pmatrix} = A \begin{pmatrix} 1 \\ 1 \end{pmatrix} e^{-bt/2m} \cos(\omega_1 t - \phi_1) + B \begin{pmatrix} 1 \\ -1 \end{pmatrix} e^{-bt/2m} \cos(\omega_2 t - \phi_2)$$

$$(12.3.10)$$

12.4. A SIMPLE DRIVEN SYSTEM

A driven mechanical system and its a-c circuit analogue are shown in Fig. 12.4.1. This is essentially the same system that is treated in Section 12.3 except that a harmonic driving force

$$F = F_0 \cos \omega t \qquad (12.4.1)$$

has been applied to one of the masses. To simplify the analysis, the masses and the springs are taken as identical. The transient solution of this system is given by Eq. (12.3.10); we now look for the steady-state solution. In particular, we will find the expression for the average power dissipated as a function of the driving frequency. According to Eq. (4.8.25), the average power for the a-c circuit can be computed from

$$\bar{P} = \tfrac{1}{2} \operatorname{Re} \frac{VV^*}{z} = \tfrac{1}{2} \operatorname{Re} \frac{\phi_0^2}{z} \qquad (12.4.2)$$

where $V = \phi_0 e^{i\omega t}$ is the complex driving voltage, z is the overall complex impedance of the system, and Re means "real part of." Alternatively, the power dissipated by the mechanical system is

$$\bar{P} = \tfrac{1}{2} \operatorname{Re} \frac{F_0^2}{z} \qquad (12.4.3)$$

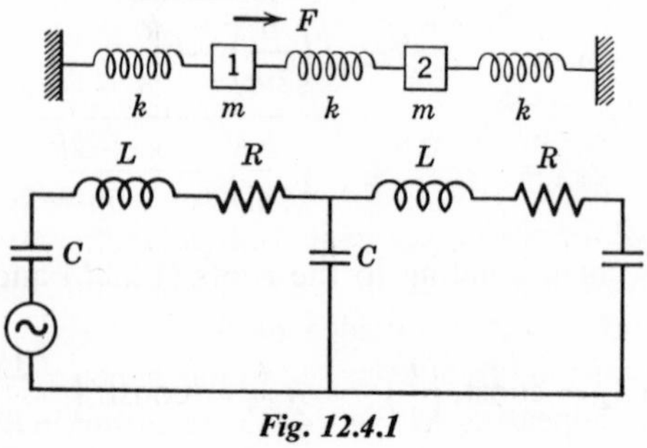

Fig. 12.4.1

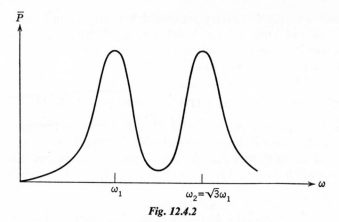

$\bar{P}$

$\omega_1 \qquad \omega_2 = \sqrt{3}\omega_1 \qquad\qquad\qquad \omega$

Fig. 12.4.2

where z is now the mechanical impedance. Essentially, then, our problem is to compute the real part of $1/z$. The impedance of the mechanical system is

$$z = b + i\omega m - \frac{ik}{\omega} + \frac{-(ik/\omega)(b + i\omega m - ik/\omega)}{b + i\omega m - 2ik/\omega} \qquad (12.4.4)$$

The reciprocal of z is found to be

$$\frac{1}{z} = \frac{b + i\omega m - 2ik/\omega}{(b + i\omega m - ik/\omega)(b + i\omega m - 3ik/\omega)}$$

$$= \frac{\frac{1}{2}}{b + i\omega m - ik/\omega} + \frac{\frac{1}{2}}{b + i\omega m - 3ik/\omega} \qquad (12.4.5)$$

$$= \frac{\frac{1}{2}[b - i(\omega m - k/\omega)]}{b^2 + (\omega m - k/\omega)^2} + \frac{\frac{1}{2}[b - i(\omega m - 3k/\omega)]}{b^2 + (\omega m - 3k/\omega)^2}$$

It is recognized that

$$\omega_1 = \sqrt{k/m} \qquad \omega_2 = \sqrt{3k/m} \qquad (12.4.6)$$

are the eigenfrequencies for the free oscillations of the system when no damping is present. The expression (4.3) for the average power, therefore, can be written

$$\bar{P} = \frac{1}{4}bF_0^2\omega^2 \left[\frac{1}{b^2\omega^2 + m^2(\omega^2 - \omega_1^2)^2} + \frac{1}{b^2\omega^2 + m^2(\omega^2 - \omega_2^2)^2} \right] \qquad (12.4.7)$$

This is just the sum of two terms which are identical in form to the expression (4.7.34) for the power when only a single mass and spring are present! The individual terms in (12.4.7) have maxima at the eigenfrequencies ω_1 and ω_2. A good idea of the response of the system to driving forces of various frequencies is found superimposing two power curves of the type appearing in Fig. (4.7.8) resulting in Fig. 12.4.2. This result can be extrapolated to systems which have larger numbers of degrees of freedom and hence a larger number of eigenfrequencies. With each eigenfrequency is to be associated a resonance in the response of the system to a periodic driving force.

12.5. PERIODIC SYSTEMS WITH f DEGREES OF FREEDOM

Very few systems with a large number of degrees of freedom can be solved exactly. One such system consists of f identical masses connected by $f + 1$ identical springs that are constrained to move in one dimension as illustrated in Fig. 12.5.1. Also shown in Fig. 12.5.1 is the electric

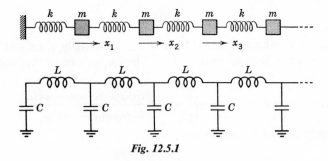

Fig. 12.5.1

circuit analogue of the mechanical system. The end-springs are assumed to be tied to fixed supports. The potential energy of the system can be expressed

$$V = \tfrac{1}{2}kx_1{}^2 + \tfrac{1}{2}k(x_2 - x_1)^2 + \cdots + \tfrac{1}{2}k(x_f - x_{f-1})^2 + \tfrac{1}{2}kx_f{}^2 \quad (12.5.1)$$

The equations of motion follow from Lagrange's equations as

$$
\begin{aligned}
m\ddot{x}_1 + 2kx_1 \quad &- kx_2 &&= 0 \\
m\ddot{x}_2 - kx_1 &+ 2kx_2 - kx_3 &&= 0 \\
&\;\;\vdots \\
m\ddot{x}_n - kx_{n-1} &+ 2kx_n - kx_{n+1} &&= 0 \\
&\;\;\vdots \\
m\ddot{x}_f - kx_{f-1} &+ 2kx_f &&= 0
\end{aligned}
\qquad (12.5.2)
$$

Another system which obeys the same differential equations consists of a light string which is stretched taut between fixed supports and then

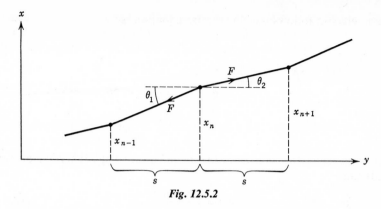

Fig. 12.5.2

loaded with point masses at equal intervals. Figure 12.5.2 shows a portion of such a system. Let the y-axis be the equilibrium position of the string. If the transverse displacements are small, the accelerations of the masses in the y-direction will be negligible, meaning that the tension F is substantially the same everywhere in the string. Then, assuming small displacements and neglecting the effect of the gravitational field, the equation of motion of mass n is

$$F\theta_2 - F\theta_1 = m\ddot{x}_n$$

$$F\frac{x_{n+1} - x_n}{s} - F\frac{x_n - x_{n-1}}{s} = m\ddot{x}_n$$

$$m\ddot{x}_n - \frac{F}{s}x_{n-1} + 2\frac{F}{s}x_n - \frac{F}{s}x_{n+1} = 0 \qquad (12.5.3)$$

which is identical in form to (12.5.2) with F/s taking the place of the spring constant.

If the solutions of (12.5.2) are assumed to be of the form

$$x_n = A_n \cos(\omega t - \phi) \qquad (12.5.4)$$

there results the system of equations

$$(-m\omega^2 + 2k)A_1 - kA_2 = 0$$

$$-kA_1 + (-m\omega^2 + 2k)A_2 - kA_3 = 0$$

$$\vdots \qquad (12.5.5)$$

$$-kA_{n-1} + (-m\omega^2 + 2k)A_n - kA_{n+1} = 0$$

from which it follows that the secular equation is

$$
\begin{vmatrix}
-m\omega^2 + 2k & -k & 0 & \cdots \\
-k & -m\omega^2 + 2k & -k & \cdots \\
0 & -k & -m\omega^2 + 2k & \cdots \\
\vdots & \vdots & \vdots &
\end{vmatrix} = 0 \quad (12.5.6)
$$

It is convenient to divide each row of the above $f \times f$ determinant by k and define

$$
c = 1 - \frac{m\omega^2}{k} \tag{12.5.7}
$$

The secular equation then reads

$$
\Delta_f =
\begin{vmatrix}
c+1 & -1 & 0 & 0 & \cdots \\
-1 & c+1 & -1 & 0 & \cdots \\
0 & -1 & c+1 & -1 & \cdots \\
0 & 0 & -1 & c+1 & \cdots \\
\vdots & \vdots & \vdots & \vdots &
\end{vmatrix} = 0 \tag{12.5.8}
$$

This is a polynomial of degree f in c and at first sight finding the roots seems formidable. The problem is made tractable as follows: Note that Δ_f obeys the following recursion formula:

$$
\Delta_f = (c + 1)\Delta_{f-1} - \Delta_{f-2} \tag{12.5.9}
$$

For example, if $f = 3$

$$
\begin{vmatrix}
c+1 & -1 & 0 \\
-1 & c+1 & -1 \\
0 & -1 & c+1
\end{vmatrix} = (c+1)
\begin{vmatrix}
c+1 & -1 \\
-1 & c+1
\end{vmatrix} - (c+1) \quad (12.5.10)
$$

Define θ by means of

$$
c + 1 = 2\cos\theta \tag{12.5.11}
$$

Then note that (12.5.9) is satisfied identically by

$$
\Delta_f = a\sin[(f+1)\theta] \tag{12.5.12}
$$

Since

$$
\Delta_1 = c + 1 = a\sin 2\theta = 2\cos\theta \tag{12.5.13}
$$

The parameter a is given by

$$a = 2 \frac{\cos \theta}{\sin 2\theta} \qquad (12.5.14)$$

Hence,

$$\Delta_f = \frac{2 \cos \theta \sin (f + 1)\theta}{\sin 2\theta} = \frac{\sin (f + 1)\theta}{\sin \theta} \qquad (12.5.15)$$

The secular equation (12.5.8) is simply

$$\frac{\sin (f + 1)\theta}{\sin \theta} = 0 \qquad (12.5.16)$$

which means

$$(f + 1)\theta = n\pi \qquad n = 1, 2, 3, \ldots, f \qquad (12.5.17)$$

Note that $n = 0$ and $n = f + 1$ are excluded since (12.5.16) is of the form $0/0$ for these values. Thus, exactly f-roots corresponding to the f-degrees of freedom have been found and there are f-possible eigenfrequencies given by

$$\omega^2 = 2 \frac{k}{m} (1 - \cos \theta) = \frac{4k}{m} \sin^2 \frac{\theta}{2}$$

$$\omega_n = 2\sqrt{\frac{k}{m}} \sin \left[\frac{n\pi}{2(f + 1)} \right] \qquad (12.5.18)$$

The eigenvectors are to be computed from (12.5.5) which we now express

$$A_2 = (c + 1)A_1$$
$$A_3 = (c + 1)A_2 - A_1$$
$$\cdot$$
$$\cdot \qquad (12.5.19)$$
$$\cdot$$
$$A_{j+1} = (c + 1)A_j - A_{j-1}$$

It is convenient to let $A_1 = \sin \theta$. Then

$$A_2 = (c + 1)A_1 = 2 \cos \theta \sin \theta = \sin 2\theta \qquad (12.5.20)$$

The jth recursion formula (12.5.19) is satisfied by

$$A_{nj} = \sin (j\theta) = \sin \left(\frac{jn\pi}{f + 1} \right) \qquad (12.5.21)$$

where, as is indicated by the double subscript, a different eigenvector is obtained for each of the f-eigenfrequencies ω_n. The orthogonality of the eigenvectors is already established; see Eq. (8.10.30). Reference to the equations of motion (12.5.2) reveals that the metric tensor is $g_{ij} = \delta_{fj}$.

The orthogonality condition, therefore, reads

$$\sum_{j=1}^{f} \sin\left(\frac{jn\pi}{f+1}\right) \sin\left(\frac{jn'\pi}{f+1}\right) = N\delta_{nn'} \qquad (12.5.22)$$

Where N is a normalization factor that is to be determined from

$$\sum_{j=1}^{f} \sin^2(j\theta) = N \qquad \theta = \frac{n\pi}{f+1} \qquad (12.5.23)$$

The above expression is summed as follows:

$$N = \sum_{j=1}^{f} (\tfrac{1}{2} - \tfrac{1}{2}\cos 2j\theta) = \frac{f}{2} - \frac{1}{2}\operatorname{Re}\sum_{j=1}^{f} e^{2ij\theta}$$

$$= \frac{f}{2} - \frac{1}{2}\operatorname{Re}\left(\sum_{j=0}^{f} e^{2ij\theta} - 1\right)$$

$$N = \frac{f}{2} + \frac{1}{2} - \operatorname{Re}\left(\frac{1 - e^{2i\theta(f+1)}}{1 - e^{2i\theta}}\right) \qquad (12.5.24)$$

$$= \frac{f+1}{2} - \operatorname{Re}\left(\frac{1 - e^{2in\pi}}{1 - e^{2in\pi/(f+1)}}\right) = \frac{f+1}{2}$$

Therefore, we may write the components of the normalized eigenvectors as

$$S_{nj} = \sqrt{\frac{2}{f+1}} \sin\left(\frac{jn\pi}{f+1}\right) \qquad (12.5.25)$$

$$S_{nj}S_{n'j} = \delta_{nn'} \qquad S_{nj}S_{nj'} = \delta_{jj'}$$

The general solution of our original mechanics problem, containing two f-arbitrary constants a_n and b_n, can be written

$$x_j = \sum_{n=1}^{f} S_{nj}(a_n \sin \omega_n t + b_n \cos \omega_n t) \qquad (12.5.26)$$

Recall that the *normal* coordinates obey

$$\ddot{q}_n + \omega^2(n)q_n = 0 \qquad \text{(no sum of } n)* \qquad (12.5.27)$$

Hence, (12.5.26) also expresses the fact that the transformation between the original coordinates x_j and the normal coordinates is

$$x_j = S_{nj}q_n \qquad q_n = S_{nj}x_j \qquad (12.5.28)$$

Since the metric tensor has the simple form $g_{ij} = \delta_{ij}$, (12.5.28) can be thought of as a rotation of a rectangular Cartesian coordinate system in the f-dimensional configuration space of the system.

---

* We write $\omega(n)$ rather than ω_n when there is to be no sum on a repeated subscript.

By setting $t = 0$ in (12.5.28) we have

$$q_n(0) = b_n = S_{nj}x_j(0) \qquad (12.5.29)$$

By differentiating (12.5.28) and then setting $t = 0$ we get

$$a_n = \frac{1}{\omega(n)} S_{nj}\dot{x}_j(0) \qquad (12.5.30)$$

These expressions allow the explicit evaluation of the constants in terms of the initial conditions.

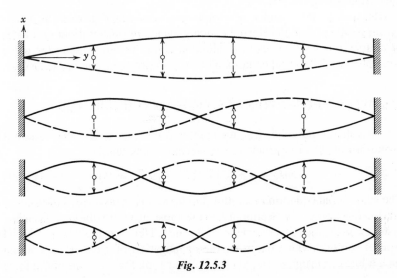

Fig. 12.5.3

As an example, consider the case $f = 4$. The normalized eigenvectors can be displayed as the rows of the transformation matrix

$$S = \sqrt{\tfrac{2}{5}} \begin{pmatrix} \sin\dfrac{\pi}{5} & \sin\dfrac{2\pi}{5} & \sin\dfrac{2\pi}{5} & \sin\dfrac{\pi}{5} \\[2mm] \sin\dfrac{2\pi}{5} & \sin\dfrac{\pi}{5} & -\sin\dfrac{\pi}{5} & -\sin\dfrac{2\pi}{5} \\[2mm] \sin\dfrac{2\pi}{5} & -\sin\dfrac{\pi}{5} & -\sin\dfrac{\pi}{5} & \sin\dfrac{2\pi}{5} \\[2mm] \sin\dfrac{\pi}{5} & -\sin\dfrac{2\pi}{5} & \sin\dfrac{2\pi}{5} & -\sin\dfrac{\pi}{5} \end{pmatrix} \qquad (12.5.31)$$

where components such as $\sin 6\pi/5$ have been written as $-\sin \pi/5$, $\sin 8\pi/5 = -\sin 2\pi/5$, etc. It is possible for the system to vibrate in any one of the four possible modes as pictured in Fig. 12.5.3. The amplitudes and relative phases of the particles can be read directly out of the matrix (12.5.31). This is facilitated by sketching the sine curves

$$\sin\frac{\pi y}{5s}, \quad \sin\frac{2\pi y}{5s}, \quad \sin\frac{3\pi y}{5s}, \quad \text{and} \quad \sin\frac{4\pi y}{5s}$$

The particles are located at $y = s$, $2s$, $3s$, and $4s$. The normal modes of vibration can actually be thought of as a type of *standing wave*. Of course, the motion of the system would generally be a mixture of the pure modes illustrated in Fig. 12.5.3.

The loaded string undergoes *transverse* vibrations whereas the mass-spring system of Fig. 12.5.1 would undergo *longitudinal* vibrations provided, of course, that the masses were all constrained to move along the same straight line. The basic equations of motion are the same in either case.

12.6. WAVE PROPAGATION IN PERIODIC STRUCTURES

Consider the periodic structures pictured in Fig. 12.6.1. The equation of motion of the nth particle of the mechanical system is

$$m\ddot{x}_n - kx_{n-1} + 2kx_n - kx_{n+1} = 0 \qquad (12.6.1)$$

The same basic equation of motion holds for the transverse vibrations of the loaded string. In Section 12.5, the most general solution of such a system consisting of f-particles was found. Here we look for a special type of solution which describes the propagation of a disturbance along such a line of particles. In particular, let us look for solutions of the form

$$x_{n-1} = a\cos(\omega t - \alpha + \beta)$$
$$x_n = a\cos(\omega t - \alpha) \qquad (12.6.2)$$
$$x_{n+1} = a\cos(\omega t - \alpha - \beta)$$

Direct substitution into the differential equation of motion and the use of

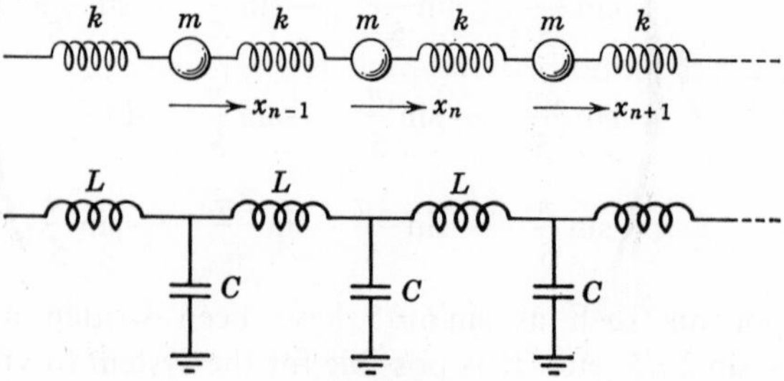

Fig. 12.6.1

the identity

$$\cos (\omega t - \alpha + \beta) + \cos (\omega t - \alpha - \beta) = 2 \cos (\omega t - \alpha) \cos \beta \quad (12.6.3)$$

shows that solutions of the form (12.6.2) do indeed exist provided that

$$\cos \beta = 1 - \frac{m\omega^2}{2k} \quad (12.6.4)$$

To get an idea of what the solution looks like, set $\alpha = 0$ and $t = 0$. Then

$$x_{n-2} = a \cos 2\beta, \qquad x_{n-1} = a \cos \beta, \qquad x_n = a,$$
$$x_{n+1} = a \cos (-\beta), \qquad x_{n+2} = a \cos (-2\beta), \qquad \text{etc.} \qquad (12.6.5)$$

For the example of the loaded string, the displacements of the particles would be as illustrated in Fig. 12.6.2. As time increases, the phases of the

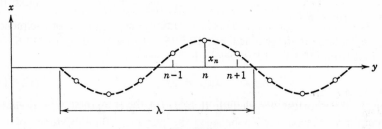

Fig. 12.6.2

displacements (12.6.2) increase, meaning that the entire disturbance moves without change of shape in the positive y-direction. This type of solution is generally referred to as a *plane wave.*

A total phase shift of 2π determines one complete wavelength. The number of particles involved in one complete wave is $2\pi/\beta$. The wavelength, therefore, is

$$\lambda = \frac{2\pi}{\beta} s \quad (12.6.6)$$

where s is the separation between particles. The velocity of propagation of the plane wave, generally known as the *phase velocity,* is

$$u_p = \frac{\lambda}{P} \quad (12.6.7)$$

where P is one complete period of the motion of a single particle as determined by

$$\omega P = 2\pi \quad (12.6.8)$$

Hence,

$$u_p = \frac{\lambda \omega}{2\pi} = \frac{s\omega}{\beta} = \frac{s\omega}{\cos^{-1}(1 - m\omega^2/2k)} \qquad (12.6.9)$$

The phase velocity depends on the material constants m and k of the medium through which the wave travels. There is also a dependence of the phase velocity on the frequency meaning that there is *dispersion*. This phenomenon is familiar from optics where is it known that plane light waves of different frequencies have different phase velocities when traveling through material media.

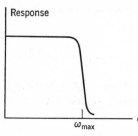

An important observation is that there is a cut-off frequency determined by setting $\cos\beta = -1$ in (12.6.4) giving

$$\omega_{max} = 2\sqrt{k/m} \qquad (12.6.10)$$

Fig. 12.6.3

This is essentially the greatest eigenfrequency of the system as determined by (12.5.18). Corresponding to (12.6.10), we find from (12.6.6) the wavelength

$$\lambda_{min} = 2s \qquad (12.6.11)$$

This is the shortest wavelength that can be propagated in a periodic structure with separation s between the particles. The existence of the maximum frequency (12.6.10) means that the response of such a system to a periodic driving force (or voltage) would be as illustrated in Fig. 12.6.3. The response could be determined, e.g., as the power delivered to the load of impedance z in Fig. 12.6.4. Such a transmission network is a *low-pass filter* in that periodic signals with frequencies above ω_{max} are not transmitted to the load.

It is also possible to construct a *high-pass filter* (Problem 12.6.7), a *band-pass filter*, or a *band-elimination filter*. A discussion of these types of filters can be found in Olson's *Dynamical Analogies* and Brillouin's *Wave Propagation in Periodic Structures*.

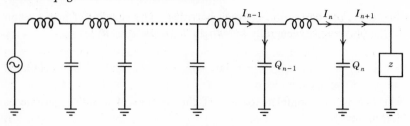

Fig. 12.6.4

12.7. CHARACTERISTIC IMPEDANCE

Consider the transmission line of Fig. 12.6.4. Suppose that a plane wave of frequency ω is traveling from the generator to the load z. With what impedance must we terminate the line in order that the plane wave shall be entirely absorbed? We will use the complex impedance method and suppose that the actual current J_n is the real part of I_n, q_n is the real part of Q_n, and so on. The terminal elements of the line obey

$$\frac{Q_n}{C} = I_{n+1}z \qquad I_n = \dot{Q}_n + I_{n+1} \tag{12.7.1}$$

Hence,

$$\frac{1}{C}(I_n - I_{n+1}) = \dot{I}_{n+1}z \tag{12.7.2}$$

We require that z be such that the traveling wave solution

$$I_n = Ae^{i(\omega t - \alpha)} \qquad I_{n+1} = Ae^{i(\omega t - \alpha - \beta)} \tag{12.7.3}$$

shall hold. The substitution of (12.7.3) into (12.7.2) yields

$$z = \frac{1}{i\omega C}[e^{i\beta} - 1] \tag{12.7.4}$$

Now, using (12.6.4) for $\cos \beta$ (with m replaced by L and k by $1/C$ we get

$$z = i\omega \frac{L}{2} + \sqrt{L/C}\sqrt{1 - \omega^2 LC/4} \tag{12.7.5}$$

This is actually the impedance that the driving force at one end of an infinite line would "see"; it is called the *characteristic impedance* of the line. If a finite line is terminated by an impedance equal to (12.7.5), which, for instance, could be an inductor of value $L/2$ and an ordinary resistor of value

$$R = \sqrt{\frac{L}{C}}\sqrt{1 - \frac{\omega^2 LC}{4}} \tag{12.7.6}$$

then a plane wave of frequency ω is completely absorbed. On the other hand, as far as the power source is concerned, we could replace the entire line by an impedance z. The power absorbed by the line, and hence the energy carried by the plane wave, is

$$\bar{P} = \tfrac{1}{2}\operatorname{Re}\left(\frac{\phi_0^2}{z}\right) = \tfrac{1}{2}\phi_0^2\sqrt{\frac{C}{L}}\sqrt{1 - \frac{\omega^2 LC}{4}} \tag{12.7.7}$$

Notice that (12.7.5) becomes purely imaginary if

$$\frac{\omega^2 LC}{4} > 1 \quad \text{or} \quad \omega > \frac{2}{\sqrt{LC}} \tag{12.8.8}$$

and no power is transmitted. Note that (12.7.8) gives again the cut-off frequency

$$\omega_{\max} = \frac{2}{\sqrt{LC}} \tag{12.7.9}$$

12.8. REFLECTION OF PLANE WAVES

What happens to a traveling wave if the load impedance z at the terminal of the line of Fig. 12.6.4 does not match the characteristic impedance? It will be shown that the plane wave incident on z is only partially absorbed, the remainder being reflected as a plane wave traveling in the opposite direction. In other words, Eq. (12.7.2) can be satisfied with z other than the characteristic impedance, provided that there is also a plane wave traveling in the opposite direction. Specifically, we assume a solution of the form

$$I_n = Ae^{i\beta+i\omega t} + Be^{-i\beta+i\omega t}$$
$$I_{n+1} = Ae^{i\omega t} + Be^{i\omega t} \tag{12.8.1}$$

The amplitudes A and B are, in general, complex—meaning that there is a definite phase difference between the waves. The substitution of (12.8.1) into (12.7.2) yields

$$B = -A \frac{e^{i\beta} - 1 - i\omega Cz}{e^{-i\beta} - 1 - i\omega Cz} \tag{12.8.2}$$

Let z_k stand for the characteristic impedance. Then, from (12.7.4),

$$e^{i\beta} - 1 = i\omega Cz_k \quad e^{-i\beta} - 1 = -i\omega Cz_k{}^* \tag{12.8.3}$$

where $z_k{}^*$ stands for the complex conjugate of z_k. Equation (12.8.2) is then

$$B = A \frac{z_k - z}{z_k{}^* + z} \tag{12.8.4}$$

This gives the amplitude of the reflected wave as a function of the terminal impedance. Note that $B = 0$ when $z = z_k$, meaning that for this special value of z there is no reflection.

The opening of the circuit loop that contains z is equivalent to $z \to \infty$. In this event,

$$B = -A \tag{12.8.5}$$

The mechanical analogue is the longitudinal motion of a series of masses and springs terminating in a rigid support as in Fig. 12.8.1. It is worth while re-deriving (12.8.5) for the mechanical system. The equation of motion for the elements nearest the support is

$$m\ddot{X}_n + 2kX_n - kX_{n-1} = 0 \tag{12.8.6}$$

As with the circuit, it is convenient to work with complex quantities and think of the actual displacement x_n as being the real part of X_n. Our

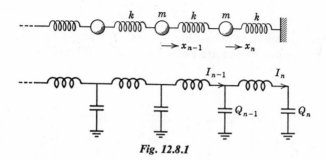

Fig. 12.8.1

solutions which consist of a superposition of waves traveling in both directions are to be taken as

$$X_n = (Ae^{i\beta} + Be^{-i\beta})e^{i\omega t}$$
$$X_{n-1} = (Ae^{2i\beta} + Be^{-2i\beta})e^{i\omega t} \tag{12.8.7}$$

We find

$$B = -A\,\frac{(-m\omega^2 + 2k)e^{i\beta} - ke^{2i\beta}}{(-m\omega^2 + 2k)e^{-i\beta} - ke^{-2i\beta}} \tag{12.8.8}$$

Since β is given by (12.6.4), (12.8.8) reduces to

$$B = -A = Ae^{i\pi} \tag{12.8.9}$$

The reflected wave is 180° out of phase with the incident wave and is equal in amplitude. The equality of the amplitudes means that the rigid support absorbs no energy from the incident wave; all is reflected.

For $B = -A$, the solution (12.8.7) is

$$X_n = A(e^{i\beta} - e^{-i\beta})e^{i\omega t}$$
$$= 2iA\sin\beta(\cos\omega t + i\sin\omega t) \tag{12.8.10}$$

The actual displacement is

$$x_n = \text{Re}\,X_n = -2A\sin\beta\sin\omega t \tag{12.8.11}$$

Similarly,

$$x_{n-1} = -2A \sin 2\beta \sin \omega t$$
$$x_{n-2} = -2A \sin 3\beta \sin \omega t \qquad (12.8.12)$$

The solutions (12.8.11) and (12.8.12) are of the form of a *standing wave*. Since they are easier to visualize, the resulting transverse vibrations of the particles on a loaded string, rather than the longitudinal vibrations of the mass-spring system are illustrated in Fig. 12.8.2. The curves in Fig. 12.8.2 are actually plots of,

$$x = -2A \sin ay \sin \omega t \qquad (12.8.13)$$

where a is a constant, at two different times. The points $ay = 0, \pi, 2\pi, \ldots$ are always stationary and are known as *nodes*. The points of maximum

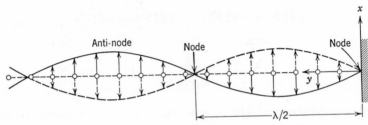

Fig. 12.8.2

amplitude of vibration are antinodes. The nodes and antinodes may not be apparent in an actual vibrating system since they do not necessarily coincide with the location of a particle.

Consider now a system of f-particles which is tied to rigid supports at both ends. This is the same system that is treated in Section 12.5. Standing wave solutions are still possible, provided that *both* ends are nodes. Since the distance between nodes is $\lambda/2$ (Fig. 12.8.2) where λ is the wavelength of the original traveling wave, only certain wavelengths are now possible. The length of the lattice is $(f + 1)s$ and the possible wavelengths are

$$(f + 1)s = \frac{\lambda}{2}, \frac{2\lambda}{2}, \frac{3\lambda}{2}, \ldots, \frac{n\lambda}{2} \qquad (12.8.14)$$

The possible wavelengths for a system consisting of only four particles are illustrated in Fig. 12.5.3. The wavelength is given by (12.6.6) so that

$$(f + 1)s = \frac{n}{2} \cdot \frac{2\pi s}{\beta}, \qquad \beta = \frac{n\pi}{f + 1} \qquad (12.8.15)$$

By using $\cos \beta$ as given by (12.6.4) there results

$$\cos \beta = \cos \left(\frac{n\pi}{f + 1} \right) = 1 - \frac{m\omega^2}{2k} \qquad (12.8.16)$$

from which follows

$$\omega_n = 2\sqrt{\frac{k}{m}} \sin \frac{n\pi}{2(f+1)} \qquad (12.8.17)$$

This result is identical to (12.5.18). Thus, corresponding to the possible wavelengths (12.8.14) are the eigenfrequencies (12.8.17).

By taking traveling wave solutions and superimposing them in the proper way so as to take into account the *boundary conditions*, i.e., the fact that the ends of the system are tied to fixed supports, we have again found the eigensolutions and eigenfrequencies of the system.

12.9. JUNCTION OF TWO LATTICES

In this section, it is shown that a plane wave incident on the junction between two periodic lattices is partially reflected and partially transmitted.

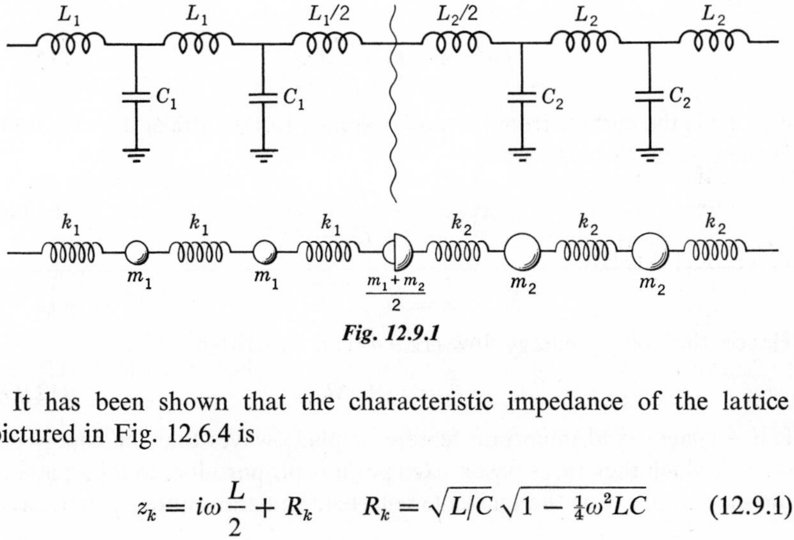

Fig. 12.9.1

It has been shown that the characteristic impedance of the lattice pictured in Fig. 12.6.4 is

$$z_k = i\omega \frac{L}{2} + R_k \qquad R_k = \sqrt{L/C}\sqrt{1 - \tfrac{1}{4}\omega^2 LC} \qquad (12.9.1)$$

Therefore, by lopping off half the first inductor, the impedance is simply R_k.

Let two lattices be joined together in the manner indicated by Fig. 12.9.1. Let the characteristic impedances of the two lattices be denoted by z_1 and z_2. The impedance of the lattice to the right of the wavy line is just

$$R_2 = \sqrt{L_2/C_2}\sqrt{1 - \tfrac{1}{4}\omega^2 L_2 C_2} \qquad (12.9.2)$$

Thus, in Eq. (12.8.4), we have

$$z_k = z_1 = i\omega \frac{L_1}{2} + R_1, \qquad z = i\omega \frac{L_1}{2} + R_2$$

$$B_1 = A_1 \frac{R_1 - R_2}{R_1 + R_2} \tag{12.9.3}$$

A plane wave of amplitude A_1 incident on the junction of the two lattices is partially reflected, the reflected amplitude being given by (12.9.3). The ratio B_1/A_1 is called the *reflection coefficient*. Note that $B_1 < 0$ if $R_2 > R_1$, meaning that the reflected wave is 180° out of phase with the incident wave. If $R_1 > R_2$, the reflected and incident waves are in phase.

It is useful to discuss the transmitted wave from the point of view of conservation of energy. The expression (12.7.7) for the average power carried by a plane wave can be written

$$\bar{P} = \tfrac{1}{2}\phi_0^2 \frac{C}{L} R_k \tag{12.9.4}$$

Since

$$\phi_0 e^{i\omega t} = I\left(i\omega \frac{L}{2} + R_k\right) \tag{12.9.5}$$

where I is the current from the power source to the lattice, the amplitude of the current has magnitude

$$|A| = \frac{\phi_0}{\sqrt{\tfrac{1}{4}\omega^2 L^2 + R_k^2}} \tag{12.9.6}$$

By (12.9.1) this is

$$|A| = \phi_0 \sqrt{C/L} \tag{12.9.7}$$

Hence, the average energy flow (12.9.4) can be written

$$\bar{P} = \tfrac{1}{2}|A|^2 R_k \tag{12.9.8}$$

It is a general and important feature of plane waves that the energy per second which they carry past a fixed point is proportional to the square of the amplitude. For the incident, reflected, and transmitted plane waves

$$\bar{P}_i = \tfrac{1}{2}|A_1|^2 R_1 \qquad \bar{P}_r = \tfrac{1}{2}|B_1|^2 R_1 \qquad \bar{P}_t = \tfrac{1}{2}|A_2|^2 R_2 \tag{12.9.9}$$

where the amplitude of the transmitted wave is denoted A_2. Conservation of energy requires

$$\tfrac{1}{2}|A_1|^2 R_1 = \tfrac{1}{2}|B_1|^2 R_1 + \tfrac{1}{2}|A_2|^2 R_2 \tag{12.9.10}$$

This combined with (12.9.3) yields

$$|A_2| = |A_1| \frac{2R_1}{R_1 + R_2} \tag{12.9.11}$$

These are magnitudes only; the phase of A_2 is still not determined. If it is recognized that the same current exists in the inductors $L_1/2$ and $L_2/2$ in Fig. 12.9.1, then

$$A_2 = A_1 + B_1 \qquad (12.9.12)$$

This is tantamount to saying that the two halves of the mass at the junction in the mechanical analogue of Fig. 12.9.1 have the same motion. Equation (12.9.12) combined with (12.9.3) leads to

$$A_2 = A_1 \frac{2R_1}{R_1 + R_2} \qquad (12.9.13)$$

which confirms (12.9.11) and shows that the transmitted wave is always in phase with the incident wave. Equation (12.9.13) gives the transmission coefficient A_2/A_1.

12.10. WAVE EQUATIONS FOR CONTINUOUS MEDIA

Consider the loaded string of Fig. 12.5.2. What happens if more and more point masses are added until they become uniformly dense so that the string is a continuous medium? Consider the equation of motion (12.5.3) to which has been added a term to account for friction:

$$\frac{F}{s}[(x_{n+1} - x_n) - (x_n - x_{n-1})] - b\dot{x}_n - m\ddot{x}_n = 0 \qquad (12.10.1)$$

The equation of motion can be expressed as

$$\frac{\frac{x_{n+1} - x_n}{\Delta y} - \frac{x_n - x_{n-1}}{\Delta y}}{\Delta y} - \frac{b}{\Delta yF}\dot{x}_n - \frac{m}{\Delta yF}\ddot{x}_n = 0 \qquad (12.10.2)$$

where the separation s between masses has been replaced by Δy. We wish to consider a limiting process whereby $\Delta y \to 0$. Notice that

$$\mu = \frac{m}{\Delta y} \qquad r = \frac{b}{\Delta y} \qquad (12.10.3)$$

become mass per unit length and viscous damping constant per unit length. In the limit, the notation $x_n \to \eta(y, t)$ will be used to denote the transverse displacement of the string from its equilibrium position as a function of the *two* variables y and t. The variable y is the position coordinate measured along the string. The limiting form of (12.10.2) is

$$\frac{\partial^2 \eta}{\partial y^2} - \frac{r}{F}\frac{\partial \eta}{\partial t} - \frac{\mu}{F}\frac{\partial^2 \eta}{\partial t^2} = 0 \qquad (12.10.4)$$

This partial differential equation is known as a *wave equation*. Its solutions give the possible motions (confined to a plane in this case) of a taught string of uniform mass density μ under a tension F. Before considering these solutions, we will derive wave equations for several other physical systems.

Consider the low-pass filter of Fig. 12.6.1. Its circuit equation is

$$\frac{I_{n+1} - I_n}{C} - \frac{I_n - I_{n-1}}{C} - R\dot{I}_n - L\ddot{I}_n = 0 \qquad (12.10.5)$$

where the ohmic resistance of the inductor has been included. Let Δy be the actual physical dimension of the cells of the filter. Write (12.10.5) as

$$\frac{\dfrac{I_{n+1} - I_n}{\Delta y} - \dfrac{I_n - I_{n-1}}{\Delta y}}{\Delta y} - rc\dot{I}_n - lc\ddot{I}_n = 0 \qquad (12.10.6)$$

where

$$r = \frac{R}{\Delta y}, \qquad c = \frac{C}{\Delta y}, \qquad l = \frac{L}{\Delta y} \qquad (12.10.7)$$

are the resistance, capacitance, and inductance per unit length. In the limit $\Delta y \to 0$,

$$\frac{\partial^2 I}{\partial y^2} - rc\frac{\partial I}{\partial t} - lc\frac{\partial^2 I}{\partial t^2} = 0 \qquad (12.10.8)$$

This wave equation is commonly known as the *telegrapher's equation* and is suitable for the discussion of the propagation of an electromagnetic signal along a thin wire or coaxial cable.

It is possible to derive the wave equation for the propagation of a compressional or longitudinal wave in a continuous elastic medium by a limiting process starting with the particle-spring system of Fig. 12.6.1. Equation (12.6.1) can be expressed by

$$\frac{\dfrac{x_{n+1} - x_n}{\Delta y} - \dfrac{x_n - x_{n-1}}{\Delta y}}{\Delta y} - \frac{m}{k(\Delta y)^2}\ddot{x}_n = 0 \qquad (12.10.9)$$

Suppose that the elastic medium is in the form of a uniform rod of crosssection σ. We write

$$\frac{m}{k(\Delta y)^2} = \frac{m}{\sigma\,\Delta y(k\,\Delta y/\sigma)} = \frac{\rho}{(k\,\Delta y/\sigma)} \qquad (12.10.10)$$

where $\rho = m/(\sigma\,\Delta y)$ is the mass density of the material. In the limit $\Delta y \to 0$, $k\,\Delta y/\sigma$ approaches a finite value which is known as *Young's modulus* for the material. Equation (12.10.10) becomes

$$\frac{\partial^2 \eta}{\partial y^2} - \frac{\rho}{Y}\frac{\partial^2 \eta}{\partial t^2} = 0 \qquad (12.10.11)$$

where $x_n \to \eta(y, t)$.

It is worthwhile re-deriving (12.10.11) by assuming at the start that the medium is continuous rather than composed of interacting particles. As the disturbance propagates in the y-direction, a given element of the rod of thickness Δy_0 is displaced to a new position. In addition to the displacement, the element also undergoes *distortion* (either compression or

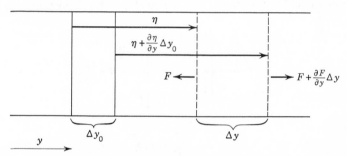

Fig. 12.10.1

elongation) so that its new thickness is Δy as indicated in Fig. 12.10.1. A convenient measure of the distortion is the *strain* which is defined to be

$$e = \frac{\Delta y - \Delta y_0}{\Delta y_0} \qquad (12.10.12)$$

Let η be the displacement from equilibrium of a cross section of the rod originally located at y. The displacement of a cross section originally at $y + \Delta y_0$ is, by a Taylor's series, $\eta + (\partial \eta/\partial y)/\Delta y_0$. Then, from the geometry of Fig. 12.10.1, the strain is

$$e = \frac{\partial \eta}{\partial y} \qquad (12.10.13)$$

Let F be the tension in the rod at the location of the distorted element. The *stress* in the rod is defined to be

$$S = \frac{F}{\sigma} \qquad (12.10.14)$$

It is consistent with the idea that the interactions between the atoms of the material under consideration can be represented approximately by harmonic oscillator potentials to assume that stress is directly proportional to strain:

$$\frac{F}{\sigma} = Y\frac{\partial \eta}{\partial y} \qquad (12.10.15)$$

The constant of proportionality is Young's modulus.

The force differential across the distorted element is $(\partial F/\partial y)\,\Delta y$. By Newton's second law,

$$\frac{\partial F}{\partial y}\Delta y = \rho\sigma\,\Delta y\,\frac{\partial^2 \eta}{\partial t^2} \qquad (12.10.16)$$

Combining (12.10.15) and (12.10.16) yields (12.10.11) again.

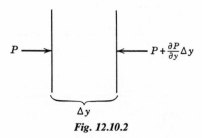

$P \longrightarrow$ $\longleftarrow P + \frac{\partial P}{\partial y}\Delta y$

Δy

Fig. 12.10.2

For the purpose of discussing the one-dimensional propagation of sound waves, it is convenient to multiply the expression (12.10.12) for the strain top and bottom by σ and write it as

$$\frac{V - V_0}{V_0} = e = \frac{\partial \eta}{\partial y} \qquad (12.10.17)$$

Where V and V_0 now refer to the displaced and original volume of the sample of gas (or liquid) under consideration. The *bulk modulus* for a substance is defined by

$$P - P_0 = -B\frac{V - V_0}{V_0} = -B\frac{\partial \eta}{\partial y} \qquad (12.10.18)$$

where P is the pressure when the volume is V, and P_0 is the pressure when the volume is V_0. Equation (12.10.18) is a little different from (12.10.15); the minus sign is included because an increase in pressure means a decrease in volume. Newton's second law for the element of gas reads:

$$-\frac{\partial P}{\partial y}V = \rho V\frac{\partial^2 \eta}{\partial t^2} \qquad (12.10.19)$$

Note the directions of the arrows which indicate pressure in Fig. 12.10.2.

Combined with (12.10.18), (12.10.19) yields

$$\frac{\partial^2 \eta}{\partial y^2} = \frac{\rho}{B} \frac{\partial^2 \eta}{\partial t^2} \tag{12.10.20}$$

The expression (12.10.18) deals with finite changes in pressure and volume and must, in this sense, be regarded as an approximation. More precisely, the bulk modulus of a given liquid or gas under pressure P_0 is

$$B = -V_0 \left(\frac{\partial P}{\partial V}\right)_0 \tag{12.10.21}$$

The compressions and rarefactions involved in the propagation of a sound wave are sufficiently rapid so that the process can be considered to be adiabatic. Thus (12.10.21) is the *adiabatic* bulk modulus.

As an example, an adiabatic process in an ideal gas is governed by

$$PV^\gamma = \text{const.} \tag{12.10.22}$$

where γ is the ratio of specific heats. Hence,

$$\left(\frac{\partial P}{\partial V}\right)_0 = -\frac{\gamma P_0}{V_0} \tag{12.10.23}$$

By the use of the ideal gas law

$$P_0 V_0 = nRT_0 \tag{12.10.24}$$

where n is the number of moles of gas, R is the gas constant, and T_0 is the absolute temperature,

$$\left(\frac{\partial P}{\partial V}\right)_0 = -\frac{\gamma nRT_0}{V_0^2} \qquad B = \frac{\gamma nRT_0}{V_0} \tag{12.10.25}$$

Moreover,

$$\frac{B}{\rho} = \frac{\gamma nRT_0}{m} = \frac{\gamma RT_0}{W} \tag{12.10.26}$$

where W is the mass of 1 mole of the gas. Thus the wave equation (12.10.20) becomes

$$\frac{\partial^2 \eta}{\partial y^2} = \sqrt{\frac{W}{\gamma RT_0}} \frac{\partial^2 \eta}{\partial t^2} \tag{12.10.27}$$

12.11. PLANE WAVE SOLUTIONS

In Section 12.10, several examples of the one-dimensional wave equation are derived—one-dimensional in the sense that propagation with respect to only one spatial coordinate is considered. It is convenient to write the wave equation, neglecting dissipation, as

$$\frac{\partial^2 \eta}{\partial x^2} - \frac{1}{c^2}\frac{\partial^2 \eta}{\partial t^2} = 0 \qquad (12.11.1)$$

where c is a constant. Mathematically speaking, this is an example of a *hyperbolic partial differential equation*. We consider in this section the simplest solution, namely, the plane wave:

$$\eta = a \sin (kx - \omega t) \qquad (12.11.2)$$

where k is a constant called the *propagation number* (not to be confused with

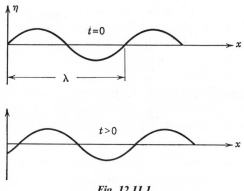

Fig. 12.11.1

the spring constant!) and ω is the circular frequency. Equation (12.11.2) is found to be a solution of (12.11.1), provided that

$$k = \frac{\omega}{c} \qquad (12.11.3)$$

Another convenient form of (12.11.2) results if new constants P and λ are introduced by means of

$$\omega = \frac{2\pi}{P} \qquad k = \frac{2\pi}{\lambda} \qquad (12.11.4)$$

where P is the period of a single particle of the medium through which the wave passes and λ is the wavelength:

$$\eta = a \sin 2\pi\left(\frac{x}{\lambda} - \frac{t}{P}\right) \qquad (12.11.5)$$

A plot of (12.11.5) at $t = 0$ and a slightly later time results in Fig. 12.11.1 and shows that the disturbance propagates in the positive x-direction. For example, Fig. 12.11.1 might represent the transverse displacements at two different times of a stretched string along which a plane wave is propagating.

To find the phase velocity, or velocity of propagation of a plane wave, observe that when the wave moves through a distance λ, a single particle has executed one period of its motion. Therefore,

$$u_p = \frac{\lambda}{P} \qquad (12.11.6)$$

Equations (12.11.3) and (12.11.4) show this to be

$$u_p = \frac{\omega}{k} = c \qquad (12.11.7)$$

Thus plane waves of all frequencies propagate with the same velocity in a medium in which there is no dissipation. There is no dispersion as is the case with wave propagation in periodic structures. Of course, all media in reality consist of particles, and dispersion can be expected at very short wavelengths. Equation (12.11.7) tells us the velocity of unattenuated waves in various types of media. From Section 12.10, the velocity of transverse waves on a string of mass μ per unit length and under a tension F is $\sqrt{F/\mu}$. Electromagnetic waves on a wire propagate at velocity $1/\sqrt{lc}$; longitudinal waves in a deformable medium of mass density ρ and Young's modulus Y propagate at velocity $\sqrt{Y/\rho}$; the velocity of sound in an ideal gas is $\sqrt{\gamma RT/W}$.

Plane waves propagating in the positive x-direction can be written in other ways, e.g.,

$$\cos(kx - \omega t), \qquad e^{i(kx - \omega t)}, \qquad \sin(kx - \omega t + \phi)$$

The argument $kx - \omega t$ or $kx - \omega t + \phi$, if the extra phase angle should be present, is called the *phase* of the wave. If the phase is of the form $kx + \omega t$, the plane wave propagates in the *negative* x-direction. The phase velocity is actually the velocity with which a point of constant phase moves. Differentiation of $kx - \omega t = $ const. gives $dx/dt = \omega/k = c$.

12.12. THE SUPERPOSITION OF PLANE WAVES

Since the wave equation (12.11.1) is linear, more complicated solutions can be built up by the superposition of plane waves. For instance, consider the superposition of two plane waves traveling in the positive x-direction:

$$\eta = Ae^{i(kx - \omega t)} + Be^{i(kx - \omega t + \phi)}$$
$$= (A + Be^{i\phi})e^{i(kx - \omega t)} \qquad (12.12.1)$$

The amplitudes are different and the two waves are out of phase with one another, but the frequencies are the same. Note that

$$A + Be^{i\phi} = Re^{i\alpha}$$

$$R = \sqrt{A^2 + B^2 + 2AB\cos\phi} \qquad \tan\alpha = \frac{B\sin\phi}{A + B\cos\phi} \qquad (12.12.2)$$

Thus, the result of the superposition is

$$\eta = Re^{i(kx-\omega t+\alpha)} \qquad (12.12.3)$$

which is a new plane wave of amplitude R and phase $kx - \omega t + \alpha$. The addition of amplitudes as given by (12.12.2) can also be accomplished by the vector addition as indicated in Fig. 12.12.1.

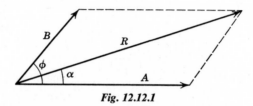

Fig. 12.12.1

Suppose two waves of the same amplitude but different frequencies are moving in the positive x-direction. Their sum is

$$\eta = Ae^{i\psi} + Ae^{i(\psi+\Delta\psi)} \qquad (12.12.4)$$

where $\psi = kx - \omega t$, $\Delta\psi = \Delta kx - \Delta\omega t$.
  We have

$$\eta = Ae^{i\psi}(1 + e^{i\Delta\psi})$$
$$= Ae^{i(\psi+\Delta\psi/2)}(e^{-i(\Delta\psi/2)} + e^{i(\Delta\psi/2)}) \qquad (12.12.5)$$
$$= 2Ae^{i(\psi+\Delta\psi/2)}\cos\frac{\Delta\psi}{2}$$

It is interesting to consider the case where the waves are only slightly different in frequency: $\Delta\psi \ll \psi$. The real part of (12.12.5) is approximately

$$\eta = 2A\cos(kx - \omega t)\cos\left(\frac{\Delta kx - \Delta\omega t}{2}\right) \qquad (12.12.6)$$

In a simple nondispersive medium, $\omega = ck$ and

$$\eta = 2A\cos k(x - ct)\cdot\cos\frac{\Delta k}{2}(x - ct) \qquad (12.12.7)$$

A plot of (12.12.7) as a function of x at $t = 0$ gives Fig. 12.12.2. The high-frequency component is sometimes called the *carrier wave*. The wave is said to be *modulated* by the low-frequency component. The modulation results in a building up of *groups* of amplitude $2A$. In acoustics, these groups are referred to as *beats*, whereas in quantum mechanics Schrödinger called them *wave packets*. As indicated by (12.12.7), the beats move with the same velocity c as the original plane waves; in fact *all* disturbances move with the same velocity in a nondispersive medium.

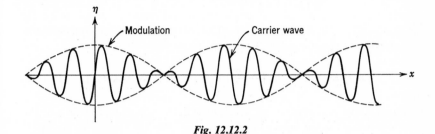

Fig. 12.12.2

Finally, let us consider two waves of the same frequency and amplitude which move in opposite directions:

$$\eta = Ae^{i(kx-\omega t)} + Ae^{i(kx+\omega t)}$$
$$= Ae^{ikx}(e^{-i\omega t} + e^{i\omega t}) \tag{12.12.8}$$
$$= 2Ae^{ikx}\cos \omega t$$

The real part is

$$\eta = 2A \cos kx \cos \omega t \tag{12.12.9}$$

and is recognized to be a *standing wave* with nodes at $kx = \pi/2, 3\pi/2, \ldots$.

12.13. THE STRING WITH BOTH ENDS FIXED

It is easy to find the eigenfunctions of a string of constant mass density μ per unit length and under a tension F. These are simply the standing waves

$$\eta(x_1 t) = \sin kx(a \sin \omega t + b \cos \omega t) \tag{12.13.1}$$

where a and b are constants and

$$k = \frac{\omega}{c} = \omega \sqrt{\frac{\mu}{F}} \tag{12.13.2}$$

If the string has a length l, then the requirement that $x = l$ be a node gives

$$k_n l = n\pi, \qquad n = 1, 2, 3, \ldots \tag{12.13.3}$$

The eigenfrequencies, therefore, are

$$\omega_n = \frac{n\pi}{l}\sqrt{\frac{F}{\mu}} = n\omega_1 \tag{12.13.4}$$

where ω_1 is the frequency of the fundamental mode of vibration and ω_2, ω_3, ... are the frequencies of the *harmonics* or *overtones*. The eigenfunctions are

$$\eta_n = \sqrt{\frac{2}{l}}\sin\left(\frac{n\pi x}{l}\right)(a_n \sin n\omega_1 t + b_n \cos n\omega_1 t) \tag{12.13.5}$$

where $\sqrt{2/l}$ is a normalizing factor included at this time for convenience. The general solution is a superposition of the eigenfunctions:

$$\eta(x, t) = \sum_{n=1}^{\infty}\sqrt{\frac{2}{l}}\sin\left(\frac{n\pi x}{l}\right)(a_n \sin n\omega_1 t + b_n \cos n\omega_1 t) \tag{12.13.6}$$

It is instructive to compare this solution with the solution (12.5.26) for the massless string loaded with f-point particles. For the loaded string, there are f-degrees of freedom, f possible eigenfunctions and eigenfrequencies, and (12.5.26) has a finite number of terms. For the continuous string, there are an infinite number of degrees of freedom and an infinite number of eigenfunctions and possible eigenfrequencies. For the loaded string, the completion of the solution required that the position and velocity of every particle be known at some time, say, $t = 0$. For the continuous string, it will be necessary to know the initial position and velocity of every point, i.e., the functions

$$\eta(x, 0) = f(x) \qquad \frac{\partial \eta}{\partial t}\bigg|_{t=0} = g(x) \tag{12.13.7}$$

must be known. At $t = 0$, (12.13.6) gives

$$f(x) = \sum_{n=1}^{\infty} b_n \sqrt{\frac{2}{l}}\sin\left(\frac{n\pi x}{l}\right) \tag{12.13.8}$$

$$g(x) = \sum_{n=1}^{\infty} a_n n\omega_1 \sqrt{\frac{2}{l}}\sin\left(\frac{n\pi x}{l}\right) \tag{12.13.9}$$

But these are just Fourier sine series for the known functions $f(x)$ and $g(x)$! The coefficients are determined in the usual way to give

$$b_n = \int_0^l f(x)\sqrt{\frac{2}{l}}\sin\left(\frac{n\pi x}{l}\right)dx \tag{12.13.10}$$

$$a_n = \int_0^l \frac{1}{n\omega_1} g(x)\sqrt{\frac{2}{l}}\sin\left(\frac{n\pi x}{l}\right)dx \tag{12.13.11}$$

The Fourier series solution obtained here can also be obtained by applying a limiting process directly to the solution for the loaded string obtained in Section 12.5. Let us write the components (12.5.25) of the normalized eigenvectors as

$$S_{nj} = \sqrt{s}\sqrt{\frac{2}{s(f+1)}} \sin\left(\frac{n\pi js}{s(f+1)}\right) \qquad (12.13.12)$$

where s is the separation between the point particles. In the limit

$$s \to \Delta x, \qquad f \to \infty, \qquad js \to x, \qquad s(f+1) \to l.$$

Hence,

$$S_{nj} \to \sqrt{s}\sqrt{\frac{2}{l}} \sin\left(\frac{n\pi x}{l}\right) \qquad (12.13.13)$$

The orthogonality condition (12.5.25) becomes

$$\sum_{j=1}^{f} S_{nj}S_{n'j} = \sum_{j=1}^{f} \frac{2}{l} \sin\left(\frac{n\pi x}{l}\right) \sin\left(\frac{n'\pi x}{l}\right) \Delta x$$

$$\to \int_{0}^{l} \frac{2}{l} \sin\left(\frac{n\pi x}{l}\right) \sin\left(\frac{n'\pi x}{l}\right) dx = \delta_{nn'} \qquad (12.13.14)$$

This establishes a direct connection between complete sets of orthogonal vectors as they are generally conceived in a space of finite dimension with the notion of orthogonality as is generally applied to complete sets of functions! The solution (12.5.26) becomes, in the limit,

$$\eta(x, t) = \sum_{n=1}^{\infty} \sqrt{\frac{2}{l}} \sin\left(\frac{n\pi x}{l}\right)(\sqrt{s}a_n \sin \omega_n t + \sqrt{s}b_n \cos \omega_n t) \quad (12.13.15)$$

For the coefficients $\sqrt{s}b_n$ we find, from (12.5.29),

$$\sqrt{s}b_n = \sum_{j=1}^{f} \sqrt{\frac{2}{s(f+1)}} \sin\left(\frac{\pi n js}{s(f+1)}\right) s x_j(0)$$

$$\to \int_{0}^{l} \sqrt{\frac{2}{l}} \sin\left(\frac{\pi n x}{l}\right) f(x) \, dx \qquad (12.13.16)$$

which is to be compared with (12.13.10).

12.14. ENERGY PROPAGATION BY PLANE WAVES

For the loaded string, the potential energy associated with the jth particle is

$$V_j = \frac{1}{2}\frac{F}{s}(x_j - x_{j-1})^2 \qquad (12.14.1)$$

If we divide through by s and interpret $\mathscr{V} = V_j/s$ as the *potential energy density* of the continuous string, then

$$\mathscr{V} = \tfrac{1}{2}F\left(\frac{\partial \eta}{\partial x}\right)^2 \tag{12.14.2}$$

where

$$\frac{x_j - x_{j-1}}{s} \to \frac{\partial \eta}{\partial x} \tag{12.14.3}$$

in the limit. The kinetic energy density is

$$\mathscr{T} = \tfrac{1}{2}\mu\dot{\eta}^2 \tag{12.14.4}$$

The total energy density, therefore, is

$$\mathscr{W} = \tfrac{1}{2}\mu\dot{\eta}^2 + \tfrac{1}{2}F\left(\frac{\partial \eta}{\partial x}\right)^2 \tag{12.14.5}$$

For the longitudinal waves propagating in one direction in an elastic solid of Young's modulus Y, the energy density (energy per unit volume) is

$$\mathscr{W} = \tfrac{1}{2}p\dot{\eta}^2 + \tfrac{1}{2}Y\left(\frac{\partial \eta}{\partial x}\right)^2 \tag{12.14.6}$$

Consider a simple plane wave solution of the wave equation:

$$\eta = a \cos(kx - \omega t) \qquad k = \frac{\omega}{c} \tag{12.14.7}$$

The energy density at any point in the medium through which such a wave is propagating is, for the example of the continuous string,

$$\mathscr{W} = a^2\mu\omega^2 \sin^2(kx - \omega t) \tag{12.14.8}$$

where $c = \sqrt{F/\mu}$ is used. The average energy density is

$$\overline{\mathscr{W}} = \tfrac{1}{2}a^2\mu\omega^2 \tag{12.14.9}$$

Since the plane wave propagates at the velocity c, the average energy transport past a given point is

$$p = \overline{\mathscr{W}}c = \tfrac{1}{2}ca^2\mu\omega^2 \tag{12.14.10}$$

Just as with energy propagation by plane waves in a periodic structure, the energy per second which a plane wave traveling in a continuous medium carries past a point is proportional to the square of the amplitude.

12.15. REFLECTION AND TRANSMISSION OF PLANE WAVES AT A BOUNDARY

Consider two uniform strings of different mass density which are joined together as in Fig. 12.15.1. Let $x = 0$ be the junction. Assume that a plane wave of frequency ω is incident on the junction from the left. Part of this wave will be transmitted to the second medium and part will

Fig. 12.15.1

be reflected. The total disturbance in the first medium consists of the sum of the original wave and the reflected wave:

$$\eta_1 = A_1 e^{i(\omega t - k_1 x)} + B_1 e^{i(\omega t + k_1 x)} \tag{12.15.1}$$

The wave transmitted to the second medium is

$$\eta_2 = A_2 e^{i(\omega t - k_2 x)} \tag{12.15.2}$$

The propagation numbers for the two strings are

$$k_1 = \frac{\omega}{c_1} = \omega \sqrt{\frac{\mu_1}{F}} \qquad k_2 = \frac{\omega}{c_2} = \omega \sqrt{\frac{\mu_2}{F}} \tag{12.15.3}$$

At $x = 0$, we obviously have $\eta_1 = \eta_2$, giving

$$A_1 + B_1 = A_2 \tag{12.15.4}$$

A second boundary condition is provided by the requirement that

$$\left(\frac{\partial \eta_1}{\partial x}\right)_{x=0} = \left(\frac{\partial \eta_2}{\partial x}\right)_{x=0} \tag{12.15.5}$$

A discontinuity in the tangent to the string would occur only if a point mass, such as a knot, were present. Equation (12.15.5) gives

$$(A_1 - B_1)k_1 = A_2 k_2 \tag{12.15.6}$$

Simultaneous solution of (12.15.4) and (12.15.6) yields the transmission and reflection coefficients

$$\frac{A_2}{A_1} = \frac{2k_1}{k_1 + k_2} \qquad \frac{B_1}{A_1} = \frac{k_1 - k_2}{k_1 + k_2} \tag{12.15.7}$$

If the second medium has the greater density, $k_2 > k_1$, the reflected wave is 180° out of phase with the incident wave; if $k_2 < k_1$ then the reflected wave is in phase with the incident wave. There phase relations are best visualized by means of the diagrams in Fig. 12.15.2 which show what happens to a single pulse which is incident on the junction.

If the second medium is infinitely dense, $k_2 \to \infty$, then $B \to -A_1$ and

$$\eta_1 = e^{i\omega t}[A_1 e^{-ik_1 x} - A_1 e^{ik_1 x}]$$
$$= -2Ae^{i\omega t} \sin (k_1 x) \qquad (12.15.8)$$

Thus the incident wave is entirely reflected and standing wave with a node at $x = 0$ results. This is identical to a string tied to a fixed support.

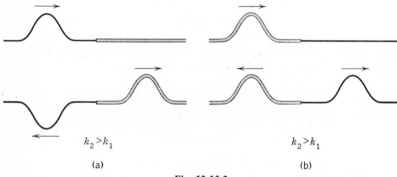

$k_2 > k_1$ $k_2 > k_1$

(a)   (b)

Fig. 12.15.2

It is possible, as in optics, to introduce a relative index of refraction by means of

$$n = \frac{c_1}{c_2} \qquad (12.15.9)$$

The transmission and reflection coefficients become

$$\frac{A_2}{A_1} = \frac{2}{1 + n} \qquad \frac{B_1}{A_1} = \frac{1 - n}{1 + n} \qquad (12.15.10)$$

These coefficients are identical to the transmission and reflection coefficients which are obtained for a plane electromagnetic wave at normal incidence on an optical medium.

12.16. THE GENERAL SOLUTION OF THE ONE-DIMENSIONAL WAVE EQUATION

It is easily verified that the wave equation (12.11.1) has the solution

$$\eta(x, t) = F(x - ct) + G(x + ct) \qquad (12.16.1)$$

where F and G are arbitrary functions of arguments $x - ct$ and $x + ct$. The plane wave solutions, e.g., $\eta = a \cos k(x - ct)$, are special cases of (12.16.1). If $F(x - ct)$ is graphed at $t = 0$ and at some later time, Fig. 12.16.1, it becomes apparent that $F(x - ct)$ represents a disturbance which propagates *without change of shape* in the positive x-direction. Similarly, $G(x + ct)$ represents a disturbance which propagates in the negative x-direction. *All* disturbances propagate with the same velocity c.

To see how the functions F and G are to be determined, suppose that $\eta(x, 0) = f(x)$ and $(\partial\eta/\partial t)_{t=0} = g(x)$ are known functions over the interval

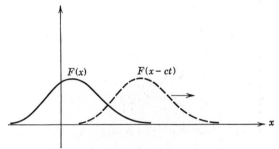

Fig. 12.16.1

$a \le x \le b$. These functions could, for instance, represent the initial position and velocity of a stretched string. Thus, at $t = 0$,

$$f(x) = F(x) + G(x) \tag{12.16.2}$$

Differentiation of (12.16.1) with respect to t yields

$$\frac{\partial\eta}{\partial t} = -cF'(x - ct) + cG'(x + ct) \tag{12.16.3}$$

where the prime stands for differentiation with respect to the argument. At $t = 0$,

$$g(x) = -c\frac{dF(x)}{dx} + c\frac{dG(x)}{dx} \tag{12.16.4}$$

If (12.16.4) is integrated between a and x,

$$G(x) - F(x) = G(a) - F(a) + \frac{1}{c}\int_a^x g(x)\,dx \tag{12.16.5}$$

This equation together with (12.16.2) can be solved for $G(x)$ and $F(x)$:

$$G(x) = \tfrac{1}{2}[G(a) - F(a)] + \tfrac{1}{2}f(x) + \frac{1}{2c}\int_a^x g(x)\,dx \tag{12.16.6}$$

$$F(x) = -\tfrac{1}{2}[G(a) - F(a)] + \tfrac{1}{2}f(x) - \frac{1}{2c}\int_a^x g(x)\,dx \tag{12.16.7}$$

Using these results, the solution (12.16.1) can be expressed

$$\eta(x, t) = \tfrac{1}{2}[f(x + ct) + f(x - ct)] + \frac{1}{2c} \int_{x-ct}^{x+ct} g(x)\, dx \quad (12.16.8)$$

The curves $x + ct = $ const. and $x - ct = $ const. are sketched in the xt-plane of Fig. 12.16.2 and are called the *characteristics* of the hyperbolic differential equation (12.11.1). Equation (12.16.8) does not actually give the solution over $a \leq x \leq b$ for all times. Suppose, for instance, that $\eta(x, t)$ is desired at the time and position indicated by point P in Fig.

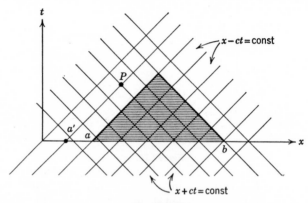

Fig. 12.16.2

12.16.2. If the characteristics are drawn through P, it is found that one of them passes through the point a' on the x-axis. But the functions f and g are not specified here so that the solution at P cannot be obtained from (12.16.8). Information about the motion propagates along the characteristic curves, so to speak. To put it another way, a disturbance which was originally at a', and hence out of the domain where initial conditions are known, has had time to propagate to the x-coordinate of P and is now influencing the motion there. For $t > 0$ the solution is entirely specified only in the triangular shaded region of Fig. 12.16.2. The solution for all times over $a \leq x \leq b$ could be specified if some boundary condition at a and b were given, such as that these are fixed points or that $\eta(a, t)$ and $\eta(b, t)$ are prescribed functions for all time.

The conditions $\eta(x, 0) = f(x)$ and $(\partial \eta / \partial t)_0 = g(x)$, which are the function and its normal derivative specified over $a \leq x \leq b$, are known as *Cauchy conditions*. It is shown in the theory of partial differential equations that Cauchy conditions specify the solution uniquely in the shaded region.

As an example, suppose that the initial displacement of a string attached to rigid supports is as illustrated in Fig. 12.16.3a. The initial velocity is zero. The solution, therefore, is

$$\eta(x, t) = \tfrac{1}{2}[f(x + ct) + f(x - ct)] \qquad (12.16.9)$$

where $f(x)$ is the initial displacement of Fig. 12.16.3a. As time progresses, $\tfrac{1}{2}f(x + ct)$ propagates in the negative x-direction and $\tfrac{1}{2}f(x - ct)$ propagates in the positive x-direction as indicated by the dashed curves in the succession of diagrams. The two dashed curves always add up to zero at the two ends of the string as is required by the boundary conditions. This is

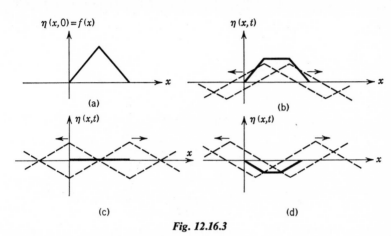

Fig. 12.16.3

actually a type of standing wave produced by two identical saw-tooth type waves moving in opposite directions! For a Fourier series solution of the same example, see Problem 12.13.17.

12.17. DISPERSION IN PERIODIC STRUCTURES: GROUP VELOCITY

When a continuous medium is approached as a limit of a periodic structure, dispersion disappears. In this section, we return to the consideration of waves in periodic structures and introduce the idea of *group velocity*, a concept which is of importance in the theory of dispersive media. A plane wave in a periodic structure can be represented as

$$\eta = Ae^{i(kx - \omega t)} \qquad (12.17.1)$$

provided that the propagation number k is defined by

$$ks = \beta \qquad \cos ks = 1 - \frac{m\omega^2}{2h} \qquad (12.17.2)$$

where h is the spring constant for the longitudinal vibration of a mass spring system and $h = F/s$ for the transverse vibrations of a loaded string. Actual particles exist only at $x = s, 2s, 3s, \ldots$. If two plane waves of slightly different frequencies are added together, the result is still Eq. (12.12.6), namely,

$$\eta = 2A \cos (kx - \omega t) \cos \left(\frac{\Delta kx - \Delta \omega t}{2} \right) \tag{12.17.3}$$

Because of the existence of dispersion in the medium, the two plane waves that have been added together to get (12.17.3) do not move at the same

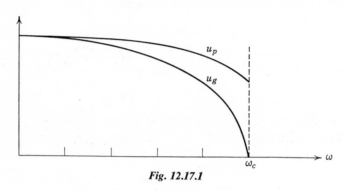

Fig. 12.17.1

velocity. Moreover, the modulations or groups move at still another velocity given by

$$u_g = \frac{\Delta \omega}{\Delta k} \rightarrow \frac{d\omega}{dk} \tag{12.17.4}$$

and called the *group velocity*. By means of (12.17.2)

$$u_g = \frac{hs}{m\omega} \sin (ks) \tag{12.17.5}$$

As ω varies from 0 up to the critical or cut-off frequency, ks varies from 0 to π. Over this range, $\sin ks > 0$. Hence,

$$u_g = s \sqrt{\frac{h}{m}} \sqrt{1 - \frac{m\omega^2}{4h}} \tag{12.17.6}$$

By contrast, the phase velocity of a single plane wave is

$$u_p = \frac{s\omega}{\cos^{-1} \left(1 - \dfrac{m\omega^2}{2h} \right)} \tag{12.17.7}$$

Graphs of u_g and u_p over the range $0 < \omega < \omega_c$ can be seen in Fig. 12.17.1.

It will now be shown that the group velocity is the velocity of energy transport in the periodic structure. Each particle can be regarded as a harmonic oscillator and as the two plane waves pass through the medium, the motion of a particle can be represented by

$$\eta = A \cos \omega_1 t + A \cos \omega_2 t \qquad (12.17.8)$$

The kinetic energy is

$$\tfrac{1}{2}m\dot{\eta}^2 = \tfrac{1}{2}mA^2(\omega_1^2 \sin^2 \omega_1 t + 2\omega_1\omega_2 \sin \omega_1 t \sin \omega_2 t + \omega_2^2 \sin^2 \omega_2 t)$$
$$(12.17.9)$$

Over a long time, the cross term will average to zero:

$$\bar{T} = \overline{\tfrac{1}{2}m\dot{\eta}^2} = \tfrac{1}{2}mA^2(\tfrac{1}{2}\omega_1^2 + \tfrac{1}{2}\omega_2^2) \qquad (12.17.10)$$

From the virial theorem it is known that for a particle moving in a harmonic oscillator potential $\bar{T} = \bar{V}$. Moreover, $\omega_1 \cong \omega_2$ so that

$$\bar{\mathscr{E}} = \bar{T} + \bar{V} = mA^2\omega^2 \qquad (12.17.11)$$

The energy density found in the medium through which the disturbance is passing is

$$\mathscr{W} = \frac{\bar{\mathscr{E}}}{s} = \frac{mA^2\omega^2}{s} \qquad (12.17.12)$$

Suppose that there are two generators operating at frequencies ω_1 and ω_2, feeding energy into the structure. By (12.9.8),

$$\bar{P} = 2 \cdot \tfrac{1}{2}A^2\omega^2\sqrt{mh}\sqrt{1 - \omega^2 m/4h} \qquad (12.17.13)$$

In (12.9.8), A is current (velocity) amplitude. Here A is displacement amplitude; hence, the extra factor of ω^2. Again, $\omega_1 \cong \omega_2 = \omega$. From (12.17.6) and (12.17.12) it is seen that (12.17.13) can be expressed by

$$\bar{P} = \mathscr{W}u_g \qquad (12.17.14)$$

Hence, the result: If two plane waves of slightly different frequency pass through a periodic structure, the velocity of energy transport is given by the group *velocity defined* as $u_g = d\omega/dk$.

12.18. ATTENUATION AND DISPERSION IN DISSIPATIVE MEDIA

In this section, brief attention is given to the propagation of plane waves in continuous dissipative media. The waves obey the differential equation

$$c^2 \frac{\partial^2 \eta}{\partial x^2} - b \frac{\partial \eta}{\partial t} - \frac{\partial^2 \eta}{\partial t^2} = 0 \qquad (12.18.1)$$

It is found that plane wave solutions of the form

$$\eta = Ae^{i(kx - \omega t)} \qquad (12.18.2)$$

exist, provided that

$$-c^2 k^2 + bi\omega + \omega^2 = 0 \qquad (12.18.3)$$

Thus, for a wave of given frequency ω,

$$k = \frac{\omega}{c}\sqrt{1 + i\frac{b}{\omega}} \cong \frac{\omega}{c}\left[1 + i\frac{b}{2\omega} + \frac{b^2}{8\omega^2} + \cdots\right] \qquad (12.18.4)$$

provided that the damping is small. The solution (12.18.2) becomes

$$\eta(x, t) = Ae^{-(b/2c)x}e^{i(k'x - \omega t)} \qquad (12.18.5)$$

where

$$k' \cong \frac{\omega}{c} + \frac{b^2}{8\omega c} \qquad (12.18.6)$$

(See Problem 12.18.25 for the exact value of k' in terms of ω.) A plane wave of given frequency ω is *attenuated* by the factor $e^{-bx/2c}$ on entering a dissipative medium. The phase velocity of the wave is

$$u_p = \frac{\omega}{k'} = c\left(1 - \frac{b^2}{8\omega^2}\right) \qquad (12.18.7)$$

Thus dissipative media are also *dispersive* because of the dependence of the phase velocity on the frequency of the wave.

12.19. THE FOURIER INTEGRAL REPRESENTATION OF A WAVE PACKET

In this section we once more consider waves in a nondispersive medium which obey the wave equation

$$\frac{\partial^2 \eta}{\partial x^2} - \frac{1}{c^2}\frac{\partial^2 \eta}{\partial t^2} = 0 \qquad (12.19.1)$$

In the development of this section, we use the Fourier integral theorem: Let $F(x)$ be a function which is defined over $-\infty < x < +\infty$. Then the Fourier transform of $F(x)$ is defined to be

$$A(k) = \frac{1}{\sqrt{2\pi}}\int_{-\infty}^{+\infty} F(x)e^{-ikx}\,dx \qquad (12.19.2)$$

provided, of course, that $F(x)$ is such that the integral exists. $A(k)$ is also called the spectral distribution of $F(x)$, or simply the spectrum of $F(x)$. We state, without formal proof, that the transform (12.19.2) can be inverted to yield

$$F(x) = \frac{1}{\sqrt{2\pi}} \int_{-\infty}^{+\infty} A(k)e^{ikx}\, dk \qquad (12.19.3)$$

Thus, if $A(k)$ is already known, $F(x)$ is determined uniquely. Most texts on advanced applied mathematics contain an account of the theory of Fourier integrals.

Another result of great usefulness follows if $A(k)$ as given by (12.19.2) is formally substituted into (12.19.3):

$$F(x) = \int_{-\infty}^{+\infty} F(x')\left\{\frac{1}{2\pi} \int_{-\infty}^{+\infty} e^{ik(x-x')}\, dk\right\} dx' \qquad (12.19.4)$$

The concept of the Dirac δ-function is introduced in Section 4.11. The comparison of (12.19.4) and (4.11.22) reveals that

$$\delta(x - x') = \frac{1}{2\pi} \int_{-\infty}^{+\infty} e^{ik(x-x')}\, dk \qquad (12.19.5)$$

which may be regarded as the Fourier integral representation of the δ-function. In the classical mathematical sense, (12.19.5) is a divergent integral. It is also possible to think of (12.19.5) as a type of orthogonality condition applicable to the "complete set" of functions e^{ikx} defined over $-\infty < x < +\infty$.

Suppose that the solution $\eta(x, t)$ of the wave equation is desired over $-\infty < x < +\infty$. The representation of $\eta(x, t)$ as a Fourier integral is

$$\eta(x, t) = \frac{1}{\sqrt{2\pi}} \int_{-\infty}^{+\infty} A(k, t)e^{ikx}\, dk \qquad (12.19.6)$$

The wave equation (12.19.1) requires that

$$\frac{1}{\sqrt{2\pi}} \int_{-\infty}^{+\infty} \left[k^2 A(k, t) + \frac{1}{c^2}\frac{\partial^2 A(k, t)}{\partial t^2}\right] e^{ikx}\, dk = 0 \qquad (12.19.7)$$

If the Fourier transform of a function is identically zero, then the function itself is zero:

$$\frac{\partial^2 A(k, t)}{\partial t^2} + k^2 c^2 A(k, t) = 0 \qquad (12.19.8)$$

The general solution of (12.19.8) is

$$A(k, t) = A(k)e^{-ikct} + B(k)e^{ikct} \qquad (12.19.9)$$

The solution (12.19.6) of the wave equation is now

$$\eta(x, t) = \frac{1}{\sqrt{2\pi}} \int_{-\infty}^{+\infty} A(k)e^{i(kx-kct)}\, dk + \frac{1}{\sqrt{2\pi}} \int_{-\infty}^{+\infty} B(k)e^{i(kx+kct)}\, dk$$

$$(12.19.10)$$

The first term in (12.19.10) represents a disturbance which propagates in the positive x-direction; the second term propagates in the negative x direction. In what follows, we will concentrate on that part of the solution which propagates in the positive x-direction:

$$\eta(x, t) = \frac{1}{\sqrt{2\pi}} \int_{-\infty}^{+\infty} A(k)e^{i(kx-kct)} \, dk \qquad (12.19.11)$$

In Section 12.12, we superimposed two waves of slightly different frequency moving in the same direction to obtain the wave packets of Fig. 12.12.2. Equation (12.19.11) is to be interpreted as the superposition of an infinite number of waves all moving in the positive x-direction and with a *continuous spectrum* of frequencies. The general distribution function $A(k)$ determines the actual shape of the disturbance or wave packet.

Suppose that at $t = 0$, $\eta(x, t)$ is known:

$$\eta(x, 0) = F(x) = \frac{1}{\sqrt{2\pi}} \int_{-\infty}^{+\infty} A(k)e^{ikx} \, dk \qquad (12.19.12)$$

The spectrum $A(k)$ is determined by the Fourier inversion theorem:

$$A(k) = \frac{1}{\sqrt{2\pi}} \int_{-\infty}^{+\infty} F(x')e^{-ikx'} \, dx' \qquad (12.19.13)$$

By combining (12.19.11) and (12.19.13),

$$\eta(x, t) = \frac{1}{2\pi} \int_{-\infty}^{+\infty} F(x') \int_{-\infty}^{+\infty} e^{ik(x-x'-ct)} \, dk \, dx' \qquad (12.19.14)$$

Now, employing (12.19.5),

$$\eta(x, t) = \int_{-\infty}^{+\infty} F(x')\delta(x - x' - ct) \, dx' = F(x - ct) \qquad (12.19.15)$$

We have merely recovered a piece of the general solution (12.16.1). In a *nondispersive* medium, a wave packet moving in the positive x-direction continues to do so without change of shape. The velocity of the wave packet is the same as the phase velocity of a single plane wave.

In a dispersive medium, it is still possible to represent a wave packet as a Fourier integral, but all the different plane wave components have different velocities! This means that such a wave packet will not retain its original shape, but will spread out as it moves. One of the most famous examples of the spreading of a wave packet occurs in connection with the wave function for a free particle in quantum mechanics. This example is also one of the simplest to treat and is taken up in the next section.

As an example of a disturbance of finite extent in space, consider

$$F(x) = e^{ik_0 x} \qquad -\Delta x < x < \Delta x$$
$$= 0 \qquad |x| > \Delta x \qquad (12.19.16)$$

The spectrum of this function is

$$A(k) = \frac{1}{\sqrt{2\pi}} \int_{-\Delta x}^{+\Delta x} e^{ix(k_0 - k)} \, dx$$

$$= \sqrt{\frac{2}{\pi}} \frac{\sin \Delta x (k_0 - k)}{k_0 - k} \qquad (12.19.17)$$

This function is sketched in Fig. 12.19.1. The wave numbers which contribute most to the wave packet lie between $\Delta x(k_0 - k) = -\pi$ and $\Delta x(k_0 - k) = +\pi$. Outside this range, $A(k)$ is very small. Thus, in

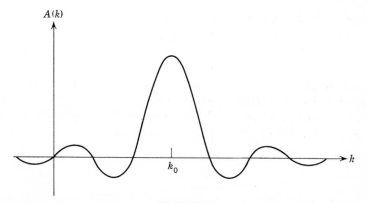

Fig. 12.19.1

order to make up the wave train of finite extent in space as given by (12.19.16), we must add plane waves, the wave numbers of which cover the range

$$\Delta x \, \Delta k = 2\pi \qquad (12.19.18)$$

As the wave train becomes longer, $\Delta x \to \infty$, the range of wave numbers Δk becomes small. The disturbance becomes more *monochromatic*. In reality, all disturbances are of finite duration and, hence, no disturbance consists of a pure frequency.

12.20. THE SCHRÖDINGER WAVE EQUATION

According to the de Broglie hypothesis, the matter waves associated with a material particle of definite momentum p have a wavelength given by

$$\lambda = \frac{h}{p} \qquad (12.20.1)$$

where h is Planck's constant. The frequency of these waves is connected with the particle energy by

$$\mathscr{E} = h\nu \qquad (12.20.2)$$

In terms of the propagation number and circular frequency

$$p = \hbar k \qquad \mathscr{E} = \hbar\omega \qquad \hbar = \frac{h}{(2\pi)} \qquad (12.20.3)$$

For free particles, the connection between energy and momentum is

$$\mathscr{E} = p^2/(2m) \qquad (12.20.4)$$

This gives the relation between ω and k:

$$\omega = \frac{\hbar k^2}{2m} \qquad (12.20.5)$$

Because of this nonlinear relation between ω and k, the matter waves associated with a material particle exhibit dispersion, i.e., the phase velocities of plane waves associated with different particle energies and momenta are different. The group velocity is given by

$$\frac{d\omega}{dk} = \frac{\hbar k}{m} = \frac{p}{m} \qquad (12.20.6)$$

and is, classically speaking, the velocity of the particle itself. Thus the group velocity is more physically meaningful than the phase velocity.

The plane waves of free-particle quantum mechanics can be expressed by

$$\psi = \psi_0 e^{i(kx - \omega t)} \qquad (12.20.7)$$

where ψ is the wave function or wave field of the particle. ψ itself has no direct physical significance but the "intensity" $|\psi|^2 = \psi\psi^*$ is interpreted as a *probability density*: $|\psi|^2\, dx$ is the probability of finding the particle between x and $x + dx$. For the plane wave, $|\psi|^2 = \psi_0^2 = $ const. Thus a plane wave, stretching from $-\infty$ to $+\infty$, is the wave function of a free particle with a precisely defined momentum and energy but completely undetermined position!

In order to construct a more localized wave packet which might represent a particle known to be in an interval Δx, it is necessary to add together many plane waves of different frequencies. According to

(12.19.18), the wave numbers must cover the range

$$\Delta k = \frac{2\pi}{\Delta x} \qquad (12.20.8)$$

In terms of the momentum as given by (12.20.3),

$$\Delta p \, \Delta x = h \qquad (12.20.9)$$

But the wave packet, once formed, will spread out due to the dispersion! Thus, more generally,

$$\Delta p \, \Delta x > h \qquad (12.20.10)$$

This is the Heisenberg uncertainty relation.

But how can it be that a whole range of momenta is associated with a single free particle? The particle cannot, of course, have more than one momentum at any one time. Equation (12.20.9) means that the price paid for knowing the position of the particle to within Δx is an unremovable uncertainty in the knowledge of its momentum. In other words, at a given instant, the exact position and momentum of a particle cannot be known, the ultimate limit on the accuracy of such measurements being given by (12.20.9).

The wave equation satisfied by plane waves of the form (12.20.7) can be easily derived. The derivatives are

$$\frac{\partial \psi}{\partial t} = -i\omega\psi \qquad \frac{\partial^2 \psi}{\partial x^2} = -k^2\psi \qquad (12.20.11)$$

By means of (12.20.5),

$$i\hbar \frac{\partial \psi}{\partial t} = -\frac{\hbar^2}{2m} \frac{\partial^2 \psi}{\partial x^2} \qquad (12.20.12)$$

The fact that the Schrödinger wave equation has a first derivative with respect to time and a second derivative with respect to x is directly related to the energy-momentum relation (12.20.4). Plane matter waves cannot be expressed by

$$\psi = \psi_0 \cos(kx - \omega t) \qquad (12.20.13)$$

because such functions obey a wave equation of the form (12.19.1) and require a linear connection between ω and k. In our previous work, the representation of various quantities in terms of complex numbers has been a mathematical convenience. In quantum mechanics, the wave-field ψ is intrinsically complex.

Let us find a more general solution of (12.20.12) by the Fourier transform method outlined in Section 12.19. If we write

$$\psi(x, t) = \frac{1}{\sqrt{2\pi}} \int_{-\infty}^{+\infty} A(k, t)e^{ikx}\, dk \qquad (12.20.14)$$

the wave equation requires that

$$\frac{\partial A(k, t)}{\partial t} + \frac{i\hbar k^2}{2m} A(k, t) = 0 \qquad (12.20.15)$$

which is solved to yield

$$A(k, t) = A(k)e^{-(i\hbar k^2/2m)t} \qquad (12.20.16)$$

Hence,

$$\psi(x, t) = \frac{1}{\sqrt{2\pi}} \int_{-\infty}^{+\infty} A(k) \exp i\left(kx - \frac{\hbar k^2}{2m} t\right) dk \qquad (12.20.17)$$

As anticipated, this general solution represents a wave packet formed by adding together many plane waves. The shape of the packet will not remain the same as time progresses, as was the case with the solution (12.19.15) of the wave equation (12.19.1).

The three-dimensional Schrödinger wave equation for a particle moving in a conservative force field given by a potential $V(x, y, z)$ is

$$i\hbar \frac{\partial \psi}{\partial t} = -\frac{\hbar^2}{2m} \nabla^2 \psi + V\psi \qquad (12.20.18)$$

12.21. A TWO-DIMENSIONAL WAVE EQUATION

Suppose that a membrane is stretched tight over an area bounded by a closed curve s as in Fig. 12.21.1. We wish to find the vibrations of the membrane perpendicular to the xy-plane. A common example is the

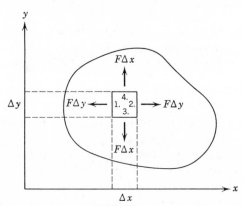

Fig. 12.21.1

motion of a drumhead. Let F be the tension in the membrane in units of force per unit length. The forces on the edges of the element of area $\Delta x\, \Delta y$ in Fig. 12.21.1 are $F\, \Delta y$ and $F\, \Delta x$. Figure 12.21.2 shows a profile of the membrane seen edge-on. The components of the forces shown in the figure in the direction of the displacement are

$$F\, \Delta y \theta_2 - F\, \Delta y \theta_1 = F\, \Delta y \left(\left.\frac{\partial \eta}{\partial x}\right|_2 - \left.\frac{\partial \eta}{\partial x}\right|_1 \right) \tag{12.21.1}$$

where small displacements and angles are assumed. If the forces on the

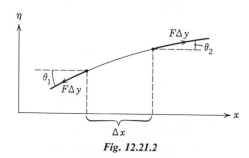

Fig. 12.21.2

other two edges of the element are similarly taken into account, the equation of motion of the element can be written

$$F\, \Delta y \left(\left.\frac{\partial \eta}{\partial x}\right|_2 - \left.\frac{\partial \eta}{\partial x}\right|_1 \right) + F\, \Delta x \left(\left.\frac{\partial \eta}{\partial y}\right|_4 - \left.\frac{\partial \eta}{\partial y}\right|_3 \right) = \rho\, \Delta x\, \Delta y\, \frac{\partial^2 \eta}{\partial t^2} \tag{12.21.2}$$

where ρ is the surface mass density (gm/cm^2) assumed constant. Equation (12.21.2) is now divided by $\Delta x\, \Delta y$ and the following limits taken:

$$\frac{\left.\frac{\partial \eta}{\partial x}\right|_2 - \left.\frac{\partial \eta}{\partial x}\right|_1}{\Delta x} \to \frac{\partial^2 \eta}{\partial x^2} \qquad \frac{\left.\frac{\partial \eta}{\partial y}\right|_4 - \left.\frac{\partial \eta}{\partial y}\right|_3}{\Delta y} \to \frac{\partial^2 \eta}{\partial y^2} \tag{12.21.3}$$

The result is

$$\frac{\partial^2 \eta}{\partial x^2} + \frac{\partial^2 \eta}{\partial y^2} - \frac{1}{c^2}\frac{\partial^2 \eta}{\partial t^2} = 0 \qquad c = \sqrt{\frac{F}{\rho}} \tag{12.21.4}$$

where c is the velocity of propagation of a disturbance on the membrane.

An important special solution of the two-dimensional wave equation is the *plane wave*, an example of which is

$$\eta = a \cos((\mathbf{k} \cdot \mathbf{r} - \omega t) \tag{12.21.5}$$

where $\mathbf{r} = \hat{\imath}x + \hat{\jmath}y$ and $\mathbf{k} = \hat{\imath}k_1 + \hat{\jmath}k_2$. It is easy to show that (12.21.5) is a solution of (12.21.4) provided that

$$\omega = kc \qquad k = \sqrt{k_1{}^2 + k_2{}^2} \tag{12.21.6}$$

The curves found by setting the phase of (12.21.5) equal to a constant,

$$\mathbf{k} \cdot \mathbf{r} - \omega t = \phi = \text{const.} \tag{12.21.7}$$

are, at a given time, straight lines. If the value of the constant is chosen to be some multiple of 2π, then η has its maximum value and we may think of the straight lines as being wave crests as illustrated in Fig. 12.21.3. In three-dimensional propagation of plane waves, the surfaces of constant phase are *planes*; hence, the term *plane* waves. The vector $\mathbf{k}$ is perpendicular to the wave crests and is called the *propagation vector*; its magnitude is

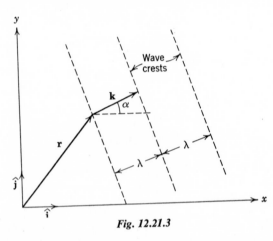

Fig. 12.21.3

the *propagation number*. It is convenient to introduce the components of $\mathbf{k}$ as $k_1 = k \cos \alpha$, $k_2 = k \sin \alpha$. The phase of the plane wave is then

$$k(x \cos \alpha + y \sin \alpha) - \omega t = \phi \tag{12.21.8}$$

At a fixed point in space, the time required for the phase to change by 2π is the period:

$$\omega P = 2\pi \tag{12.21.9}$$

By differentiating (12.21.8) and setting $d\phi/dt = 0$ we find that a point of constant phase, e.g., a point on the crest of a wave, moves according to

$$\dot{x} \cos \alpha + \dot{y} \sin \alpha = c \tag{12.21.10}$$

Thus the waves advance in the direction of the propagation vector $\mathbf{k}$ at a speed c. Moreover, if ωt increases by 2π, $x \cos \alpha + y \sin \alpha$ increases by one wavelength:

$$k\lambda - 2\pi = 0 \qquad k = \frac{2\pi}{\lambda} \tag{12.21.11}$$

Consider the three-dimensional Cartesian coordinate system of Fig. 12.21.4 which is constructed out of the three variables x, y, and t. The surfaces

$$r - ct = \text{const.} \qquad (12.21.12)$$

are *characteristic surfaces* of the two-dimensional wave equation. If we write $r = \sqrt{(x - x_p)^2 + (y - y_p)^2}$ then the characteristic surface $r - ct = r_0$ intersects the xy-plane in a circle of radius r_0 centered at x_p, y_p. The surface is actually a cone with vertex at x_p, y_p, $t_p = r_0/c$ as shown in

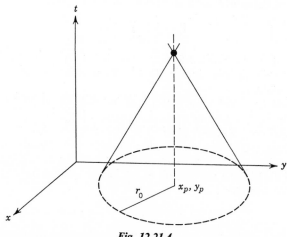

Fig. 12.21.4

Fig. 12.21.4. The significance of the characteristic surface is that all waves which arrive at the point x_p, y_p of the membrane at $t = t_p$ were on the circle of radius r_0 at $t = 0$. Conversely, if the membrane is struck a sharp blow, circular waves go out at velocity c and generate a characteristic surface in the xyt-space. In the theory of the propagation of electromagnetic waves, the characteristic surfaces are called *light cones*.

PROBLEMS

12.1.1   Verify Eq. (12.1.7).

12.2.2   A uniform thin rod of mass M and length s is suspended at its ends by two identical springs. Find the linear approximation to the motion if the rod is constrained to move in the xy-plane. Let x and y be the coordinates of the center of mass and θ be the angle of rotation, all measured with respect to the equilibrium position of the rod. What are the normal coordinates?

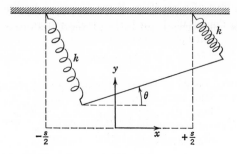

12.3.3   Find the general solution of the system (12.3.2) if

(a) $\dfrac{k}{m} < \left(\dfrac{b}{2m}\right)^2 < \dfrac{k + 2k_3}{m}$ (b) $\left(\dfrac{b}{2m}\right)^2 = \dfrac{k}{m}$

12.4.4   Find the Lagrangian and the Rayleigh dissipation function for the mechanical system pictured in Fig. 12.4.1. Verify that they give the correct equations of motion.

12.4.5   Find the steady-state velocity of the mass to which the force F is applied in Fig. 12.4.1. Use the method of complex impedances.

12.5.6   Consider the system pictured in Fig. 12.5.1 for the case where there are four masses. Write out the complete solution if the initial conditions are $x_1 = x_0, x_2 = x_3 = x_4 = 0,\ \dot{x}_1 = \dot{x}_2 = \dot{x}_3 = \dot{x}_4 = 0.$

12.6.7   Show that the nth element of the illustrated transmission line obeys the differential equation

$$L(2\ddot{q}_n - \ddot{q}_{n-1} - \ddot{q}_{n+1}) + \frac{1}{C}q_n = 0$$

and that it acts as a *high-pass filter*. Find the expression for the lowest frequency that the net work will transmit.

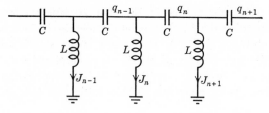

12.7.8

Consider the network and the periodic input in the illustration. The input is

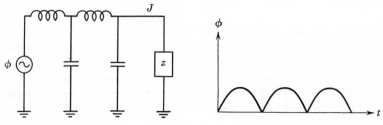

typical of the output of a rectifier. If the frequency of the input voltage is somewhat greater than $2/\sqrt{LC}$, predict, without actually making a calculation, what the out-put current J to the load z will be like. *Hint:* Consider the Fourier series expansion of $\phi(t)$.

12.8.9 Show that $z_k e^{-i(\beta/2)} = z_k * e^{i(\beta/2)}$.

12.8.10 Consider the longitudinal vibrations of the chain of masses and springs which terminates with a mass M as illustrated. What is the electric circuit analogue? What is the impedance of the terminal element of the chain? Write down the differential equation of motion of the mass M. Derive the reflection coefficient for plane waves by assuming solutions of the form

$$x_n = (Ae^{i\beta} + Be^{-i\beta})e^{i\omega t}$$
$$x_{n+1} = (A + B)e^{i\omega t}$$

Check the result against Eq. (12.8.4).

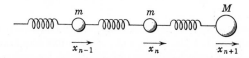

12.8.11 Refering to the results of Problem 12.8.10, show that if $M \to \infty$, the system becomes identical to a chain terminating on a fixed support. Show that if $M = m/2$, the incident and reflected wave add up to a standing wave with an *antinode* at the location of the terminal mass. Where is the antinode located if $M = m$? *Hint:* Use the result of Problem 12.8.9. Find all the eigenfrequencies for the longitudinal vibrations of a chain of identical particles which is *free* at both ends.

12.9.12 Show that if $\omega > 2/\sqrt{L_2 C_2}$, the incident wave undergoes *total reflection* at the boundary of the two lattices in Fig. 12.9.1.

12.9.13 Let ω be greater than the critical frequency in the periodic structure of Fig. 12.6.1. Show that a solution of (12.6.1) exists which is of the form of an *attenuated wave:*

$$x_{n-1} = Ae^{i\omega t}e^{\beta}, \qquad x_n = Ae^{i\omega t},$$
$$x_{n+1} = Ae^{i\omega t}e^{-\beta}, \qquad x_{n+2} = Ae^{i\omega t}e^{-2\beta} \cdots$$

where, if $\beta = \gamma + i\pi$,

$$\cosh \gamma = \frac{m\omega^2}{2k} - 1$$

Suppose a plane wave is incident on the junction of the two lattices in Fig. 12.9.1 and that in the second lattice $\omega = 2\omega_{crit.}$. Estimate how far the disturbance penetrates the second lattice by computing how much the amplitude of the wave has decreased by the time it reaches the second full inductor.

12.9.14 A blower pumps air through the acoustical filter or muffler as shown in the illustration. The smaller pipe has a cross sectional area $\sigma = 1$ ft^2 and the short segments connecting the large chambers have a length of $s = 2$ ft. Find the correct volume V of the large chambers if the device is to cut out blower noise above 100 cycles per second. See Section 4.9. Take the velocity of sound in air as 1100 ft/sec.

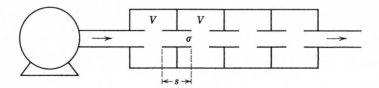

12.10.15 The telegrapher's equation, which takes into account not only the ohmic resistance of the wire but also leakage from the wire to ground, is

$$\frac{\partial I}{\partial t}(ls + rc) + Irs + \frac{\partial^2 I}{\partial t^2}lc = \frac{\partial^2 I}{\partial y^2}$$

where s is *leakage conductance* per unit length to ground, i.e., the reciprocal of the leakage resistance of 1 cm of the wire. Derive this equation as a limiting case of the equation obeyed by the periodic structure shown in the figure. The value of the leakage resistance is denoted $1/s$.

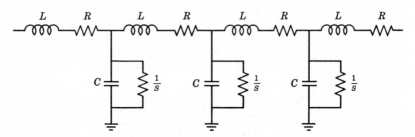

12.12.16 Show that the addition of the $N + 1$ plane waves indicated by

$$\eta = \cos \psi + \cos (\psi + \phi) + \cos (\psi + 2\phi) + \cdots + \cos (\psi + N\phi)$$

where $\psi = kx - \omega t$ gives the result

$$\eta = \cos \left(\psi + \frac{N\phi}{2}\right) \frac{\sin \frac{N+1}{2}\phi}{\sin \frac{\phi}{2}}$$

If $N = 7$, sketch the amplitude of η as a function of ϕ.

12.13.17 A string is stretched between fixed supports and is given the initial displacement shown in the figure. The initial velocity is zero. Compute all the coefficients in Eq. (12.13.6). Write out the first three terms of $\eta(x, t)$.

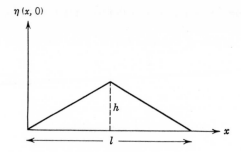

$\eta(x, 0)$

h

l

12.13.18   Show that the eigenfrequencies (12.5.18) for the loaded string become identical to (12.13.4) in the limit of a continuous string. Recall that the spring constant must be replaced by F/s for the loaded string.

12.13.19   Find the eigenfrequencies for the propagation of acoustical waves along a hollow pipe closed at one end. As boundary conditions, assume that $x = 0$ is a displacement node and $x = l$ is a pressure node.

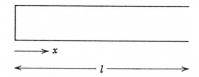

x

l

12.14.20   Derive (12.14.10) by a limiting process starting with (12.9.8). The amplitude A in (12.9.8) is current amplitude. In the analogous mechanical system it is *velocity amplitude*. In (12.14.10), a is *displacement amplitude*. Can you identify the characteristic impedance of a continuous infinite string?

12.15.21   Consider the energy flow in the incident, transmitted, and reflected waves and show that the coefficients (12.15.7) satisfy the condition that energy is conserved.

12.15.22   A string of uniform mass density has a knot tied in it which can be considered as a point mass. What happens when a plane wave is incident on the knot?

12.16.23   A stretched string with both ends fixed is struck a sharp blow in its center so that its initial shape is as illustrated. Show on a series of diagrams how the motion of the string progresses in time.

12.17.24   Show that the phase velocity and the group velocity become the same at low frequencies. Is this to be expected? What is the limiting value of u_p and u_g as $\omega \to 0$?

12.18.25 Show that the precise value of k' in Eq. (12.18.5) is given by

$$k' = \frac{\omega}{c\sqrt{2}} \sqrt{\sqrt{1 + \frac{b^2}{\omega^2}} + 1}$$

12.19.26 A function $F(x)$ is defined over $0 < x < \infty$. (a) The Fourier cosine transform of $F(x)$ is defined to be

$$A(k) = \sqrt{\frac{2}{\pi}} \int_0^\infty F(x) \cos kx \, dx$$

Show that

$$F(x) = \sqrt{\frac{2}{\pi}} \int_0^\infty A(k) \cos kx \, dk$$

$$\delta(x - x') = \frac{2}{\pi} \int_0^\infty \cos kx \cos kx' \, dk$$

Hint: Extend $F(x)$ into $-\infty < x < 0$ in such a way that $F(x)$ is even: $F(x) = F(-x)$. Use (12.19.2) and (12.19.3). (b) The Fourier sine transform of $F(x)$ is defined to be

$$A(k) = \sqrt{\frac{2}{\pi}} \int_0^\infty F(x) \sin kx \, dx$$

Show that

$$F(x) = \sqrt{\frac{2}{\pi}} \int_0^\infty A(k) \sin kx \, dk$$

$$\delta(x - x') = \frac{2}{\pi} \int_0^\infty \sin kx \sin kx' \, dk$$

12.19.27 Suppose that the solution of the wave equation (12.19.1) is desired over $-s < x < +s$ subject to some boundary conditions at $x = \pm s$. Assume a complex Fourier solution of the form

$$\eta(x, t) = \sum_{n=-\infty}^{+\infty} A_n(t) e^{inkx} \qquad k = \frac{\pi}{s}$$

and determine $A_n(t)$ from the wave equation. Note the resemblance of the Fourier series solution so obtained to (12.19.10). As $s \to \infty$, we can think of the Fourier series solution as approaching the Fourier integral representation of the solution in the limit.

12.19.28 Find the spectrum of the step function

$$F(x) = F_0 \qquad -\Delta x < x < +\Delta x$$
$$F(x) = 0 \qquad \text{otherwise}$$

Does (12.19.18) hold in this case? Show from your result that

$$\int_{-\infty}^{+\infty} \frac{\sin (k\Delta x) \cos (kx)}{k} \, dk = \frac{\pi}{2} \qquad -\Delta x < x < +\Delta x$$
$$= 0 \qquad |x| > \Delta x$$

12.20.29 Suppose that through some measurement the wave function of a free particle moving in the x-direction is known to be $\psi(x, 0) = F(x)$ at $t = 0$. Show that the wave function for later times is

$$\psi(x,t) = \frac{1}{2m} \int_{-\infty}^{+\infty} F(x') \int_{-\infty}^{+\infty} \exp i\left[k(x - x') - \frac{\hbar^2 k^2}{2m} t \right] dk \, dx'$$

12.21.30 Show that a solution of the two-dimensional wave equation is

$$\eta = f_1(\hat{\mathbf{n}} \cdot \mathbf{r} + ct) + f_2(\hat{\mathbf{n}} \cdot \mathbf{r} - ct)$$

where $\hat{\mathbf{n}}$ is a unit vector and f_1 and f_2 are arbitrary functions.

12.21.31 Show that a solution of the three-dimensional wave equation,

$$\nabla^2 \eta - \frac{1}{c^2} \frac{\partial^2 \eta}{\partial t^2} = 0$$

which depends only on $r = \sqrt{x^2 + y^2 + z^2}$ is

$$\eta = \frac{f_1(r - ct)}{r} + \frac{f_2(r + ct)}{r}$$

where f_1 and f_2 are arbitrary functions. *Hint:* Write the Laplacian in spherical coordinates. It is possible to show that this is the most general solution which depends only on r. What does the solution represent?

REFERENCES

1. Page and Adams, *Principles of Electricity*, Van Nostrand, New York, 1948.
2. Morse, Philip M., *Vibration and Sound*, McGraw-Hill Book Co., New York, 1948.
3. Pauling, Linus, and Wilson, E. Bright, *Introduction to Quantum Mechanics*, McGraw-Hill Book Co., New York, 1935.

13

Special Relativity

In Chapter 8, the Galilean covariance of Newtonian mechanics was demonstrated. In Chapter 10, the transformation theory of accelerated reference frames was developed. We now show that Galilean transformations are valid only at velocities which are much less than the velocity of light. The program of special relativity is essentially the generalization of the concepts of Galilean transformations to include velocities arbitrarily near the velocity of light. Should we try to extend the transformation theory of accelerated frames to include arbitrary velocities and accelerations, we would be led to the concepts of general relativity. It might be said that special relativity is the most general covariant theory of inertial frames of reference.

13.1. THE MICHELSON-MORLEY EXPERIMENT

The propagation of light waves is governed by a wave equation of the same form as that satisfied by waves propagating in an elastic medium. The velocity of propagation of waves in an elastic medium is determined by the properties of the medium, e.g., tension and density in the example of the stretched string. Suppose an observer is moving with a velocity v parallel to a string along which a pulse is propagating. No one would doubt that relative to the moving observer the pulse is moving at a velocity $c' = c - v$. It is natural to try to extend this idea to light waves. Must it not be true that the velocity of light waves is somehow dependent on the intrinsic properties of the space through which the waves travel?

When electromagnetic waves were first investigated, it was actually assumed that space contained a kind of idealized elastic substance called the *ether*. Since the earth is obviously moving with respect to this ether, the velocity of light relative to the earth must be different in different directions and different at different times during the year. The Michelson-Morley experiment was an attempt to measure this effect.

In this experiment, light goes from a source, S, Fig. 13.1.1a, to a half-silvered mirror, P. Part of the light is transmitted and part of it is reflected.

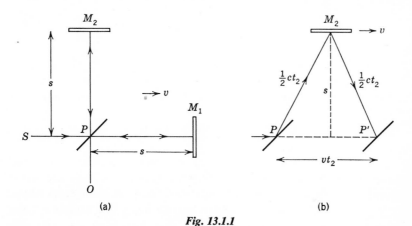

(a)   (b)

Fig. 13.1.1

The two light beams so formed reflect from mirrors M_1 and M_2 and recombine at P. Finally, the two beams are observed at O. The apparatus is fixed to the earth and moves through the alleged ether at a velocity v. It will be assumed that the direction of the motion is parallel to one of the light beams as indicated in Fig. 13.1.1a.

Figure 13.1.1a shows the experiment as it appears to an observer on the earth. To a conceptual observer who is fixed in space, i.e., who is stationary with respect to the alleged ether, the apparatus moves a distance vt_2 during the time t_2 that the light beam requires to go from P to M_2 and back again. Thus the ether observer would see the experiment as pictured in Fig. 13.1.1b. The time t_2 is computed as

$$(\tfrac{1}{2}ct_2)^2 = s^2 + (\tfrac{1}{2}vt_2)^2,$$

$$t_2 = \frac{2s}{c\sqrt{1 - v^2/c^2}} \qquad (13.1.1)$$

where s is the distance from P to M_2 and $c = 2.99 \times 10^{10}$ cm/sec is the speed of light with respect to the ether. The light beam which goes parallel to v has a velocity of $v - c$ relative to the apparatus in going from P to M_1 and velocity $v + c$ in returning from M_1 to P. The total time is

$$t_1 = \frac{s}{c - v} + \frac{s}{c + v} = \frac{2s}{c} \frac{1}{1 - v^2/c^2} \qquad (13.1.2)$$

Thus, on account of the motion of the apparatus, the two beams considered as plane waves are out of phase with one another, when observed at O, by an amount

$$\omega(t_1 - t_2) \cong \omega \frac{2s}{c}\left[1 + \frac{v^2}{c^2} - \left(1 + \frac{1}{2}\frac{v^2}{c^2} \right) \right] = \frac{\omega s v^2}{c^3} \qquad (13.1.3)$$

The two beams produce an interference pattern at O which can be optically observed. If the apparatus is slowly rotated so that PM_2 becomes coincident with v, there will be a phase shift between the two beams given by

$$2\frac{\omega s v^2}{c^3} \qquad (13.1.4)$$

and, hence, a shift in the interference pattern. By using for v the known velocity of the earth as it moves around the sun, it can be determined that the effect (13.1.4) is large enough to be detected. The experiment was performed at different times during the year to remove the possibility that the earth's velocity with respect to the ether might accidentally be zero at any one time.

The Michelson-Morley experiment yields a negative result. No actual phase shift exists. In the consideration of this experiment, and many others, Einstein concluded that the most satisfactory assumption to make is that the speed of light is the same in all inertial frames of reference. If an observer on the earth measures the speed of the light coming to him from a star known to be receding at great speed from the earth, he measures the same constant value c as if the star were stationary. The alleged ether, being undetectable, does not exist. Only *relative* motions are significant, e.g., as far as measurements of the speed of light are concerned, "the source is moving away from the observer" and "the observer is moving away from the source" are equivalent.

The velocity of light does not obey the Galilean law for the transformation of velocities. In the next section, another manifestation of the breakdown of Galilean transformations is demonstrated when it is shown that the Lorentz force transforms incorrectly, leading to a basic inconsistency between classical Newtonian mechanics and the Maxwellian theory of electricity and magnetism.

13.2. GALILEAN TRANSFORMATION OF THE LORENTZ FORCE

Classical mechanics and Galilean relativity predict that the law of transformation of force is

$$\mathbf{F'} = \mathbf{F} \tag{13.2.1}$$

It will now be shown that the Lorentz force acting on a charged particle has a different transformation law.

Only some simple solutions of Maxwell's equations are needed. Consider two parallel line charges of charge μ per unit length (stat-coulombs/cm). The electric field of either line charge is

$$E = \frac{2\mu}{r} \tag{13.2.2}$$

There is a force of repulsion between the two line charges given by

$$F_E = \frac{2\mu^2}{r} \text{ dynes/cm} \tag{13.2.3}$$

Consider an inertial frame F' moving at a velocity v parellel to the line charges. An observer in F' would see electric currents of magnitude $J = \mu v$ and, hence, magnetic fields. The magnetic field of either line charge as seen in F' is

$$B = \frac{2J}{rc} = \frac{2\mu v}{rc} \tag{13.2.4}$$

The force of attraction between two wires carrying equal currents is

$$F_B = \frac{1}{c} JB = \frac{2J^2}{rc^2} = \frac{2\mu^2 v^2}{rc^2} \tag{13.2.5}$$

per unit length. Thus, in frame F', a total force

$$F' = F_E - F_B = \frac{2\mu^2}{r}\left(1 - \frac{v^2}{c^2}\right) \tag{13.2.6}$$

is observed between the two line charges. Our development thus leads to the law of transformation of force

$$F' = F\left(1 - \frac{v^2}{c^2}\right) \tag{13.2.7}$$

in clear violation of the law (13.2.1) as obtained from Newtonian mechanics. The speed of light in vacuum has the value

$$c = 2.9979 \times 10^{10} \text{ cm/sec} \tag{13.2.8}$$

so that the difference between (13.2.7) and (13.2.1) would not ordinarily be observed. There is, none the less, an inconsistency in principle. This inconsistency is removed in the theory of special relativity. One of the basic postulates of special relativity is that the laws of mechanics and electrodynamics must be covariant with respect to transformations among all possible inertial frames. In particular, as required by the Michelson-Morley experiment, the velocity of light has the same constant value in all inertial frames. Since Newtonian mechanics and Galilean transformations are known to be correct at velocities small compared to the velocity of light, it is necessarily true that these principles shall remain valid in the limit $v \to 0$ when the new theory is developed. In particular, when the "correct" law of transformation of force is derived [which (13.2.7) is not], it must be true that $\mathbf{F}' = \mathbf{F}$ at low velocities.

By definition, an inertial frame is one in which a neutral test particle experiences no acceleration when released. Thus accelerated frames and gravitational phenomena are, strictly speaking, not part of special relativity, but belong to the domain of general relativity. Since acceleration is an intrinsic part of the theory of particle motion, it will be necessary to lay down some hypotheses concerning accelerated frames as the theory is developed.

13.3. SIMULTANEITY

Suppose point A is equidistant from B and C in frame F, Fig. 13.3.1. Light signals from A arrive simultaneously at B and C according to observers in F. Suppose that, on receiving the signal from A, both B and C emit secondary signals. In F, these signals meet half-way between B and C. Consider points B' and C' of frame F' which were instantaneously coincident with B and C when the signals arrived from A. Since B' and C' are moving with respect to F at a constant velocity v, the secondary signal does not meet half-way between B' and C'. However, according to our

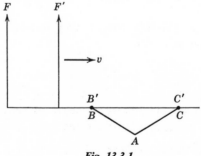

Fig. 13.3.1

basic postulate, the speed of each signal is c with respect to both F and F'. Therefore, it must be that the secondary signal from C' starts earlier than the one from B'. We conclude that two events which are simultaneous in one inertial frame are *not* simultaneous in another. The simple time transformation $\Delta t = \Delta t'$ is no longer valid.

13.4. THE LORENTZ TRANSFORMATION

Recall (Section 8.1) that the Galilean transformation connecting the distance and time intervals between two events as seen in frames F and F' is

$$\Delta x' = \Delta x - v\,\Delta t \qquad \Delta y' = \Delta y$$
$$\Delta z' = \Delta z \qquad\qquad \Delta t' = \Delta t \qquad\qquad (13.4.1)$$

where, for simplicity, F' is assumed to be moving in the x-direction. For a light ray propagating in the x-direction (13.4.1) predicts a transformation

$$c' = c - v \qquad (13.4.2)$$

in direct violation of the result of the Michelson-Morley experiment.

Assume that there exists a relativistically correct transformation of the form

$$x_i' = x_i'(x_j, t) \qquad t' = t'(x_j, t) \qquad (13.4.3)$$

where the Latin subscripts can take on the values 1, 2, and 3. Our failure to find an "ether" or an "absolute rest frame" leads us to assume that all inertial frames are equivalent, no one being singled out over the others for preferential treatment. This guarantees the existence of the inverse transformation:

$$x_i = x_i(x_j', t') \qquad t = t(x_j', t') \qquad (13.4.4)$$

Moreover, the transformations are differentiable. It is best to work with the transformations in differential form:

$$dx_i' = \frac{\partial x_i'}{\partial x_j}\,dx_j + \frac{\partial x_i'}{\partial t}\,dt$$
$$dt' = \frac{\partial t'}{\partial x_j}\,dx_j + \frac{\partial t'}{\partial t}\,dt \qquad (13.4.5)$$

Then, just as with Galilean transformations, (dx_i, dt) can be thought of as the distance and time intervals between events an infinitesimal distance apart as measured in F, and (dx_i', dt') are the distance and time intervals between the *same* two events as would be measured in F'.

In the special case where F' moves in the x-direction of F and the x- and x'-axes remain always parallel, it is possible to specialize (13.4.5) to

$$dx' = \alpha\, dx + \beta\, dt \qquad dy' = dy$$
$$dt' = \gamma\, dt + \epsilon\, dx \qquad dz' = dz$$

$$(13.4.6)$$

where α, β, γ and ϵ are written for the partial derivatives.

Suppose that a mirror M is fixed in F' as in Fig. 13.4.1. A light signal goes from a point on the x'-axis and, traveling parallel to the y'-axis, is

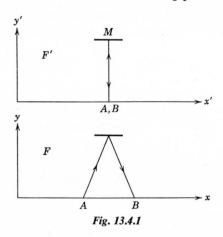

Fig. 13.4.1

reflected back to the same point. This same experiment as it would be seen by observers in F is also illustrated in Fig. 13.4.1. The two events which we want to consider are the departure of the light signal from A and its arrival at B. In F',

$$dx' = 0 \qquad dt' = \frac{2\, dy'}{c} = \frac{2\, dy}{c}$$

$$(13.4.7)$$

whereas, for these same two events as seen in F,

$$\frac{dx}{2} = \sqrt{\left(\frac{c\, dt}{2}\right)^2 - dy^2} = \frac{v\, dt}{2}$$

$$(13.4.8)$$

In (13.4.7) and (13.4.8), $dy = dy'$ refers to the perpendicular distance between M and the x- and x'-axes. By solving (13.4.7) and (13.4.8) for dt we find

$$dt' = dt\sqrt{1 - v^2/c^2}$$

$$(13.4.9)$$

The substitution of (13.4.9) and (13.4.7) into the transformation equations (13.4.6) yields

$$0 = \alpha v + \beta \qquad \sqrt{1 - v^2/c^2} = \gamma + \epsilon v$$

$$(13.4.10)$$

In order to determine the coefficients completely, two more conditions are needed. These are provided by the converse of the experiment just performed. A mirror is held stationary in F and a light ray is considered which goes parallel to the y-axis. The result is

$$\gamma = \frac{1}{\sqrt{1 - v^2/c^2}} \qquad \beta = -\frac{v}{\sqrt{1 - v^2/c^2}} \tag{13.4.11}$$

The solution of (13.4.10) for α and ϵ yields

$$\alpha = \frac{1}{\sqrt{1 - v^2/c^2}} \qquad \epsilon = -\frac{v}{c^2\sqrt{1 - v^2/c^2}} \tag{13.4.12}$$

The transformation (13.4.6) can now be written

$$\begin{aligned} dx' &= \gamma(dx - v\, dt) & dy' &= dy \\ dt' &= \gamma(dt - v\, dx/c^2) & dz' &= dz \end{aligned} \tag{13.4.13}$$

This result will be referred to as the *special* or *restricted* Lorentz transformation. In the limit $v/c \to 0$, $\gamma \to 1$ and (13.4.13) becomes identical to the Galilean transformation (13.4.1) as it must.

To find the inverse of (13.4.13), note that F moves relative to F' at velocity $-v$. The equivalence of all inertial frames requires that the inverse transformation be identical to (13.4.13) with v replaced by $-v$:

$$dx = \gamma(dx' + v\, dt')$$
$$dt = \gamma\left(dt' + \frac{v\, dx'}{c^2}\right) \tag{13.4.14}$$

The Lorentz transformation is *linear*, i.e., the coefficients depend on only v and c, and not on x and t. It is thus possible to integrate to get

$$\begin{aligned} x' - x_0' &= \gamma[x - x_0 - v(t - t_0)] \\ t' - t_0' &= \gamma[t - t_0 - v(x - x_0)/c^2] \end{aligned} \tag{13.4.15}$$

Thus we may consider two events which are as far separated in space and time as we please. If one of the events is the coincidence of the origins of F and F' at $t_0 = t_0' = 0$, then

$$x' = \gamma[x - vt] \qquad t' = \gamma\left[t - \frac{vx}{c^2}\right] \tag{13.4.16}$$

which is the *homogeneous Lorentz transformation*.

13.5. THE LORENTZ CONTRACTION

In applications of the special Lorentz transformation (13.4.13), (dx, dt) and (dx', dt') are to be identified as *intervals* between events or experiments as would be observed from two different reference frames. Due to the linearity of the transformation, in many instances the differential intervals can be replaced by finite intervals. Suppose, for instance, that the x'-axis of frame F' is a moving measuring rod. In frame F, Fig. 13.5.1, observers at A and B make *simultaneous* readings on the rod in their frame. The two events under discussion are these two readings. In F, they are separated

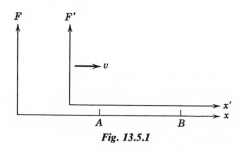

Fig. 13.5.1

by intervals $dt_{AB} = 0$ and $dx_{AB} = x_B - x_A$. To see what are the space-time intervals between these *same* two events, as seen by observers in F', (13.4.13) is used:

$$dx'_{AB} = x_B{}' - x_A{}' = \gamma(x_B - x_A) \tag{13.5.1}$$

$$dt'_{AB} = t_B{}' - t_A{}' = -\gamma \frac{v}{c^2}(x_B - x_A) = -\frac{v}{c^2}(x_B{}' - x_A{}') \tag{13.5.2}$$

Since $\gamma > 1$, the distance-interval between the two events is greater in F' than it is in F. The rod "appears" shorter in F; hence, the name Lorentz contraction for this particular phenomena. The statement "the rod appears shorter" must be interpreted only in the context of an experiment performed exactly as just described. Another experiment, such as the actual photographing of a moving object, is a different experiment involving the recording of light rays coming from the object; it will not necessarily lead to the same results. Each experiment presents a different set of circumstances which must be analyzed anew, using the Lorentz transformation.

Equation (13.5.2) shows that the two events are not simultaneous in F'. Since $t_B{}' - t_A{}' < 0$, the measurement at B occurs earlier than does the one at A. This does not matter since F' is the rest frame of the measuring rod and observers in F' would therefore not question the fact that the readings as taken by the observers in F represents the actual distance between A and B.

13.6. THE TIME DILATION

Consider a clock which is *fixed* in frame F' and is compared sequentially with two clocks at different points A and B on the x-axis of F.

Since the clock remains always at the same point of F', the events which happen at A and B are characterized by $dx' = 0$. From Eq. (13.4.14),

$$dt' = dt\sqrt{1 - v^2/c^2} \qquad (13.6.1)$$

The time interval between the event-points A and B is *less* in frame F' than it is in F. This effect is known as the time dilation.

One frequently sees the statement made, "the moving clock is running slower than the stationary clock." This statement is correct *only* in the context of the experiment as just outlined, namely, the comparison of a single moving clock with two stationary clocks. Observers in the moving frame F' are not aware of any slowing down except as they might deduce from observations made on the clocks stationed at A and B in frame F. All conditions, laws, and processes are quite normal in F' as is required by the general covariance principle.

We now describe an experiment that is a direct verification of the time dilation. Cosmic ray bombardment of the atmosphere produces high-velocity μ-mesons which have a characteristic life time. These particles are then effectively moving clocks. In a recent experiment,* μ-mesons incident on top of Mt. Washington, New Hampshire, were selected to have speeds in the range $0.9950C$ to $0.9954C$. The number of these which survived to reach sea level was measured in Cambridge, Massachusetts. The number to be expected without time dilation was calculated from the known mean life of stationary μ-mesons. From the number actually to be observed and the known distance of descent, a time dilation factor of 8.8 ± 0.08 was deduced. This compares with $\gamma = (1 - v^2/c^2)^{-1/2} = 8.4 \pm 0.2$ as deduced from the known particle velocities.

Suppose that a *single* stationary clock in F is compared to two moving clocks at A' and B' in frame F' as they go past. It is now true that $dx = 0$ and from (13.4.13)

$$dt = dt'\sqrt{1 - v^2/c^2} \qquad (13.6.2)$$

which is just opposite from (13.6.1). There is, however, no contradiction or paradox. The two equations apply to different experiments.

---

* D. H. Frisch, J. H. Smith, "Measurement of the Relativistic Time Dilation Using μ-Mesons," *American Journal of Physics*, **31**, 342 (1963).

13.7. THE METRIC OF SPECIAL RELATIVITY

By using the transformation (13.4.13) it is shown that

$$dx'^2 + dy'^2 + dz'^2 - c^2\,dt'^2 = dx^2 + dy^2 + dz^2 - c^2\,dt^2 \quad (13.7.1)$$

Thus the quantity

$$ds^2 = dx_i\,dx_i - c^2\,dt^2 \qquad (13.7.2)$$

is a *scalar invariant* with respect to the special Lorentz transformation. Recall that, in Chapter 1, the development of the theory of orthogonal transformations was centered on the idea that the three-dimensional distance $r_i r_i$ between two points is an invariant. In Chapter 2, the development of the theory of arbitrary coordinate transformations involved the invariance of the three-dimensional line element which is written as

$$ds^2 = g_{ij}\,dq^i\,dq^j \qquad (13.7.3)$$

in the most general kind of notation where the difference between covariant and contravarient quantities is taken into account.

The situation is similar in special relativity. The invariant (13.7.2) is the four-dimensional space-time separation or metric between two events. This metric is peculiar in that it is *indefinite* due to the presence of the minus sign; ds^2 can be positive, negative, or zero. At this stage, a formal definition of Lorentz transformations can be given. *Any* transformation which leaves the quantity (13.7.2) an invariant,

$$dx_i'\,dx_i' - c^2\,dt'^2 = dx_i\,dx_i - c^2\,dt^2 \qquad (13.7.4)$$

is said to be a Lorentz transformation. The special Lorentz transformation (13.4.13) is merely one example. Other simple examples are *space inversion*

$$dx_1' = -dx_1, \qquad dx_2' = dx_2, \quad dx_3' = dx_3, \quad dt' = dt \quad (13.7.5)$$

and time inversion

$$dx_i' = dx_i \qquad dt' = -dt \qquad (13.7.6)$$

which are examples of *improper* Lorentz transformations.

Mathematically, the situation as regards the indefinite metric can be handled in two ways. One method is to introduce the imaginary coordinate

$$dx_4 = ic\,dt \qquad (13.7.7)$$

The metric then reads

$$ds^2 = dx_i\,dx_i + dx_4\,dx_4 = dx_\lambda\,dx_\lambda \qquad (13.7.8)$$

where the repeated Greek subscripts is understood to run from 1 to 4. The theory of Lorentz transformations then becomes formally identical to the theory of orthogonal transformations. The most general Lorentz transformation can be expressed by

$$dx_\mu' = S_{\mu\lambda}\, dx_\lambda \tag{13.7.9}$$

where the coefficients obey the orthogonality conditions

$$S_{\lambda\mu}S_{\lambda\nu} = \delta_{\mu\nu} \qquad S_{\mu\lambda}S_{\nu\lambda} = \delta_{\mu\nu} \tag{13.7.10}$$

Of course, the $S_{\mu\nu}$ are complex numbers so that in this notation the Lorentz transformations are *complex orthogonal* transformations in a four-dimensional space. The inverse of (13.7.9) is

$$dx_\mu = S_{\lambda\mu}\, dx_\lambda' \tag{13.7.11}$$

Any set of four quantities which obeys the transformation equations (13.7.9) and (13.7.11) is called a *four-vector*.

As an example, the coefficients $S_{\mu\nu}$ for the special Lorentz transformation (13.4.13) can be constructed by first writing the transformation in the form

$$dx' = \gamma\, dx + i\,\frac{\gamma v}{c}\,(ic\, dt)$$

$$(ic\, dt') = -i\,\frac{\gamma v}{c}\, dx + \gamma(ic\, dt) \tag{13.7.12}$$

The $S_{\mu\nu}$ are seen to be the components of the matrix

$$S = \begin{pmatrix} \gamma & 0 & 0 & i\gamma\dfrac{v}{c} \\ 0 & 1 & 0 & 0 \\ 0 & 0 & 1 & 0 \\ -i\gamma\dfrac{v}{c} & 0 & 0 & \gamma \end{pmatrix} \tag{13.7.13}$$

Just as with orthogonal transformations that represent rotations in three-dimensional Cartesian space, the inverse of (13.7.13) is found by taking the transpose:

$$S^{-1} = \tilde{S} \tag{13.7.14}$$

The second method for dealing with the indefinite metric which avoids the use of complex numbers involves the use of the metric tensor

$$G = \begin{pmatrix} 1 & 0 & 0 & 0 \\ 0 & 1 & 0 & 0 \\ 0 & 0 & 1 & 0 \\ 0 & 0 & 0 & -1 \end{pmatrix} \tag{13.7.15}$$

The four-dimensional line element is then expressed by

$$ds^2 = g_{\lambda\mu}\, dx^\lambda\, dx^\mu \tag{13.7.16}$$

where $g_{\lambda\mu}$ are the components of the matrix (13.7.15). In this notation,

$$dx^1 = dx, \qquad dx^2 = dy, \qquad dx^3 = dz, \qquad dx^4 = c\, dt \tag{13.7.17}$$

the superscript indicates a contravariant transformation law:

$$dx^{\lambda'} = P_\mu^{\lambda'}\, dx^\mu \tag{13.7.18}$$

The coefficients $P_\mu^{\lambda'}$ can, of course, be interpreted as the partial derivatives:

$$P_\mu^{\lambda'} = \frac{\partial x_{\lambda'}}{\partial x_\mu} \tag{13.7.19}$$

If the special Lorentz transformation is written out in the form

$$dx^{1'} = \gamma\, dx^1 - \gamma\frac{v}{c}\, dx^4$$
$$\tag{13.7.20}$$
$$dx^{4'} = -\frac{v\gamma}{c}\, dx^1 + \gamma\, dx^4$$

the coefficients $P_\mu^{\lambda'}$ are seen to be the components of the matrix

$$P = \begin{pmatrix} \gamma & 0 & 0 & -\dfrac{v}{c}\gamma \\[2mm] 0 & 1 & 0 & 0 \\[1mm] 0 & 0 & 1 & 0 \\[2mm] -\dfrac{v}{c}\gamma & 0 & 0 & \gamma \end{pmatrix} \tag{13.7.21}$$

The comparison of (13.7.21) and (13.7.13) reveals the connection between the transformation coefficients in the two notations

$$P_k^{j'} = S_{jk} \qquad P_4^{j'} = iS_{j4} \qquad P_j^{4'} = -iS_{4j} \qquad P_4^{4'} = S_{44} \tag{13.7.22}$$

where j and k take on the values 1, 2, and 3.

The inverse transformation is conveniently indicated by lowering the prime:

$$P_\mu^{\lambda'} = \frac{\partial x_\lambda}{\partial x_{\mu'}} \tag{13.7.23}$$

Thus

$$P_{\mu'}^{\lambda}\, P_\nu^{\mu'} = \delta_\nu^{\ \lambda} \qquad P_\mu^{\lambda'} P_{\nu'}^{\mu} = \delta_\nu^{\ \lambda} \tag{13.7.24}$$

For the special Lorentz transformation, the coefficients $P_{\nu'}^{\mu}$, are given by

$$P^{-1} = \begin{pmatrix} \gamma & 0 & 0 & \dfrac{v}{c}\gamma \\ 0 & 1 & 0 & 0 \\ 0 & 0 & 1 & 0 \\ \dfrac{v}{c}\gamma & 0 & 0 & \gamma \end{pmatrix} \tag{13.7.25}$$

The reader can easily verify that

$$PP^{-1} = P^{-1}P = I \tag{13.7.26}$$

The theory is formally identical to the theory of arbitrary coordinate transformations in three-space. The metric tensor transforms as

$$g'_{\lambda\mu} = P_{\lambda}^{\alpha} P_{\mu'}^{\beta} g_{\alpha\beta} \tag{13.7.27}$$

In matrix notation,

$$G' = \widetilde{P^{-1}} G P^{-1} \tag{13.7.28}$$

A Lorentz transformation is then formally defined as any transformation which leaves the metric tensor invariant. In other words, P is a Lorentz transformation provided that G' comes out to be the same as G. The contravariant form of the metric tensor is found from

$$g^{\lambda\mu} g_{\nu\mu} = \delta_{\nu}^{\lambda} \tag{13.7.29}$$

The metric is so simple that the components $g^{\lambda\mu}$ come out to be the same as $g_{\lambda\mu}$. The components $g^{\lambda\mu}$ transform as

$$g^{\lambda\mu'} = P_{\alpha}^{\lambda'} P_{\beta}^{\mu'} g^{\alpha\beta} \tag{13.7.30}$$

The covariant form of the displacement vector (or any vector) is related to the contravariant form by the raising and lowering of indices:

$$dx_{\lambda} = g_{\lambda\mu} dx^{\mu} \qquad dx^{\mu} = g^{\mu\lambda} dx_{\lambda} \tag{13.7.31}$$

Note that

$$dx_{\lambda'} = P_{\lambda'}^{\mu} dx_{\mu} \tag{13.7.32}$$

It should be evident to the reader that ordinary orthogonal transformations in three-space are included as special cases of Lorentz transformations. Moreover, it is possible to include arbitrary coordinate transformations in the three-space, i.e., we can work with a metric of the form

$$G = \begin{pmatrix} g_{11} & g_{12} & g_{13} & 0 \\ g_{21} & g_{22} & g_{23} & 0 \\ g_{31} & g_{32} & g_{33} & 0 \\ 0 & 0 & 0 & -1 \end{pmatrix} \tag{13.7.33}$$

and still stay within the framework of the transformation theory so far developed. In working mechanics problems, this freedom will allow us to pick the most advantages curvilinear coordinate system in the three-space.

The reader may be annoyed by having to learn two different ways of doing the same thing. Both notations are in widespread use at the present time and both notations have advantages. The imaginary notation is a little simpler and is probably the most convenient for the development of "classical" relativity theory, i.e., the relativistic theory of Newtonian mechanics and Maxwellian electrodynamics. On the other hand, the real notation is preferred by quantum field theorists and is also the notation required if the theory is to be broadened to encompass gravitation and accelerated reference frames.

Quantum field theorists generally write the metric in the form

$$G = \begin{pmatrix} 1 & 0 & 0 & 0 \\ 0 & -1 & 0 & 0 \\ 0 & 0 & -1 & 0 \\ 0 & 0 & 0 & -1 \end{pmatrix} \tag{13.7.34}$$

and allow the Greek subscripts to take on the values 0, 1, 2, 3. The g_{00} component then plays the same role as the g_{44} component in the notation of this text. There is superimposed on this the various possible systems of units in use in electrodynamics. The author can only deplore this situation. It seems as though every one who develops the relativistic version of some theory feels that he must finally discover the "best" notation and that all previous notations are somehow inferior.

13.8. THE GENERAL LORENTZ TRANSFORMATION WITHOUT ROTATION

In Fig. 13.8.1, the origin of F' moves in the xy plane of F at an angle θ with respect to the x-axis. There is no motion in the z-direction. Frame F' is not rotated with respect to F. We wish to know the Lorentz transformation between F and F'. This is done by first rotating F until the x-axis is parallel to $\mathbf{v}$

$$\bar{X} = RX \tag{13.8.1}$$

where R denotes the rotation by the angle θ and X is a column matrix which represents any four vector.

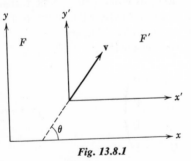

Fig. 13.8.1

Now transform to a frame which is moving at the speed v with respect to F by means of the special Lorentz transformation:

$$\bar{X} = P\bar{X} = PRX \qquad (13.8.2)$$

Now rotate in the opposite sense by the angle θ so that the final frame F' is not rotated with respect to F:

$$X' = R^{-1}\bar{\bar{X}} = (R^{-1}PR)X = TX \qquad (13.8.3)$$

Written out in full, using real notation, this is

$$T = \begin{pmatrix} \cos\theta & -\sin\theta & 0 & 0 \\ \sin\theta & \cos\theta & 0 & 0 \\ 0 & 0 & 1 & 0 \\ 0 & 0 & 0 & 1 \end{pmatrix} \begin{pmatrix} \gamma & 0 & 0 & -\dfrac{v}{c}\gamma \\ 0 & 1 & 0 & 0 \\ 0 & 0 & 1 & 0 \\ -\dfrac{v}{c}\gamma & 0 & 0 & \gamma \end{pmatrix}$$

$$\times \begin{pmatrix} \cos\theta & \sin\theta & 0 & 0 \\ -\sin\theta & \cos\theta & 0 & 0 \\ 0 & 0 & 1 & 0 \\ 0 & 0 & 0 & 1 \end{pmatrix} \qquad (13.8.4)$$

By writing $v_x = v\cos\theta$, $v_y = v\sin\theta$ the reader can verify that the final result is

$$T = \begin{pmatrix} 1+(\gamma-1)\dfrac{v_x^2}{v^2} & (\gamma-1)\dfrac{v_xv_y}{v^2} & 0 & -v_x\dfrac{\gamma}{c} \\ (\gamma-1)\dfrac{v_xv_y}{v^2} & 1+(\gamma-1)\dfrac{v_y^2}{v^2} & 0 & -v_y\dfrac{\gamma}{c} \\ 0 & 0 & 1 & 0 \\ -v_x\dfrac{\gamma}{c} & -v_y\dfrac{\gamma}{c} & 0 & \gamma \end{pmatrix} \qquad (13.8.5)$$

Suppose now that F' moves at a completely arbitrary direction with respect to F. It is easy to see how (13.8.5) is to be extended to this more general situation:

$$
T = \begin{pmatrix}
1 + (\gamma - 1)\dfrac{v_1^2}{v^2} & (\gamma - 1)\dfrac{v_1 v_2}{v^2} & (\gamma - 1)\dfrac{v_1 v_3}{v^2} & -v_1\dfrac{\gamma}{c} \\[2ex]
(\gamma - 1)\dfrac{v_1 v_2}{v^2} & 1 + (\gamma - 1)\dfrac{v_2^2}{v^2} & (\gamma - 1)\dfrac{v_2 v_2}{v^2} & -v_2\dfrac{\gamma}{c} \\[2ex]
(\gamma - 1)\dfrac{v_1 v_3}{v^2} & (\gamma - 1)\dfrac{v_2 v_3}{v^2} & 1 + (\gamma - 1)\dfrac{v_3^2}{v^2} & -v_3\dfrac{\gamma}{c} \\[2ex]
-v_1\dfrac{\gamma}{c} & -v_2\dfrac{\gamma}{c} & -v_3\dfrac{\gamma}{c} & \gamma
\end{pmatrix}
$$

$$(13.8.6)$$

This is called the general Lorentz transformation without rotation. The inverse, T^{-1}, is obtained by simply reversing the signs of all the velocity components.

Let $(\mathbf{r}, \Delta t)$ and $(\mathbf{r}', \Delta t')$ be the space-time intervals connecting two events as would be observed in the two frames F and F'. By means of (13.8.6) it is possible to express the general Lorentz transformation without rotation in the notation of ordinary three-dimensional vectors:

$$
\mathbf{r}' = \mathbf{r} + \mathbf{v}\left[\frac{\mathbf{r} \cdot \mathbf{v}}{v^2}(\gamma - 1) - \gamma\Delta t\right]
\tag{13.8.7}
$$

$$
\Delta t' = \gamma\left[\Delta t - \frac{\mathbf{r} \cdot \mathbf{r}}{c^2}\right]
\tag{13.8.8}
$$

13.9. CAUSALITY

The Lorentz transformation becomes undefined as $v \to c$:

$$
\lim_{v \to c} \gamma = \infty
\tag{13.9.1}
$$

This result in itself does not mean that velocities equal to or larger than c are meaningless or impossible. The concept of velocity comes up in many ways. We can talk about particle velocities, the velocity of propagation of signals bearing information, and, in the theory of wave motion, of phase and group velocity. Each of these types of velocity requires special treatment and we will concern ourselves here with the propagation of information.

The following basic assumption is added to the theory: If an event A causes an event B, then in no frame of reference can the event B occur *before* event A. Suppose that a neutron moves from a point A to a point B and then decays. According to the causality assumption just made, there is no frame of reference in which the neutron is observed to decay at B before it leaves A. More generally, if A causes B, then some sort of information-bearing signal must propagate from A to B. It is shown in this section that such signals cannot propagate at a speed greater than c with respect to any frame of reference.

Assume, as indicated in Fig. 13.9.1, that our neutron leaves A, travels along the x-axis at a speed ϵc to B where it decays. We will assume that

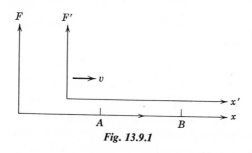

Fig. 13.9.1

$\epsilon > 1$ and show that it leads to a violation of causality. In the second frame F', the time interval between events A and B is

$$\Delta t'_{AB} = \gamma\left(\Delta t_{AB} - \frac{v}{c^2}\Delta x_{AB}\right) \tag{13.9.2}$$

But $\Delta x_{AB} = \epsilon c\, \Delta t_{AB}$ so that

$$\Delta t'_{AB} = \gamma\frac{\Delta x_{AB}}{c}\left(\frac{1}{\epsilon} - \frac{v}{c}\right) \tag{13.9.3}$$

Since $1/\epsilon < 1$, v can be chosen such that

$$\frac{1}{\epsilon} - \frac{v}{c} < 0 \tag{13.9.4}$$

even without making v greater than c. Thus, a frame of reference F' exists in which $\Delta t'_{AB} < 0$ meaning that the decay at B occurs *before* the departure of the neutron from A in violation of the causality hypothesis. It must be concluded that the assumption that the neutron could travel at a speed greater than c with respect to the original frame F is false.

13.10. THE TRANSFORMATION OF PARTICLE VELOCITIES

In the special Lorentz transformation,

$$dx' = \gamma(dx - v\, dt) \qquad dt' = \gamma(dt - v\, dx/c^2) \tag{13.10.1}$$

let (dx, dt) and (dx', dt') be the space-time intervals between two events taken very close together on the trajectory of a particle. The x-component of the particle's velocity as determined by observers in F and F' is

$$u_x = \frac{dx}{dt} \qquad u_x' = \frac{dx'}{dt'} \qquad (13.10.2)$$

By means of (13.10.1),

$$u_x' = \frac{u_x - v}{1 - \frac{vu_x}{c^2}} \qquad (13.10.3)$$

Since $dy = dy'$ and $dz = dz'$, the y and z components of particle velocity transform according to

$$u_y' = \frac{1}{\gamma} \frac{u_y}{1 - \frac{vu_x}{c^2}} \qquad u_z' = \frac{1}{\gamma} \frac{u_z}{1 - \frac{vu_x}{c^2}} \qquad (13.10.4)$$

If all velocities are small compared to the velocity of light, then (13.10.3) and (13.10.4) become $u_x' = u_x - v$, $u_y' = u_y$, and $u_z' = u_z$ which is just the Galilean addition of velocities.

Suppose that a particle is traveling in the x-direction. Then (13.10.3) is

$$u' = \frac{u - v}{1 - \frac{vu}{c^2}} \qquad (13.10.5)$$

We note that if $u \rightarrow c$, then $u' \rightarrow c$ also. Thus the particle has the same limiting value c for its velocity in all frames of reference.

The formulas of this section apply to instantaneous particle velocities; the particles can be undergoing accelerations and following arbitrary trajectories.

13.11. THE DOPPLER EFFECT

An observer stationed in a moving frame F' receives signals sent to him from an observer at O in frame F. Let the receipt of the signals be the two events A and B in Fig. 13.11.1. Since the receiver is stationary in his own frame F', $\Delta x' = 0$ and from the special Lorentz transformation (13.4.14)

$$\Delta x = \gamma v \, \Delta t' \qquad \Delta t = \gamma \, \Delta t' \qquad (13.11.1)$$

Let P be the period of the signals sent from O. Since the second signal travels a distance $\Delta x \cos \alpha$ farther than the first signal, the actual time interval between the events A and B as measured in F is

$$\Delta t = P + \frac{\Delta x \cos \alpha}{c} \qquad (13.11.2)$$

where it is assumed that the signal propagates at velocity c. Writing $P' = \Delta t'$ as the period of the signals as observed in F' and combining (13.11.1) and (13.11.2)

$$\gamma P' = P + \frac{\gamma v P'}{c} \cos \alpha \qquad (13.11.3)$$

In terms of circular frequencies as given by $\omega = 2\pi/P$, $\omega' = 2\pi/P'$

$$\omega' = \omega \gamma \left(1 - \frac{v}{c} \cos \alpha \right) \qquad (13.11.4)$$

This is one form of the Doppler formula.

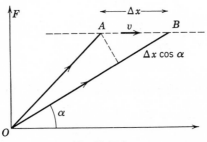

Fig. 13.11.1

The same experiment can be analyzed another way. Regard F' as "stationary" and the observer at O as "moving." Let the two events to which the Lorentz transformation is to be applied be the departure of signals from two different locations of O as he moves in the negative x'-direction, Fig. 13.11.2. Now $\Delta x = 0$ and the special Lorentz transformation gives

$$\Delta x' = \gamma v \, \Delta t = -\gamma v P$$
$$\Delta t' = \gamma \, \Delta t = \gamma P \qquad (13.11.5)$$

The second signal travels a distance $|\Delta x'| \cos \alpha'$ farther than the first so that the period observed at A, B is

$$P' = \Delta t' + \frac{|\Delta x'| \cos \alpha'}{c} \qquad (13.11.6)$$

Hence,

$$P' = \gamma P + \frac{\gamma v P \cos \alpha'}{c},$$
$$\omega = \omega' \gamma \left(1 + \frac{v}{c} \cos \alpha' \right) \qquad (13.11.7)$$

Note that α' is the direction of the signals as observed in F' whereas in

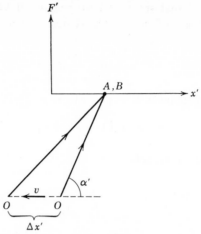

Fig. 13.11.2

(13.11.4) α is the direction of the signal as seen in F. By combining (13.11.4) and (13.11.7) we get

$$\cos \alpha' = \frac{\cos \alpha - v/c}{1 - (v/c)\cos \alpha} \qquad (13.11.8)$$

which is sometimes called the relativistic aberration formula.

13.12. PHOTOGRAPHING A MOVING OBJECT

Suppose that a camera is pointed directly along the y-axis as in Fig. 13.12.1a. The idea is to photograph a high-speed object moving in the x-direction. Let the object be in the shape of a sphere and suppose that its radius, r_0, is much less than the distance between camera and sphere. The appropriate form of the relativistic aberation formula is

$$\cos \alpha' = \frac{\cos \alpha + v/c}{1 + (v/c)\cos \alpha} \qquad (13.12.1)$$

where α is the direction the light must travel to reach the camera and α' is the direction of these same light rays with respect to the rest frame F' of the sphere. Since the light rays must travel nearly down the y-axis in order to be recorded by the camera, $\cos \alpha \cong 0$ and

$$\cos \alpha' = \frac{v}{c} \qquad \sin \alpha' = \sqrt{1 - \frac{v^2}{c^2}} = \frac{1}{\gamma} \qquad (13.12.2)$$

The shutter of the camera is opened only for a brief instant so that light rays must arrive simultaneously at the camera in order to be recorded.

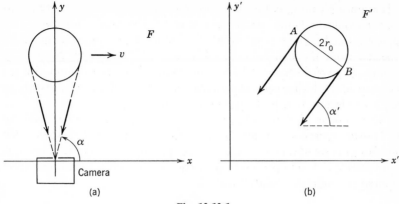

Fig. 13.12.1

Consider light rays which come from the extremities A and B of a diameter as in Fig. 13.12.1b. It is evident that some of the light which reaches the camera actually comes from behind the sphere. The sphere, moving at high velocity, gets out of the way of this light. Let the emission of light rays from A and B be the two events to which the Lorentz transformation is to be applied. From the geometry of Fig. 13.12.1b,

$$\Delta x' = +2r_0 \sin \alpha' = + \frac{2r_0}{\gamma}$$

$$\Delta y' = -2r_0 \cos \alpha' = -2r_0 \frac{v}{c} = \Delta y \tag{13.12.3}$$

The light ray emitted from A has a longer way to travel down the y-axis of frame F than does the light ray emitted from B and so it must be emitted earlier in order to arrive at the camera at the same time as light from B. The time interval between events A and B, therefore, is

$$\Delta t = -\frac{\Delta y}{c} = \frac{2r_0 v}{c^2} \tag{13.12.4}$$

By substituting into the Lorentz transformation $\Delta x' = \gamma(\Delta x - v \, \Delta t)$ we get

$$\frac{2r_0}{\gamma} = \gamma \, \Delta x - \frac{2r_0 \gamma v^2}{c^2}$$

which is solved for Δx with the result

$$\Delta x = 2r_0 \tag{13.12.5}$$

The amazing thing is that the apparent diameter of the sphere is just what it would be if the sphere were at rest! The sphere, however, does appear rotated. There is no visible flattening of the sphere as might be expected from the conventional derivation of the Lorentz contraction formula as given in Section 13.5. This example serves to emphasize the necessity of analyzing each experiment in terms of the events exactly as they take place in that particular experiment.

The phenomenon that we have just discussed is called the *invisibility of the Lorentz contraction*. It means, for instance, that distant galaxies in the universe will give normal (i.e., undistorted) photographs in spite of possible high velocities. The Lorentz transformation is precisely the transformation required to minimize visual distortion.

13.13. THE CLOCK PARADOX

Much confusion has resulted from misinterpretations of the time dilation discussed in Section 13.6. Suppose that a space ship goes out from the earth and eventually returns. On board the ship is a clock and we may suppose that it is compared with a succession of clocks that have been placed in its path and that are at rest in the reference frame of the earth. Equation (13.6.1) applies. By integrating over the entire path of the ship

$$t' = \oint \sqrt{1 - v^2/c^2}\, dt \qquad (13.13.1)$$

On the other hand, consider a reference frame which is rigidly attached to the ship. Synchronized clocks are placed throughout this frame. According to observers on the ship, it is the earth that moves away and eventually returns to the ship. A fixed clock on the earth is compared with a succession of clocks in the reference frame of the ship. Equation (13.6.2) applies, leading to

$$t' = \oint \frac{dt}{\sqrt{1 - v^2/c^2}} \qquad \text{(incorrect)} \qquad (13.13.2)$$

The discrepancy between Eqs. (13.13.1) and (13.13.2) is the famous clock paradox of special relativity.

In our derivation of (13.13.1) and (13.13.2), the tacit assumption was made that reference frames attached to the ship and to the earth are equivalent. This is not true for the simple reason that in order to make a round trip the ship must undergo acceleration over at least some portion of the trip and, therefore, is not an inertial frame of reference.

Before the paradox can be removed, some assumptions about accelerated reference frames must be made. Specifically, in transforming from an inertial frame F to an accelerated frame F', the basic invariance of the four-dimensional line element is assumed to hold:

$$ds^2 = dx_i\, dx_i - c^2\, dt^2 = g'_{\mu\nu}\, dx^{\mu'}\, dx^{\nu'} \qquad (13.13.3)$$

The metric tensor $g'_{\mu\nu}$ of the accelerated frame is *not* identical to the simple metric (13.7.15) of an inertial frame! However, consider two events which are the comparison of a single clock that is at a fixed point in F' with two clocks a differential distance apart in F. Since $dx^{1'} = dx^{2'} = dx^{3'} = 0$,

$$ds^2 = (v^2 - c^2)\, dt^2 = g'_{44}(dx^{4'})^2 \qquad (13.13.4)$$

where $v^2 = dx_i\, dx_i/dt^2$ is the *instantaneous* velocity of the moving clock. Of course, v is now not a constant. Moreover, other clocks in frame F' *do not* necessarily have the same velocity with respect to F at a given time. The actual time interval recorded on the accelerated clock is called its *proper time* and is given by

$$-c^2\, d\tau^2 = g'_{44}(dx^{4'})^2 \qquad (13.13.5)$$

Due to its importance, the Greek symbol τ is given to the proper time. The so-called *coordinate* time interval in frame F' is given by

$$dx^{4'} = c\, dt' \qquad (13.13.6)$$

The rate of a coordinate clock can be chosen to be anything, just as different scales can be used for measuring ordinary distances. Equation (13.13.5) then becomes a definition of g'_{44}. Eqs. (13.13.4) and (13.13.5) give

$$d\tau = dt\sqrt{1 - v^2/c^2} \qquad (13.13.7)$$

This is identical to (13.6.1) with the exception that dt' is replaced by $d\tau$. An important aspect of the proper time is that it is an *invariant* which we may associate with the accelerated clock, particle, spaceship, or whatever else is under discussion:

$$-c^2\, d\tau^2 = ds^2 \qquad (13.13.8)$$

It is possible to place synchronized clocks as far apart as we please in the inertial frame F; it is *not* possible to do this in the accelerated frame F' unless the effects of the *gravitational potential difference* existing between the clocks is taken into account. It is shown in general relativity that a gravitational field existing between clocks at different locations affects their relative rates. Equation (13.13.2), therefore, is incorrect since in its derivation it is assumed that clocks can be placed arbitrarily far apart in the reference frame of the ship. The correct result is Eq. (13.13.1) which we now write in terms of the proper time of the ship as

$$\tau = \oint \sqrt{1 - v^2/c^2}\, dt \qquad (13.13.9)$$

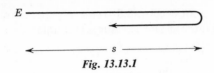

Fig. 13.13.1

The passengers on board the ship return to earth younger than if they had stayed at home!

Due to the controversial nature of the time dilation effect, it is worthwhile working through a specific simple example. A ship goes out from the earth E (Fig. 13.13.1) along a straight line path. At a distance s from the earth, it turns around and comes back. The time required to turn around is assumed to be small compared to the total time for the trip and the velocity of the ship is assumed to be constant over the other parts of the trajectory. Suppose that a transmitter is placed on the ship and that its natural frequency is v_0. The frequency received on the earth emitted by the ship as it moves away is

$$v_1 = v_0 \gamma \left(1 - \frac{v}{c} \right) \tag{13.13.10}$$

whereas the frequency received from the ship as it moves toward the earth is

$$v_2 = v_0 \gamma \left(1 + \frac{v}{c} \right) \tag{13.13.11}$$

The frequency received on earth does not shift from v_1 to v_2 immediately when the ship turns around because of the time required for the signal to propagate from the turning point to the earth. The frequency v_1 is received for a time

$$\frac{s}{v} + \frac{s}{c} \tag{13.13.12}$$

whereas v_2 is received for a time

$$\frac{s}{v} - \frac{s}{c} \tag{13.13.13}$$

The number of signals received on earth for the whole trip is

$$N_E = v_1 \left(\frac{s}{v} + \frac{s}{c} \right) + v_2 \left(\frac{s}{v} - \frac{s}{c} \right) = \frac{2 v_0 s}{v} \sqrt{1 - \frac{v^2}{c^2}} \tag{13.13.14}$$

The ship time is $t_s = N_E / v_0$ and the earth time is $t_E = 2s/v$. Hence,

$$t_s = t_E \sqrt{1 - v^2/c^2} \tag{13.13.15}$$

Suppose now that the ship receives signals from the earth. The frequencies ν_1 and ν_2 are received for just half the journey. This is because the frequency shift seen by observers on the ship occurs just as they turn around. It is here that explicit account is taken of the nonequivalence of the two frames of reference. The number of signals received by the ship is

$$N_s = \nu_1 \frac{t_s}{2} + \nu_2 \frac{t_s}{2} = \nu_0 \gamma t_s \tag{13.13.16}$$

Observers on the ship infer that earth time is $t_E = N_s/\nu_0$. Hence,

$$t_s = t_E \sqrt{1 - v^2/c^2} \tag{13.13.17}$$

Since (13.13.15) and (13.13.17) are identical, there is no paradox.

13.14. SPACE-TIME DIAGRAMS

It is useful to represent the trajectories of particles and light rays on a *space-time diagram*. For instance, a light ray which starts at the origin and propagates in the x-direction has the equation

$$x = ct \tag{13.14.1}$$

and is represented by the straight line OAB in Fig. 13.14.1 where ct is plotted against x. Possible trajectories of the space ship discussed in Section 13.13 in connection with the clock paradox are the solid curves inside the triangle OAC. These trajectories are sometimes called *world lines*. The slope of these lines must always exceed $45°$; otherwise a velocity greater than c is implied.

It is the proper time of the spaceship which is a measure of "distance" in the four-dimensional space-time continuum:

$$-c^2 \, d\tau^2 = dx^2 - c^2 \, dt^2 = ds^2 \tag{13.14.2}$$

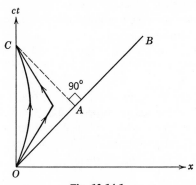

Fig. 13.14.1

It is interesting that the longer the world lines become in the sense of *ordinary* geometry, the shorter is the path in the sense

$$\tau = \oint d\tau \qquad (13.14.3)$$

The limiting case is the path OAC which would require that the ship move at the velocity c. Then $d\tau = 0$ and the time for the entire trip as measured by the ship's passengers is zero!

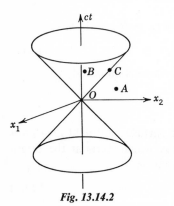

If all the light rays that pass through the origin and move in the $x_1 x_2$-plane are plotted, a cone is obtained as pictured in Fig. 13.14.2. This is called a *light cone*. (See also the discussion in the last part of Section 12.21.) The length s of a displacement vector that projects from the origin of space-time is given by

$$s^2 = g_{\mu\nu} x^\mu x^\nu = r^2 - c^2 t^2 \quad (13.14.4)$$

Fig. 13.14.2

where $r^2 = x_1{}^2 + x_2{}^2 + x_3{}^2$. In the discussion that follows, reference is made to Fig. 13.14.2; the conclusions also apply to the four-dimensional x_1, x_2, x_3, ct-space. If the tip of the four-vector, the length of which is given by (13.14.4), is at point A, then $s^2 > 0$. Since s^2 is an invariant, if a transformation to another frame F' is made

$$s^2 = r^2 - c^2 t^2 = r'^2 - c^2 t'^2 > 0 \qquad (13.14.5)$$

Thus in *all* frames of reference, the point A lies *outside* the light cone. In particular, it would be possible to choose the frame F' such that $t' = 0$, but it is *not* possible to choose F' such that $r' = 0$. In *all* reference frames, the two events connected by this four vector are separated by a *spatial* distance. Four vectors for which $s^2 > 0$ are said to be space-like. Events which are separated by a space-like four vector can have no causal connection with one another.

If the tip of the four-vector lies at the point B, then

$$s^2 = r^2 - c^2 t^2 = r'^2 - c^2 t'^2 < 0 \qquad (13.14.6)$$

No frame of reference exists in which $t' = 0$, but it *is* possible to make $r' = 0$ so that the vector points in the "time" direction. This class of four-vectors is said to be time-like. Causal connections can exist between events which are separated by time-like four-vectors.

If the four-vector joins O and C, then

$$r^2 - c^2 t^2 = r'^2 - c^2 t'^2 = 0 \qquad (13.14.7)$$

Such a vector is a *null vector*. Events separated by null vectors can be joined by light signals; in fact, the light cone is frequently called the *null cone*.

13.15. THE USE OF HYPERBOLIC FUNCTIONS

Let θ be a function of velocity defined by

$$\tanh \theta = \frac{v}{c} \tag{13.15.1}$$

Recall that $-1 < \tanh \theta < 1$ so that (13.15.1) includes all possible velocities. The function θ is called the *rapidity*. By means of the identities

$$\cosh^2 \theta - \sinh^2 \theta = 1 \qquad \tanh \theta = \frac{\sinh \theta}{\cosh \theta} \tag{13.15.2}$$

it is readily shown that

$$\cosh \theta = \frac{1}{\sqrt{1 - v^2/c^2}} = \gamma \qquad \sinh \theta = \frac{v}{c} \gamma \tag{13.15.3}$$

By means of (13.15.1) and (13.15.3) it is possible to find numerical values of various functions of v/c from a table of hyperbolic functions.

The special Lorentz transformation can be expressed in the interesting form

$$\begin{aligned} dx' &= dx \cosh \theta - c \, dt \sinh \theta \\ c \, dt' &= -dx \sinh \theta + c \, dt \cosh \theta \end{aligned} \tag{13.15.4}$$

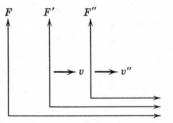

Thus the special Lorentz transformation is like a rotation with hyperbolic functions replacing the trigonometric functions!

Consider the three reference frames of Fig. 13.15.1. F' moves at velocity v rela-

Fig. 13.15.1

tive to F, F'' moves at velocity v' relative to F' and at velocity v'' relative to F. The transformation from F to F' is given by (13.15.4); the transformation from F' to F'' is

$$\begin{aligned} dx'' &= dx' \cosh \theta' - c \, dt' \sinh \theta' \\ c \, dt'' &= -dx' \sinh \theta' + c \, dt' \cosh \theta' \end{aligned} \tag{13.15.5}$$

The transformation *directly* from F to F'' follows from

$$\begin{aligned} \begin{pmatrix} \cosh \theta' & -\sinh \theta' \\ -\sinh \theta' & \cosh \theta' \end{pmatrix} & \begin{pmatrix} \cosh \theta & -\sinh \theta \\ -\sinh \theta & \cosh \theta \end{pmatrix} \\ &= \begin{pmatrix} \cosh (\theta' + \theta) & -\sinh (\theta' + \theta) \\ -\sinh (\theta' + \theta) & \cosh (\theta' + \theta) \end{pmatrix} \end{aligned} \tag{13.15.6}$$

where use is made of

$$\cosh (\theta' + \theta) = \cosh \theta' \cosh \theta + \sinh \theta' \sinh \theta$$
$$\sinh (\theta' + \theta) = \sinh \theta' \cosh \theta + \cosh \theta' \sinh \theta \qquad (13.15.7)$$

Thus a succession of special Lorentz transformations is again a special Lorentz transformation with rapidity

$$\theta'' = \theta' + \theta \qquad (13.15.8)$$

Rapidities add like Galilean velocities! Obviously,

$$\tanh \theta'' = \frac{v''}{c} \qquad (13.15.9)$$

13.16. FOUR-VECTORS

So far the only four-vector which has been discussed is the space-time interval between events:

$$dx^\mu = (dx', dx^2, dx^3, dx^4) \qquad (13.16.1)$$

where $dx^4 = c \, dt$. In general, a four-vector is a set of four quantities which has the same transformation property as the prototype four-vector (13.16.1). In this section, several other examples of four-vectors are presented.

Let (13.16.1) represent the differential space-time interval between two points on the trajectory of a particle. The *proper* time interval, i.e., the actual time that the particle's own clock records between these two points, is an invariant:

$$d\tau = \frac{1}{c} \sqrt{- g_{\mu\nu} \, dx^\mu \, dx^\nu} = dt\sqrt{1 - u^2/c^2} \qquad (13.16.2)$$

By dividing (13.16.1) by the invariant $d\tau$ another four-vector results which is called the four-velocity of the particle:

$$U^\mu = \frac{dx^\mu}{d\tau} \qquad (13.16.3)$$

It is convenient to display the spatial and temporal components of U^μ as

$$U^\mu = \left(\frac{\mathbf{u}}{\sqrt{1 - u^2/c^2}}, \frac{c}{\sqrt{1 - u^2/c^2}} \right) \qquad (13.16.4)$$

where $\mathbf{u}$ is just the ordinary three-velocity of the particle, the components of which are

$$\mathbf{u} = \left(\frac{dx^1}{dt}, \frac{dx^2}{dt}, \frac{dx^3}{dt} \right) \qquad (13.16.5)$$

Since

$$-c^2\, d\tau^2 = g_{\mu\nu}\, dx^\mu\, dx^\nu = dx^i\, dx^i - (dx^4)^2 \qquad (13.16.6)$$

it follows that the "square of the length" of the four-velocity is

$$-c^2 = g_{\mu\nu}U^\mu U^\nu = U^i U^i - (U^4)^2 \qquad (13.16.7)$$

Recall the rule (13.7.31) for relating covariant and contravariant components of a vector:

$$U_\lambda = g_{\lambda\mu}U^\mu \qquad U^\mu = g^{\mu\lambda}U_\lambda \qquad (13.16.8)$$

For the simple metric (13.7.15) these relations are

$$U_i = U^i \qquad U_4 = -U^4 \qquad (13.16.9)$$

It is sometimes convenient to introduce the particle *rapidity*, $\tanh\theta = u/c$, and display the components of the four-velocity as

$$U^\mu = (\hat{\boldsymbol{\lambda}} c \sinh\theta,\, c \cosh\theta) \qquad (13.16.10)$$

where $\hat{\boldsymbol{\lambda}}$ is a unit vector tangent to the particle trajectory in the three-space. If the particle moves in the x-direction, (13.16.10) is

$$U^\mu = (c \sinh\theta,\, 0,\, 0,\, c \cosh\theta) \qquad (13.16.11)$$

The *four-acceleration* of the particle is

$$A^\mu = \dot{U}^\mu \qquad (13.16.12)$$

where the dot stands for differentiation with respect to proper time. By differentiating (13.16.7),

$$0 = g_{\mu\nu}U^\mu A^\nu \qquad (13.16.13)$$

Thus, in the sense of the geometry of four-space, the four-velocity and the four-acceleration are "perpendicular" to one another. This result, of course, does not have the same simple geometric interpretation as it would for ordinary three-vectors.

Consider the propagation of a plane electromagnetic wave, the electric field of which is

$$\mathbf{E} = \mathbf{E}_0 \cos(\mathbf{k}\cdot\mathbf{r} - \omega t) \qquad (13.16.14)$$

Assume that the direction of propagation is in the $x_1 x_2$-plane and at an angle α with the x_1-axis as in Fig. 12.21.3. We do not know yet how to transform electric fields, but whatever the rule is, both $\mathbf{E}$ and $\mathbf{E}_0$ transform the same way. Therefore, $\cos(\mathbf{k}\cdot\mathbf{r} - \omega t)$ is just a scalar factor and the phase $\mathbf{k}\cdot\mathbf{r} - \omega t$ is an invariant. Since $\mathbf{r}$, ct is the displacement four-vector from the origin, the propagation vector $\mathbf{k}$ and the frequency make up another four-vector:

$$k^\mu = \left(\mathbf{k}, \frac{\omega}{c}\right) \qquad (13.16.15)$$

The phase of the plane wave can be expressed by

$$\mathbf{k} \cdot \mathbf{r} - \omega t = g_{\mu\nu}k^{\mu}x^{\nu} \qquad (13.16.16)$$

Consider the transformation of (13.16.15) by means of the special Lorentz transformation:

$$k_1' = \gamma\left(k_1 - \frac{v}{c}\frac{\omega}{c}\right) \qquad (k_1 = k^1) \qquad (13.16.17)$$

$$\frac{\omega'}{c} = \gamma\left(\frac{\omega}{c} - \frac{v}{c}k_1\right) \qquad (13.16.18)$$

Remember (12.21.8) that

$$k_1 = k\cos\alpha = \frac{2\pi}{\lambda}\cos\alpha \qquad k_1' = k'\cos\alpha' = \frac{2\pi}{\lambda'}\cos\alpha' \quad (13.16.19)$$

Moreover,

$$\omega = ck \qquad \omega' = ck' \qquad (13.16.20)$$

Thus (13.16.18) can be expressed by

$$\omega' = \omega\gamma\left(1 - \frac{v}{c}\cos\alpha\right) \qquad (13.16.21)$$

This is just the Doppler effect, Eq. (13.11.4). If (13.16.17) is divided by (13.16.18), the result can be expressed by

$$\cos\alpha' = \frac{\cos\alpha - v/c}{1 - (v/c)\cos\alpha} \qquad (13.16.22)$$

which is the relativistic aberration formula (13.11.8).

13.17. CHARGE AND CURRENT AS A FOUR-VECTOR

A new basic assumption will now be added to the development of our theory: The electric charge on an elementary particle is an absolute invariant. There is ample experimental evidence to support this hypothesis. For instance, the atoms of the heavier elements contain orbital electrons moving at sufficiently high velocities that any increase or decrease of the total charge as a result should be apparent.

Consider a beam of identically charged particles moving in the x-direction of F. Let F' move at the same velocity as the particles, Fig. 13.17.1. Consider two events, A and B, that are simultaneous in F. Suppose, for example, that these events are the closing of gates in the path of the particles. Since $\Delta t = 0$, the special Lorentz transformation gives $\Delta x' = \gamma \, \Delta x$. If there are N-particles between A and B, the charge density in F will be

$$\rho = \frac{Nq}{\Delta x \sigma} = \gamma \frac{Nq}{\Delta x' \sigma} \qquad (13.17.1)$$

where σ is the cross-sectional area of the beam. Events A and B are not

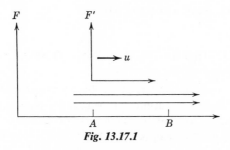

Fig. 13.17.1

simultaneous in F', but it does not matter since the particles do not move relative to F'. Since q is an invariant, the charge density in F' is

$$\rho_0 = \frac{Nq}{\Delta x' \sigma} \qquad (13.17.2)$$

Comparison of (13.17.1) and (13.17.2) reveals

$$\rho = \frac{\rho_0}{\sqrt{1 - u^2/c^2}} \qquad (13.17.3)$$

The current density (charge per second per unit area) in F is

$$j = \rho u = \rho_0 \frac{u}{\sqrt{1 - u^2/c^2}} \qquad (13.17.4)$$

The set of four quantities

$$j^\mu = (\rho u, 0, 0, \rho c) \qquad (13.17.5)$$

constitutes a four-vector. To see this, write (17.5) as

$$j^\mu = \left(\frac{\rho_0 u}{\sqrt{1 - u^2/c^2}}, 0, 0, \frac{\rho_0 c}{\sqrt{1 - u^2/c^2}} \right) = \rho_0 U^\mu \qquad (13.17.6)$$

where U^μ are the components of the four-velocity of the charged particles.

In general, if charged matter is moving in an arbitrary way through space, we may, at a given point, interpret the three spatial components of $j^\mu = \rho_0 U^\mu$ as the ordinary current density and the temporal component as c times the charge density; ρ_0 must then be the value of the charge density in the instantaneous rest frame of the moving matter.

Charge invariance with the consequent interpretation of charge and current as a four-vector is basic to the development of the transformation theory of the electromagnetic field and, finally, the relativistically correct equations of motion of a charged particle.

13.18. THE ELECTROMAGNETIC FIELD TENSOR

The transformation properties of the electromagnetic field can be deduced from simple solutions of Maxwell's equations. Consider an infinite line charge parallel to the x_1-axis of frame F and moving at a velocity u in the positive x-direction. Let F' be another frame of reference moving at velocity v relative to F. One of the basic postulates of special relativity is the universal validity of the laws of physics in all inertial frames of reference. Hence, the electric and magnetic fields of the line charge in F and F' are

$$E = \frac{2\mu}{r} \qquad B = \frac{2J}{rc} \qquad (13.18.1)$$

$$E' = \frac{2\mu'}{r} \qquad B' = \frac{2J'}{rc} \qquad (13.18.2)$$

where (J, μ) are current and charge per unit length as determined by observers in F and (J', μ') are these quantities as determined by observers in F'. Since r is the distance from the line charge measured perpendicular to v, it is the same in both F and F'. In Section 13.17, we learned that $(J, 0, 0, \mu c)$ are the components of a four-vector. Hence,

$$J' = \gamma(J - v\mu) \qquad \mu' = \gamma\left(\mu - \frac{v}{c^2}J\right) \qquad (13.18.3)$$

The combination of (13.18.1), (13.18.2), and (13.18.3) yields

$$B' = \gamma\left(B - \frac{v}{c}E\right) \qquad E' = \gamma\left(E - \frac{v}{c}B\right) \qquad (13.18.4)$$

The lines of B are circles around the x_1-axis; the lines of E are radical

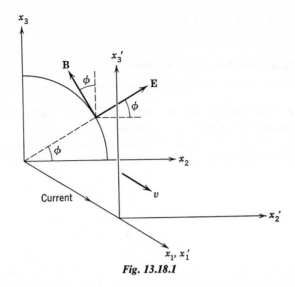

Fig. 13.18.1

as indicated in Fig. 13.18.1. We wish to know the transformation properties of the components of E and B. Thus multiply (13.18.4) by $\cos \phi$:

$$B' \cos \phi = \gamma \left(B \cos \phi - \frac{v}{c} E \cos \phi \right)$$

$$E' \cos \phi = \gamma \left(E \cos \phi - \frac{v}{c} B \cos \phi \right)$$

(13.18.5)

Since $x_2' = x_2$ and $x_3' = x_3$, the angle ϕ looks the same in both F and F'. From the geometry of Fig. 13.18.1,

$$B_3' = \gamma \left(B_3 - \frac{v}{c} E_2 \right) \qquad E_2' = \gamma \left(E_2 - \frac{v}{c} B_3 \right) \qquad (13.18.6)$$

Similarly, the multiplication of (13.18.4) by $\sin \phi$ leads to

$$B_2' = \gamma \left(B_2 + \frac{v}{c} E_3 \right) \qquad E_3' = \gamma \left(E_3 + \frac{v}{c} B_2 \right) \qquad (13.18.7)$$

Now consider a charged conducting sheet with a surface charge density ρ stat-coulombs/cm² which is perpendicular to the x_1- and x_1'-axes as indicated in Fig. 13.18.2. The electric fields produced in the x_1- and x_1'-directions are

$$E_1 = 4\pi\rho \qquad E_1' = 4\pi\rho' \qquad (13.18.8)$$

But, since $\Delta x_2 = \Delta x_2'$ and $\Delta x_3 = \Delta x_3'$, the surface charge density looks the same in both F and F'.   Hence,

$$E_1' = E_1 \tag{13.18.9}$$

The reader can construct a proof leading to

$$B_1' = B_1 \tag{13.18.10}$$

for example, by considering the magnetic field of a surface current flowing in the x_3-direction of the $x_1 x_3$-plane.

Equations (13.18.6), (13.18.7), (13.18.9) and (13.18.10) are a complete set of equations for transforming the components of the electromagnetic

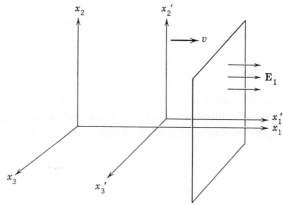

Fig. 13.18.2

field by means of the special Lorentz transformation. Even though simple solutions of Maxwell's equations have been used, the results are general and apply to any electromagnetic field.   It is obvious that the six components of the electromagnetic field do not in any way comprise a four-vector. It is, however, possible to construct a second rank antisymmetric tensor, called the field tensor, the transformation of which correctly gives the transformation of the electromagnetic field.   The contravariant components of this tensor are

$$(F^{\mu\nu}) = \begin{pmatrix} 0 & B_3 & -B_2 & -E_1 \\ -B_3 & 0 & B_1 & -E_2 \\ B_2 & -B_1 & 0 & -E_3 \\ E_1 & E_2 & E_3 & 0 \end{pmatrix} \tag{13.18.11}$$

Its transformation is

$$F^{\mu\nu'} = P_\alpha^{\mu'} P_\beta^{\nu'} F^{\alpha\beta} \tag{13.18.12}$$

where now $P_\alpha^{\mu'}$ can be the coefficients of any Lorentz transformation.   In particular, if $P_\alpha^{\mu'}$ are the coefficients of the special Lorentz transformation given by (13.7.21), (13.18.12) can be conveniently represented by the matrix equation $(F^{\mu\nu'}) = P(F^{\mu\nu})\tilde{P}$:

$$(F^{\mu\nu})' = \begin{pmatrix} \gamma & 0 & 0 & -\gamma\dfrac{v}{c} \\ 0 & 1 & 0 & 0 \\ 0 & 0 & 1 & 0 \\ -\gamma\dfrac{v}{c} & 0 & 0 & \gamma \end{pmatrix} \begin{pmatrix} 0 & B_3 & -B_2 & -E_1 \\ -B_3 & 0 & B_1 & -E_2 \\ B_2 & -B_1 & 0 & -E_3 \\ E_1 & E_2 & E_3 & 0 \end{pmatrix}$$

$$\times \begin{pmatrix} \gamma & 0 & 0 & -\gamma\dfrac{v}{c} \\ 0 & 1 & 0 & 0 \\ 0 & 0 & 1 & 0 \\ -\gamma\dfrac{v}{c} & 0 & 0 & \gamma \end{pmatrix} \qquad (13.18.13)$$

The reader should confirm the correctness of (13.18.13) by carrying out the indicated matrix multiplications.

There is, of course, a covariant form of the field tensor:

$$F_{\mu\nu} = g_{\mu\alpha}g_{\nu\beta}F^{\alpha\beta} \qquad (13.18.14)$$

It is easily verified that

$$(F_{\mu\nu}) = \begin{pmatrix} 0 & B_3 & -B_2 & E_1 \\ -B_3 & 0 & B_1 & E_2 \\ B_2 & -B_1 & 0 & E_3 \\ -E_1 & -E_2 & -E_3 & 0 \end{pmatrix} \qquad (13.18.15)$$

Moreover,

$$F'_{\mu\nu} = P_{\mu'}^{\alpha} P_{\nu'}^{\beta} F_{\alpha\beta} \qquad (13.18.16)$$

where $P_{\mu'}^{\alpha}$ are given by (13.7.25) in the case of the special Lorentz transformation. In matrix notation,

$$(F_{\mu\nu})' = \widetilde{P^{-1}}(F_{\alpha\beta})P^{-1} \qquad (13.18.17)$$

13.19. MAXWELL'S EQUATIONS

The Maxwell field equations in conventional vector notation are

$$\mathbf{\nabla} \times \mathbf{E} + \frac{1}{c}\frac{\partial \mathbf{B}}{\partial t} = 0 \qquad \mathbf{\nabla} \cdot \mathbf{E} = 4\pi\rho$$

$$\mathbf{\nabla} \times \mathbf{B} - \frac{1}{c}\frac{\partial \mathbf{E}}{\partial t} = \frac{4\pi}{c}\mathbf{j} \qquad \mathbf{\nabla} \cdot \mathbf{B} = 0 \qquad (13.19.1)$$

It is easy to verify that all the Maxwell equations are contained in the tensor equations

$$\partial_\mu F^{\nu\mu} = \frac{4\pi}{c} \rho_0 U^\nu \tag{13.19.2}$$

$$\partial_\mu F_{\nu\lambda} + \partial_\lambda F_{\mu\nu} + \partial_\nu F_{\lambda\mu} = 0 \tag{13.19.3}$$

where the abbreviation $\partial_\mu = \partial/\partial x_\mu$ is used. The symbol ∂_μ can be transformed formally like a covariant vector in expressions in which it appears. Notice the cyclic permutation of indices in (13.19.3).

Consider, for example, $\nu = 1$ in (13.19.2):

$$\partial_2 F^{12} + \partial_3 F^{13} + \partial_4 F^{14} = \frac{4\pi}{c} \rho_0 U^1 \tag{13.19.4}$$

Writing $j^1 = \rho_0 U^1$ as the first component of the current density and using the contravariant form of the field tensor (13.18.11),

$$\partial_2 B_3 - \partial_3 B_2 + \frac{1}{c} \frac{\partial}{\partial t}(-E_1) = \frac{4\pi}{c} j^1 \tag{13.19.5}$$

As another example, let $\mu, \nu, \lambda = 1, 2, 3$ in (13.19.3):

$$\begin{aligned} \partial_1 F_{23} + \partial_3 F_{12} + \partial_2 F_{31} &= 0 \\ \partial_1 B_1 + \partial_3 B_3 + \partial_2 B_2 &= 0 \end{aligned} \tag{13.19.6}$$

Equations (13.19.2) and (13.19.3) are tensor equations and are valid in all coordinate systems. Thus the covariance of Maxwell's equations with respect to the group of all possible Lorentz transformations is established.

The vector potential is easily introduced in a covariant fashion into the theory. Let the field tensor be expressed in terms of a four-vector by means of

$$F_{\mu\nu} = \partial_\mu A_\nu - \partial_\nu A_\mu \tag{13.19.7}$$

This way of writing the field tensor automatically satisfies (13.19.3); the reader should verify this statement. A gauge transformation

$$A_\nu' = A_\nu + \partial_\nu \psi \tag{13.19.8}$$

where ψ is any scalar function, does not affect the field tensor, i.e.,

$$F_{\mu\nu} = \partial_\mu A_\nu' - \partial_\nu A_\mu' \tag{13.19.9}$$

It is the general practice to restrict the possible vector potentials to those which obey the *Lorentz gauge condition*

$$\partial^\mu A_\mu = 0 \qquad (13.19.10)$$

where

$$\partial^\mu = g^{\mu\nu}\partial_\nu \qquad (13.19.11)$$

is the contravariant form of the gradient operator; (13.19.10) can, of course, be expressed in other ways such as

$$\partial_\mu A^\mu = 0, \qquad g_{\mu\nu}\partial^\mu A^\nu = 0, \qquad \text{etc.} \qquad (13.19.12)$$

In terms of the vector potential, the field equations (13.19.2) are

$$\partial_\mu(\partial^\nu A^\mu - \partial^\mu A^\nu) = \frac{4\pi}{c}\rho_0 U^\nu \qquad (13.19.13)$$

The first term vanishes due to the Lorentz gauge condition leaving

$$(\partial_\mu\partial^\mu)A^\nu = -\frac{4\pi}{c}\rho_0 U^\nu \qquad (13.19.14)$$

In conventional terms,

$$\partial_\mu\partial^\mu = \nabla^2 - \frac{1}{c^2}\frac{\partial^2}{\partial t^2} \qquad (13.19.15)$$

In (13.19.14) the equations for $\nu = 1, 2$ and 3 can be expressed

$$\left[\nabla^2 - \frac{1}{c^2}\frac{\partial^2}{\partial t^2}\right]\mathbf{A} = -\frac{4\pi}{c}\mathbf{j} \qquad (13.19.16)$$

Writing $A^4 = \psi$ as the conventional scalar potential gives, for $\nu = 4$,

$$\left[\nabla^2 - \frac{1}{c^2}\frac{\partial^2}{\partial t^2}\right]\psi = -4\pi\rho \qquad (13.19.17)$$

These are the familiar wave equations for the vector and scalar potentials.

13.20. POINT CHARGE MOVING AT CONSTANT VELOCITY

As an application of the theory which we have developed, we will derive the expressions for the electric and magnetic fields associated with a point charge moving at constant velocity.

Consider a point charge q moving at a constant velocity u in the x-direction of F. Let F' be the rest frame of the particle, Fig. 13.20. If q is at the origin of F', the components of the electric field in the $x_1'x_2'$-plane are

$$E_1' = \frac{qx_1'}{r'^3} \qquad E_2' = \frac{qx_2'}{r'^3} \qquad (13.20.1)$$

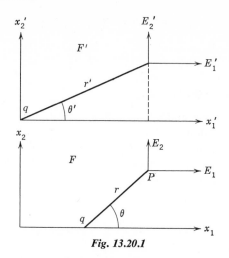

Fig. 13.20.1

There is no magnetic field in F'. In frame F, the fields at point P and the location of the charge are to be measured *simultaneously*. These measurements *do not* appear simultaneous in F', but this does not matter since charge and fields are static in F'. The transformation of these two events is

$$x_1' = \gamma(\Delta x_1 - u\,\Delta t) = \gamma\,\Delta x_1 = \gamma r \cos\theta$$
$$x_2' = x_2 = r \sin\theta$$
$$r' = (x_1'^2 + x_2'^2)^{1/2} = (\gamma^2 r^2 \cos^2\theta + r^2 \sin^2\theta)^{1/2} \qquad (13.20.2)$$
$$= r(1 + \cos^2\theta[\gamma^2 - 1])^{1/2}$$

For the transformation of the electric field we may write

$$E_1' = E_1 \qquad E_2 = \gamma\left(E_2' + \frac{u}{c}B_3'\right) = \gamma E_2' \qquad (13.20.3)$$

Combining these results yields

$$E_1 = \frac{q \cos\theta}{r^2}\,\frac{\gamma}{(1 + \cos^2\theta[\gamma^2 - 1])^{3/2}} \qquad (13.20.4)$$

$$E_2 = \frac{q \sin\theta}{r^2}\,\frac{\gamma}{(1 + \cos^2\theta[\gamma^2 - 1])^{3/2}} \qquad (13.20.5)$$

The transformation of the magnetic field is

$$B_3 = \gamma\left(B_3' + \frac{u}{c}E_2'\right) = \gamma\frac{u}{c}E_2'$$

$$B_2 = \gamma\left(B_2' - \frac{u}{c}E_3'\right) = 0 \qquad (13.20.6)$$

$$B_1 = B_1' = 0$$

Hence,

$$B_3 = \frac{u}{c}\frac{q\sin\theta}{r^2}\frac{\gamma}{(1 + \cos^2\theta[\gamma^2 - 1])^{3/2}} \tag{13.20.7}$$

13.21. PARTICLE MECHANICS

In this section, we address ourselves to the problem of finding the necessary modifications of the laws of Newtonian particle mechanics which will make them consistent with the requirements of special relativity. The *only* force which is legitimately the domain of special relativity in this respect is the Lorentz force acting on a charged particle. The treatment

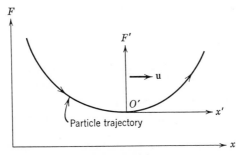

Fig. 13.21.1

of gravitational forces requires general relativity; theories of the remaining two types of forces which are at present known, namely, weak interactions and nuclear forces, can presumably be made relativistically covariant, but, of course, Newtonian mechanics is inapplicable.

Consider a charged particle moving along an arbitrary trajectory with respect to a frame F. At a given point of the path, the particle has a definite velocity u. Relative to an inertial frame F', also moving at the velocity u, the particle is instantaneously (though not permanently) at rest. At *low* velocities we know how to write down the equations of motion of a particle. At the instant that the particle is at rest relative to F', its equation of motion is

$$q\mathbf{E}' = m\frac{d\mathbf{u}'}{dt'} \tag{13.21.1}$$

where m is the mass of the particle. In order to find out what the equation of motion is with respect to frame F relative to which the particle's velocity may be anything from 0 to c, it is necessary merely to transform (13.21.1). Let us suppose that at the point O' (Fig. 13.21.1) under consideration the particle is instantaneously moving in the x-direction. The calculation of the transformation for acceleration begins with

$$u_1 = \frac{u_1' + u}{1 + uu_1'/c^2} \qquad u_2 = \frac{1}{\gamma}\frac{u_2'}{1 + uu_1'/c^2} \qquad u_3 = \frac{1}{\gamma}\frac{u_3'}{1 + uu_1'/c^2} \quad (13.21.2)$$

where u is the *constant* velocity of F' which is *instantaneously* the same as the velocity of the particle. The differentiation of the equation for u_1 yields

$$\frac{du_1}{dt} = \frac{du_1'/dt}{1 + uu_1'/c^2} - \frac{u_1' + u}{(1 + uu_1'/c^2)^2}\frac{u}{c^2}\frac{du_1'}{dt} \qquad (13.21.3)$$

Setting $u_1' = 0$ results in

$$a_1 = \frac{du_1}{dt} = \left(1 - \frac{u^2}{c^2}\right)\frac{du_1'}{dt} \qquad (13.21.4)$$

But $dt = \gamma(dt' + u\,dx'/c^2) = \gamma\,dt'$ so that

$$a_1 = \left(1 - \frac{u^2}{c^2}\right)\frac{1}{\gamma}a_1' = \frac{1}{\gamma^3}a_1' \qquad (13.21.5)$$

By a similar calculation,

$$a_2' = \gamma^2 a_2 \qquad a_3' = \gamma^2 a_3 \qquad (13.21.6)$$

With the aid of the transformation equations for the fields, the components of (13.21.1) are easily found to be

$$qE_1 = m\gamma^3 a_1$$

$$q\left(E_2 - \frac{u}{c}B_3\right) = m\gamma a_2$$

$$q\left(E_3 + \frac{u}{c}B_2\right) = m\gamma a_3 \qquad (13.21.7)$$

where now all quantities are expressed in frame F. It is possible to express (13.21.7) in the form

$$q\left(\mathbf{E} + \frac{1}{c}\mathbf{u} \times \mathbf{B}\right) = m\gamma^3 \mathbf{a}_{\|} + m\gamma \mathbf{a}_{\perp} \qquad (13.21.8)$$

where $\mathbf{a}_{\|}$ and $\mathbf{a}_{\perp}$ refer to the components of the particle acceleration *parallel* and *perpendicular* to the particle trajectory. The particle is seen to behave as though it had a velocity dependent mass! Moreover, the effective mass is different, depending on whether we are talking about acceleration parallel or perpendicular to the trajectory. We define a *longitudinal* and a *transverse* mass by

$$m_{\|} = m\gamma^3 \qquad m_{\perp} = \gamma m \qquad (13.21.9)$$

The left side of (13.21.8) represents the familiar Lorentz force. An immediate consequence of this result is that force and acceleration are no longer in the same direction. It is relatively harder to accelerate a particle in the same direction in which it is already going than perpendicular to this direction.

Equation (13.21.8) is not a convenient form of the equation of motion for most purposes. In order to rewrite the equations of motion, we prove the identity

$$\frac{d}{dt}(\gamma\mathbf{u}) = \gamma\mathbf{a}_\perp + \gamma^3\mathbf{a}_{\parallel} \tag{13.21.10}$$

Note that

$$\frac{d}{dt}(\gamma\mathbf{u}) = \gamma\frac{d\mathbf{u}}{dt} + \mathbf{u}\frac{d\gamma}{dt} = \gamma\mathbf{a} + \mathbf{u}\frac{\gamma^3}{c^2}u\frac{du}{dt} \tag{13.21.11}$$

By differentiation of $\mathbf{u}\cdot\mathbf{u} = u^2$,

$$\mathbf{u}\cdot\mathbf{a} = u\frac{du}{dt} = ua_{\parallel} \tag{13.21.12}$$

Moreover, $\mathbf{u}a_{\parallel} = ua_{\parallel}$. Hence, (13.21.11) is

$$\frac{d}{dt}(\gamma\mathbf{u}) = \gamma(\mathbf{a}_{\parallel} + \mathbf{a}_\perp) + \gamma^3\frac{u^2}{c^2}\mathbf{a}_{\parallel} \tag{13.21.13}$$

It is easily shown that

$$\gamma + \gamma^3\frac{u^2}{c^2} = \gamma^3 \tag{13.21.14}$$

Hence, (13.21.10) follows. The equation of motion (13.21.8), therefore, is

$$\mathbf{F} = q\left(\mathbf{E} + \frac{1}{c}\mathbf{u}\times\mathbf{B}\right) = \frac{d}{dt}(\gamma m\mathbf{u}) \tag{13.21.15}$$

We identify

$$\mathbf{p} = \gamma m\mathbf{u} \tag{13.21.16}$$

as the appropriate relativistic generalization of the linear momentum of a particle.

As in nonrelativistic mechanics, the concept of energy can be introduced by considering the rate at which the net force does work on a particle:

$$\mathbf{F}\cdot\mathbf{u} = m(\gamma\mathbf{a}_\perp + \gamma^3\mathbf{a}_{\parallel})\cdot\mathbf{u}$$
$$= m\gamma^3 u\frac{du}{dt} = \frac{d}{dt}(m\gamma c^2) = \frac{dW}{dt} \tag{13.21.17}$$

Thus $\mathbf{F}\cdot\mathbf{u}$ can be expressed as the total time rate of change of the function of the velocity $W(u) = m\gamma c^2$. The kinetic energy of a particle is simply the difference between $W(u)$ and $W(0)$:

$$T = W(u) - W(0) = m\gamma c^2 - mc^2 \qquad (13.21.18)$$

If $u \ll c$,

$$\gamma = \left(1 - \frac{u^2}{c^2}\right)^{-1/2} \cong 1 + \frac{1}{2}\frac{u^2}{c^2} + \frac{3}{8}\frac{u^4}{c^4} + \cdots \qquad (13.21.19)$$

$$T \cong \tfrac{1}{2}mu^2\left(1 + \frac{3}{4}\frac{u^2}{c^2} + \cdots\right) \qquad (13.21.20)$$

The rate at which the Lorentz force does work on the particle can also be expressed in terms of the electric field:

$$\mathbf{F}\cdot\mathbf{u} = q\left(\mathbf{E} + \frac{1}{c}\mathbf{u}\times\mathbf{B}\right)\cdot\mathbf{u} = q(\mathbf{E}\cdot\mathbf{u}) \qquad (13.21.21)$$

For static fields,

$$\mathbf{E} = -\boldsymbol{\nabla}\psi, \qquad \mathbf{F}\cdot\mathbf{u} = -q\frac{d\psi}{dt} \qquad (13.21.22)$$

Combining this result with (13.21.17) yields

$$\frac{d}{dt}(W + q\psi) = 0 \qquad (13.21.23)$$

Thus,

$$\mathscr{E} = W + q\psi \qquad (13.21.24)$$

can be identified as the total energy of the particle and is a constant of the motion if the fields through which the particle moves are static.

13.22. THE FOUR-DIMENSIONAL FORMULATION OF PARTICLE MECHANICS

Consider the four-vector which results if the mass of a particle is multiplied by its four-velocity:

$$P^\mu = mU^\mu = (\gamma m\mathbf{u}\ , \gamma mc) = \left(\mathbf{p}, \frac{W}{c}\right) \qquad (13.22.1)$$

The three spatial components are identical to the linear momentum (13.21.16) while the temporal components are essentially the function W given by (13.21.17). This four-vector is called the energy-momentum four-vector, or sometimes just the four-momentum. It follows directly, from (13.16.7) that

$$P_\mu P^\mu = p^2 - \frac{W^2}{c^2} = -m^2c^2 \qquad (13.22.2)$$

The following tensor equation contains the equation of motion (13.21.15):

$$\frac{q}{c} F^{\mu\nu} U_\nu = m\dot{U}^\mu \tag{13.22.3}$$

The dot stands for differentiation with respect to the proper time of the particle. The reader should verify that $\mu = 1$, 2, and 3 gives the three components of (13.21.15). For $\mu = 4$ (13.22.3) is

$$\frac{q}{c} (F^{41}U_1 + F^{42}U_2 + F^{43}U_3) = m\frac{dU^4}{d\tau} \tag{13.22.4}$$

Referring to the contravariant form of the field tensor (13.18.11), and remembering that $d\tau = dt/\gamma$,

$$\frac{q}{c} (E_1\gamma u_1 + E_2\gamma u_2 + E_3\gamma U_3) = m\gamma\frac{d}{dt}(\gamma c)$$

$$q(\mathbf{E} \cdot \mathbf{u}) = \frac{d}{dt}(m\gamma c^2) \tag{13.22.5}$$

This is recognized as the expression for the rate at which the Lorentz force does work on the particle. In other words, the four-dimensional form of the equations of motion contains the expression for the time rate of change of the energy as well as the time rate of change of momentum.

The four vector

$$K^\mu = m\dot{U}^\mu \tag{13.22.6}$$

is called the four-force, or sometimes the Minkowski force, acting on the particle. In this form, the equations of motion resemble the familiar Newtonian equations of nonrelativistic particle mechanics. The components of the Minkowski force are related to the ordinary force by

$$K^\mu = \left(\gamma\mathbf{F}, \frac{\gamma}{c}\mathbf{F} \cdot \mathbf{u}\right) \tag{13.22.7}$$

13.23. HYPERBOLIC MOTION*

We consider here the solution of the equations of motion for a constant force; for example, a charged particle moving in a uniform electric field. We define a vector $\mathbf{g}$ by means of

$$\mathbf{F} = m\mathbf{g} = \frac{d}{dt}(\gamma m\mathbf{u}) \tag{13.23.1}$$

---

* So named because the particle trajectory on a space-time diagram is a hyperbola. See Problem 13.24.23.

Let the x-direction coincide with $\mathbf{F}$.   Then

$$g = \frac{d}{dt}(\gamma u) \qquad (13.23.2)$$

It is convenient to introduce the rapidity and the proper time of the particle:

$$\frac{u}{c} = \tanh\theta \qquad dt = \gamma\, d\tau = \cosh\theta\, d\tau \qquad (13.23.3)$$

The equation of motion reduces to

$$g = \dot\theta c \qquad (13.23.4)$$

where, as usual, the dot refers to differentiation with respect to proper time.   The rapidity has the advantage in one-dimensional problems that it reduces the differential equation to a very simple one.

The components of the four-velocity of the particle are

$$\dot x = c \sinh\theta \qquad c\dot t = c \cosh\theta \qquad (13.23.5)$$

The integration of (13.23.4) yields

$$\theta = \frac{g\tau}{c} \qquad \dot x = c \sinh\frac{g\tau}{c} \qquad c\dot t = c \cosh\frac{g\tau}{c} \qquad (13.23.6)$$

where the initial condition $\dot x = 0$ at $\tau = 0$ is used.   A second integration yields

$$x = \frac{c^2}{g}\left[\cosh\frac{g\tau}{c} - 1\right] \qquad t = \frac{c}{g}\sinh\frac{g\tau}{c} \qquad (13.23.7)$$

which expresses x and t as parametric equations in the proper time of the particle.

The vector $\mathbf{g}$ has a simple physical interpretation.   If our particle were a space ship, $\mathbf{g}$ would be the gravitational field experienced by the passengers due to the acceleration of the ship.

13.24. THE EQUIVALENCE OF MASS AND ENERGY

Consider a system of N isolated point particles which, for the moment, are far enough apart so that they do not interact.   In a reference frame F' which is moving relative to the laboratory at a velocity $\mathbf{v}$, the fourth component of the energy-momentum four-vector is

$$W' = \sum_{i=1}^{N} \frac{m_i c^2}{\sqrt{1 - u_i'^2/c^2}} \qquad (13.24.1)$$

Let F' be the center of mass coordinate system defined by the condition that the total linear momentum of the system is zero with respect to F':

$$\mathbf{p'} = \sum_{i=1}^{N} \frac{m_i \mathbf{u}_i'}{\sqrt{1 - u_i'^2/c^2}} = 0 \qquad (13.24.2)$$

In transforming back to the laboratory coordinates, it is convenient to use the general Lorentz transformation without rotation in vector form as given by (13.8.7) and (13.8.8). By remembering that $(\mathbf{p}, W/c)$ is a four-vector in the same sense as $(\mathbf{r}, c\,\Delta t)$ we can write the transformation in inverted form as

$$\mathbf{p} = \mathbf{p'} + \mathbf{v}\left[\frac{\mathbf{p'} \cdot \mathbf{v}}{v^2}(\gamma - 1) + \gamma \frac{W''}{c^2}\right] \qquad W = \gamma(W' + \mathbf{v} \cdot \mathbf{p'}) \qquad (13.24.3)$$

Since F' is such that $\mathbf{p'} = 0$,

$$\mathbf{p} = \gamma \mathbf{v} \frac{W'}{c^2} \qquad W = \gamma W' \qquad (13.24.4)$$

Here, $\mathbf{v}$ can be interpreted as the velocity of the center of mass of the system. The total mass of the system of noninteracting particles relative to the laboratory can be found by means of

$$W = M\gamma c^2 = \gamma W' \qquad (13.24.5)$$

Then, by (13.24.1),

$$M = \sum_{i=1}^{N} \frac{m_i}{\sqrt{1 - u_i'^2/c^2}} \qquad (13.24.6)$$

The apparent mass of the system relative to the laboratory is not just the sum of the individual masses of the particles! The kinetic energy of the individual particles relative to F' can be introduced by means of (13.21.18):

$$M = \sum_{i=1}^{N} m_i + \sum_{i=1}^{N} \frac{T_i'}{c^2} \qquad (13.24.7)$$

Since the system is isolated, the total energy remains constant. If, after a time, the particles come together to form a composite system, some of the kinetic energy is converted into other forms, such as potential energy or heat energy. Then (13.24.7) must be replaced by

$$M = \sum_{i=1}^{N} m_i + \sum_{i=1}^{N} \frac{T_i}{c^2} + \sum_{i=1}^{N} \frac{V_i}{c^2} \qquad (13.24.8)$$

The internal energy of a system, whatever its form, causes an increase in the total mass. For example, compressing a spring makes it a little heavier; a beaker of hot water weighs more than an identical beaker of cold water. These effects are, of course, extremely small.

In the case of a bound system, such as an atomic nucleus, there is a force of attraction between the nucleons and the potential energy function is negative. Such a stable bound system can be expected to have a *smaller* mass than the sum of the masses of the individual particles making it up. As an example, consider the nuclear reaction

$$_3Li^7 + {_1}H^1 \rightarrow {_2}He^1 + {_2}He^1$$

The masses of the various atomic particles in atomic mass units are

$$_3Li^7 = 7.0166 \qquad _1H^1 = 1.0076 \qquad _2He^4 = 4.0028$$

The so called *mass defect* or loss of mass during the reaction is

$$\Delta m = 7.0166 + 1.0076 - 2 \times 4.0028 = 0.0186 \qquad (13.24.9)$$

Thus an amount of energy $\Delta m c^2$ appears in the form of the kinetic energy of the product particles. Nuclear forces are strong and the mass defect is easily measured, giving an important confirmation of the mass-energy equivalence principle.

As mentioned earlier, nuclear forces do not fit into the framework of Newtonian mechanics and hence our development of relativistic mechanics is inapplicable to them. However, one of the most important confirmations of special relativity involves nuclear forces. The principles of energy and momentum are very basic and remain valid even when the concept of force in the Newtonian sense no longer applies requiring that classical mechanics be replaced by another theory.

If an isolated particle moves at a velocity **u**, it is the general practice to refer to $W = \gamma m c^2$ as the total energy of the particle. If the particle is at rest, $W(0) = mc^2$. This is called the *rest energy* of the particle. There is ample experimental evidence to show that this rest energy can be made available as energy in the conventional sense. For example, if an electron and a positron come together, they annihilate with the production of γ-rays of combined energy $2mc^2$. If a particle moves in a force-field derivable from a simple scalar potential, its total energy must be written as $\mathscr{E} = W + V$. We could even subtract out the rest energy if it was convenient in a given problem, and interpret $T + V$ as the total mechanical energy of the particle just as in nonrelativistic mechanics.

13.25. BETA DECAY OF NEUTRONS

A simple example of the application of the principles of energy and momentum conservation is provided by the decay of a neutron into a proton, an electron, and a neutrino. The fact that three particles appear makes the problem indeterminate, and so we will calculate the maximum possible energy of the ejected electron. This occurs when no energy and momentum is carried away by a neutrino. If the neutron is at rest when it decays, conservatism of energy and momentum require

$$m_n c^2 = W_e + W_p \tag{13.25.1}$$

$$p_e = p_p \tag{13.25.2}$$

The invariant (13.22.2) can be applied to both the proton and the electron after the decay:

$$p_e^{\,2} - \left(\frac{W_e}{c}\right)^2 = -m_e^{\,2} c^2 \tag{13.25.3}$$

$$p_p^{\,2} - \left(\frac{W_p}{c}\right)^2 = -m_p^{\,2} c^2 \tag{13.25.4}$$

The above equations are solved for W_e to yield

$$W_e = \frac{(m_n^{\,2} - m_p^{\,2} + m_e^{\,2}) c^2}{2 m_n} \tag{13.25.5}$$

Actually, a spectrum of energies is observed due to the varying amounts of energy which the neutrino carries away. Equation (13.25.5) represents the *upper limit* on the possible electron energy.

Suppose, now, that the neutron is initially moving at a velocity v and it is desired to know the maximum electron energy as a function of its direction after the decay takes place. Let W_e' stand for the electron energy in the center of mass coordinates, i.e., the rest frame of the original neutron; W_e' is, of course, just (13.25.5). The transformation equations for the energy and momentum of the electron can be written

$$W_e' = \gamma(W_e - v p_e \cos \theta)$$

$$p_e' \cos \theta' = \gamma\left(p_e \cos \theta - \frac{v}{c^2} W_e\right) \tag{13.25.6}$$

These equations, combined with the useful expression (13.25.3) which, being an invariant, is valid in *all* frames, leads to

$$W_e = \frac{W_e' + \dfrac{v}{c} \cos \theta \sqrt{W_e'^2 - \gamma^2 m_e^{\,2} c^4 \left(1 - \dfrac{v^2}{c_2} \cos^2 \theta\right)}}{\gamma \left[1 - \dfrac{v^2}{c^2} \cos^2 \theta\right]} \tag{13.25.7}$$

13.26. THE COMPTON EFFECT

A basic assumption of quantum theory is that electromagnetic radiation can interact with matter only in units of $h\nu$ where ν is the frequency of the radiation and h is Planck's constant. These quanta of radiation carry a definite energy and momentum and we assume that, when radiation interacts with matter, energy and momentum are conserved. A free electron will scatter radiation. The classical picture of this process is

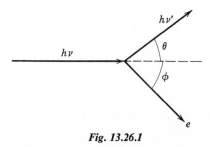

Fig. 13.26.1

that the electron is driven by the oscillating electric field and hence becomes a secondary source of electromagnetic radiation of the *same frequency* as the driving field.

Here we will treat the problem by assuming that the photon encounters the electron, interacts with it after which the electron and a photon emerge. In other words, it is treated just like a collision problem between two particles. The mass of a photon is zero. Hence, the relation between its energy and its momentum is

$$p^2 - \left(\frac{W_p}{c}\right)^2 = 0 \tag{13.26.1}$$

Since $W_p = h\nu$,

$$p = \frac{h\nu}{c} \tag{13.26.2}$$

The scattering process is illustrated in Fig. 13.26.1. A photon of frequency ν strikes an electron, initially at rest, after which the electron and the photon move off at angles ϕ and θ. The frequency of the scattered photon must be less because the electron carries away some of the energy. The equations expressing conservation of energy and momentum are

$$p = p' \cos \theta + p_e \cos \phi \tag{13.26.3}$$

$$0 = p' \sin \theta - p_e \sin \phi \tag{13.26.4}$$

$$h\nu + mc^2 = h\nu' + W_e \tag{13.26.5}$$

where p and p' are initial and final momenta of the photon, p_e and W_e are the momentum and energy of the electron after scattering and m is the mass of the electron. Write (13.26.3) and (13.26.4) as

$$(p - p' \cos \theta)^2 = p_e{}^2 \cos^2 \phi$$
$$(p' \sin \theta)^2 = p_e{}^2 \sin^2 \phi$$

add to get

$$p^2 - 2pp' \cos \theta + p'^2 = p_e{}^2$$

Now use (13.25.3) and (13.26.5) to get

$$p^2 - 2pp' \cos \theta + p'^2 = (p - p' + mc)^2 - m^2c^2 \qquad (13.26.6)$$

Use has been made of (13.26.2). The above reduces to

$$mc(p - p') = pp'(1 - \cos \theta) \qquad (13.26.7)$$

The wavelength can be introduced by means of

$$p = \frac{h\nu}{c} = \frac{h}{\lambda} \qquad p' = \frac{h}{\lambda'} \qquad (13.26.8)$$

The result is

$$\lambda' - \lambda = \frac{h}{mc}(1 - \cos \theta) \qquad (13.26.9)$$

The quantity h/mc is called the *Compton wavelength*.

13.27. RELATIVISTIC ROCKET

The day may not be far off when space craft will move at high enough velocities to require relativistic mechanics for a description of their motion. Let P^μ be the energy-momentum four-vector of the ship. The ship ejects an increment of matter in the form of burned fuel, the energy-momentum four-vector of which is dQ^μ. Then, by conservation of energy and momentum,

$$P^\mu = dQ^\mu + (P^\mu + dP^\mu) \qquad (13.27.1)$$

where $P^\mu + dP^\mu$ is the new four-momentum of the ship. We have

$$dQ^\mu = \delta m W^\mu \qquad dP^\mu = d(mU^\mu) \qquad (13.27.2)$$

where δm is the mass and W^μ is the four-velocity of the ejected matter *with respect to an inertial frame*. We combine (13.27.1) and (13.27.2) and divide by $d\tau$ where τ is the proper time of the ship:

$$0 = \frac{\delta m}{d\tau} W^\mu + \frac{dm}{d\tau} U^\mu + m \frac{dU^\mu}{d\tau} \qquad (13.27.3)$$

Here, $\delta m/d\tau$ is the rate of increase of the mass of the ejected matter and $dm/d\tau$ is the rate of change of the mass of the ship. It is *not* true that $\delta m = -dm$ for exactly the same reason that the masses of the fragments in a nuclear decay are not equal to the mass of the original nucleus! Let $\lambda = \delta m/d\tau$. Then

$$\dot{m}U^\mu + m\dot{U}^\mu = -\lambda W^\mu \qquad (13.27.4)$$

Multiply (13.27.4) by U_μ and recall that

$$U_\mu U^\mu = -c^2, \quad U_\mu \dot{U}^\mu = 0$$

Then

$$\lambda = \dot{m}\frac{c^2}{U_\mu W^\mu} \qquad (13.27.5)$$

Therefore, the equation of motion of a relativistic rocket, or, more generally of a particle of variable mass, is

$$\frac{d}{d\tau}(mU^\mu) = -\left(\frac{\dot{m}c^2}{U_\nu W^\nu}\right)W^\mu \qquad (13.27.6)$$

The solution of (13.27.6) for the special case of one-dimensional motion is not very difficult. As usual in one-dimensional problems, a simplification results if rapidities are used:

$$\frac{w}{c} = \tanh\phi \qquad \frac{u}{c} = \tanh\theta \qquad (13.27.7)$$

Here, w and u are the ordinary velocities of ejected matter and ship with respect to an inertial frame. Since rapidities add like Galilean velocities, we may define

$$\alpha = \theta - \phi \qquad (13.27.8)$$

and interpret

$$h = c\tanh\alpha \qquad (13.27.9)$$

as the ordinary exhaust velocity of the ejected matter relative to the ship. Moreover,

$$U_\nu W^\nu = c^2(\sinh\theta\sinh\phi - \cosh\theta\cosh\phi)$$
$$= -c^2\cosh(\theta - \phi) = -c^2\cosh\alpha \qquad (13.27.10)$$

For $\mu = 1$, (13.27.6) gives

$$\frac{d}{d\tau}(mc\sinh\theta) = \frac{\dot{m}c^2}{c^2\cosh\alpha}c\sinh\phi \qquad (13.27.11)$$

which reduces to the simple differential equation

$$m\dot{\theta} + \dot{m}\frac{h}{c} = 0 \qquad (13.27.12)$$

The actual force on the ship in the conventional sense is

$$F = m \frac{d}{dt}(\gamma u) \qquad (13.27.13)$$

If, as in the example of hyperbolic motion treated in Section 13.23, we write $F = mg$, we find $g = \dot{\theta}c$. The significance of g is that it is the actual gravitational field that would be registered by an accelerometer placed on the ship.

If h is a constant, the solution of (13.27.12) for θ in terms of the mass of the ship is

$$\theta = \log\left(\frac{m}{M}\right)^{-h/c} \qquad (13.27.14)$$

where M is a constant of integration which can be interpreted as the initial mass of the ship. The velocity of the ship as a function of its mass is

$$\frac{u}{c} = \tanh\theta = \frac{1 - e^{-2\theta}}{1 + e^{-2\theta}} = \frac{1 - \left(\dfrac{m}{M}\right)^{2h/c}}{1 + \left(\dfrac{m}{M}\right)^{2h/c}} \qquad (13.27.15)$$

The optimum condition for obtaining the largest possible speed is $h \to c$. It does no good to raise the energy of ejected particles to enormous values. For instance, if $h \simeq c$ and one-half of the initial mass of the ship is ejected,

$$\frac{u}{c} = \frac{1 - (0.5)^2}{1 + (0.5)^2} = \frac{3}{5} \qquad (13.27.16)$$

13.28. SPACE SHIP DRIVEN BY PHOTONS

Of possible interest is a propulsion system for a space craft which consists of a device for ejecting a beam of electromagnetic radiation. Assume an equation of the form

$$\frac{d}{d\tau}(mU^{\mu}) = -\lambda P^{\mu} \qquad (13.28.1)$$

where P^{μ} is the four-momentum of the ejected radiation. As in Section 13.27, we multiply (13.28.1) by U_{μ} and arrive at

$$\lambda = \frac{\dot{m}c^2}{U_{\mu}P^{\mu}} \qquad (13.28.2)$$

Thus,

$$\frac{d}{d\tau}(mU^{\mu}) = -\left(\frac{\dot{m}c^2}{U_{\nu}P^{\nu}}\right)P^{\mu} \qquad (13.28.3)$$

The thing that makes the photon rocket different is the fact that the momentum four-vector of the photons is a *null vector*:

$$P_\mu P^\mu = 0 \tag{13.28.4}$$

As a specific example, consider again one-dimensional motion. Then (13.28.4) is

$$(P^1)^2 - (P^4)^2 = 0 \qquad P^1 = \pm P^4 \tag{13.28.5}$$

If the ship is going in the positive x-direction, the momentum of the photons must necessarily be negative: $P^1 = -P^4$. Specifically, we may think of the energy of a single photon as being $W = h\nu$. Then $P^4 = h\nu/c$, $P^1 = -h\nu/c$. Thus

$$\begin{aligned} U_\nu P^\nu &= U^1 P^1 - U^4 P^4 \\ &= (U^1 + U^4)P^1 = -(U^1 + U^4)P^4 \end{aligned} \tag{13.28.6}$$

and (13.28.3) gives for $\mu = 1$ and $\mu = 4$

$$(U^1 + U^4) \frac{d}{d\tau}(mU^1) = -\dot{m}c^2 \tag{13.28.7}$$

$$(U^1 + U^4) \frac{d}{d\tau}(mU^4) = \dot{m}c^2 \tag{13.28.8}$$

The addition of the two equations gives

$$\frac{d}{d\tau}(mU^1 + mU^4) = 0 \tag{13.28.9}$$

Hence,

$$mU^1 + mU^4 = Mc \tag{13.28.10}$$

where M is the initial mass of the ship, i.e., the mass when $U^1 = 0$, and $U^4 = c$. In terms of the rapidity

$$U^1 = c \sinh \theta \qquad U^4 = c \cosh \theta$$
$$m(\sinh \theta + \cosh \theta) = M$$
$$m = Me^{-\theta} \tag{13.28.11}$$
$$\frac{u}{c} = \tanh \theta = \frac{1 - (m/M)^2}{1 + (m/M)^2}$$

There is no dependence of the final speed on the frequency of the photons. No advantage in this respect is obtained by using hard X rays instead of radiation of lower frequency. Design difficulties lie in constructing a photon drive with sufficient thrust. The comparison of (13.28.11) and (13.27.15) shows that the photon drive has no advantage over an ion propulsion system which ejects material particles near the speed of light.

13.29. LAGRANGE'S AND HAMILTON'S EQUATIONS

By the same procedure used in Section 7.6, a Lagrangian which gives the relativistically correct equations of motion of a charged particle in an electromagnetic field can be established to be

$$L = \tfrac{1}{2}mU_\mu U^\mu + \frac{q}{c} A_\mu U^\mu \qquad (13.29.1)$$

where A_μ are the components of the four-vector potential introduced in Eq. (13.19.7). There are *four* Lagrange equations

$$\frac{d}{d\tau}\frac{\partial L}{\partial \dot{q}^\mu} - \frac{\partial L}{\partial q^\mu} = 0 \qquad (13.29.2)$$

where q^1, q^2, and q^3 are a suitable set of generalized coordinates in the three-space and $q^4 = ct$. Should we try to use the equations of motion in the Newtonian form either as given by (13.21.15) or (13.22.3) in a curvilinear coordinate system, we would run into a familiar problem; namely, that of expressing the curvilinear components of acceleration correctly. The Lagrange formulation neatly avoids this difficulty. Note that

$$A_\mu U^\mu = \gamma(\mathbf{A} \cdot \mathbf{u}) - \gamma c\psi \qquad (13.29.3)$$

where A and ψ are the familiar vector and scalar potential of the electromagnetic field through which the particle moves.

Maxwell's equations themselves as given in tensor form by (13.19.2) and (13.19.3) are covariant *only* with respect to Lorentz transformations; transformations to curvilinear coordinates, or more generally to a four-dimensional space of arbitrary metric are not included. The modification required turns out not to be drastic. In fact (13.19.3) is all right the way it is, but (13.19.2) must be replaced by

$$\frac{1}{\sqrt{-g}}\frac{\partial}{\partial q^\mu}(\sqrt{-g}\, F^{\nu\mu}) = \frac{4\pi\rho_0}{c} U^\nu \qquad (13.29.4)$$

where g is the determinant of $g_{\mu\nu}$.

Let us write the Lagrangian in the form

$$L = \tfrac{1}{2}mg_{\mu\nu}\dot{q}^\mu\dot{q}^\nu + \frac{q}{c} g_{\mu\nu}\dot{q}^\mu A^\nu \qquad (13.29.5)$$

Then the generalized momenta are seen to be

$$p_\alpha = \frac{\partial L}{\partial \dot{q}^\alpha} = mg_{\alpha\mu}\dot{q}^\mu + \frac{q}{c} g_{\alpha\mu}A^\mu = m\dot{q}_\alpha + \frac{q}{c} A_\alpha \qquad (13.29.6)$$

As in nonrelativistic mechanics, p_α is a constant of the motion if q^α itself does not appear in the Lagrangian. For example, consider

$$p_4 = m\dot{q}_4 + \frac{q}{c} A_4 = -\left(m\dot{q}^4 + \frac{q}{c} A^4\right) \tag{13.29.7}$$

But $\dot{q}^4 = c\,dt/d\tau = c\gamma$ and $A^4 = \psi$ so that

$$cp_4 = -(mc^2\gamma + q\psi) = -\mathscr{E} \tag{13.29.8}$$

The total energy of the particle appears as the canonical momentum conjugate to the coordinate ct! If the electromagnetic field through which the particle moves is static so that $q^4 = ct$ does not appear explicitly in the Lagrangian, then $\mathscr{E}$ is a constant of the motion.

The Hamiltonian is given by

$$\begin{aligned}
H &= p_\mu \dot{q}^\mu - L \\
&= \left(m\dot{q}_\mu + \frac{q}{c} A_\mu\right)\dot{q}^\mu - \tfrac{1}{2}m\dot{q}_\mu\dot{q}^\mu - \frac{q}{c}\dot{q}_\mu A^\mu \tag{13.29.9} \\
&= \tfrac{1}{2}m\dot{q}_\mu\dot{q}^\mu = -\tfrac{1}{2}mc^2
\end{aligned}$$

The Lagrangian is always independent of the *proper* time of the particle and the Hamiltonian is always a constant of the motion. In order to use Hamilton's canonical equations, it is necessary to express the Hamiltonian in terms of coordinates and generalized momenta. Thus,

$$\dot{q}_\mu = \frac{1}{m}\left[p_\mu - \frac{q}{c} A_\mu\right]$$

$$\begin{aligned}
H &= \frac{1}{2m}\left[p_\mu - \frac{q}{c} A_\mu\right]\left[p^\mu - \frac{q}{c} A^\mu\right] \\
&= \frac{1}{2m}\left[g^{\mu\nu}p_\mu p_\nu - \frac{2q}{c} A^\mu p_\mu + \frac{q^2}{c^2} A_\mu A^\mu\right] \tag{13.29.10}
\end{aligned}$$

The equations of motion of the charged particle then follow from

$$\dot{p}_\mu = -\frac{\partial H}{\partial q^\mu} \qquad \dot{q}^\mu = \frac{\partial H}{\partial p_\mu} \tag{13.29.11}$$

13.30. THE TWO-BODY CENTRAL FORCE PROBLEM

As an application of the Lagrange formulation, we will find the motion of a point charge in the field of a *stationary* point charge. The generalization to the more realistic situation in which both particles move around a common center of mass is very difficult as we must take into account the finite time required for electromagnetic signals to propagate from one particle to the other. This problem has never been solved completely. If, however, one of the particles is very heavy so that it is stationary, the relativistic problem is hardly more difficult than the nonrelativistic problem.

The four-vector potential through which our particle moves is

$$A^\mu = (0, 0, 0, q'/r) \tag{13.30.1}$$

where q' is the charge of the stationary particle which is setting up the field. If polar coordinates are used in the plane in which the particle moves, the appropriate metric tensor is

$$(g_{\mu\nu}) = \begin{pmatrix} 1 & 0 & 0 & 0 \\ 0 & r^2 & 0 & 0 \\ 0 & 0 & 1 & 0 \\ 0 & 0 & 0 & -1 \end{pmatrix} \tag{13.30.2}$$

Hence,

$$U^\mu = \dot{q}^\mu = (\dot{r}, \dot{\theta}, 0, \dot{x}^4)$$
$$g_{\mu\nu}U^\mu U^\nu = \dot{r}^2 + r^2\dot{\theta}^2 - (\dot{x}^4)^2$$
$$g_{\mu\nu}U^\mu A^\nu = \frac{-\dot{x}^4 q'}{r} \tag{13.30.3}$$
$$L = \tfrac{1}{2}m(\dot{r}^2 + r^2\dot{\theta}^2 - (\dot{x}^4)^2) - \frac{qq'\dot{x}^4}{rc}$$

There are two cyclic coordinates, θ and x^4. Corresponding to θ, we find

$$p_\theta = \frac{\partial L}{\partial \dot{\theta}} = mr^2\dot{\theta} = mr^2\gamma \frac{d\theta}{dt} = l = \text{const.} \tag{13.30.4}$$

which is just the angular momentum. Corresponding to x^4,

$$p_4 = \frac{\partial L}{\partial \dot{x}^4} = -\frac{\mathscr{E}}{c} = -\left(m\dot{x}^4 + \frac{qq'}{rc}\right) \tag{13.30.5}$$

where $\mathscr{E}$ is the total energy of the particle. There is a third integral of the motion namely,

$$\dot{q}_\mu \dot{q}^\mu = \dot{r}^2 + r^2\dot{\theta}^2 - (\dot{x}^4)^2 = -c^2 \tag{13.30.6}$$

The radial equation follows from

$$\frac{d}{d\tau}\frac{\partial L}{\partial \dot{r}} - \frac{\partial L}{\partial r} = 0 \tag{13.30.7}$$

and is

$$m(\ddot{r} - r\dot{\theta}^2) - \frac{qq'}{r^2 c}\dot{x}^4 = 0 \tag{13.30.8}$$

As it turns out, all *four* of the equations (13.30.4), (13.30.5), (13.30.6), and (13.30.8) are *not* independent. Any one of them can be obtained from the other three. The elimination of $\dot{x}^4$ between (13.30.5) and (13.30.8) gives the result

$$m(\ddot{r} - r\dot{\theta}^2) - \frac{qq'}{r^2c}\left[\frac{\mathscr{E}}{mc} - \frac{qq'}{mrc}\right] = 0 \qquad (13.30.9)$$

Through using the angular momentum equation (13.30.4), we can write an orbit equation in terms of the variable $x = 1/r$:

$$\frac{d^2x}{d\theta^2} + \alpha^2 x = -\frac{qq'\mathscr{E}}{l^2c^2} \qquad \alpha^2 = 1 - \left(\frac{qq'}{lc}\right)^2 \qquad (13.30.10)$$

The solution is

$$\frac{1}{r} = -\frac{qq'\mathscr{E}}{l^2c^2\alpha^2} + A\cos(\alpha\theta) \qquad (13.30.11)$$

Since the parameter α is not unity, the orbit does not close on itself but is a precessing figure. Equation (13.30.6) can be expressed in the form

$$\left(\frac{dx}{d\theta}\right)^2 + x^2 - \left(\frac{\mathscr{E}}{lc} - \frac{qq'x}{lc}\right)^2 = -\left(\frac{mc}{l}\right)^2 \qquad (13.30.12)$$

By substituting the solution (13.30.11) into (13.30.12), the constant of integration A can be found in terms of other constants. The result is

$$A = \sqrt{\left(\frac{\mathscr{E}}{lc\alpha^2}\right)^2 - \left(\frac{mc}{l\alpha}\right)^2} \qquad (13.30.13)$$

Notice that *nonperiodic* solutions of (13.30.10) exist when $\alpha^2 < 0$. Let us suppose that an electron in an atom has an angular momentum of order $\hbar$. Let $q' = Zq$ where Z is the atomic number of the atom. Now if

$$\alpha^2 = 1 - \left(\frac{q^2Z}{\hbar c}\right)^2 < 0 \qquad (13.30.14)$$

then

$$Z > \frac{\hbar c}{q^2} \cong 137 \qquad (13.30.15)$$

Thus atoms of greater atomic number than 137 presumably could not exist because of a kind of *relativistic collapse*. The quantity

$$\frac{q^2}{\hbar c} \cong \frac{1}{137} \qquad (13.30.16)$$

is the famous *fine structure constant*.

PROBLEMS

13.4.1   Verify Eq. (13.4.11).

13.4.2   It is desired to place synchronized clocks at number of different places in a given inertial frame. How might this be done? Devise a method which does not depend on a prior knowledge of the value of c.

13.5.3   A measuring rod moves at a velocity v past a single observer O in frame F. O makes two observations on the rod as it goes past. Based on this experiment, how long does O think the rod is as compared to the measurements of an observer O' in the rest frame of the rod? Assume for the purposes of visualization that O very quickly marks the rod at two different places (and times) as it goes past. O' can then, at his leisure, measure the distance between these marks.

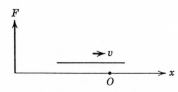

13.6.4   A box at rest in the laboratory contains N_0 radio active particles that decay according to $N = N_0 e^{-t/\tau}$ where τ is a constant. If the box, originally containing N_0 particles, is moving at a velocity v with respect to the laboratory, how many particles are left after t seconds as measured by laboratory clocks?

13.7.5   Verify that the matrix (13.7.13) obeys $S\tilde{S} = \tilde{S}S = I$.

13.7.6   Carry out the matrix multiplications indicated by (13.7.28) for the special Lorentz transformation (13.7.25) and confirm that $G' = G$. What is the matrix representation of equation (13.7.30)?

13.8.7   Carry out the steps required to go from (13.8.4) to (13.8.5). Write out the imaginary version of (13.8.6).

13.10.8   For the general Lorentz transformation without rotation, show that particle velocities transform according to

$$\mathbf{u}' = \frac{\dfrac{1}{\gamma}\mathbf{u} + \mathbf{v}\left[\dfrac{1}{v^2}\left(1 - \dfrac{1}{\gamma}\right)(\mathbf{v}\cdot\mathbf{u}) - 1\right]}{1 - \dfrac{\mathbf{v}\cdot\mathbf{u}}{c^2}}$$

13.10.9   A particle moves in the xy-plane at an angle θ with respect to the x-axis. If frames F and F' are connected by the special Lorentz transformation, show that the particle moves at an angle θ' with respect to the x'-axis given by

$$\tan\theta' = \frac{1}{\gamma}\frac{\sin\theta}{\cos\theta - v/u}$$

13.11.10   A light signal travels between two points. Consider the departure and arrival of the signal as the two events. Obtain (13.11.8) by a direct application of the Lorentz transformation.

13.13.11   A space ship moves on a circular path at a speed of $0.9c$. The radius of the path is such that the acceleration of the ship is 980 cm/sec^2. What is the radius of the path in light-years? How long in years does it take the ship to make one complete round trip (a) as recorded by clocks in the ship, (b) as recorded by clocks in the inertial frame relative to which the ship moves?

13.14.12   Consider two events separated by intervals t and x ($y = z = 0$) where $t > 0$ and $x > 0$. (a) If $x^2 - c^2t^2 > 0$, show, using the special Lorentz transformation directly that a frame F' with velocity $v < c$ exists such that $t' \leq 0$. (b) If $x^2 - c^2t^2 < 0$, show that *no* frame F' exists with $v < c$ such that $t' \leq 0$.

In part (b), the two events are *ordered* with respect to time. Considering 0 (Fig. 13.14.2) to be one event-point, the other event *always* lies in the future in all reference frames. The displacement vector lies *inside* the light cone for $t > 0$, and this region is sometimes referred to as the *future*. Similarly, the region inside the light cone for $t < 0$ is the *past*.

As illustrated in part (a), event points which lie outside the light cone can happen in the past, present, or future, depending on which frame one happens to be in. This is all right since the event at point O can in no way have a causal influence on events outside the light cone. In some texts, we find the three invariant subregions, into which the light cone divides the space, called "the absolute past, the absolute future, and the absolute elsewhere."

13.15.13   Show that (13.15.8) is the equivalent of

$$v'' = \frac{v' + v}{1 + vv'/c^2}$$

Note that this is just a special case of (13.10.5).

13.17.14   A plane electromagnetic wave propagates in a material medium of refractive index n_0 so that the phase velocity is given by

$$u_0 = \frac{\omega_0}{k_0} = \frac{c}{n_0}$$

Suppose that the medium moves with a velocity v in the same direction that the electromagnetic wave is propagating. Show that the apparent index of refraction and phase velocity of the wave relative to the laboratory are

$$n = \frac{n_0 + v/c}{1 + n_0 v/c} \qquad u = \frac{c/n_0 + v}{1 + v/(n_0 c)}$$

If $v/c \ll 1$,

$$u = \frac{c}{n_0} + v\left(1 - \frac{1}{n_0^2}\right)$$

The quantity $1 - 1/n_0^2$ is known as the *Fresnell dragging coefficient*.

13.17.15   How are the coefficients $P^\mu_{\nu'}$ related to the $S_{\mu\nu}$? Show that the coefficients of a general Lorentz transformation obey the conditions

$$\delta_{ij} = P^k_{i'} P^k_{j'} - P^4_{i'} P^4_{j'}$$
$$-\delta_{4\mu} = P^k_{4'} P^k_{\mu'} - P^4_{4'} P^4_{\mu'}$$

Show that the above conditions are the equivalent of

$$\delta_{\mu\nu} = S_{\mu\alpha}S_{\nu\alpha}$$

13.19.16   There are two other versions of the field tensor, each of which is of the form of a mixed tensor:

$$F_\mu{}^\nu = g_{\alpha\mu}F^{\alpha\nu} \qquad F^\nu{}_\mu = g_{\alpha\mu}F^{\nu\alpha}$$

Write out the matrices which give the components of these tensors in terms of field quantities. How do the mixed tensors transform?

13.19.17   By the rules of general tensor analysis, the quantity

$$\phi = F_{\alpha\beta}F^{\alpha\beta}$$

is a scalar invariant. What is this invariant in terms of field quantities?

13.19.18   Show that the correct form of the field tensor for use with complex notation is

$$(F_{\mu\nu}) = \begin{pmatrix} 0 & B_3 & -B_2 & -iE_1 \\ -B_3 & 0 & B_1 & -iE_2 \\ B_2 & -B_1 & 0 & -iE_3 \\ iE_1 & iE_2 & iE_3 & 0 \end{pmatrix}$$

meaning that the correct transformation of the electromagnetic field is given by

$$F'_{\mu\nu} = S_{\mu\alpha}S_{\nu\beta}F_{\alpha\beta}$$

The determinant of a second rank tensor is an invariant under orthogonal transformations. By calculating the determinant of the above tensor, show that this invariant is essentially

$$\psi = \mathbf{B} \cdot \mathbf{E}$$

By using the special transformations (13.18.6), (13.18.7), (13.18.9), and (13.18.10), verify directly that

$$\mathbf{B'} \cdot \mathbf{E'} = \mathbf{B} \cdot \mathbf{E}$$

13.20.19   Assume that (13.20.2) is valid in an inertial frame F. Prove, by transforming directly all quantities which appear in the equation, that in a second inertial frame F'

$$\partial_\alpha' F^{\beta\alpha'} = \frac{4\pi}{c} \rho_0 U^{\beta'}$$

The important thing to remember is that the primed quantities have the same physical meaning to observers in F' as do the unprimed quantities to observers in F.

13.21.20   What do the electric and magnetic fields of a point charge moving at a constant velocity look like in the limit $u \to c$?

13.22.21   An electron is moving in a region of space where there is an electric field of 10^4 stat-volts/cm in the x_1-direction and a magnetic field of 10^4 gauss in the negative x_3-direction. The electron moves in the x_1x_2-plane. Calculate the direction of the force and the direction of the acceleration at an instant when the electron is moving in the x_1-direction at a velocity of $0.8c$.

13.24.22 What do the equations of motion (13.23.7) look like in the non-relativistic limit $g\tau/c \to 0$?

13.24.23 From (13.23.7), obtain x explicitly as a function of t and sketch the trajectory on a space-time diagram. Locate the asymptotes. Show that if a space ship goes directly out from the earth under constant acceleration (i.e., g = const.) there is a time after which communication with the ship is no longer possible.

13.24.24 A space ship travels straight out from the earth under the influence of a constant force. The force is such that $g = 980$ cm/sec². What is the speed of the ship relative to earth after the ship's clocks have recorded a time $\tau = 1$ year? How much time has elapsed on earth?

13.26.25 Verify Eq. (13.25.7).

13.26.26 Antiprotons are to be produced by bombarding stationary protons with protons out of an accelerator. Conservation of nucleon number and total charge allows two possible reactions:

$$p + p \to \bar{n} + n + p + p$$
$$p + p \to \bar{p} + 3p$$

where $\bar{p}$ and $\bar{n}$ stand for antiproton and anti-neutron. What minimum kinetic energy, in units of $m_p c^2$, must the bombarding protons have in order to make the reaction go? The masses of all the particles can be considered to be the same. Solve the problem first in the center of mass coordinates.

13.26.27 A block of mass m_2 at rest on a frictionless surface is struck by a bullet of energy W, momentum p, and mass m_1. The bullet is initially traveling horizontally and remains embedded in the block after the collision. What is the final mass of the system? How much heat is there from the collision? What is the final kinetic energy of block plus bullet?

13.27.28 Show that the kinetic energy of the recoil motion of the electron in the Compton effect is

$$T = h\nu \, \frac{2\left(\dfrac{h}{mc\lambda}\right)\sin^2\dfrac{\theta}{2}}{1 + 2\left(\dfrac{h}{mc\lambda}\right)\sin^2\dfrac{\theta}{2}}$$

13.28.29 How must the mass of a rocket moving in one dimension vary as a function of the proper time of the ship in order to maintain g = const. = 980 cm/sec²? Assume that the exhaust velocity is $h = c$. A trip to a star about four light years away is contemplated. The plan is to accelerate uniformly for one year (ship time), coast at constant velocity for whatever time is necessary, and then decelerate uniformly for one year (again as measured by the ship's clocks), at which point the destination is reached. Assume that during the periods of acceleration $g = 980$ cm/sec². Compute the total time for the trip as measured by both the ship's clocks and earth clocks. What fraction of the ship's mass must be expended by the time the destination is reached?

13.30.30   Verify that the Lagrangian (13.29.1) gives the correct equations of motion (13.22.3) for a charged particle. You may assume that the spatial coordinates are Cartesian.

13.30.31   A charged particle moves at right angles to a uniform magnetic field. Show that the orbit is circular and find an expression for the circular frequency in terms of the radius. (The equations of motion in the Newtonian form (13.21.15) are the most convenient here.) Find an expression for rB, where r is the radius of the orbit, in terms of the energy of the particle.

13.30.32   A charged particle moves in an axially symmetric static magnetic field, (i.e., as discussed in Section 7.11). Find a Lagrangian and all the constants of the motion including the relativistic version of Busch's theorem equation (7.11.7). Find differential equations for r and z as functions of the proper time of the particle.

APPENDICES

APPENDIX E

Appendix A

The following trigonometric identities find application in the solution of problems in non-linear mechanics.

1. $\cos^3 \theta = \frac{3}{4} \cos \theta + \frac{1}{4} \cos 3\theta$
2. $\cos^4 \theta = +\frac{3}{8} + \frac{1}{2} \cos 2\theta + \frac{1}{8} \cos 4\theta$
3. $\cos^5 \theta = \frac{5}{8} \cos \theta + \frac{5}{16} \cos 3\theta + \frac{1}{16} \cos 5\theta$
4. $\cos \theta \cos 2\theta = \frac{1}{2} \cos 3\theta + \frac{1}{2} \cos \theta$
5. $\cos \theta \cos 3\theta = \frac{1}{2} \cos 4\theta + \frac{1}{2} \cos 2\theta$
6. $\cos \theta \cos 4\theta = \frac{1}{2} \cos 3\theta + \frac{1}{2} \cos 5\theta$
7. $\cos \theta \cos 5\theta = \frac{1}{2} \cos 4\theta + \frac{1}{2} \cos 6\theta$
8. $\cos^2 \theta \cos 2\theta = \frac{1}{4} + \frac{1}{2} \cos 2\theta + \frac{1}{4} \cos 4\theta$
9. $\cos^2 \theta \cos 3\theta = \frac{1}{4} \cos \theta + \frac{1}{2} \cos 3\theta + \frac{1}{4} \cos 5\theta$
10. $\cos^2 \theta \cos 4\theta = \frac{1}{4} \cos 2\theta + \frac{1}{2} \cos 4\theta + \frac{1}{4} \cos 6\theta$
11. $\cos^2 \theta \cos 5\theta = \frac{1}{4} \cos 3\theta + \frac{1}{2} \cos 5\theta + \frac{1}{4} \cos 7\theta$
12. $\cos \theta \cos^2 2\theta = \frac{1}{2} \cos \theta + \frac{1}{4} \cos 3\theta + \frac{1}{4} \cos 5\theta$
13. $\cos \theta \cos^2 3\theta = \frac{1}{2} \cos \theta + \frac{1}{4} \cos 5\theta + \frac{1}{4} \cos 7\theta$
14. $\cos 2\theta \cos^3 \theta = \frac{1}{2} \cos \theta + \frac{3}{8} \cos 3\theta + \frac{1}{8} \cos 5\theta$
15. $\cos 2\theta \cos 3\theta = \frac{1}{2} \cos \theta + \frac{1}{2} \cos 5\theta$
16. $\cos^2 2\theta \cos 3\theta = \frac{1}{4} + \frac{1}{4} \cos 2\theta + \frac{1}{4} \cos 4\theta + \frac{1}{4} \cos 6\theta$

Further identities can generally be established as needed, e.g., by substituting 2θ for θ or by combining the above identities in various ways.

Appendix B

Maxwell's Equations in Gaussian and mks Units:

Gaussian

$$\nabla \times \mathbf{E} + \frac{1}{c}\frac{\partial \mathbf{B}}{\partial t} = 0$$

$$\nabla \times \mathbf{B} - \frac{1}{c}\frac{\partial \mathbf{E}}{\partial t} = \frac{4\pi}{c}\mathbf{j}$$

$$\nabla \cdot \mathbf{E} = 4\pi\rho$$

$$\nabla \cdot \mathbf{B} = 0$$

mks

$$\nabla \times \mathbf{E} + \frac{\partial \mathbf{B}}{\partial t} = 0$$

$$\nabla \times \mathbf{B} - \mu_0\epsilon_0\frac{\partial \mathbf{E}}{\partial t} = \mu_0\mathbf{j}$$

$$\nabla \cdot \mathbf{E} = \frac{1}{\epsilon_0}\rho_0$$

$$\nabla \cdot \mathbf{B} = 0$$

Field of a Point Charge:

$$\mathbf{E} = q\frac{\mathbf{r}}{r^3}$$

$$\mathbf{E} = \frac{q}{4\pi\epsilon_0}\frac{\mathbf{r}}{r^3}$$

Integral Relations for Static Fields:

Gauss' Law:

$$\oint \mathbf{E} \cdot d\boldsymbol{\sigma} = 4\pi q$$

$$\oint \mathbf{E} \cdot d\boldsymbol{\sigma} = \frac{1}{\epsilon_0}q$$

Ampere's Law:

$$\oint \mathbf{B} \cdot d\mathbf{s} = \frac{4\pi}{c}J$$

$$\oint \mathbf{B} \cdot d\mathbf{s} = \mu_0 J$$

Biot-Savart Law:

$$\mathbf{B} = \frac{J}{c}\oint \frac{d\mathbf{s} \times \mathbf{r}}{r^3}$$

$$\mathbf{B} = \frac{\mu_0 J}{4\pi}\oint \frac{d\mathbf{s} \times \mathbf{r}}{r^3}$$

Vector and Scalar Potentials:

$$\mathbf{B} = \nabla \times \mathbf{A}$$

$$\mathbf{E} = -\nabla\psi - \frac{1}{c}\frac{\partial \mathbf{A}}{\partial t}$$

$$\mathbf{B} = \nabla \times \mathbf{A}$$

$$\mathbf{E} = -\nabla\psi - \frac{\partial \mathbf{A}}{\partial t}$$

Lorentz Gauge Condition:

$$\nabla \cdot \mathbf{A} + \frac{1}{c}\frac{\partial\psi}{\partial t} = 0$$

$$\nabla \cdot \mathbf{A} + \frac{1}{c^2}\frac{\partial\psi}{\partial t} = 0$$

Lorentz Force on a Charged Particle:

$$\mathbf{F} = q\mathbf{E} + \frac{1}{c}q(\mathbf{u} \times \mathbf{B})$$

$$\mathbf{F} = q\mathbf{E} + q(\mathbf{u} \times \mathbf{B})$$

Field Tensor:

$$F^{\mu\nu} = \begin{pmatrix} 0 & B_z & -B_y & -E_x \\ -B_z & 0 & B_x & -E_y \\ B_y & -B_x & 0 & -E_z \\ E_x & E_y & E_z & 0 \end{pmatrix} \qquad F^{\mu\nu} = \begin{pmatrix} 0 & cB_z & -cB_y & -E_x \\ -cB_z & 0 & cB_x & -E_y \\ cB_y & -cB_x & 0 & -E_z \\ E_x & E_y & E_z & 0 \end{pmatrix}$$

Covariant Form of Maxwell's Equations:

$$\partial_\mu F^{\nu\mu} = \frac{4\pi}{c}j^\nu$$

$$\partial_\mu F_{\nu\lambda} + \partial_\lambda F_{\mu\nu} + \partial_\nu F_{\lambda\mu} = 0$$

$$\partial_\mu F^{\nu\mu} = \mu_0 c j^\nu$$

$$\partial_\mu F_{\nu\lambda} + \partial_\lambda F_{\mu\nu} + \partial_\nu F_{\lambda\mu} = 0$$

Fundamental Quantities and Conversion Factors:

Name	Symbol	mks	Gaussian
Charge	q	1 coulomb	3×10^9 statcoulomb
Current density	$\mathbf{j}$	1 ampere m^{-2}	3×10^5 statamp cm^{-2}
Electric field	$\mathbf{E}$	1 volt m^{-1}	$\frac{1}{3} \times 10^{-4}$ statvolt cm^{-1}
Magnetic field	$\mathbf{B}$	1 weber m^{-2}	10^4 gauss
Potential	ψ	1 volt	$\frac{1}{300}$ statvolt
Resistance	R	1 ohm	$\frac{1}{9} \times 10^{-11}$ sec cm^{-1}
Capacitance	C	1 farad	9×10^{11} cm
Inductance	L	1 henry	$\frac{1}{9} \times 10^{-11}$ sec^2 cm^{-1}

Fundamental Constants:

$$\mu_0 = 4\pi \times 10^{-7} \text{ weber amp}^{-1}\text{ m}^{-1}$$

$$\epsilon_0 = 8.854 \times 10^{-12} \text{ farad m}^{-1}$$

$$c = 1/\sqrt{\epsilon_0\mu_0} = 2.998 \times 10^8 \text{ m sec}^{-1}$$

Appendix C

1. $y = snu$ $\quad\quad (y')^2 = (1 - y^2)(1 - k^2y^2) \quad\quad\quad y' = cnu\,dnu$

$\quad\quad\quad\quad\quad\quad y'' + y(1 + k^2) - 2k^2y^3 = 0$

2. $y = cnu$ $\quad\quad (y')^2 = (1 - y^2)(1 - k^2 + k^2y^2) \quad\quad y' = -snu\,dnu$

$\quad\quad\quad\quad\quad\quad y'' + y(1 - 2k^2) + 2k^2y^3 = 0$

3. $y = dnu$ $\quad\quad (y')^2 = (1 - y^2)(y^2 - 1 + k^2) \quad\quad y' = -k^2snucnu$

$\quad\quad\quad\quad\quad\quad y'' + y(k^2 - 2) + 2y^3 = 0$

4. $y = tnu$ $\quad\quad (y')^2 = (1 + y^2)[1 + (1 - k^2)y^2] \quad y' = dnu/cn^2u$

$\quad\quad\quad\quad\quad\quad y'' - y(2 - k^2) - 2y^3(1 - k^2) = 0$

5. $y = 1/snu$ $\quad (y')^2 = (y^2 - 1)(y^2 - k^2) \quad\quad\quad y' = -cnu\,dnu/sn^2u$

$\quad\quad\quad\quad\quad\quad y'' + y(1 + k^2) - 2y^3 = 0$

6. $y = 1/cnu$ $\quad (y')^2 = (y^2 - 1)[(1 - k^2)y^2 + k^2], \quad y' = snu\,dnu/cn^2u$

$\quad\quad\quad\quad\quad\quad y'' + y(1 - 2k^2) - 2y^3(1 - k^2) = 0$

7. $y = dnutnu$ $\quad y' = \sqrt{(1 + y^2)^2 - 4k^2y^2}$

$\quad\quad\quad\quad\quad\quad y'' + 2y(2k^2 - 1) - 2y^3 = 0$

8. $y = sn^2u$ $\quad\quad (y')^2 = 4y(1 - y)(1 - k^2y)$

$\quad\quad\quad\quad\quad\quad y'' - 2 + 4y(k^2 + 1) - 6k^2y^2 = 0$

9. $y = cn^2u$ $\quad\quad (y')^2 = 4y(1 - y)(1 - k^2 + k^2y)$

$\quad\quad\quad\quad\quad\quad y'' - 2(1 - k^2) - 4y(2k^2 - 1) + 6k^2y^2 = 0$

10. $y = dn^2u$ $(y')^2 = 4y(1 - y)(y - 1 + k^2)$
$y'' - 2(k^2 - 1) - 4y(2 - k^2) + 6y^2 = 0$

11. $y = 1/sn^2u$ $(y')^2 = 4y(y - 1)(y - k^2)$
$y'' - 2k^2 + 4y(k^2 + 1) - 6y^2 = 0$

12. $y = 1/cn^2u$ $(y')^2 = 4y(y - 1)[(1 - k^2)y + k^2]$
$y'' + 2k^2 - 4y(2k^2 - 1) - 6y^2(1 - k^2) = 0$

13. $y = dn^2utn^2u$ $(y')^2 = 4y[(1 + y)^2 - 4k^2y]$
$y'' - 2 - 8y(1 - 2k^2) - 6y^2 = 0$

14. $y = tn^2u$ $(y')^2 = 4y(1 + y)[1 + (1 - k^2)y]$
$y'' - 2 - 4(2 - k^2)y - 6(1 - k^2)y^2 = 0$

Appendix D

MISCELLANEOUS PHYSICAL CONSTANTS

Charge on the Electron

$q = 1.602 \times 10^{-19}$ coulombs $= 4.803 \times 10^{-10}$ stat-coulombs

Planck's Constant

$h = 6.625 \times 10^{-27}$ erg sec

$\hbar = 1.0544 \times 10^{-27}$ erg sec

Mass of the Proton, Electron, and Neutron

$m_e = 9.11 \times 10^{-28}$ gm

$m_p = 1.6724 \times 10^{-24}$ gm

$m_n = 1.6747 \times 10^{-24}$ gm

Magnetic Dipole Moment of the Earth's Magnetic Field

$\mu = 8.1 \times 10^{25}$ stat-coulomb cm

Gravitational Constant

$G = 6.670 \times 10^{-8}$ cm^3 gm^{-1} sec^{-2}

$\quad = 3.42 \times 10^{-8}$ ft^3 slug^{-1} sec^{-2}

Velocity of Light

$c = 2.998 \times 10^{10}$ cm sec^{-1}

Data in Regard to the Earth

mean radius: 6.371×10^6 m

mass: 5.983×10^{27} gm

angular velocity of rotation: 7.27×10^{-5} sec^{-1}

Astronomical Data

Mass of the moon: 0.01228 earth masses

Mean earth-moon distance: 38×10^4 km

Mean earth-sun distance: 149×10^6 km

Mass of the sun: 329390 earth masses

Period of the moon: 27.3 days

Inclination of moon's orbit to equatorial plane of the earth: $5°$

Periods of rotation of Uranus and Neptune:

Uranus: 10.7 hr

Neptune: 14.0 hr

Radii of Uranus and Neptune:

Uranus: 26,700 km

Neptune: 24,850 km

Index